LAGOMORPHS

Lagomorphs
Pikas, Rabbits, and Hares of the World

EDITED BY Andrew T. Smith
Charlotte H. Johnston
Paulo C. Alves
Klaus Hackländer

JOHNS HOPKINS UNIVERSITY PRESS | BALTIMORE

© 2018 Johns Hopkins University Press
All rights reserved. Published 2018
Printed in China on acid-free paper
9 8 7 6 5 4 3 2 1

Johns Hopkins University Press
2715 North Charles Street
Baltimore, Maryland 21218-4363
www.press.jhu.edu

Library of Congress Cataloging-in-Publication Data

Names: Smith, Andrew T., 1946-, editor.
Title: Lagomorphs : pikas, rabbits, and hares of the world / edited
by Andrew T. Smith, Charlotte H. Johnston, Paulo C. Alves, Klaus
Hackländer.
Description: Baltimore : Johns Hopkins University Press, 2018. | Includes
bibliographical references and index.
Identifiers: LCCN 2017004268| ISBN 9781421423401 (hardcover) |
ISBN 1421423405 (hardcover) | ISBN 9781421423418 (electronic) | ISBN
1421423413 (electronic)
Subjects: LCSH: Lagomorpha. | BISAC: SCIENCE / Life Sciences / Biology /
General. | SCIENCE / Life Sciences / Zoology / Mammals. | SCIENCE /
Reference.
Classification: LCC QL737.L3 L35 2018 | DDC 599.32—dc23
LC record available at https://lccn.loc.gov/2017004268

A catalog record for this book is available from the British Library.

Frontispiece, top to bottom: courtesy Behzad Farahanchi, courtesy
David E. Brown, and © Alessandro Calabrese.

*Special discounts are available for bulk purchases of this book. For more
information, please contact Special Sales at 410-516-6936 or specialsales
@press.jhu.edu.*

Johns Hopkins University Press uses environmentally friendly book
materials, including recycled text paper that is composed of at least 30
percent post-consumer waste, whenever possible.

CONTENTS

PREFACE

The IUCN Species Survival Commission Lagomorph Specialist Group (LSG) and the World Lagomorph Society (WLS) are pleased to present this comprehensive compendium of all the lagomorphs in the world. This work is designed to expand coverage of the world's lagomorphs and update the 1990 LSG Lagomorph Action Plan (*Rabbits, Hares and Pikas: Status Survey and Conservation Action Plan*, compiled and edited by Joseph A. Chapman and John E. C. Flux). The Action Plan has served as the most thorough single source of information on lagomorphs for biologists, but it was never widely available to the public and it has become outdated. In this book we present updated range maps of all lagomorph species, high-quality images of most species, as well as current information on identification, systematics, ecology, behavior, reproduction, genetics, physiology, and conservation and management of the pikas, rabbits, and hares of the world. The book also summarizes key components of topics of broad interest across all lagomorph species: evolution, systematics, lagomorph diseases, introduced lagomorphs, and conservation and management. Despite several ongoing controversies in lagomorph taxonomy, we have maintained a conservative systematic approach. Nevertheless, we highlight relevant taxonomic issues that require attention.

This work has been a team effort, with 82 specialists contributing to species accounts. We especially thank Rachel Fadlovich for her work on the references, and Aryn Musgrave for constructing all the species range maps. ATS thanks Harriet Smith for her insightful editorial work on his chapters. We are grateful to all the photographers who provided their work for free. We appreciate the meticulous copy editing by Maria E. denBoer. Finally, the collaboration with our editorial team from Johns Hopkins University Press, Vincent Burke, Tiffany Gasbarrini, Debby Bors, and Meagan M. Szekely, is highly appreciated.

CONTRIBUTORS

Pelayo Acevedo
Instituto de Investigación en Recursos Cinegéticos
Ciudad Real, Spain

Maria Altemus
School of Natural Resources and the Environment
University of Arizona
Tucson, Arizona, United States

Sergio Ticul Álvarez-Castañeda
Centro de Investigaciones Biológicas del Noroeste
La Paz, Baja California Sur, Mexico

Paulo C. Alves
Faculdade de Ciências & CIBIO
Universidade do Porto, Campus de Vairão
Vairão, Vila do Conde, Portugal

Nguyen The Truong An
Leibniz Institute for Zoo and Wildlife Research
Berlin, Germany

Anders Angerbjörn
Department of Zoology
Stockholm University
Stockholm, Sweden

Asma Awadi
Faculty of Sciences of Tunis
University El Manar
Tunis, Tunisia

Fernando Ballesteros
Sistemas Naturales
c/ Santa Susana 15 3° C
Oviedo, Spain

Ron Barry
P. O. Box 471
Lewiston, Maine, United States

Amando Bautista
Centro Tlaxcala de Biología de la Conducta
Universidad Autónoma de Tlaxcala
Tlaxcala, Mexico

Penny A. Becker
Washington Department of Fish and Wildlife
Olympia, Washington, United States

Erik A. Beever
U.S. Geological Survey
Northern Rocky Mountain Science Center
Bozeman, Montana, United States

Hichem Ben Slimen
High Institute of Biotechnology of Béja
Béja, Tunisia

Joel Berger
Department of Fish, Wildlife and Conservation Biology
Colorado State University
Fort Collins, Colorado, United States

Leah K. Berkman
Cooperative Wildlife Research Laboratory
Southern Illinois University
Carbondale, Illinois, United States

Sabuj Bhattacharyya
Molecular Ecology Laboratory
Centre for Ecological Sciences
Indian Institute of Science
Bangalore, Karnataka, India

Jorge Bolaños
El Colegio de la Frontera Sur
Departamento de Conservación de la Biodiversidad
San Cristóbal de Las Casas, Chiapas, Mexico

Never Bonino
Instituto Nacional de Tecnologica Agropecuaria
Bariloche, Argentina

Christy J. Bragg
Drylands Conservation Programme
The Endangered Wildlife Trust
South Africa

David E. Brown
School of Life Sciences
Arizona State University
Tempe, Arizona, United States

Arturo Carrillo-Reyes
Oikos: Conservación y Desarrollo Sustentable
San Cristóbal de Las Casas, Chiapas, Mexico

Fernando A. Cervantes
Instituto de Biología, UNAM
Colección Nacional de Mamíferos
Distrito Federal, Mexico

Kai Collins
Mammal Research Institute
Department of Zoology & Entomology
University of Pretoria
Pretoria, South Africa

Brian D. Cooke
Institute for Applied Ecology
University of Canberra
Canberra, Australia

Mayra de la Paz
Centro de Investigaciones Biológicas del Noroeste
La Paz, Baja California Sur, Mexico

Miguel Delibes-Mateos
CIBIO, Universidade do Porto
Campus Agrario de Vairão
Vairão, Vila do Conde, Portugal

Robert C. Dowler
Department of Biology, Angelo State University
San Angelo, Texas, United States

Ricardo Farrera-Muro
Instituto Tecnológico de Estudios Superiores de
    Monterrey
Campus Puebla, Puebla, Mexico

Craig Faulhaber
Florida Fish and Wildlife Conservation Commission
Ocala, Florida, United States

John E. C. Flux
23 Hardy Street, Waterloo
Lower Hutt, Wellington, New Zealand

Johnnie French
Department of Biology
Portland State University
Portland, Oregon, United States

Antonio García-Méndez
El Colegio de la Frontera Sur
Departamento de Conservación de la Biodiversidad
San Cristóbal de Las Casas, Chiapas, Mexico

Fernando Gopar-Merino
Centro de Investigaciones en Geografia Ambiental
Universidad Nacional Autonoma de Mexico
Morelia, Michoacán, Mexico

Thomas Gray
WWF—Greater Mekong
Phnom Penh, Cambodia

Klaus Hackländer
University of Natural Resources and Life Sciences Vienna
Institute of Wildlife Biology and Game Management
Vienna, Austria

David C. D. Happold
Research School of Biology
Australian National University
Canberra, A.C.T., Australia

Jeremy Holden
Flora and Fauna International
Cambridge, United Kingdom

Yeong-Seok Jo
Division of Animal Research
National Institute of Biological Resources
Incheon, South Korea

Charlotte H. Johnston
School of Life Sciences
Arizona State University
Tempe, Arizona, United States

Patrick A. Kelly
Department of Biological Sciences
California State University, Stanislaus
Turlock, California, United States

Howard Kilpatrick
391 Rt. 32
North Franklin, Connecticut, United States

Adrienne Kovach
University of New Hampshire
Durham, New Hampshire, United States

Charles J. Krebs
Department of Zoology
University of British Columbia
Vancouver, British Columbia, Canada

Hayley C. Lanier
Department of Zoology and Physiology
University of Wyoming
Casper, Wyoming, United States

John W. Laundré
James San Jacinto Mountains Natural Reserve
Idyllwild, California, United States

Antonio Lavazza
Virology Department, IZSLER
Brescia, Italy

Weidong Li
Natural Ecological Protection Studio
Ürümqi, Xinjiang, China

Andrey Lissovsky
Zoological Museum, Moscow State University
Moscow, Russia

John A. Litvaitis
Department of Natural Resources and the Environment
University of New Hampshire
Durham, New Hampshire, United States

Marian Litvaitis
Department of Natural Resources and the Environment
University of New Hampshire
Durham, New Hampshire, United States

Shaoying Liu
Sichuan Academy of Forestry
Chengdu, Sichuan, China

Consuelo Lorenzo
El Colegio de la Frontera Sur
Departamento Conservación de la Biodiversidad
San Cristóbal de Las Casas, Chiapas, Mexico

Debbie Martyr
Fauna & Flora International-Indonesia Programme
Ragunan, Jakarta, Indonesia

Conrad A. Matthee
Department of Botany and Zoology
Stellenbosch University
Stellenbosch, South Africa

Jennifer L. McCarthy
University of Delaware, Associate of Arts Program
Wilmington, Delaware, United States

Robert McCleery
Department of Wildlife Ecology and Conservation
University of Florida
Gainesville, Florida, United States

José Melo-Ferreira
CIBIO, Centro de Investigação em Biodiversidade e
    Recursos Genéticos
InBIO Laboratório Associado & Faculdade de Ciências
Universidade do Porto
Porto, Portugal

José M. Mora
Instituto Internacional en Conservación y Manejo de Vida
    Silvestre
Universidad Nacional
Heredia, Costa Rica

Sanjay Molur
Zoo Outreach Organization
Coimbatore, Tamil Nadu, India

Dennis L. Murray
Department of Biology
Trent University
Peterborough, Ontario, Canada

P. O. Nameer
Centre for Wildlife Studies
Kerala Agricultural University
Thrissur City, Kerala, India

Clayton K. Nielsen
Cooperative Wildlife Research Laboratory and
    Department of Forestry
Southern Illinois University
Carbondale, Illinois, United States

Janet L. Rachlow
Department of Fish and Wildlife Sciences
University of Idaho
Moscow, Idaho, United States

Juan Pablo Ramírez-Silva
Programa Académico de Biología
Unidad Académica de Agricultura
Universidad Autónoma de Nayarit
Xalisco, Nayarit, Mexico

Chris Ray
Institute for Arctic and Alpine Research
University of Colorado
Boulder, Colorado, United States

Tamara Rioja-Paradela
Sustentabilidad y Ecología Aplicada
Universidad de Ciencias y Artes de Chiapas
Tuxtla Gutiérrez, Chiapas, Mexico

Terry J. Robinson
Department of Botany and Zoology
Stellenbosch University
Matieland, South Africa

Luisa Rodríguez-Martínez
Centro Tlaxcala de Biología de la Conducta
Universidad Autónoma de Tlaxcala
Tlaxcala, Mexico

Luis A. Ruedas
Department of Biology
Portland State University
Portland, Oregon, United States

Eugenia C. Sántiz-López
El Colegio de la Frontera Sur
Departamento de Conservación de la Biodiversidad
San Cristóbal de Las Casas, Chiapas, Mexico

Stéphanie Schai-Braun
University of Natural Resources and Life Sciences Vienna
Institute of Wildlife Biology and Game Management
Vienna, Austria

Lisa A. Shipley
School of the Environment
Washington State University
Pullman, Washington, United States

Andrew T. Smith
School of Life Sciences
Arizona State University
Tempe, Arizona, United States

Adia Sovie
Department of Wildlife Ecology and Conservation
University of Florida
Gainesville, Florida, United States

Andrew Tilker
Leibniz Institute for Zoo and Wildlife Research
Berlin, Germany, and
Global Wildlife Conservation
Austin, Texas, United States

Zelalem Gebremariam Tolesa
Department of Biology
Hawassa University
Hawassa, Ethiopia

Myles B. Traphagen
Westland Resources, Inc.
Tucson, Arizona, United States

Julieta Vargas
Instituto de Biología, UNAM
Colección Nacional de Mamíferos
Distrito Federal, Mexico

Jorge Vázquez
Centro Tlaxcala de Biología de la Conducta
Universidad Autónoma de Tlaxcala
Tlaxcala, Mexico

Alejandro Velázquez
Centro de Investigaciones en Geografia Ambiental
Universidad Nacional Autonoma de Mexico
Morelia, Michoacán, Mexico

Rafael Villafuerte
IESA-CSIC
Campo Santo de los Mártires 7
Códoba, Spain

Fumio Yamada
Department of Wildlife Biology
Forestry and Forest Products Research Institute
Matsunosato 1, Tsukuba, Ibaraki, Japan

1

Introduction

ANDREW T. SMITH

"Lagomorph"—the colloquial word used to describe members of the mammalian order Lagomorpha—is not as familiar in everyday usage as are other common names of mammal groups such as "carnivore" or "rodent." Even the definition of "lagomorph"—"hare shaped"—is perplexing due to its circularity. But the rabbits, hares, and pikas, which constitute the order, are certainly well known to most people. Children grow up with the white rabbit and March hare characters in Lewis Carroll's *Alice's Adventures in Wonderland* or Hazel the rabbit of Richard Adams's *Watership Down*. The European rabbit (*Oryctolagus cuniculus*) has been domesticated, and many rabbit breeds are found around the globe. As a result, many children have had rabbits as pets, and in some circles rabbits are considered a food delicacy. Mark Twain humorously captured the physique and character of hares (genus *Lepus*) in *Roughing It*:

> As the sun was going down, we saw the first specimen of an animal known familiarly over two thousand miles of mountain and desert—from Kansas clear to the Pacific Ocean—as the "jackass rabbit." He is well named. He is just like any other rabbit, except that he is from one third to twice as large, has longer legs in proportion to his size, and has the most preposterous ears that ever were mounted on any creature but a jackass. When he is sitting quiet, thinking about his sins, or is absent-minded or unapprehensive of danger, his majestic ears project above him conspicuously; but the breaking of a twig will scare him nearly to death, and then he tilts his ears back gently and starts for home. (Twain 1962:37)

Comprehensive treatments of how rabbits and hares form an integral part of our culture are given in Evans and Thomson (1974), Carnell (2010), Lumpkin and Seidensticker (2011), and Dickenson (2014).

The pikas (genus *Ochotona*) are the least well known of the lagomorphs, although recently they have been getting increased attention. Gift shops within the range of the American pika (*O. princeps*) now stock an assortment of pika paraphernalia. In Japan, the Pika Fan Club of Hokkaido is one of the largest green groups in the country. And in spring 2015, an image of the first Ili pika (*O. iliensis*) to be seen in 20 years (portrayed in that account) grabbed the attention of the world; for two days the so-called teddy-bear pika was the number one trending online story in the world.

Not all views of lagomorphs are favorable; they are considered pests in some regions. For example, the European rabbit is viewed as an alien invasive species in Australia, New Zealand, and other areas of the world, where it has caused calamitous economic damage (see "Introduced Lagomorphs" and "Diseases of Lagomorphs"). Many farmers, both industrial and artisanal, are plagued by invasions of rabbits eating their crops. The white-tailed jackrabbit (*L. townsendii*) remains classified in Wyoming as a "varmint" (definition: an objectionable or undesirable animal), even though some conservationists suggest its numbers could be at risk (Berger et al. 2005). For more than four decades, the plateau pika (*O. curzoniae*) has been subject to what is likely the most massive poisoning campaign directed at a native mammal species because it is considered (falsely) to be an agricultural pest on the high alpine grasslands of the Qinghai-Tibetan Plateau (Delibes-Mateos et al. 2011).

An underappreciated aspect of lagomorph biology is the remarkable diversity of monotypic genera (a genus consisting of a single species) of rabbits. Few realize that there are black rabbits in Japan (Amami rabbit *Pentalagus furnessi*), montane rabbits living just outside Mexico City (volcano rabbit *Romerolagus diazi*), and dwarf rabbits in the inter-montane west sagebrush flats of North America (pygmy rabbit *Brachylagus idahoensis*). These are just three of seven monotypic genera of rabbit found across the globe, and by definition, each of these species represents a unique genetic lineage.

Use of Lagomorphs

Commercial use of pikas has been very limited. In the past, steppe-dwelling pikas in Asia were harvested to make high-quality felt, but this activity has ceased. The state of Alaska recently reclassified the collared pika (*O. collaris*) as a fur-bearer, but there is no evidence of any subsequent harvest. In Central Asia the soft feces of pikas are distilled and filtered to make "mumeo," a local remedy for rheumatism and a wide variety of other ills.

Leporids (rabbits and hares) have been used consumptively or in sport for millennia. These species provide high-quality meat, and there is evidence that many ancient peoples relied heavily on them. As long as 50,000 years ago, rabbits contributed to a consistently high proportion of the game consumed by early hominins on the Iberian Peninsula (Fa et al. 2013). Similarly, in the American Southwest both cottontail rabbits and jackrabbits figured prominently in the diet of the Anasazi, who were even known to engage in stylized rabbit drives (Cordell 1977). Around the globe today, rabbits and hares as bushmeat form a consistent part of the human diet wherever they are found (Lumpkin and Seidensticker 2011), and the eastern cottontail (*Sylvilagus floridanus*) is the most common game animal in the United States. The most recent rabbit to be discovered by science, the Annamite striped rabbit (*N. timminsi*), was initially found in a game market.

From Europe to North America, rabbits and hares are a desired game species. Millions of hares and rabbits are shot in Europe each year. In areas of unsustainable use, wild hares and rabbits have been translocated across Europe, or even imported from Argentina, to keep hunting bags high. However, hunting bags have declined over the past decades, mainly due to the intensification of agricul-

ture. This development has resulted in both habitat management activities and strategies to implement sustainable harvest rates (Marboutin et al. 2003).

An example of the difficulties in understanding the status of cottontail rabbits and hares is the manner in which these species are managed as game species by state resource agencies in the United States. Hares, and particularly cottontail rabbits, have long been favored species by hunters. The goal in management should be the sustainability of these species as a game resource, but often the hunt statistics of multiple leporid species have been lumped together, making it difficult to understand long-term trends. This is further complicated by natural fluctuations in the density of these species. Last, data on the status of cottontail rabbits and hares are rarely shared between state agencies. Analyses are now under way to examine the trends across all western states, and preliminary results are alarming: over the past 10–50 years most leporid species show declining population indices, and hunt success for these species has similarly declined dramatically (D. Brown and G. Beatty, personal communication). These declines may be due to changes in land use and habitat quality, extended drought, and possibly increased predation. This example highlights the need to understand better even those lagomorph species that most consider common, or they may spiral into decline.

A similar situation can be found in the European Alps, where alpine mountain hares (*Lepus timidus varronis*) and European hares (*L. europaeus*) are often lumped together in the hunting statistics. Thus, a potential decline of alpine mountain hares due to global warming might be camouflaged by stable hunting statistics that include increasing numbers of European hares.

Historically, one of the most stylized forms of pursuing rabbits was "coursing"—or the hunting of hares with dogs. This practice, with competition and betting on the greyhounds engaged in the chase, dates back to the time of Elizabeth I. Times have changed, and hunting of hares with dogs was outlawed by the British parliament in 2004 (Carnell 2010).

Ecosystem Services Provided by Lagomorphs

Lagomorphs provide critical ecosystem services; many are considered keystone species in the environment they occupy, and others are ecosystem engineers. Many of the species accounts in this book highlight the role that lagomorphs play as critical prey species for a wide variety of carnivores. The Canadian lynx (*Lynx canadensis*) subsists almost entirely on the snowshoe hare (*L. americanus*). On the Tibetan Plateau, the plateau pika is the primary prey species targeted by nearly every avian and mammalian carnivore; when the pikas are indiscriminately poisoned these carnivores disappear (either starving to death or moving to where pikas have not been poisoned; Smith and Foggin 1999; Badingiuying et al. 2016). Two of the world's most endangered carnivores, the Spanish imperial eagle (*Aquila heliaca*) and the Iberian lynx (*L. pardinus*), rely almost exclusively on the European rabbit to survive. Thus, the status of these carnivores is linked to the declining numbers of European rabbits, which in turn may be negatively impacted by disease or habitat loss (Delibes-Mateos et al. 2011).

There are a myriad number of ways in which lagomorphs play a positive role in natural ecosystems. They increase vertebrate and invertebrate species richness, increase plant species richness, improve soil organic content and soil moisture, increase available nitrogen in soil, enhance nutrient cycling, promote increased plant growth and biomass, actively disperse seeds, improve seed germination rate and success, and help stabilize vegetative communities (Delibes-Mateos et al. 2011). One recent study highlights the global impact that a lagomorph can have. The plateau pika is a burrowing form, and its burrows have been shown to increase infiltration of water and minimize local erosion from seasonally heavy monsoonal rains on the Qinghai-Tibetan Plateau. When this species is poisoned, its burrows collapse, and the potential for downstream flooding increases. As the rivers that originate in the range of the plateau pika impact downstream approximately 20% of the world's human population, one can see the importance of this one small species (Wilson and Smith 2015).

How to Use This Book

We begin each account with the complete scientific name, common name, and any additional common names for each species. Each account includes a contemporary geographic range map of the species and a description of its current range occupancy. A full description of the appearance and unique morphological characteristics of each species is accompanied by a range of standard measurements of adult specimens in mm or g: (1) head and body length; (2) tail length; (3) hind foot length; (4) ear length; (5) greatest length of skull (or condylobasal length); and (6) weight. When available, an image of each species is presented, sometimes accompanied by a contrasting image of the species in winter or engaged in a particular behavior. Habitat features are shown for some representative species.

Subsequent sections discuss any known paleontological data concerning the species and the current state of its taxonomy and geographic variation. This latter section includes subspecies and any relevant historical or currently unresolved controversies in the systematics of the form. This section, iterated over all the species accounts, highlights that the state of lagomorph systematics is far from resolved; significant ongoing and future work will be necessary to solidify our understanding of the systematics of many lagomorph species.

Narratives covering the ecology, habitat and diet, behavior, genetics and physiology, reproduction and development, and parasites and diseases follow. Some lagomorph species are well known, but the information that we have on several species is rudimentary or lacking. In some accounts we combine some of these topics. It is clear that significant new studies need to be initiated on nearly all species of lagomorph, and we hope that this volume clearly points to our gaps in knowledge and provides the most current information available. We need an army of new lagomorph biologists to expand our knowledge of lagomorph biology, particularly with intensive field investigations.

As many as one-quarter of all lagomorph species are threatened with extinction. In each species account we present the International Union for Conservation of Nature (IUCN) Red List status (see "Lagomorph Conservation"); additionally, we present the national threatened status for those species so listed. We intertwine our discussion of conservation threats and initiatives with other aspects of management in a single concluding narrative in each species account. This approach was taken because conservation cannot be considered in isolation from other interactions that the species has with local people.

Each species account concludes with a list of key references; the full bibliographic citation for each of these appears in a master reference list.

2

Evolution of Lagomorphs

LUIS A. RUEDAS, JOSÉ M. MORA, AND
HAYLEY C. LANIER

What, if anything, is a rabbit?

(Wood 1957:417)

As currently understood, the order Lagomorpha has two living families: the Ochotonidae (pikas) and the Leporidae (cottontails and hares or jackrabbits). There are 29 extant species of pika in the genus *Ochotona*, whereas the Leporidae includes 63 species in 11 genera. The Sardinian pika, †*Prolagus sardus*, represents a third, recently extinct family of pika relatives, the †Prolagidae. However, and notwithstanding this apparent simplicity, both the taxonomic and the evolutionary histories of the Lagomorpha are complex and have been extensively debated.

The original taxonomic understanding of the Lagomorpha (Linnaeus 1758) was unusual by modern standards. Linnaeus recognized eight orders of mammals, ranging taxonomically from what we might today consider a family (the whales, "Cetae") to strange collections of subclasses (his "Bestiæ" and "Glires"). Lagomorphs were in the order Glires, which included six genera: *Rhinoceros* (rhinoceroses, two species), *Hystrix* (porcupines, five species), *Lepus* (hares and cottontails, four species), *Castor* (beaver, two species), *Mus* (mice and allies, 15 species), and *Sciurus* (squirrels, seven species). Thus the first taxonomic association of Lagomorpha was in what we might today consider a group of several distinct orders including rodents and, incongruously, an odd-toed ungulate: the rhinoceros.

Linnaeus's "genera" in Glires are in fact akin to modern families, if not orders. Linnaeus did not include pikas in the definitive tenth edition of his nomenclatural treatise; the four species included in *Lepus* were *L. timidus*, *L. capensis*, "*Lepus*" (= *Oryctolagus*) *cuniculus*, and "*Lepus*" (= *Sylvilagus*) *brasiliensis*. As a bit of trivia, the legendary modern "jackalope" appears to have been known to Linnaeus, who noted (1758:57) that the horned hare, if not a hybrid, was almost certainly mythical.

In the particular instance of *Lepus* in the tenth edition, Linnaeus's family concept was identical to what today we would call Leporidae. However, his diagnosis of the genus *Lepus* based on the duplicated first pair of upper incisors is in fact the modern diagnosis of the order Lagomorpha (Gidley 1912). Illiger (1811) used that character to erect the family Duplicidentata (still within the rodents), including two genera: *Lepus* and *Lagomys* (= *Ochotona*). Fischer called this same family-level group of *Lepus* and *Lagomys* the Leporini (Fischer 1817:372) and Leporinorum (Fischer 1817:409). The first use of "Leporidae" was by Gray (1821:304); however, he used the term in the same sense as Illiger and Fischer, to include *Lepus* and *Lagomys* (thus corresponding to Lagomorpha) in his order Rosores. Waterhouse (1839) later split "rodents" into three higher categories, or sections, the third of which, Leporina, corresponded to the contemporary concept of Lagomorpha.

It was Brandt (1855:295) who eventually coined for the group the now commonly used term "Lagomorpha" ("hare shaped"), also at the subordinal rank within rodents (along with Sciuromorpha, Myomorpha, and Hystricomorpha). In addition, although he clearly called lagomorphs a suborder ("Subordo IV. Lagomorphi seu

Lagomorpha"), Brandt (1855:295) began his discussion of the groups with the words "Ordo *Leporinus* . . . ," underscoring their stark distinction from rodents, based on the presence of four upper incisors in lagomorphs. Thus, while it could be argued that Brandt was the first to suggest ordinal status for lagomorphs, it was not until 1912 that J. W. Gidley formally called for ordinal rank for the Lagomorpha.

Evolutionary History and Chronology of the Lagomorpha—Interordinal Relationships

The early evolutionary diversification of Glires, the group including both rodents and lagomorphs, has been hypothesized to have occurred in the Late Cretaceous, immediately preceding the Cretaceous-Paleogene boundary (Asher et al. 2005), some 66 mya (near the time that non-avian dinosaurs went extinct). Evidence for this early diversification comes from the extinct †Mimotonidae, a non-natural grouping (in this instance, not including all descendants from a last common ancestor) of Glires forming with Lagomorpha the Duplicidentata, and the acknowledged closest relatives to Lagomorpha (Asher et al. 2005). Successive sister groups included †*Gomphos elkema* and also the non-natural †Eurymylidae (Asher et al. 2005).

Within Lagomorpha, as currently understood, the earliest known fossil dates to at least 53 mya and is known from Early Eocene Cambay Shale deposits in west-central India from the Vastan lignite mine (Rose et al. 2008). Prior to that report, the oldest Lagomorpha were dated to the Irdinmanhan (48.6–37.2 mya; Middle Eocene) of Asia, and late Early or early Middle Eocene of Central Asia (Li et al. 2007).

Our understanding of the early evolutionary relationships with Lagomorphs has also been impacted by Rose et al. (2008), who carried out a phylogenetic analysis of the fossils using 71 taxa and 228 characters based on a matrix previously published by Asher et al. (2005). Using those characters, the Vastan fossils were indicated to be a sister group to *Oryctolagus*, to the exclusion of *Lepus* and *Sylvilagus* (figure 1), leading Rose et al. (2008) to hypothesize that the modern families Leporidae and Ochotonidae had already diverged by the Early Eocene. Molecular analyses by Springer et al. (2003) support this deep division, proposing dates for the split between these two lagomorph families to range from 41 to 71 mya, depending on the methods employed. In a similar approach that included

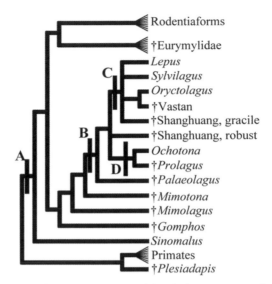

Fig. 1. Detail of the Glires portion of the phylogenetic tree of Rose et al. (2008:1205, Fig. 3) based on 228 morphological characters (characters from Asher et al. 2005, with 20 additional tarsal characters from Rose et al. 2008 [electronic supplementary material]). Taxa daughter to node A include all Glires; daughters to node B include Lagomorpha; daughters to node C include Leporidae; and daughters to node D, Ochotonidae.

more fossil taxa and a much greater number of morphological characters, O'Leary et al. (2013) also supported a clade age for lagomorphs of 53 mya, while cautioning that only one early lagomorph was included.

Some recent authors have used earlier dates based on the work of Stucky and McKenna (1993), who extended the fossil record of lagomorphs into the Late Cretaceous (93.5–89.3 mya). However, that assertion relies on fossils of †Zalambdalestidae and †Pseudictopidae being considered lagomorphs, a hypothesis that does not appear to be supported by current data (e.g., McKenna and Bell 1997; O'Leary et al. 2013). Additional molecular estimates for the ancestry of the Lagomorpha likewise support a deep origin: e.g., 86.4 mya (range 79.7–94.1; Murphy et al. 2007) or 79.5 mya (range 71.5–94.1; Meredith et al. 2011).

Disagreement among estimates for the origins of the Lagomorpha may partially be explained by the use of differing datasets and with missing data. While the Rose et al. (2008) data enable identification of their fossils as potentially lagomorph in nature, it may be insufficient to provide strong taxonomic placement within Lagomorpha, particularly since some of the characters are primitive and shared across all early placental mammals (e.g., Asher et al. 2005), and hence are not suitable for determination of sister taxon status.

Based on the above, a conservative position might be that definitive lagomorphs were present around the late Early Eocene (~ 55–50 mya), based on fossils of that age that can be assigned to Leporidae (Stucky and McKenna 1993), the Rose et al. (2008) fossils, and combined molecular and morphological analyses (Asher et al. 2005; O'Leary et al. 2013). Similarly, †*Heomys*, originally described as a eurymyloid from the Early to Late Paleocene, has been controversial: McKenna and Bell (1997) considered it a rodent, but Asher et al. (2005) supported its inclusion in †Eurymylidae, as transitional between lagomorphs and rodents, but fully neither (Carroll 1988). Indeed, although a relationship with rodents has been proposed for eurymylids (Li and Ting 1985), the analysis of Asher et al. (2005) suggests that eurymylids are sister to lagomorphs. Benton and Donoghue (2007) indicated that eurymylids may be an outgroup to both Rodentia and Lagomorpha, and proposed a minimal estimate for the split between the two orders of 65.5 to 61.7 mya (Danian, Early Paleocene of China). In contrast, Asher et al. (2005) suggested that eurymylids were not a natural group, with †*Synomylus* sister to Rodentia + (†Eurymylidae + Lagomorpha).

Another, recently described, potential ancestral lagomorph taxon, †*Dawsonolagus antiquus*, is known from the late Early Eocene of Asia, and can be taxonomically assigned based on cranial, dental, and ankle bone materials (Li et al. 2007). However, Li et al. (2007) explicitly indicated that †*Dawsonolagus* is represented by partial elements and exhibits a mosaic of mimotonid and lagomorph features. O'Leary et al. (2013) also showed †*Dawsonolagus* as sister to Lagomorpha. At issue would therefore be whether †*Mimotona* (dating from the Early to Late Paleocene; Meng and Wyss 2005) and mimotonids (†Mimotonidae: †Mixodontia?) are in fact lagomorphs. Certainly mimotonids appear to be sister to the modern concept of Lagomorpha (Asher et al. 2005; figure 2). Likewise, if the broader definition of Lagomorpha were employed, as used by Simpson (1945:75), then eurymylids are surely lagomorphs, and hence by definition are mimotonids (but see below). In that case, Lagomorpha would indeed date to the very earliest Eocene, and have origins in the Late Cretaceous, likely the Maastrichtian, about 65.5–70.6 mya, as suggested by the analyses of Asher et al. (2005).

Although mimotonids (particularly smaller ones) share some synapomorphies with true lagomorphs, the relationships among mimotonids, eurymylids, and lagomorphs often differ based on the characters analyzed (López-

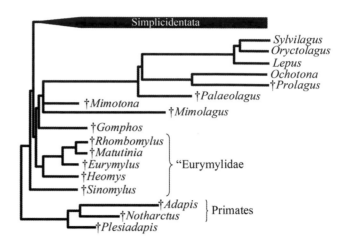

Fig. 2. Detail of the Glires portion of the hypothesis of phylogenetic relationships from Asher et al. (2005; Fig. 3), redrawn to reflect the order of the taxa as displayed in figure 1 for ease of comparison. The topology is based on combined analysis of unordered morphological characters and cytB, A2AB, IRBP, vWF, and GHR genes. Asher et al. (2005:1093 and Fig. S7) note that the morphological data alone results in a sister group relationship between eurymylids and Rodentia, with †*Sinomylus* basal to dupicidentates + {simplicidentates + eurymylids}. They also note that placing †*Palaeolagus* within crown Lagomorpha as sister taxon to leporids requires just two additional steps in the combined (morphological + molecular) analysis; neither alternative for both sets of topological relationships could be decisively rejected based on Wilcoxon rank sum tests.

Martínez 2008). In support of this position, Averianov (1994a) argued that †*Mimotona* in particular lack important shared characters and the evolutionary tendencies of later lagomorphs (most notably a complex first lower molar). Averianov (1994a) concluded that †Mixodontia (which he construed as †Mimotonidae + †Eurymylidae) constitutes a grade (i.e., a transitional taxon) rather than a clade (a natural group defined by ancestor-descendant relationships) because although †Mimotonidae and †Eurymylidae are apparently sister taxa, the latter appears to be most closely related to Rodentia, while the former is more closely related to Lagomorpha.

"True" (modern) lagomorphs may therefore have split from the mimotonid lineage at any time between the K-Pg boundary (65 mya) and the Middle Eocene given that definitive lagomorphs are not known until the Middle Eocene ("Selandian"; ~ 58.7–61.7 mya; McKenna and Bell 1997; Meng et al. 2005) of both Asia, the likely center of origin of the group (Meng et al. 2003; Asher et al. 2005; Li et al. 2007), and North America. Other genera previously assigned to Leporidae by Stucky and McKenna (1993) and McKenna and Bell (1997) are instead Lagomorpha *incertae sedis* as "lagomorphs of modern aspect" (Li et al. 2007:97).

Evolutionary History and Chronology of the Lagomorpha—Intraordinal Relationships

Given that there only are two extant families in Lagomorpha (Leporidae Fischer de Waldheim, 1817, and Ochotonidae Thomas, 1897), they are sister taxa, or each other's closest relatives, by definition. However, numerous fossil families of Lagomorpha also have been named, including †Paleolagidae Dice, 1929; †Mytonolagidae Burke, 1941; †Desmatolagidae Burke, 1941; †Prolagidae Gureev, 1962; †Mimolagidae Erbajeva, 1986; and †Strenulagidae Averianov and Lopatin, 2005. As noted by Li et al. (2007), not only is the rank of these taxa unstable, but further, with respect to †Strenulagidae, may be diagnosed by primitive features, suggesting that †Strenulagidae may represent a grade rather than a clade (Li et al. 2007). Erbajeva et al. (2011) recognized five families within Lagomorpha: Mimolagidae, Leporidae, Paleolagidae, Prolagidae, and Ochotonidae.

Fossil data suggest that the divergence between Ochotonidae and Leporidae occurred before about 42 mya, based on the first occurrence of Ochotonidae in the Eocene (Uintan North American faunal stage; 42–46.2 mya). This occurrence is based on a record of a species of †*Desmatolagus*, †*D. vusillus*, in the Swift Current Creek fauna of Saskatchewan (Storer 1984). Storer (1984) argued that the specimen, described as a †*Procaprolagus* at the time, represented a leporid. However, †*Procaprolagus* and †*Desmatolagus* have since been synonymized (Meng and Hu 2004). A Uintan date significantly pushes back the Late Eocene results obtained by Asher et al. (2005), based on combined analyses of morphology and molecules, as well as the proposal of Erbajeva et al. (2015), who hypothesized that the first Ochotonidae are from the Asian early Late Oligocene (Early Chattian, < 28.1 mya) while not considering †*Desmatolagus*. Fostowicz-Frelik and Meng (2013) hypothesized †*Desmatolagus* to be a stem lagomorph outside crown group lagomorphs rather than an early ochotonid.

Springer et al. (2003) suggested a range of 41 to 71 mya for the ochotonid-leporid split: the lower estimate is congruent with the foregoing fossil data. However, the upper estimate is more congruent with the origin of Lagomorpha. In contrast, Matthee et al. (2004) suggested a somewhat younger divergence date for this familial divergence of 28.96 ± 3.80—22.44 to 37.15 mya (estimate ± standard deviation, 95% credibility interval) based on an analysis of molecular data from combined nuclear and mitochondrial genes. Using a 37 mya divergence between Ochotonidae and Leporidae, based on fossil data, resulted in a Middle to Late Miocene diversification among *Ochotona* (Lanier and Olson 2009), a range of dates that is in agreement with the first fossil occurrence of the genus. On the basis of family-level sampling, a 26-gene fragment supermatrix, and multiple fossil calibrations, Meredith et al. (2011) recovered a somewhat more recent date of 50.2 mya (range 47.4–56.9 mya). Given that the earliest lagomorph fossils, even accepting eurymylids and mimotonids as lagomorphs, barely extend the fossil record to the K-Pg boundary, and that early definitive modern lagomorphs display characters common to both ochotonid and leporid lineages until the Eocene, the 50.2 mya date of Meredith et al. (2011) may represent the earliest plausible date for the divergence between the Ochotonidae and the Leporidae.

Conclusions

There is a broad perception that the morphological conservatism of craniodental and skeletal features in Lagomorpha makes this an intractable taxonomic group. Nevertheless, and with particular respect to these taxonomic issues, Meng et al. (2003, 2004), Asher and colleagues (Asher et al. 2005), and in particular Wible (2007), have shown that lagomorph crania and dentitions are character rich, waiting to be mined and analyzed within the context of the phylogenetic framework (O'Leary et al. 2013). Identifying species limits and relationships between species, long considered among the most challenging remaining problem of lagomorphs (Alves and Hackländer 2008), should now be based on both morphological and molecular work. Joint consideration of morphological and molecular characters may allow for better placement of fossil taxa (e.g., Asher et al. 2005), better resolution of ambiguous dental and cranial characters, and potentially stronger estimates of major evolutionary transitions in the group.

With respect to the issue of basic alpha taxonomy, we indicated in the taxonomic summary above that Lagomorpha currently comprises some 92 recognized extant species in 12 genera and 2 families. However, this process of species detection, description, and classification remains one of the more pressing areas of investigation in this group, and is of particular importance for studies of evolution, conservation, and ecology. It is also of particular importance insofar as the speciose genera are concerned: new species have recently been described or taxonomically resurrected in *Ochotona*, *Lepus*, and *Sylvilagus*. More undescribed and unrecognized species-level taxa undoubtedly exist in these genera, and their revelation will undoubtedly help in conservation efforts. Furthermore, the presence of widespread hybridization and introgression between certain species, as well as geographic variation, underscores the importance of validating proposed classifications with multiple lines of molecular and morphological evidence.

What, If Anything, Is a Lagomorph?

We began this chapter by quoting the classic title of Albert Elmer Wood's 1957 paper, "What, if anything, is a rabbit?" Wood expanded on the question in the text: first, that the question had to do not with rabbits in particular (Leporidae), but rather that there was no specific common name for all members of Lagomorpha, and that he referred to rabbits as lagomorphs in general. Thus today, the question is more aptly phrased as "what, if anything, is a lagomorph?" Following Simpson, who had suggested that rodents and lagomorphs were distinctive enough that "they have nothing at all to do with each other aside from being eutherian mammals" (Simpson 1945:196), Wood used

dental features to assert that there was no valid basis for considering that rodents and lagomorphs have a common ancestral basis (Wood 1957).

However, the vastly expanded morphological datasets in use today (e.g., Asher et al. 2005; Meng et al. 2003, 2004; Wible 2007) as well as the molecular data (e.g., Asher et al. 2005; 2009; Murphy et al. 2007; Meredith et al. 2011) all point to one conclusion: Rodentia and Lagomorpha are each other's sister taxon and jointly constitute a monophyletic group currently known as Glires. Indeed, the construct that Simpson (1945) called cohort Glires and Linnaeus called order Glires (absent rhinoceros) has if anything been strengthened as it has increasingly become restricted to Lagomorpha and Rodentia. The new burning questions therefore are more concerned with the fossil taxa: taxa that remain poorly defined and analyzed, with numerous characters still missing in particular instances. Is †*Sinomylus* really a eurymylid? Are eurymylids—*sensu stricto* or *sensu lato*—the sister taxon to Lagomorpha? Are mimotonids lagomorphs, or do they constitute a distinct, ordinal-level taxon? As a consequence of the massive datasets being developed and analyzed today, Wood might today rather ask: What is Glires, and where in Glires do we draw the line between Lagomorpha and non-Lagomorpha?

3

Systematics of Lagomorphs

JOSÉ MELO-FERREIRA AND PAULO C. ALVES

Taxonomy and systematics are often considered to be synonymous. However, taxonomy is a branch of biology concerned with identifying, naming, and classifying organisms, whereas systematics includes both taxonomy and the relatedness over time in the form of a tree of life (phylogeny). Thus, systematics aims to classify and name biodiversity components according to evolutionary patterns. One of the major challenges of systematics is species delimitation as the basic unit of the classification of biodiversity. Because evolution is a continuous process and species as taxonomic entities are discrete units, this task remains fairly controversial among specialists. From the initial naturalist perspective, when morphology was generally used as the sole guide for species classification, to the present, when our ability to collect massive amounts of information from natural populations is unprecedented, this debate has resulted in numerous attempts to create simple definitions using objective criteria (see, e.g., Avise 2004 for a summary of species concepts). While for deep levels of divergence different concepts tend to agree in the identification of species, for closely related natural populations that lie in the so-called gray zone of species divergence, different criteria result in distinct classifications (De Queiroz 2007), generating obvious difficulties for taxonomy. In addition, even within species concepts, the continuous makeup of population diversification and the peculiarities of each situation create substantial difficulties inherent to the evolutionary process. For example, extensive sharing of characters among sister species is expected in cases of speciation during rapid radiations, or of secondary hybridization between populations that maintain some or high levels of reproductive isolation, the key criterion of the biological species concept (Mayr 1942; Coyne and Orr 2004). Another common difficulty is geographic separation, which tends to foment classification of distinct species even if the common history is very recent.

The complex nature of species delimitation, coupled with our ability to merge important amounts of information from different disciplines, led to the creation of the notion of integrative taxonomy to define species (Dayrat 2005). This approach seeks consensus among distinct disciplines and attempts to use the evolutionary perspective to understand mechanisms of species divergence and resolve disagreements (Pante et al. 2015). The pragmatic definition of species thus remains complex, multidimensional, and provisional, and corresponds to the best understanding of the biology and evolution of the organisms under study.

Difficulties and Progress in Lagomorph Systematics

Lagomorphs comprise more than 90 species, distributed in 2 families, Ochotonidae and Leporidae. The family Ochotonidae is composed of a single genus, *Ochotona*, known as pikas. The leporids are composed of 11 genera, 10 known as rabbits and 1 as hares (*Lepus*). Most rabbit genera (*Brachylagus*, *Bunolagus*, *Caprolagus*, *Oryctolagus*, *Pentalagus*, *Poelagus*, and *Romerolagus*) are monotypic.

Thus, most lagomorph species are included in only 3 genera: *Ochotona* (29), *Sylvilagus* (18), and *Lepus* (32). While there are some inconsistencies in the phylogenetic relationships at a generic level, namely, among the rabbit group (Matthee et al. 2004), the main systematic concerns are among the species within genera. These controversies are mainly related to evolutionary history (rapid radiation, local adaption, and hybridization) and ecological features (inhabiting a great variety of environments, including the arctic, deserts, grasslands, or tropical and boreal forests), coupled with extensive lack of information on many of the species within *Ochotona*, *Sylvilagus*, and *Lepus*.

Rapid Radiations

During most of the past 45 million years, lagomorphs were much more common on the planet than today (López-Martínez 2008). The fossil record shows that around 78 genera and more than 200 species inhabited both the Old World and North America, suggesting that lagomorphs underwent numerous explosive radiations during their evolution. This led to complete replacement and diversification of genera and species over very short periods of time. At the generic level, lagomorphs initially spread rapidly in Asia and North America during the Eocene, and later to Europe during Late Eocene–Early Oligocene, and then to Africa in the Late Oligocene–Early Miocene. The most recent diversification occurred during the Pliocene, with a subsequent decline during the Quaternary (López-Martínez 2008). A similar pattern of rapid radiations is seen at the species level. At the end of Miocene, e.g., the formation of the steppe zone in Eurasia allowed the rapid diversification of the genus *Ochotona* (see, e.g., Erbajeva and Zheng 2005). Likewise, diversification of leporids was explosive (Hibbard 1963), and within the genus *Lepus* extant species diversified recently and very rapidly (Matthee et al. 2004). The fast nature of speciation events during rapid diversification processes leads to extensive sharing of characters among the resulting species, which complicates the correct inference of species splits (i.e., the phylogenetic patterns) and consequently the delimitation of species. In leporids, e.g., efforts to resolve the phylogeny using morphological characters have generally resulted in poorly resolved estimates mostly at recent evolutionary scales, such as infra-genus or among closely related genera, but also provided conflicting results even at deeper evolutionary scales (Hibbard 1963; Dawson 1981; Corbet 1983; Ave-

rianov 1999; Fostowicz-Frelik and Meng 2013; Ge et al. 2015). Note, however, that morphological characters are particularly prone to convergent evolution due to similar adaptive pressures, which may explain at least part of the lack of phylogenetic signal (e.g., Ge et al. 2015). Chromosome painting analyses also suggest a rapid radiation of the most recent common ancestor of lagomorphs with 2n = 48, with only punctual chromosomal rearrangement in some genera (Robinson and Matthee 2005; Robinson et al. 2002). At the infra-generic level, when investigated, karyotypes tend to be extremely conserved, as, e.g., within *Lepus*, where all species have 2n = 48 chromosomes (Robinson et al. 1983a).

At the molecular level, rapid radiations are characterized by extensive sharing of lineages due to retention of ancestral variation, which is one of the strongest confounding effects for phylogenetic reconstruction. Importantly, this effect varies along the genome, and phylogenetic trees based on a single gene may not correspond to the correct speciation history (Maddison 1997). From the use of single gene phylogenies to attempts to reconstruct speciation histories in lagomorphs (e.g., Biju-Duval et al. 1991; Halanych et al. 1999), multilocus phylogenies have emerged as the most powerful inference methods, initially using supermatrix approaches (Matthee et al. 2004) and then coalescent-based species-tree reconstruction methods (Melo-Ferreira et al. 2012; Melo-Ferreira et al. 2015). However, low phylogenetic resolution was still generally found at lower scales of divergence, namely, at the genus level (*Ochotona*, *Sylvilagus*, and *Lepus*), which reflects the effect of rapid radiations. Going from a dozen to hundreds or thousands of loci holds great promise to resolve many uncertainties (see Posada 2016 and references therein).

Local Adaptation and Cryptic Speciation

Rapid radiations are often associated with fast local adaptations, and when these are coupled with patterns of global colonization, they make lagomorphs an excellent model for studying the genomic signature of local adaptations. However, rapid local adaptations may enhance phenotypic differences among species that diverged recently, creating additional difficulties for systematics. Lagomorphs have a worldwide distribution, covering arctic to desert regions, open to dense forest landscapes, and altitudinal gradients (sea level to > 6,000 m asl). Changes in the environment create new resources, eliminate existing resources, or open new environmental niches, thus promoting physiological and phenotypic adaptations. Divergent lagomorph species, namely, pikas, but also some hare and rabbit species, clearly demonstrate interesting cases of local adaptation, and have very restricted ranges (e.g., Ili pika *O. iliensis*; riverine rabbit *B. monticularis*; Amami rabbit *Pentalagus furnessi*; Manzano Mountain cottontail *S. cognatus*; Omiltemi rabbit *S. insonus*; broom hare *L. castroviejoi*). On the contrary other species span an impressive range of different environments (e.g., mountain hare *L. timidus*; Cape hare *L. capensis*; eastern cottontail *S. floridanus*), suggesting extreme physiological plasticity. However, even though neighboring species can potentially interbreed, some are too distantly distributed to allow for genetic exchange. L. capensis provides an example of this situation, where there is no evidence of gene flow between the South African and northern populations, and several subspecies have been described. Nevertheless, other evidence suggests a close phylogenetic relationship between *L. capensis* and the European hare (*L. europaeus*), and some authors even consider that *L. europaeus* belongs in the *L. capensis* complex (Petter 1959).

The complexity of species delimitation also becomes problematic when evolutionary forces are able to maintain intrinsic divergence with little morphological differentiation, which is known as cryptic speciation. This phenomenon implies that species may accumulate genetic divergence over time, but environmental and ecological forces maintain morphological conservatism without detectable diagnostic differences. Molecular markers to assess phylogenetic relationships have been used to illustrate several cases of cryptic divergence in lagomorphs. For example, a recent study using a multilocus coalescent-based phylogeny shows that the snowshoe hare (*L. americanus*) is composed of three major evolutionary units, and one shows deep genetic divergence (~ 2.0 mya), comparable to what is found among other species such as the black-tailed jackrabbit (*L. californicus*) and the white-tailed jackrabbit (*L. townsendii*; Cheng et al. 2014; Melo-Ferreira et al. 2014a). This divergence was apparently not accompanied by morphological differentiation and is not associated with subspecies classification. Similar cases of cryptic diversity have been reported in the collared pika (*O. collaris*) and the tapetí (*S. brasiliensis*; see that account).

Hybridization

Hybridization between diverging populations and subsequent backcross lead to secondary transmission of derived characters, which brings obvious complications for taxonomy and systematics. Lagomorph species have been shown to be particularly prone to hybridization and secondary introgression. This tendency puts the group at the forefront of models for research focused on the formation of species and the causes of reticulate evolution (Melo-Ferreira et al. 2012). However, it complicates the delimitation of species. Because genetic introgression may affect different loci along the genome, the use of a single or only a few loci to reconstruct species phylogenies may lead to erroneous inferences of the evolutionary relationships among taxa, as is the case of mtDNA in hares (Alves et al. 2006). In the European rabbit (*Oryctolagus cuniculus*) multilocus (and even genomic) approaches have been fundamental to our understanding of the evolutionary history and systematics of the species, which is composed of two parapatric subspecies (*O. c. algirus* and *O. c. cuniculus*) that occupy the Iberian Peninsula. While most of the genome shows little differentiation between the subspecies, due to extensive gene flow, some other genomic regions show exceptionally high levels of differentiation, suggesting that genes in these regions are strongly involved in the maintenance of reproductive isolation between the forms (see Carneiro et al. 2014a). This mosaic pattern of genomic differentiation situates the rabbit subspecies in the gray zone of species divergence and complicates simple assessments of their taxonomic status.

Hares provide numerous examples of hybridization and secondary introgression between species. Interestingly, instances of gene flow between species not only reflect recent secondary contacts, such as between the Iberian hare (*L. granatensis*) and *L. europaeus* or between *L. timidus* and *L. europaeus* (Fredsted et al. 2006; Thulin et al. 1997; Melo-Ferreira et al. 2014c), but also historical hybridization in former areas of distribution of extant species (Melo-Ferreira et al. 2005; Melo-Ferreira et al. 2012). *L. granatensis* hybridized with *L. timidus* presumably by the end of the last glacial period, before the latter went extinct in the Iberian Peninsula. Traces of *timidus* are still found in the genome of *granatensis*, but generally at low frequencies, except for mtDNA, which remains at high frequency (Melo-Ferreira et al. 2005; Melo-Ferreira et al. 2009). Similar situations affected the Corsican hare (*L. corsicanus*) in Italy, *L. castroviejoi* in northern Iberia, and *L. europaeus* in its Iberian range. These results suggest that these species are well delimited from *L. timidus*, but that either demographic or selective processes cause high frequencies of introgression at particular loci (Alves et al. 2008; Melo-Ferreira et al. 2011; Melo-Ferreira et al. 2014b). A similar pattern is found in North American

L. americanus that has been massively affected by mtDNA introgression from *L. californicus* (Melo-Ferreira et al. 2014a). Hybridization and introgression are also phenomena of concern for conservation because the introduction of species in non-natural ranges may lead to pollution of the gene pool of native species and to its replacement with hybridization, as has been documented in Sweden (introduced *L. europaeus* overcoming *L. timidus*; Thulin 2003) and Corsica (introduced *L. capensis*, *L. europaeus*, and *L. granatensis* hybridizating with *L. corsicanus*; Pietri et al. 2011). The apparent predominance of hybridization within *Lepus* may be related to recent diversification and chromosome structure conservatism (Robinson et al. 1983a), which differs in other genera, such as *Sylvilagus* (Robinson et al. 1983b). However, species delimitation must take the possibility of hybridization into account, even if sometimes it is apparently unlikely. At finer scales, between sister taxa, this demands reconstructing the speciation parameters and inference of genetic migration rates using multilocus approaches that allow taking into account the effects of incomplete lineage sorting (Melo-Ferreira et al. 2012; Melo-Ferreira et al. 2014a). Genomic approaches will allow characterizing in detail the pervasiveness of gene flow between species and assess the nature of the speciation process (e.g., Carneiro et al. 2014a).

Allopatry

Allopatric species often retain close genetic similarities, which mostly occurs when allopatry is a result of recent fragmentation of historical larger ranges or recent colonization. These situations complicate the assessment of species status because one cannot assess the degree of reproductive isolation, and species classification must rely on quantifications of amounts of morphological or genetic differentiation or divergence. *L. timidus* perfectly illustrates this situation. It contains several isolated populations, in the Alps and Ireland, but these express a relatively low degree of differentiation (Hamill et al. 2007). They are therefore considered conspecific. Conversely, the Alaskan hare (*L. othus*) and the Arctic hare (*L. arcticus*), distributed in Alaska and Canada, respectively, also show a close genetic relationship with *L. timidus* (Melo-Ferreira et al. 2012) but, on the contrary, are currently considered

distinct species. Some other cases in hares with allopatric ranges are well identified and deserve further reflection about possible taxonomic revisions, such as the similarity of the black jackrabbit (*L. insularis*) and *L. californicus*, and that of *L. castroviejoi* and *L. corsicanus*. Similar situations may occur in *Sylvilagus*, which demand careful assessments.

Perspectives and Implications for Conservation

Many uncertainties persist in lagomorph taxonomy and systematics, namely, within those genera with many species: *Ochotona*, *Sylvilagus*, and *Lepus*. These ambiguities reflect an intricate evolutionary history, which couples adaptive and non-adaptive radiations, hybridization and range fragmentation, and emphasizes the challenge of species delimitation. New genomic-based approaches hold the promise to provide unprecedented resolution to solve these complex questions. The recent initiative LaGomiCs—the Lagomorph Genomics Consortium, an international collaborative effort for sequencing the genomes of all lagomorph species—will potentiate genomic resources, and thus contribute to clarifying some of the existing systematic uncertainties in this mammalian order (Fontanesi et al. 2016).

While the delimitation and classification of species will remain a complex and laborious scientific process, most conservation planning, such as global and regional assessments, and International Union for Conservation of Nature (IUCN) Red Lists and action plans, are based on species status. This adds a nonscientific dimension to the complicated nature of species delimitation: the political species concept. Certainly much needs to be done to better match species classification to the biological and evolutionary nature of populations. But it is our understanding that conservation planning should focus on preserving uniqueness, be it genetic, ecological, or of any other nature, regardless of the species status. That is what matters the most to preserve biodiversity and ecosystems at the regional level. Shifting conservation efforts from lagomorph species to subspecies, based on the simple criterion of uniqueness, may contribute to the resolution of much of the political debate.

4

Introduced Lagomorphs

BRIAN D. COOKE, JOHN E. C. FLUX, AND
NEVER BONINO

According to Long (2003) in his widely cited book *Introduced Mammals of the World*, nine species of lagomorph have been deliberately introduced by humans to areas beyond their historically known geographical ranges. Of these species, three have become established in such numbers that they have caused major economic damage or environmental disruption. The European rabbit (*Oryctolagus cuniculus*), the European hare (*Lepus europaeus*), and the eastern cottontail (*Sylvilagus floridanus*) are now considered invasive species in several countries. However, we know today that Long (2003) could not have been aware of other introductions, particularly in the Mediterranean region, which are still coming to light because of modern genetic research (Pierpaoli et al. 1999; Pietri et al. 2011). It turns out that the Corsican hare (*L. corsicanus*), endemic to peninsular Italy and Sicily, was actually introduced to Corsica (despite its specific name), where it was recently rediscovered (Scalera and Angelici 2003; Angelici and Spagnesi 2008a). The European hare was also introduced into Corsica in the twentieth century (from France and other countries), as well as the Iberian hare (*L. granatensis*; from Spain), which was also introduced in southern France. Finally, the Cape hare (*L*. cf. *capensis*) has apparently been introduced into Sardinia with the subspecies *L. c. mediterraneus*, probably introduced from North Africa, although its real systematic and phylogeographic status is yet to be determined (Scandura et al. 2007; Angelici and Spagnesi 2008b).

Despite the frequency with which lagomorphs may have been moved from one locality to another, perhaps for at least 2,000 years, few have become established widely and perhaps those on islands are often mere indicators of a wider traffic to mainland areas where releases did not persist because of more complex ecosystems. European rabbits, nevertheless, became established in countries such as Australia, New Zealand, Chile, and Argentina as well as hundreds of islands throughout the world (Flux 1994), ranging from sub-Antarctic islands such as Kerguelen to tropical islands such as Phoenix Island in the Pacific within 3° of the equator. European hares are now found in many of the same countries as rabbits, including Australasia and South America, but have also become established in North America around the Great Lakes area. The eastern cottontail, a North American species, was introduced comparatively recently into Italy, France, and Spain, and populations in Italy are spreading apparently at the expense of local European hares. Other introductions of lagomorphs have been confined to extensions of range, establishment of populations elsewhere on the same continent, or small numbers of islands.

History of Introductions and Spread

Not all introductions of lagomorphs were carried out for the same reason. European rabbits were systematically spread through northeastern Europe in medieval times with monasteries and Norman dukes and kings playing a major role in the spread into Britain in the thirteenth century. Rabbit hunting and rights to keep rabbits in a garenne or warren were at times used to cultivate political allies or to reward for past services. Rabbits were spread more widely as various European countries became colonial powers and releases of rabbits in far-flung localities became increasingly common. Rabbits were released in the Azores and Canary Islands soon after these localities became strategically important for Portuguese and Spanish ships traveling to the Americas in the sixteenth century.

In many instances rabbits and hares were introduced for hunting (e.g., rabbits and hares into Corsica and Sardinia), with some introductions possibly being made as recently as the sixteenth century (Angelici and Spagnesi 2008a; Pietri et al. 2011).

European rabbits and European hares were introduced into Australia and New Zealand mainly for hunting, but also because these animals were reminders of a faraway "home" for nostalgic colonists. These introductions were partly to re-create opportunities for sport shooting, coursing with dogs, falconry, or, in the case of New Zealand, hunting on horseback with rabbits and hares as substitutes for foxes. Indian hares (*L. nigricollis*) were taken from India to Mauritius by European colonists too, although those in the Seychelles were apparently introduced by plantation workers as a source of food (Long 2003).

Eastern cottontails, found mostly throughout the eastern and southern United States, but extending into the northernmost parts of South America, were also widely introduced into other parts of North America by hunters seeking to have more game animals. In the 1960s, they were also introduced by European hunters as substitutes for European rabbits after the viral disease myxomatosis

was introduced into France and reduced the numbers of European rabbits dramatically (see "Diseases of Lagomorphs").

Current Distributions and Status of Species Considered Invasive
The European Rabbit

European rabbits are native to the Iberian Peninsula, but were well established on Mediterranean islands in Roman times. Historical documents show that even in the fifteenth century Spanish and Portuguese navigators transported European rabbits during colonization of the Americas (Delibes and Delibes-Mateos 2015); rabbits were established in the Azores, Madeira, and Canary Islands by the sixteenth century, and these and similar introductions persist today. European rabbits were successfully introduced into Australia and New Zealand in the nineteenth century, and there was a contemporaneous spread of European rabbits into parts of northern Europe. European rabbits only became established in Sweden and parts of Russia in the early twentieth century.

In South America European rabbits were introduced into several regions of Chile, from where they invaded the southwestern part of Argentina. Rabbits are now present in three regions of Argentina: Tierra del Fuego, Santa Cruz Province, and Neuquén and Mendoza Provinces (Bonino and Soriguer 2009), and they are also present on the Falkland (Malvinas) Islands. In Santa Cruz and Neuquén the rabbit is currently undergoing a process of dispersion and invasion of new areas (Bonino and Gader

Table 1. Lagomorphs that have been introduced to regions beyond their natural distribution

Common name	Scientific name	Origin	Introduced
European rabbit*	*Oryctolagus cuniculus*	SW Europe	Worldwide
European hare*	*Lepus europaeus*	Europe and Asia	Sweden, Great Lakes area USA/Canada, New Zealand, Australia, South America, Ireland, Elba, Corsica, some Aegean islands
Corsican hare	*Lepus corsicanus*	Peninsular Italy, Sicily	Corsica
Iberian hare	*Lepus granatensis*	Iberian peninsula	Corsica, S France
Sardinian hare	*Lepus* cf. *capensis mediterraneus*	North Africa?	Sardinia
Eastern cottontail*	*Sylvilagus floridanus*	E USA	Other parts of North America, Italy, France
Snowshoe hare	*Lepus americanus*	Canada, N USA including Alaska	Elsewhere in North America
Arctic hare	*Lepus arcticus*	NE Canada, Greenland	Newfoundland and Anticosti Island, Quebec
Black-tailed jackrabbit	*Lepus californicus*	W USA	Several states in E USA
Indian hare	*Lepus nigricollis*	Indian subcontinent	Islands: Indian Ocean, Indonesia
Mountain hare	*Lepus timidus*	N Europe, Asia	Faroes, other northern islands
White-tailed jackrabbit	*Lepus townsendii*	Central W USA, S Canada	Michigan and Wisconsin USA

*Species considered invasive

1987; Bonino and Soriguer 2004). It is expected that these rabbits in central Argentina will eventually spread to join up with populations in Santa Cruz well to the south and may colonize temperate areas to the north in Bolivia along the Andean Cordillera.

The European Hare

European hares were originally distributed through Europe, the Middle East, and Central Asia. Recent documented liberations were made in the Falkland Islands about 1740, Barbados 1842, New Zealand 1851 (failed) and 1867, Ireland 1852, Tasmania 1854, Australia 1859, Sweden 1886, South America 1888, the United States 1888, Canada 1912, Siberia 1935, and Irkutsk 1938 (references in Flux 1990a). The Swedish introduction was probably earlier, when hares introduced to Ven Island crossed the ice and spread on the mainland from 1870 (Fraguglione 1962–1964).

In South America European hares are found throughout Argentina, Chile and Uruguay, southeastern Peru, southwestern Bolivia, southeastern Paraguay, and the central part of southern Brazil (Bonino et al. 2010). This is most of the zone expected to be favorable for them based on the climate within their original native range. In southeastern Brazil their spread has been associated with land clearing for cattle production, indicating the importance of habitat modification as well as a suitable climate. European hares have conspicuously moved north along the Andean Cordillera through southwestern Bolivia and into southern Peru, where they are found at altitudes of up to 4,400 m asl (Cossios 2004).

European hares occupy wide areas of southeastern Australia from the central Queensland tablelands to semi-arid regions of South Australia, but are most prominent in farming areas and are absent from the arid interior. They have not yet been able to bridge the arid Nullarbor Plain to reach Western Australia.

The Eastern Cottontail

The native range of eastern cottontails is extensive, ranging from areas with winter snow into Neotropical parts of Central America. However, they were deliberately introduced into Washington, Oregon, and British Columbia on the western side of the continent and have expanded into New England with land alteration.

Releases of eastern cottontails are largely associated with hunting industries, to increase hunting resources in areas where local rabbit species are seen as providing a less sustainable resource. Their introduction to Europe was also driven by hunting interests after myxomatosis, first reported in France in 1953, had severely reduced rabbit abundance and game bags. Because myxoma viruses circulate naturally in cottontails in both North and South America, it was known that American hares and their relatives were resistant to severe forms of the disease.

Eastern cottontails have established permanent populations in parts of Italy and are extending their range (Vidus Rosin et al. 2008). They occur sympatrically with native European hares (Vidus Rosin et al. 2008; Vidus Rosin et al. 2011).

Why Some Releases Have Not Been Successful

Although European rabbits became established across the southern two-thirds of Australia, they have never succeeded in colonizing the monsoonal tropics. The reasons for this are complex, but investigations have resulted in the development of a climate-based model that closely predicts the rabbit's current distribution (Cooke 2012a). Extending this model to other continents indicates that large areas of most continents, except Antarctica, would in fact be suitable for rabbits if climate were the only constraint. Such models have immediate use in anticipating the potential distribution limits of European rabbits in South America, where they are still spreading.

Yet, despite well-documented, large-scale attempts to introduce European rabbits into North America, they have become established permanently only on offshore islands. Likewise, in southern Africa introductions were forbidden and conditions for illegal release were possibly less suitable because of a lack of European pasture species. However, rabbits thrive on Robben Island, confirming that the climate in adjacent mainland areas must be suitable, as modeling suggests.

Among the many factors that could explain why rabbits have never become established in continental southern Africa and North America, one in particular can be immediately considered. This is the presence of native hares and rabbits that would compete directly or carry parasites and diseases that are lethal for newcomers. For example, there are 28 species of lagomorph in North America and eight native lagomorphs in southern Africa, so there is a wide range of potential competitors or parasites that could have obstructed successful introductions. There are no native lagomorphs in Australia or New Zealand or in those parts of Chile and Argentina where European rabbits have become established in the wild. Nonetheless, countering this suggestion as a broad principle, both the European hare and the eastern cottontail have been able to establish in areas where other lagomorphs are present.

There is one clear example where disease in one lagomorph species seems to have limited the rapid spread of another. This is myxomatosis, previously mentioned, which originates in the Americas. One form mildly affects the tapetí (*Sylvilagus brasiliensis sensu stricto*), its natural host, but causes high mortality when it spreads into European rabbits. Indeed, it was this high level of lethality that led to the virus being used as a biological agent for the control of introduced rabbits in Australia. The presence of myxomatosis in South America may explain why European rabbits are spreading much more slowly than they did when first introduced into Australia. Bonino and Soriguer (2009) estimated that rabbits are spreading in Argentina at a maximum rate of about 9 km per year. This is only 12% of the rate of initial spread of rabbits across Australia, estimated at 70 km per year, long before myxoma virus was introduced (Stodart and Parer 1988). Because European hares are not susceptible to acute myxomatosis, they may have been able to spread across Argentina far more readily than rabbits. Hares have extended their range faster than rabbits at estimated rates of between 10 and 37 km/year (Bonino et al. 2010).

Weighing Costs versus Benefits of Lagomorph Introductions
Economic Considerations

When European rabbits were introduced into northwestern Europe in medieval times, they were regarded as beneficial by the ruling classes, but often at the expense of poorer peasant farmers. Rabbits ate their crops! Nevertheless, they were seen as one way of using poorly productive sandy heathland to advantage. Arguments about rabbits and mixed benefits for farmers or hunters persist to this day.

In Australia and New Zealand plentiful rabbits were exploited for skins and meat, no doubt contributing to the economy of both countries. However, governments also realized that supplies of wild game products and markets were highly erratic and not a basis for stable industry. Uncontrolled rabbits interfere badly with livestock industries, generally reducing productivity through competition with sheep and cattle. Similar debate over benefits and costs of introduced lagomorphs persists today in Argen-

tina. Considering that some areas occupied by the rabbit in Neuquén and Mendoza Provinces are very important for the provincial economies, the situation may worsen. According to inhabitants of the region, rabbits compete with cattle for food and cause considerable damage to alfalfa crops, orchards, and coniferous forest plantations (Bonino and Soriguer 2009). Live European hares sent from Argentina to Europe for restocking hunting areas are seen as contributing to the export market, and hare meat has at times also been valuable with exports to Germany in the 1990s that were worth many millions of dollars (Roth and Merz 1997), further complicating the wider debate.

Conservation Implications

There is increasing awareness that European rabbits have had a major impact on the native flora and fauna of Australia. It is now recognized that even at densities of less than one rabbit per hectare, seedlings of highly palatable native trees and shrubs can be eliminated. Rabbits also selectively graze pastures, leading to changes in vegetation composition. Rabbits in Australia have enabled increases in predators such as dingoes, eagles, and introduced cats and red foxes (*Vulpes vulpes*), which in turn put extra predation pressure on smaller native mammals leading to extinctions. Since rabbit hemorrhagic disease was introduced into Australia to reduce rabbit abundance, four rare native mammal species have made extensive recoveries (Pedlar et al. 2016). There is no evidence that prey switching by introduced rabbit-specialist predators detracted from native species' recovery or that native predators like wedge-tailed eagles (*Aquila audax*) have become so rabbit-dependent that they cannot survive without them (Olsen et al. 2014). In New Zealand, with a different suite of predators and rare prey species, researchers are less sure about environmental benefits from rabbit hemorrhagic disease.

In Argentina the establishment of the European rabbit in some protected areas (Lanín and Laguna Blanca National Parks) and its potential invasion into others (Nahuel Huapi and Los Glaciares National Parks) pose a threat for the wild fauna of these areas. The presence of this exotic herbivore, with its semi-fossorial habits and high reproductive rate, could harm some native species directly, through competition for food and shelter, or indirectly, by favoring increases in the populations of native carnivores (Bonino and Soriguer 2009).

In Ireland the introduction of European hares has led to a conservation issue of a different kind. There it is believed that they are displacing or hybridizing with Irish hares (*L. timidus hibernicus*; Caravaggi et al. 2016).

Conclusions

Of 12 lagomorph species listed as being introduced to areas outside their normal range (see table 1), only 3 can be regarded as "invasive." All three are obviously problematic enough to cause concern, but neither the European hare nor the eastern cottontail has caused the extreme economic and ecological damage that followed the introduction of European rabbits to countries such as Australia.

We have models for exploring climatic constraints on the distribution of many lagomorph species and predicting their likely geographic distribution should they be introduced into new regions, countries, and continents (Bomford 2003). We also have ways of measuring the risk that species impose based on previous, within region, capacity for spread. Eastern cottontails, e.g., had already

shown the capacity to spread naturally in the 1870s and 1880s and had been deliberately and successfully introduced into new areas in the United States in the 1920s and 1930s well before it was introduced to Europe in the 1950s. Likewise, when rabbits were introduced to Australia, their potential as an agricultural pest was already well understood despite being ignored.

What is perhaps more alarming is the fact that all three invasive lagomorphs have shown some involvement in the transfer of parasites and diseases to other lagomorph species (see "Diseases of Lagomorphs"). That should be enough to advocate extreme caution in introducing rabbits or hares to new regions, especially where potential ranges overlap with those of existing rabbit or hare populations. This theme is also raised when considering why European rabbits have not become widely established in North America or southern Africa. Put simply, from the point of view of lagomorph conservation, deliberately spreading lagomorphs to areas outside their historic, native range is not a good idea without careful prior study.

5

Diseases of Lagomorphs

ANTONIO LAVAZZA AND BRIAN D. COOKE

Diseases in the European rabbit (*Oryctolagus cuniculus*) have always been of wide interest, largely because of the rabbit's importance as a domestic species and its use for meat and fur production. It is also a common laboratory animal and increasingly used as a model for exploring human diseases (Duranthon et al. 2012), especially with the development of transgenic tools that enable enhanced gene expression or silencing. Rosell (2000) lists and discusses a wide range of naturally occurring diseases in domestic European rabbits, including dietary deficiencies; bacterial respiratory, reproductive, and enteric diseases; parasites; and viral diseases. Some of these diseases, in particular, those caused by viruses, are confined to a single host species within the order Lagomorpha, whereas there are other multi-host diseases, like tularemia, that affect not only many lagomorph species, but a wide range of other taxa as well.

With the exclusion of multifactorial conditioned diseases, many of these same diseases and conditions also afflict wild lagomorphs, but are often neglected in terms of their impacts on natural populations. Nevertheless, in this chapter the aim is not to produce an exhaustive list of all diseases of wild lagomorphs, but rather to give some basic information on select diseases and to suggest precautionary procedures for carefully managing their inadvertent introduction and spread that could compromise lagomorph management and conservation.

Some lagomorph diseases are well studied because lethal viruses such as myxoma virus (MYXV) have been deliberately used to reduce numbers of introduced European rabbits where they have become serious pests. However, the introduction of MYXV within the European rabbit's native range (Spain, Portugal, and France) in the mid-1950s also caused intense conservation concern (Angulo and Bárcena 2007; Ross and Tittensor 1986) and, on becoming endemic in wild rabbits, have recurrent annual impacts on rabbit meat farms. Other viruses such as rabbit hemorrhagic disease virus (RHDV) and European brown hare syndrome virus (EBHSV) have apparently emerged spontaneously within Asia and Europe, respectively, and RHDV has been used as a biological control agent in Australia as well.

When MYXV was introduced into Australia more than 60 years ago, research by Fenner in particular (Fenner and Fantini 1999; Fenner and Ratcliffe 1965) not only showed why a relatively benign MYXV from a South American lagomorph (the tapetí *Sylvilagus brasiliensis sensu stricto*) caused lethal disease on transfer into European rabbits, but also tracked the course of evolution of subsequent disease resistance in rabbits. This work has since become the foundation for wider theory and has been fundamental in understanding subsequent diseases such as rabbit hemorrhagic disease (RHD).

RHD and European brown hare syndrome (EBHS) are similar diseases caused by two highly related but phylogenetically distinct RNA viruses belonging to the genus *Lagovirus* of the family Caliciviridae. RHD and EBHS have restricted host specificity both naturally and experimentally (Bergin et al. 2009; Lavazza et al. 1996; Lenghaus et al. 1994). In fact, they were initially considered genus-specific, the former infecting only wild and domestic European rabbits and the latter mainly European hares (*Lepus europaeus*).

Nevertheless, there is increasing evidence that lago-viruses such as RHDV are rapidly co-evolving to match increased disease resistance in rabbits. A new virus sero-type (RHDV2) has emerged (Le Gall-Reculé et al. 2011; Le Gall-Reculé et al. 2013), and it is apparently replacing those strains (RHDV and RHDVa) that initially spread.

The origin of RHDV2 is still under study, but the pre-liminary results indicate that it could not have evolved from RHDV, but it rather represents a new viral emer-gence from an unknown source (Le Gall-Reculé et al. 2013); the hypothesis of a species jump has also been proposed (Esteves et al. 2015). The genomic and anti-genic profiles of RHDV2 are quite different from those of RHDV, to such an extent that it could be considered a distinct serotype, even able to infect rabbits with specific immunity against classical RHDV. This advantage enabled RHDV2 to rapidly spread in Europe, causing significant losses in farmed and wild rabbits.

This finding has wider implications, not simply in terms of the need for improved vaccines to protect do-mestic rabbits, but also because RHDV2 is not confined to European rabbits and has shown a capacity to cause dis-ease in other lagomorphs, notably Cape hares (*L. capensis mediterraneus*; Puggioni et al. 2013), and, more sporad-ically, as a probable result of spillover events, Corsican hares (*L. corsicanus*; Camarda et al. 2014) and even Euro-pean hares (*L. europaeus*; Capucci et al., personal observa-tions).

Moreover, the epidemiological pattern and distribution of the disease is complicated by the fact that, in addition to virulent strains of RHDV that progressively formed at least six geno-groups that appeared and sometimes disap-peared as it evolved over time (Le Gall-Reculé et al. 2003), nonpathogenic rabbit RHDV-like viruses (rabbit calicivi-ruses, or RCVs) have been identified in wild and domestic rabbits in both Europe (Capucci et al. 1996; Marchandeau et al. 2005) and Australia (Strive et al. 2009). The extent of genetic diversity among these RCVs is so wide—from non-protective, partially protective, to protective strains—that there is a gradient of cross-protection among domes-tic and wild rabbit populations due to their circulation.

Thus, where MYXV provides an example of a virus jumping hosts, RHDV may have initially evolved from a nonpathogenic calicivirus, but has since shown changes in virulence, enabling it to spread into a wider range of lagomorphs. Likewise, the transfer of EBHSV has been recorded between European hares and both mountain hares (*L. timidus*) in Scandinavia (Syrjälä et al. 2005) and Corsican hares in Italy (Guberti et al. 2000) as well as, and occasionally, eastern cottontail rabbits (*S. floridanus*; Lavazza et al. 2015).

Populations of native lagomorphs in Europe have been significantly reduced by these three important viral diseases (myxomatosis and RHD in rabbits, and EBHS in hares), and there has been considerable effort made to support populations by managing lagomorph popula-tions and by improving hunting and restocking strategies (Delibes-Mateos et al. 2014). These include the release of farm-bred rabbits and hares and the translocation of rab-bits and hares from localities where they are numerous to areas where they have become rare. In Italy wild hares for release are sourced mainly from eastern European coun-tries and sometimes from as far away as South America, which implies long-distance translocations and even the possibility of transfer of pathogens from different coun-tries or continents.

Nonetheless, the direct effects of diseases on rabbit and hare populations are only part of the story, and specific case histories point to wider, more complex problems that can result when lagomorphs are incautiously moved to new habitats. Some recent examples are reported here with reference to the most severe lagomorph diseases.

Rabbit Hemorrhagic Disease (RHD) in European Rabbits

European rabbits of the subspecies *O. c. algirus* have been established in the Canary Islands for many centuries, probably since the islands became strategically impor-tant in reaching European colonies in South America. Nonetheless, introductions of additional rabbits in 1988, without quarantine or certification of disease-free status, apparently led to the introduction of RHDV (Foronda et al. 2005). Rabbit populations fell markedly at the time, and the response seems to have been the introduction of additional rabbits, increasingly *O. c. cuniculus* rather than *O. c. algirus*, because populations of the former appear to be recovering from RHDV more rapidly. The conse-quences of these actions will not be known for some time, but potentially interbreeding between the two rabbit sub-species could either enhance or decrease the rate of development of disease resistance depending on its genetic basis. Alternatively, disease-induced natural selection could see one subspecies of *Oryctolagus* at an advantage over the other.

Portuguese researchers recently raised the possibility that RHDV strains that have evolved in *O. c. algirus* in the Azores over the past 17 years might be returned to Portugal and Spain if the islands are used as a source for replenishing declining *O. c. algirus* populations on the Iberian Peninsula (Esteves et al. 2014; Almeida et al. 2015). With a new form of the virus (RHDV2) dominating in Portugal and Spain in recent years, genetic resistance in European rabbits is almost certainly shifting to counter the present virus. As a result, it is argued, resistance to RHDV strains similar to the one in the Azores may be declining. Under those circumstances bringing rabbits and their locally adapted viruses to peninsular Portugal and Spain from the Azores could be counterproductive.

An enhanced decline of wild European rabbits could have a dramatic effect on biodiversity and nature conservation since they are essential keystone elements for the native Mediterranean ecosystem. In fact, more than 30 predator species prey on rabbits, including the critically endangered Iberian lynx (*Lynx pardinus*) and the Iberian imperial eagle (*Aquila adalberti*), the decline of which has been partly due to rabbit decline. Rabbits are also "landscape modelers" with important impacts on plant communities, and their burrows provide habitats for many invertebrates. Therefore, the control of RHD for reversing rabbit decline becomes one of the biggest challenges for nature conservation in Europe. Unauthorized movement of rabbits to restock hunting areas could potentially undo gains in rabbit resistance to disease rather than improving conservation prospects.

European Brown Hare Syndrome (EBHS) in Different Hare Species

Since the first description of EBHS in Sweden in the early 1980s, its distribution has been restricted to Europe, where it is common in the European hare, the main host species. However, EBHS has also been reported with lower frequency in the mountain hare and Corsican hare, but never in other European species such as the Iberian hare (*L. granatensis*) and the broom hare (*L. castroviejoi*), both present mainly on the Iberian Peninsula, and the Cape hare in Sardinia (Puggioni et al. 2013).

European hares were introduced into Sweden in 1886 and established widely across the south of the country. They intermingled with mountain hares within their new range. However, after EBHS first appeared among Sweden's European hares, it was subsequently found that it also affected mountain hares, although only within the range where the two species overlapped (Gavier-Widen and Mörner 1993). It seemed therefore that mountain hares only became infected where there was an associated disease reservoir in the European hare. Results of similar surveys in Finland are, however, less encouraging. There, EBHS appears in mountain hares well beyond the zone of overlap between the two species, implying that introduced European hares could have facilitated the transfer of this new disease into mountain hares (Syrjälä et al. 2005).

A study aimed at evaluating the temporal dynamics of infection and estimating the yearly effects of population density on EBHS prevalence (Chiari et al. 2014) revealed that population density influences virus maintenance in the hare population (sero-prevalence was 3.3 times higher in high-density areas—those with > 15 hares/km^2). Since the eradication of EBHS in wild populations is not feasible, the strategy of promoting the endemic stability of the virus through density-dependent mechanisms (populations > 15 hares/km^2) is the only way to minimize EBHS impact.

The Role of Eastern Cottontails in the Epidemiology of Rabbit and Hare Diseases

In 1966, the eastern cottontail (*S. floridanus*) was introduced for hunting purposes to France and Italy, where it is currently widespread. Its ecological niche is similar to that of native rabbits and hares, and increasing overlap in distribution brings these species into ever-closer contact.

With the translocation of this alien species, at least four new parasite species have been brought into Europe (Tizzani et al. 2014): three nematodes (*Obeliscoides cuniculi*, *Trichostrongylus calcaratus*, *Passalurus nonannulatus*) and the flea *Euphoplopsyllus glacialis affinis* (Baker 1904). *O. cuniculi* has subsequently been reported from European hares (Tizzani et al. 2011), and there is speculation that if this weakens the European hares it could give eastern cottontails a competitive advantage.

In turn, Eastern cottontails are considered at risk of infection with both the lagoviruses RHDV and EBHSV, endemically present in Italy. Field and experimental data have recently demonstrated that the eastern cottontail is susceptible to infection with EBHSV (but not with

RHDV), and this occasionally evolves into an EBHS-like disease. At this stage the eastern cottontail can be considered a spillover or dead-end host for EBHSV, unless further evidence is found to confirm that it plays an active role in epidemiology (Lavazza et al. 2015).

Tularemia—an Important Zoonosis and Multispecies Infection of Wildlife

Tularemia is the perfect example of a disease whose importance goes beyond the impact it has on lagomorph populations. It is one of the major known zoonoses. It is a multi-host, contagious, and vector-borne zoonosis caused by *Francisella tularensis*, and it is considered to be an emerging and reemerging infection causing new outbreaks in Europe and North America in the past decades. In fact, it is widely distributed in humans, wildlife, and arthropod vectors, from the most northerly European countries to the most southern, and from the east to the west. Its epidemiology is extraordinarily complex and varies with ecosystem and geographical region. One of its constant epidemiological characteristics is its local emergence and reemergence in wildlife and humans, where it has a clear seasonal pattern. Nevertheless, there are fundamental gaps in the knowledge of the role of arthropod vectors, wildlife, and maintenance of natural habitats. In fact, besides affecting rodents and lagomorphs (in Europe mainly European hares and rabbits), tularemia, as classically described, can cause disease and death in many other domestic and wild animal species. It is recognized that wildlife reservoirs and arthropods are important sources of infection for humans and other animal species, some of which have been poorly considered, like the red fox (*Vulpes vulpes*), the raccoon dog (*Nyctereutes procyonoides*), the wild boar (*Sus scrofa*), and wild ruminants.

Conclusions

To conclude, because MYXV transferred readily from *S. brasiliensis* to *O. cuniculus*, and because RHDV could have had an origin outside Europe (Esteves et al. 2014), there is little doubt that serious diseases can spread between closely related lagomorph species. We also know that there are other viruses known from North American cottontails (Trivittatus virus: Bunyaviridae) in the eastern cottontail and desert cottontail (*S. audubonii*; Hal 2012) that may not be in European rabbits, and no doubt rabbits and hares in Europe could be hosts for unique diseases that are not present in cottontails and jackrabbits in the Americas. Even with prior testing for known diseases and parasites, incautious transfer of lagomorphs from one continent to another is contraindicated because we know that benign, nonpathogenic viruses in one lagomorph species can sometimes become a lethal disease in another. In addition, there are also some diseases, like tularemia, that can easily be transmitted from rodents and lagomorphs to other wild animals, humans, and even arthropods.

This overall information sounds a warning for disease management and conservation of lagomorphs in general, considering that when any animal is translocated its parasites and pathogens are moved at the same time. The introduction of lagomorphs to areas beyond their normal geographical range should be avoided where possible. After all, preventing disease introduction into susceptible populations is the most efficient and cost-effective method of disease management.

6

Conservation of Lagomorphs

ANDREW T. SMITH

Most people have come to view lagomorphs through the window of the order's best-known species, the European rabbit (*Oryctolagus cuniculus*). This species has successfully spread (as an invasive in many cases) throughout the world, and the phrase "they breed like rabbits!" stems from this species' adaptability and fecundity. It is less widely known, however, that many species of lagomorphs are among the most globally threatened of all the mammals (Smith 2008). Here I examine the vulnerabilities of these threatened or endangered species and highlight the conservation efforts that are being undertaken to ensure their survival.

The global standard for understanding the degree of vulnerability of species to extinction is the International Union for Conservation of Nature (IUCN) Red List (http://www.iucnredlist.org/initiatives/mammals), which is continuously updated by the all-volunteer army of specialists of the IUCN Species Survival Commission (SSC). The status of each lagomorph species is evaluated by the Lagomorph Specialist Group (LSG) of the SSC. The Red List captures and evaluates quantitative data for measures that could lead a species to extinction. These evaluations are free from politics and represent the least biased of all threatened species assessments. Species not under threat are considered to be of *Least Concern* (LC). Threatened forms are classified in a rank order according to their susceptibility to extinction from *Vulnerable* (VU), to *Endangered* (EN), to *Critically Endangered* (CR). Species for which we attempt evaluation, but for which there are insufficient data, are termed *Data Deficient* (DD). *Not Evaluated* (NE) is the designation used for species that have not yet been assessed. A special category of *Near Threatened* (NT) is given to LC species that do not meet the stringent criteria for being designated in one of the threatened categories, but for which there are ominous signs that the species may be in trouble in the near future.

The last comprehensive review of all lagomorphs was undertaken in 2008 (see Schipper et al. 2008). At the time 61 of 92 species of lagomorph (66%) were considered to be LC (free of threat), while the fate of the remaining 31 species (34%) could have been in jeopardy. Of this latter category, 19 (21% of lagomorph species) were placed in one of the three threatened categories, 8 species (9%) were DD, and 4 (4%) were NT. Since 2008, systematic changes within the order and reevaluation of some species have somewhat altered these percentages, but they still reflect the general level of threat faced today by many lagomorph species. And the large number of DD species indicates, as is also apparent in many of the species accounts in this book, that much remains to be learned about the ecology and status of many lagomorph species.

Red List evaluations are a first step in the conservation process, followed by conservation actions. The conservation measures taken to ensure the persistence of species may take many forms. With lagomorphs, there is no one-size-fits-all model; actions to conserve threatened lagomorphs run the gamut of conservation interventions.

Taxonomic Changes and Not Evaluated (NE) Species

Since the 2008 review of the status of all lagomorphs several new taxa have been identified. Some of these retain their nomenclature from the 2008 review, but now include forms that were previously considered independent. For example, the Gaoligong pika (*Ochotona gaoligongensis*—DD) and black pika (*O. nigritia*—DD), initially described as independent species, are now considered to be melanistic forms of Forrest's pika (*O. forresti*—LC) and are included in that species. Similarly, the Muli pika (*O. muliensis*—DD) is now included in Glover's pika (*O. gloveri*—LC), and the Himalayan pika (*O. himalayana*—LC) is now included in Royle's pika (*O. roylei*—LC). These changes eliminate three of the former DD lagomorph species.

Other pika forms have been split, such as Pallas's pika (*O. pallasii*—LC) into *O. pallasii* and the Kazakh pika (*O. opaca*), and the Korean pika (*O. coreana*) and the Manchurian pika (*O. mantchurica*) splitting from the northern pika (*O. hyperborea*). A good example of how changes in systematics present challenges to accurate classification of the status of species can be found in the tapetí (*Sylvilagus brasiliensis sensu stricto*). In 2008, this species was considered to range from southern Mexico through most of northern South America, from the east coast, through interior tropical forests, to the high mountains of the Andes in the west. Recently Gabb's cottontail (*S. gabbi*), the Páramo cottontail (*S. andinus*), and the Rio de Janeiro dwarf cottontail (*S. tapetillus*), each with restricted geographic ranges, have been identified within the former range of *brasiliensis*, and *S. brasiliensis* itself appears to be confined to a small area in easternmost Brazil. And work is under way to evaluate the status of rabbits found throughout the remainder of the former broad distribution of *brasiliensis*. These changes present situations that challenge our ability to conduct status surveys of these forms, as prior evaluations within the former (larger) species ranges cannot apply to the new taxa (unless an evaluation was site-specific and clearly identified as occurring within the range of a newly identified species). Thus, the next generation of Red Listing of lagomorphs may include several NE forms until biologists undertake new surveys to determine their status.

Extremely Isolated Species

Another challenge facing lagomorph biologists is that several forms are highly range restricted or occur in localities that are extremely remote. The Omiltemi rabbit (*S. insonus*—EN) may be one of the rarest mammals on earth (Cervantes et al. 2004). Despite several dedicated surveys by Mexican biologists, this species is known from only four complete museum specimens plus one skin. It already occurs in a natural reserve area (Omiltemi State Ecological Park, in the Sierra Madre del Sur, Guerrero, Mexico), so protection is not a problem. Still, it is difficult to engage in proscriptive conservation measures when so little is known about a species.

Similarly, several pika species have very restricted distributions, such as Hoffmann's pika (*O. hoffmanni*—EN), which is found on two isolated mountains in southern Russia and northern Mongolia, and the Helan Shan pika (*O. argentata*—CR), which is found within a narrow range in the Helanshan (mountains) in central China. These populations are vulnerable to stochastic extinction due to their isolation, and protection of their habitat may be the best hope of ensuring their survival.

Another isolated species is the Ili pika (*O. iliensis*—EN), which is found on two spurs of the Tianshan (mountains) in the Xinjiang-Uyghur Autonomous Region, China. The story of the Ili pika highlights the difficulty of conducting conservation in remote areas (Li and Smith 2005, 2015). The species was discovered in 1983 and named in 1986. It was then studied for about a decade, leading to a basic understanding of its biology and distribution. Only 27 living Ili pikas were seen or captured during this interval. Then 10 years passed before an attempt was made to reassess the status of the species. The results were chilling: based on the characteristic sign left by pikas (haypiles, scat piles) it was determined that fewer than half of the previously known localities contained pikas, and no living pikas were observed. Several thorough expeditions subsequently failed to find any living Ili pikas until summer 2014, when two individuals were observed, the first Ili pikas seen in 20 years. The challenge we face is planning meaningful conservation strategies for a species in severe decline, but for which very little information is available. First we must try to identify the root cause(s) of decline. Additional information on the Ili pika is now being gathered with infrared camera traps, and we hope soon to have a clearer understanding of the actual distribution and extent of decline of this species. We have ruled out disease as an explanation for the observed decline, leaving climate change or the increasing pressure on the land by pastoralists (and their mastiff dogs) as likely candidates. Once we have a better understanding of the root causes of decline, specific conservation actions to preserve the species can be put in place. The establishment of protected areas in regions occupied by the Ili pika in which grazing is restricted seems to be the most viable option for the Ili pika at the present time.

Ghost Rabbits

The two striped rabbits, the Sumatran striped rabbit (*Nesolagus netscheri*—VU) of Sumatra and the Annamite striped rabbit (*N. timminsi*—DD) of Vietnam, have been ghosts to biologists. The Annamite striped rabbit was first described only in 2000, after it was discovered in a local market, and decades have elapsed between sightings of the Sumatran striped rabbit. Now biologists are focusing on these forest animals using modern technology. As in the case of the Ili pika, camera traps are for the first time revealing the distribution and abundance of these species (McCarthy et al. 2012). In addition, the range of the Annamite striped rabbit is being assessed using the clever technique of analyzing leeches for the presence of rabbit blood (Schnell et al. 2012). Habitat loss is a key threat to both species, particularly the Sumatran striped rabbit; from 1990 to 2010, Sumatra lost roughly 70% of its total forest cover. The Annamite striped rabbit, however, appears to be primarily threatened by the bushmeat trade and illegal hunting. Interestingly, while closely related, the meat of the Sumatran striped rabbit is considered to taste bad, so that hunting is not a threat to this species. In both cases, however, technology is allowing conservationists to understand more of the biology of these species, which should lead to improved interventions to arrest their decline. We should have enough data on the Annamite striped rabbit in the near future to remove it from the DD classification.

Threats by Invasive Species

The Amami rabbit (*Pentalagus furnessi*—EN) is found on only two small islands (Amami-ohshima and Tokuno-shima) in the Ryukyu Archipelago in southern Japan. This species faces two main threats: habitat loss due to logging (in the past four decades from 70% to 90% of old growth forests, the primary habitat of the Amami rabbit, have been lost) and predation by introduced non-native carnivores (Yamada 2008). Feral dogs and cats, followed by the introduction of the Indian mongoose (*Herpestes javanicus*), led to initial losses of Amami rabbits. The Japanese Ministry of the Environment responded with a mongoose eradication program that has been highly successful; the

mongoose population and distribution on Amami-oshima Island decreased dramatically followed by an increase in native fauna, including the Amami rabbit. The most severe threat at present is predation by feral cats, which has greatly reduced the population of Amami rabbits, especially on Tokuno-shima Island. Initial culls of cats have been halted, however, due to protests by animal rights activists. Negotiations are under way to allow further feral cat control, without which the rabbit is likely to go extinct on the island.

Reintroduction as a Conservation Tool

The wild population of the pygmy rabbit (*Brachylagus idahoensis*—LC) from the isolated Columbia Basin in the central portion of the state of Washington (United States) went extinct in 2004. This population was genetically distinct from rabbits in the remaining distribution of the species and was considered to be an endangered distinct population segment (DPS) under the U.S. Endangered Species Act. In 2001, 16 adult pygmy rabbits were captured and transferred to an *ex situ* captive breeding program. There were initial problems with husbandry and inbreeding depression, and ultimately the few remaining purebred rabbits were crossed with rabbits from the extant population (largely from Idaho). Husbandry issues (primarily disease) continued to plague the breeding program. Finally, in 2011, the captive breeding program was phased out and animals were translocated to large outdoor enclosures in native sagebrush habitat within the historic range of the species. Additional wild-caught pygmy rabbits supplemented this breeding stock, which was highly successful. In 2014 and 2015, 1,408 rabbits, including 164 adults, were released into the wild. Young with high amounts of Columbia Basin genetic ancestry are retained each year for future breeding. Monitoring of rabbits tracked with VHF transmitters coupled with winter burrow surveys inform future release efforts. Overall, this program demonstrates an adaptive strategy for reintroduction, and highlights the steps that must be taken to ensure a successful reintroduction program (Becker and DeMay 2016).

What about Species That Occur Only on Private Land?

The riverine rabbit (*Bunolagus monticularis*—CR) is a flagship species of one of South Africa's conservation priority areas, the Karoo (Ahlmann et al. 2000). This species is a biological indicator for river zones in the Karoo that are important for groundwater storage and livestock grazing; these areas also serve as a buffer zone during periods of drought. The riverine rabbit faces many threats from habitat loss (as much as 50% to 80% over the past century), hunting, predation by dogs, misuse of gin traps, and road kills. On the horizon are the potential threats of fracking and climate change. Protection of this critically endangered species presents special challenges, as the linear nature of the remaining habitat fragments it occupies are difficult to manage. In addition, the riverine rabbit almost exclusively occurs on private land; thus establishment of protected areas is not a viable option. The solution utilized by the Endangered Wildlife Trust and its Riverine Rabbit Conservation Project is to partner with local farmers, who in the past may have been antagonistic toward rabbits, and form Riverine Rabbit Conservancies. These conservancies promote a conservation stewardship concept across the landscape, and in turn, allow for the development and implementation of conservation management plans. A "Conservation Farmer of the Year" award has been developed to reward those who go the extra mile to promote sustainable farming practices as well as protecting the rabbits. This overarching program includes support of Eco-Schools and a nursery that grows native plants for revegetation projects. The revegetation program, in turn, employs local people, rewarding them for their efforts to protect the rabbits. Overall, this program shows how conservation of an endangered species can be integrated within the fabric of local culture for the betterment of both.

Sometimes You Have to Try Everything

Since the 1960s, the distribution of New England cottontails (*Sylvilagus transitionalis*—VU) has declined by more than 80% (Litvaitis et al. 2006). Rather than delay recovery until a decision to federally list New England cottontails as threatened or endangered was made, several federal agencies joined forces with state fish and wildlife agencies within the current range of the New England cottontail and nongovernmental organizations (NGOs) to develop a comprehensive strategy to expand their existing populations (Fuller and Tur 2012). Current efforts that are under way include restoration and management of more than 20,000 ha of habitat and the development of captive breeding facilities. Initial accomplishments have shown some success, but substantial challenges remain and must be overcome. Most land containing suitable

habitat is in private ownership (Tash and Litvaitis 2007), and federal cost-share programs are recruiting landowners to develop habitat restoration projects. Additionally, the federal government is attempting to establish a network of publicly owned refuges that will serve as core habitat in recovery focal areas (USFWS 2016). Efforts to expand populations of New England cottontails in landscapes that have been subsequently colonized by non-native eastern cottontails (*S. floridanus*) will need to be more nuanced as eastern cottontails directly compete with New England cottontails for food and cover (Probert and Litvaitis 1996). Because restoration of suitable habitats and generating a sufficient supply of captive-bred rabbits for stocking will take time, enhancing the survival of all remaining populations of New England cottontails should be a high priority. Remnant populations are often small and isolated and thus vulnerable to high rates of predation from foxes (*Vulpes vulpes*, *Urocyon cinereoargenteus*), coyotes (*Canis latrans*), and free-ranging house cats. Positioning supplemental feeders in dense patches of understory vegetation among habitats occupied by cottontails has been shown to reduce rabbit vulnerability to predators and substantially increase overwinter survival rates (Weidman and Litvaitis 2011). Combined, these efforts were recently considered sufficient enough to permit the U.S. Fish and Wildlife Service to forgo listing New England cottontails as threatened or endangered.

Threats from Land Conversion and Development

Although the brush rabbit (*Sylvilagus bachmani*—LC) is still thought to be present or even common in brushy habitat throughout most of its historic range along much of the western seaboard of North America, there is cause for concern about its welfare because of the growing human population in much of that coastal region. There is further concern in the context of climate change and its impact on ecosystems, especially because so little is known about the status of subspecies and populations throughout most of that wide distribution.

Conversion of habitat to agricultural use resulted in the listing of the riparian brush rabbit (*S. b. riparius*) as endangered by the state of California and by the U.S. Fish and Wildlife Service (Williams et al. 2008). However, a number of significant conservation and recovery actions have been implemented since 2000 to establish new populations, restore and create new habitat, and protect populations from flooding (Hamilton et al. 2010; Kelt et al. 2014). Flooding and wildfire are major threats to riparian brush rabbit populations. Although the subspecies is still considered endangered, its status has significantly improved since 2000.

The status of brush rabbits in Baja California appears to be of even greater concern than that of those in the United States. Lorenzo et al. (2013) fear that *S. b. peninsularis* may even be extinct, and they attribute this to tourism development. Also, they suggest that *S. b. cerrosensis* and *S. b. exiguus* are threatened. As with most other species of lagomorphs, a range-wide assessment of brush rabbits is warranted, but given their habitat preferences, such an assessment will be particularly challenging.

A City on the March

Mexico City is one of the world's largest metropolises, and as it continues to expand, it encroaches against the volcanoes to the south and east of the city. These volcanoes are the sole habitat of the volcano rabbit (*Romerolagus diazi*—EN), or zacatuche, thus the spreading of Mexico City increases fragmentation of the rabbit's habitat and decreases the area that it can occupy. Currently volcano rabbits occur in 15 isolated patches of varying size. Like the work done to engage local people in South Africa, the future of the volcano rabbit is in the hands of local communities that for generations have utilized the landscape of pines and zacaton bunchgrasses that also function as the rabbit's primary habitat. Conservation scientists from Mexico City have conducted workshops with local people to explore the long-term sustainability of this fragile habitat, with the goal of mutual benefit to people and rabbits. The way ahead is to enhance corridors between currently isolated populations of volcano rabbits by promoting scientific landscape ecology principles for the management of this species (Velázquez et al. 2003).

No More Roads and Tunnels

While the northern pika (LC) has the largest distribution of any pika, the subspecies that occurs on Hokkaido Island (*O. h. yesoensis*; listed as NT by the Ministry of the Environment of Japan) is very special to the people of Japan. There the Pika Fan Club is one of the largest green groups in the country, and they represent a true grassroots movement for conservation action (Ichikawa 1999). The government views building roads and tunnels as an important economic driver, but these projects sometimes seriously jeopardize the habitat of critical populations of pikas. The Pika Fan Club has successfully stopped many of these proposed projects, and additionally has engaged in significant public awareness measures to highlight the importance of pikas and all native species on Hokkaido.

Conservation of a Common Species

The plateau pika (*Ochotona curzoniae*—LC) does not immediately strike one as a species in need of conservation action; in fact, it is so common that it is considered a pest species and poisoned widely across its entire distributional range on the Qinghai-Tibetan Plateau (Smith and Foggin 1999). However, that is the reason that this species is of conservation concern. The plateau pika is a keystone species for biodiversity and an ecosystem engineer on the Qinghai-Tibetan Plateau. When pikas are poisoned much of the functioning of the ecosystem on the plateau is compromised. For example, after pikas are poisoned their burrows collapse, the ground becomes more compacted, infiltration of water following summer monsoonal rains is inhibited, and the potential for local erosion and downstream flooding increases (Wilson and Smith 2015). Thus in this instance, preventing the widespread poisoning of plateau pikas is a conservation action that enhances biodiversity and sustainability of the Qinghai-Tibetan Plateau grassland ecosystem for the benefit of local people.

Summary

These examples of conservation action directed to protect lagomorph species demonstrate much of the complexion of conservation in the world today. Each situation demands a unique and targeted approach, informed by science, but also incorporating local circumstances and the needs of people as well as nature. These examples also highlight the important and state-of-the art work being conducted by lagomorph conservation biologists worldwide. Their work is encouraging, but much remains to be done to protect lagomorph species and the environments they occupy. As we have seen, lagomorphs are integral species in ecosystems across the world. We lose those species at our own risk. In the words of the American forester, ecologist, and writer Aldo Leopold: "To keep every cog and wheel is the first precaution of intelligent tinkering" (Leopold 1966:177).

SPECIES ACCOUNTS

Order Lagomorpha Brandt, 1855

The Lagomorpha consists of two families: the pikas (Ochotonidae; 29 species) and the rabbits and hares (Leporidae; 63 species). The Lagomorpha and the Rodentia constitute the superorder Glires. While lagomorphs and rodents each have large diastemas (the gap between the incisors and the molar teeth), the lagomorph's small peg-like incisors without a cutting edge that lie immediately behind the larger first incisors distinguish them from rodents. The general morphology of lagomorphs includes large ears and elongated hind limbs. The feet of lagomorphs are fully furred. Lagomorphs are generalized herbivores and possess a very large gut that houses bacteria to aid in the digestion of cellulose. All lagomorphs produce two types of feces: small, dry, round pellets and soft, black cecal pellets, the latter of which are normally eaten directly from the anus (this process of re-ingestion is termed "coprophagy"). Coprophagy allows lagomorphs to optimize their uptake of essential vitamins (particularly B-complex vitamins) and to enhance their assimilation of energy from their vegetarian diet.

Family Ochotonidae Thomas, 1897

Pikas (genus *Ochotona*) are small, egg-shaped lagomorphs without a conspicuous tail (tail measurements are usually not recorded for pikas). They have short rounded ears and their fur is normally soft and dense. Pikas have forefeet with five digits and hind feet with four digits. Their dental formula is 2.0.3.2/ 1.0.2.3 = 26.

Most pika species tend to be diurnally active, and none hibernate. Ecologically there are two basic habitats occupied by pikas: rock/ talus (non-burrowing forms) and meadow/ steppe (burrowing forms). Talus-dwelling pikas normally are long-lived, have low reproductive rates, and live at low densities, whereas burrowing pikas are short-lived, have high reproductive rates, and occur at high densities. Many pika species are highly vocal, and most construct food caches, called haypiles, that serve as a source of food during winter.

All pikas are in a single genus *Ochotona*; 27 of the 29 species occur in Asia and two species are found in the mountains of western North America.

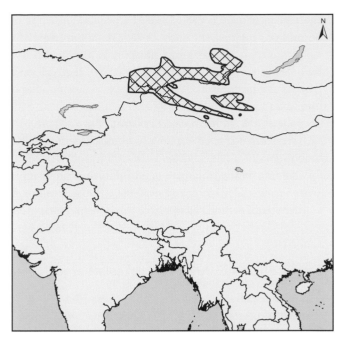

Ochotona alpina. **Photo courtesy Andrey Lissovsky**

Ochotona alpina (Pallas, 1773)
Alpine Pika

OTHER COMMON NAMES: Altai pika; Gaoshan shutu (Chinese); Tagiin ogdoi (Mongolian); Altaiskaya pischuha (Russian)

DESCRIPTION: The dorsal pelage of the alpine pika in summer is variable, ranging from light yellowish gray ochraceous to bright ochraceous, rusty ochraceous, black-brown ochraceous, or reddish brown. The flanks are tinged with a rusty color. The ventral pelage is pale yellowish ochraceous or pale reddish brown. In winter the dorsal pelage is lighter, either an ashy gray with pale yellowish brown flanks or a dark reddish brown on the back fading to an ochre tone on the sides and belly. The black or black-brown vibrissae range in length from 60 to 70 mm.

The stoutly built skull is long and narrow in contrast to that of the northern pika (*O. hyperborea*). No fenestrae are present in the frontals, and the incisive foramina are small and rounded. The anterior palatal foramina are completely separate from the incisive. The tympanic bullae are of moderate size and narrow in aspect. The rear of the cranium is flexed downward.

SIZE: Head and body 152–251 mm; Hind foot 26–38 mm; Ear 17–26 mm; Greatest length of skull 41–54 mm; Weight 226–426 g

CURRENT DISTRIBUTION: The alpine pika occurs in the mountainous regions of Central Asia. The range extends from the easternmost region of Kazakhstan, through NW China (Xinjiang), W Mongolia, and S Russian Federation (Tuva, Altai Krai, Khakassia). It occupies elevations from 400 to 3,000 m.

TAXONOMY AND GEOGRAPHIC VARIATION: Subgenus *Pika*. Five subspecies: *O. a. alpina* (includes *ater*; E and C Altai Mountains); *O. a. changaica* (Mongolia); *O. a. nanula* (S Tuva); *O. a. nitida* (includes *sushkini*; E Altai and W Sayan Mountains); *O. a. sayanica* (E Sayan Mountains). The forms *svatoshi* and *cinereofusca*, once included in *O. alpina*, are now considered to be in *O. hyperborea*, and *scorodumovi* is a junior synonym of the Mantchurian pika (*O. mantchurica*). In earlier treatments the Helan Shan pika (*O. argentata*) was considered a form of *O. alpina*; in reality it is more closely related to Pallas's pika (*O. pallasii*).

ECOLOGY: Alpine pikas live in colonies with familial territories consisting of one adult pair and their young. Population density varies with location, but tends to be low at 10–12/ha, although some colonies may reach a density of 50–60/ha. Where alpine pikas occur sympatrically with the northern pika, there is partial segregation either altitudinally or by microclimate. Annual mortality of alpine pikas varies from 41% to 53%. Overall, there appear to be no significant changes in population size from year to year within colonies of alpine pikas. Vegetation composition and diversity is influenced by the presence of the alpine pika; the proportion of flowering plants and seeds deposited in the soil decreases and succession is slowed in areas occupied by the pikas. Potential ecosystem benefits provided by alpine pikas are facilitated plant colonization due to haypile decomposition and nitrophilic vegetation appearance in areas where fecal pellets accumulate at high quantities. In winter, several native and domestic animals will concentrate in areas where the pika has cached food. Siberian red deer (*Cervus elaphus sibiricus*), reindeer (*Rangifer tarandus*), cows, and horses will opportunistically feed on the haypiles. The density of voles (*Myodes*) and shrews (*Sorex*) is four times higher in areas with pika haypiles than in the surrounding forest. Alpine pikas are an important prey base for sables (*Martes zibellina*), as well as other native carnivores. Up to 87% of sable scats contain pikas, and in some regions with pikas the density of sables is 10 times higher than in areas without pikas.

HABITAT AND DIET: The alpine pika occurs in the alpine or sub-alpine regions of the taiga zone. Talus (scree) or rocky regions are the preferred habitat, where it occupies crevices or empty spaces between rocks. It will occasionally occupy old moss-covered scree or burrows found under tree roots. When associated with great expanses of rocky terrain, it will be most dominant on the margins nearer vegetation. Preference is given to rocky areas with stones that range between 0.5 and 1.0 m in diameter.

Alpine pikas are generalized herbivores that will forage on nearly all vegetation found within their habitat. Berries, seeds, shrubs, buds, grasses, mushrooms, lichens, leaves, and shoots are consumed when they are available, with green parts of available vegetation constant in their diet year-round. Vegetation is cached in haypiles for over-winter survival, with the most intensive collection occurring in August. Haypiles are typically stored under large stones or in crevices. In one region more than 100 species of plant were recorded in pika haypiles, and in general the composition of plants in a haypile reflects that of the local vegetation. At the same time, some plant species that are very common in the environment are ignored by alpine pikas; thus they show selectivity in their foraging behavior. Haypiles can be quite large, reaching a height of 2 m and a diameter at the base of 1.5 m; the weight of some haypiles reaches 30 kg.

BEHAVIOR: Alpine pikas are diurnally active. Their activity level is highest on cool days, and activity decreases when the ambient temperature rises to 18°C. Animals maintained in a captive setting have died when temperatures reached 25°C. During midday summer heat and in winter, outside activity decreases and they will retreat to their shelters. Alpine pikas will often bask in the sun on a stone.

Like other rock-dwelling pikas in Asia, alpine pikas occupy family territories consisting of an adult male and female and their offspring. Adult residents store food together and aggressively defend their family territory from neighboring pikas. Scars and scratches resulting from aggressive encounters are common, particularly among adult males during the reproductive season (up to 62% of adult males in some populations), and in all animals during late summer (when juveniles are becoming settled). Overall the boundaries of family territories within a colony tend to remain constant from year to year, although territory size may vary depending on local environmental conditions. Territory size averages 1,500 m², with a range from 150 to 3,000 m².

Alpine pikas have a large repertoire of calls, including the long call (song) produced by males during the reproductive season, an alarm call, and a territorial call. The acoustic nature of the alarm call varies depending on context and the species of animal eliciting the call. Alpine pikas also communicate their presence by rubbing their neck (buccal gland) on rocks. This behavior is most often seen during the dispersal phase or when an individual claims a new territory.

PHYSIOLOGY AND GENETICS: Diploid chromosome number = 42

REPRODUCTION AND DEVELOPMENT: The reproductive season occurs from February to mid-July. Two litters are produced each year, with a postpartum estrus, and a third litter is possible. The average litter size is 3 (2–4) for the first litter and 4.4 (3–6) for the second litter. Gestation time is approximately 30 days. Young are born with their eyes and auditory meatuses closed, but they grow very fast. Animals that first become surface active in May reach adult size by August. Longevity can be up to 6 years, although most animals live only to the age of 3.

PARASITES AND DISEASES: A wide variety of ectoparasites are found on alpine pikas, including many species of flea. Several gamasid mites (*Haemogamasus kitanoi*, *Hirstionyssus isabellinus*, *Poecilochirus necrophori*, *Laelaps hilaris*, *Parasitus* spp.) and ixodid (*Ixodes persulcatus*, *Dermacentor nuttalli*) ticks are found on alpine pikas. They arc also very susceptible to infections by the larvae of red mites. Throughout summer the larvae of several species of western warble flies parasitize alpine pikas, and up to 30–40 larvae may be found on a single pika. Many species of helminth (cestodes, nematodes) are also found in alpine pikas.

CONSERVATION STATUS:

IUCN Red List Classification: Least Concern (LC)

MANAGEMENT: There are no active conservation actions directed toward alpine pikas, although they are known to occur in several protected areas within their geographic range. Habitat fragmentation coupled with global climate change may pose a threat to the species. Terrain between suitable habitats is likely to become inhospitable for this species because it does not tolerate temperatures above 25°C very well, making dispersal difficult or impossible. In the 1970s, when a catastrophic decline in population numbers occurred in the Sayan Mountains, predation was identified as a threat to the species.

ACCOUNT AUTHOR: Andrew T. Smith

Key References: Lissovsky 2003, 2014; Lissovsky et al. 2007, 2008; Naumov 1974; Nikolskii and Mukhamediev 1997; Ognev 1966; Orlov 1983; Sludskii and Strautman 1980; Smith et al. 1990; Sokolov et al. 2009.

Ochotona argentata Howell, 1928
Helan Shan Pika

OTHER COMMON NAMES: Silver pika; Ningxia shutu, Helanshan shutu (Chinese)

DESCRIPTION: The Helan Shan pika is a relatively large pika. It was originally described based on winter specimens in which the entire back appears to sport a "striking silvery color" with hairs a pale grayish steel with fine black tips. The original common name, from the species epitaph *argentata*, was "silver pika." Even the feet are white (with a hint of buff), and the underparts are white tinged with buff. The head, face, and rump are yellowish. In contrast, the summer pelage is a bright rusty red over the entire back and head, while the ventral surface is a light silver gray. The distinct rufous coloration distinguishes *O. argentata* from Pallas's pika (*O. pallasii*), which has a gray-brown dorsal coloration. The ears are not reddish, and the white stripe along their border is indistinct.

The skull is slightly convex, and the nasals are relatively long. The interorbital region is broad and flat, and the posterior palatal foramina are pear-shaped. The auditory bullae are large and widely spaced.

SIZE: Head and body 208–235 mm; Hind foot 31–35 mm; Ear 22–25 mm; Greatest length of skull 45–52 mm; Weight 176–236 g

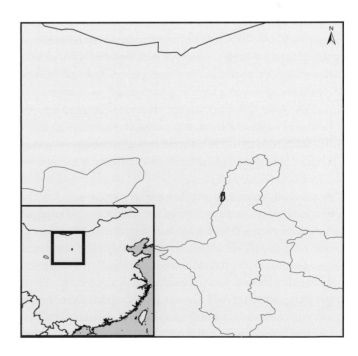

Ochotona argentata summer. Photo courtesy Zhaoding Wang

Ochotona argentata winter. Photo courtesy Zhaoding Wang

CURRENT DISTRIBUTION: The Helan Shan pika is known to occur in three areas in the Helan Shan (mountains) of Ningxia and Nei Mongol. These areas encompass a north-south distance of about 13 km, with a width of about 5 km. However, no comprehensive surveys have been undertaken on this species, and its distribution may be more extensive, but still confined within this isolated mountain range.

TAXONOMY AND GEOGRAPHIC VARIATION: Subgenus *Pika.* No subspecies (*helanshanensis* is a synonym). *O. argentata* was originally described as a subspecies of the alpine pika (*O. alpina*). It varies from *O. alpina* in behavior, chromosome number, morphology, and molecular characteristics, and thus is independent of that form. Recent molecular analyses indicate that *O. argentata* is closely allied with *O. pallasii*, although *O. argentata* appears independent based on a clear separation of morphological measurements between the two forms.

ECOLOGY: This is a rock-dwelling pika, and it is assumed that much of its natural history parallels that of other rock-dwelling pikas. They occur as high as 3,000 m asl.

HABITAT AND DIET: The Helan Shan pika has been found living among rocks in open or forested regions. They may also occupy artificial rock slides associated with mining, and even can be found living as far as 20 m inside mine tunnels. Some occupy artificial piled up stones (obo) used in local Mongolian rituals. The Helan Shan pika, like all pikas, appears to be a generalized herbivore. It constructs a typical pika haypile (cache of vegetation).

BEHAVIOR: The Helan Shan pika appears to be relatively quiet in comparison with the frequent vocalizations given by the similar Pallas's and alpine pikas. However, it does vocalize in the spring breeding season.

PHYSIOLOGY AND GENETICS: Diploid chromosome number = 38

CONSERVATION STATUS:
IUCN Red List Classification: Critically Endangered (CR)—B1ab(i,ii)+2ab(i,ii)
National-level Assessments: China Red List (Critically Endangered (CR)—A2c)

MANAGEMENT: The Helan Shan pika occupies a very restricted range in an isolated mountain chain, and its habitat throughout the extent of its geographic range has been severely altered in past decades.

ACCOUNT AUTHORS: Weidong Li and Andrew T. Smith

Key References: Erbajeva and Ma 2006; Formozov 1997; Formozov et al. 2004; Howell 1928; Jiang et al. 2016; Lissovsky 2014; Lissovsky et al. 2007; Smith and Xie 2008.

Ochotona cansus Lyon, 1907
Gansu Pika

OTHER COMMON NAMES: Jianlu shutu (Chinese)

DESCRIPTION: The Gansu pika resembles the Moupin pika (*O. thibetana*), but is smaller in size and generally lighter in pelage coloration. The summer coat can be quite variable, ranging from dark russet, tea brown, or dark brown to dull grayish buff. The belly is white with trace amounts of buffy hairs. There is a pale stripe that extends from the chest to the abdomen, which is not easily seen. The ears are blackish gray with white edges. The upper surface of the forefeet and hind feet is cream-buff, and the thickly furred soles are a sooty brown. The winter pelage becomes a homogeneous drab gray to a gray-russet color.

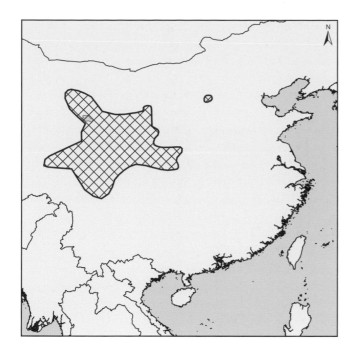

Ochotona cansus. Photo courtesy Jon Hall www.mammalwatching.com

Ochotona cansus habitat. Photo courtesy Shaoying Liu

The skull is large proportional to the body size, although compared with the skull of the Moupin pika is shorter (< 37 mm versus > 37 mm), and the zygomatic arch (< 17.2 mm versus > 17.3 mm) and interorbital (< 4 mm versus > 4.2 mm) widths are narrower. The measurements of *O. cansus morosa* are close to those of *O. thibetana*, but *morosa* is only distributed in the Qingling Mountains, where there are no *thibetana*. The general outline of the braincase is pear-shaped. The auditory bullae are well developed.

SIZE: Head and body 116–165 mm; Hind foot 22–29 mm; Ear 14–24 mm; Greatest length of skull 33–37 mm; Weight 50–99 g

CURRENT DISTRIBUTION: The Gansu pika is endemic to China. It ranges from E Qinghai and Gansu Provinces, south across W Sichuan, and isolated populations are found in Shanxi, S Shaanxi, and N Chongqing Provinces.

TAXONOMY AND GEOGRAPHIC VARIATION: Subgenus *Ochotona*. Four subspecies: *O. c. cansus* (E Gansu, E Qinghai, NW Sichuan); *O. c. morosa* (S Shaanxi, N Chongqing); *O. c. sorella* (N Shanxi); *O. c. stevensi* (W Sichuan). *O. cansus* was originally included in *O. roylei*, but it is quite unlike that species. It frequently has been included in *O. thibetana*, with which it shares many morphological similarities, but detailed genetic and morphological analyses confirm the separation of these forms. Some *O. cansus* specimens in China are labeled as the form "annectens"; however, that form is actually included as a synonym within *O. dauurica huangensis*.

ECOLOGY: The Gansu pika is a characteristic burrowing species and most aspects of the species life history parallel those of other meadow/steppe burrowing forms. The burrow system of the Gansu pika is one of the least complex of all the pikas. Burrows consist of only a few entrances, several branching tunnels, and a single nest chamber, and tunnel length normally extends only 1 m. Population density is low in spring (5–10 animals/ha) and higher in fall (30–50/ha).

HABITAT AND DIET: Gansu pikas primarily inhabit shrub habitat consisting of deciduous shrubs such as *Potentilla fruticosa* and *Caragana jubata* adjacent to alpine meadows. On the Qinghai-Xizang Plateau, the Gansu pika occupies steep hillsides, the primary zone where these shrubs grow. This pika prefers slightly open areas and avoids the thickest areas of brush. In regions where these pikas are sympatric with the meadow-dwelling *O. curzoniae* they appear to be excluded from open alpine meadows. The Gansu pika occurs between 2,700 and 5,000 m asl, the highest distribution of pikas in the subgenus *Ochotona*.

The Gansu pika is considered a generalized herbivore that feeds on moss, grasses, and forbs. Among 40 types of vegetation consumed by Gansu pikas in one study, they showed significant selectivity. Favored species included grasses and sedges (*Kobresia pygmaea*, *Carex* spp.), as well as forbs known to be toxic range plants (*Astragalus floridus*, *Oxytropis* spp.; *Pedicularis longiflora*).

BEHAVIOR: The Gansu pika is one of the most purely diurnal of all pika species; they become surface active at first light and seldom enter their burrows throughout their daytime activity period. They remain surface active even under conditions of wind and rain. Average time above ground each day exceeds 11 hours.

Home ranges of paired adult males and females overlap considerably and the spatial arrangement of home ranges is relatively constant. The utilization of a family home range does vary seasonally. In spring the home range of females may be small (500 m²), whereas that of males is considerably larger (1,200 m²), perhaps due to exploratory mating forays. Home range movements by both males and females enlarge during summer once young become surface active, and then ranges of both species contract considerably in fall and throughout winter. Natal burrows are located near the border of the home range, and ultimately it is in these areas that juveniles become established. While the primary social organization is that of pairs of adults and their young, neighboring males may extend their influence to include the home ranges of additional females following the death or disappearance of a male. Thus the mating system is flexible, normally resulting in facultative monogamy with occasional polygyny. Males show heightened aggression when they encounter each other, particularly near the border of a family home range. In contrast, behavioral interactions among adult females are rare. Almost all affiliative interactions (consisting of mouth-nose rubbing, allogrooming, sitting in contact) occur between paired animals.

Gansu pikas are quite vocal and they utter four different types of call. As in most pika species, the long call (song) is given only by males and primarily during the breeding season. Short calls and trills are uttered by both sexes, and these calls are normally returned in kind by family members. Juveniles utter whines that appear to function as a method of mother-infant contact.

Cheek glands located about 5 mm below the base of the ear are used in scent marking; only males exhibit this behavior. Cheek rubbing takes place during the mating season and apparently serves as a warning signal for territory maintenance.

REPRODUCTION AND DEVELOPMENT: The breeding season of the Gansu pika is from early April to September, depending on the locality. There is little synchrony in reproduction among females in a population. Litter size varies from one to six young per litter, and up to three litters per year (normally two) are produced with a postpartum estrus. Gestation lasts for 20 days. Mothers remain separate from their young except for a short time allocated for nursing (usually a half hour, but as short as 10 minutes). Young of the year can become reproductively active.

PARASITES AND DISEASES: The Gansu pika is host to the nematodes *Ohbayashinema patriciae* and *Murielus tjanschaniensis*, each a characteristic parasite within *Ochotona*.

CONSERVATION STATUS:

IUCN Red List Classification: Least Concern (LC)

MANAGEMENT: There are no known major threats to the Gansu pika, and no active management plans have been directed toward this species. Populations of the Gansu pika may be reduced by poisoning in portions of its distributional range because of perceived indications that it may be a rangeland pest. It is recommended that research be conducted to determine population status of this species, particularly for the subspecies *O. c. sorella* (not seen in the wild for 70 years) and the isolated population of *O. c. morosa*.

ACCOUNT AUTHORS: Andrew T. Smith and Shaoying Liu

Key References: Allen 1938; Durette-Desset et al. 2010; Jiang and Wang 1981; Lissovsky 2014; Smith and Xie 2008; Smith et al. 1990; Su 2001.

Ochotona collaris (Nelson, 1893)
Collared Pika

OTHER COMMON NAMES: Cony (Aboriginal); Pica à collier (French)

DESCRIPTION: The pelage of the collared pika is brownish gray dorsally with a light gray to cream-colored venter, a characteristic white collar extends from the throat to below the ears, and a buffy patch covers the facial gland. Ears are round, with a dark interior and a white margin.

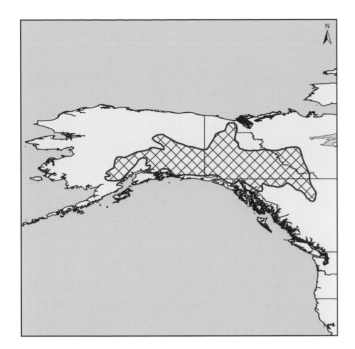

Ochotona collaris. Photo courtesy Moose Peterson/WRP

Young of the year are a light gray with less distinction between dorsal and ventral coloration and a less pronounced collar than the adults; a dark molt line is visible on adults throughout most of the summer. The feet have hairy plantar and palmar surfaces and black claws.

The skull is medium to large in size and relatively flat. The skull is slightly broader than that of the American pika (*O. princeps*), with larger auditory bullae and significant differences in tooth morphology. The incisive and the palatal foramina are separate.

SIZE: Head and body 149–205 mm; Hind foot 25–36 mm; Ear 17–26 mm; Greatest length of skull 38–46 mm; Weight 101–188 g

PALEONTOLOGY: Fossil evidence demonstrates that pikas were intermittently present on the cold, dry mammoth steppe of Alaska and NW Canada throughout the Pleistocene. These fossils include a larger bodied species (*O. whartoni*), considered to be more of a grassland specialist, and a smaller species, similar to *O. collaris* or *O. princeps*. Although pikas originated in Asia and migrated through Beringia (the ice-free corridor between Alaska and Russia), parasitological evidence indicates that the lineage that gave rise to collared pikas diverged from a common ancestor with *O. princeps* in North America (likely before the Pleistocene) and subsequently migrated north. The collared pika lineage subsequently diverged geographically, with molecular data identifying a larger stable population in the unglaciated region near the north-central portion of their range. A mummified specimen is known from Pleistocene deposits near Fairbanks, Alaska, and feces and teeth have been found in Pleistocene deposits in the Yukon Territory (Old Crow Basin, Bluefish Caves). Although this fossil material is difficult to assign to either *O. collaris* or *O. princeps*, molecular evidence suggests that populations of collared pika were regionally present and had already begun to diversify in Alaska at the last interglacial (~ 148 kya [range 126–323 kya]).

CURRENT DISTRIBUTION: Known from Alaska and NW Canada, collared pikas can be found from the Chigmit Mountains in W Alaska, north to the Richardson Mountains in the Yukon Territory, east into the MacKenzie Mountains and Kotaneelee Range of the Northwest Territories, Canada, and south to the area around Atlin in British Columbia. In Alaska they have not been found on the Kenai Peninsula or north of the Yukon River, and no populations or specimens are known from the Brooks Range. They are known from elevations ranging from 300 to 2,000 m, and are separated from American pikas by a 800 km gap.

TAXONOMY AND GEOGRAPHIC VARIATION: Subgenus *Pika*. No subspecies. *O. collaris* was once considered to be conspecific with *O. princeps* and the Asian alpine pika (*O. alpina*), but it is morphologically and phylogenetically distinct. Recent molecular analyses indicate that at least three geographically and genetically distinct lineages exist within *O. collaris*.

ECOLOGY: Collared pikas inhabit solitary territories on talus slopes, with one large or several small haypiles of cached food for the winter. They rarely forage more than 10 m from talus slopes, and an obvious grazing gradient can be seen in the vegetation by the end of the summer. Latrine sites are primarily located within the talus, and can be identified for several years after the territory has been vacated. Haypile locations within a territory tend to be static regardless of occupant, and abandoned territories can be identified by old haypiles.

Population densities for collared pikas are low, with estimates ranging from 1.5 animals/ha in the Ruby Range of Canada up to 7.2 animals/ha in Denali National Park, Alaska. Inter-annual variability in population size is also relatively low, as is characteristic for talus-dwelling species of pika, ranging from 0.6 to 2.3 animals/ha over a 15-year period in the Ruby Range. The majority of the mortality occurs during winter, especially in the first year of life. Although few observations of successful predation exist, potential predators of collared pikas include ermine (*Mustela erminea*) and other weasels, red fox (*Vulpes vulpes*), and raptors.

While collared pikas are primarily non-burrowing, some small amount of tunneling occurs around the edges of talus slopes, presumably increasing safe access to vegetation. Collared pikas share their territories with hoary marmots (*Marmota caligata*), Arctic ground squirrels (*Spermophilus parryii*), and several species of vole (*Microtus* spp. and *Myodes rutilus*) as well as collared and brown lemmings (*Dicrostonyx groenlandicus* and *Lemmus trimucronatus*). They minimize competition with other species by spatially partitioning the available ecological niches.

HABITAT AND DIET: Collared pikas prefer talus and block fields adjacent to open alpine meadows and can also be found in small populations on nunataks (exposed rocky ridges surrounded by glacier) in the Wrangell–St. Elias region. Sizable patches of talus habitat, with large patch perimeter, and the presence of *Dryas* spp. and *Carex* spp. are the best predictors of habitat occupancy. Generally population densities are higher on south-facing slopes, which will tend to be snow free earlier in the season. This species has also been occasionally documented in unusual low-elevation or forested habitats, such as along the shores of Kluane Lake and in isolated populations in a spruce-birch-willow forest of C Alaska.

Collared pikas are generalist herbivores. They mostly consume vegetation but, like rabbits, will also re-consume their own soft feces. Proximity to talus seems to be the most important factor driving vegetation selection, although pikas distinguish between grasses with high toxin loads (such as *Festuca*) and those that are safe to consume.

They are also thought to cache and consume scats from hoary marmots and ermines. Individuals on nunataks are known to collect and store the carcasses of dead birds (presumably scavenged from the surrounding glaciers) in their haypiles, and to consume the brains from these individuals.

BEHAVIOR: The collared pika is a diurnal and fairly asocial species, with each individual controlling a small territory (16–20 m²). They are kleptoparasitic, meaning they will steal cached vegetation from the haypiles of their neighbors. The majority of the interactions between individuals are tolerant or aggressive (few affiliative interactions), with scent marking, calls, and chasing behavior to demarcate and defend territories. Territories generally remain fairly static from year to year, and are inherited by individuals of the same sex as the former occupant, maintaining an alternating male-female pattern on the landscape. Overlap among adjacent territories is greatest between opposite-sex pairs, and appears to be greatest during the fall. Their mating system is promiscuous, with both males and females taking multiple mates. Young of the year are philopatric and will tend to disperse less than 1 km (500 m average) from their natal site, often to the closest available territory. Dispersal is not sex-biased. Collared pikas are very vocal and respond to the calls of conspecifics as well as calls of hoary marmots and Arctic ground squirrels. They exhibit less call variability than their sister species, the American pika, and do not give a long call. Although calls contain individual-specific information, individuals react similarly (calling, approaching, or yawning) to calls of near neighbors and strange pikas.

PHYSIOLOGY AND GENETICS: Diploid chromosome number = 68. Collared pikas show lower genetic diversity within populations and less variation between populations than is observed in co-occurring alpine species (hoary marmots, Arctic ground squirrels, singing voles, or brown lemmings) or in other species of pika (the American pika or the plateau pika, *O. curzoniae*). This appears to be due to historical (i.e., post-glacial movements) as well as ecological factors. Genetic evidence for dispersal distance indicates that most collared pikas disperse around 500 m, dispersal distances of 2 km or farther are possible, but populations at greater distances are unlikely to exchange migrants. Even so, there is little evidence for inbreeding, even when population sizes are small. Little work has been done on the physiology of collared pikas, but what is known suggests adaptation to lower altitudes and cooler temperatures than American pikas. Oxygen affinity of collared pika hemoglobin is uniformly lower than that of

American pikas. The body temperature of collared pikas is 39°C, lower than that of the American pika (40.1°C).

REPRODUCTION AND DEVELOPMENT: The reproductive rate of collared pikas is relatively low. Breeding occurs at 1 year of age from mid-May to early June, and young are born approximately 30 days later. Most females initiate only one litter per year. Although embryo counts and placental scars suggest that litter size at conception ranges from two to six offspring, studies in the Ruby Range indicate that females successfully wean only one offspring per litter per summer on average. Young pikas develop rapidly, reaching adult size within 40–50 days. After weaning they disperse to new territories and begin to rapidly cache food before winter. Collared pikas generally live around 4 years, but have been known to live up to 7 years.

PARASITES AND DISEASES: Collared pikas are hosts to multiple species of parasites. Although this parasite fauna represents only a subset of the species diversity found on American pikas or Asian species, most collared pikas harbor fleas in addition to several endoparasites. Endoparasites include tapeworms (*Schizorchis*), pinworms (*Labiostomum* and *Cephaluris*), and coccidea (*Eimeria*, *Isospora*, and *Sarcocystis*). Ectoparasites include fleas (*Amphalius*, *Ctenophyllus*, and *Monopsyllus*), mites, and botflies.

CONSERVATION STATUS:

    IUCN Red List Classification: Least Concern (LC)

    National-level Assessments: Canada (Special Concern)

MANAGEMENT: In general, collared pikas experience little direct anthropogenic disturbance throughout most of their range. Hunting and trapping of this species for meat and fur are allowed throughout the year in some regions of Alaska, likely representing a small number of actual individuals taken. A larger concern is the loss of suitable habitat and other environmental changes associated with global warming. Aberrant midwinter rains and freeze-thaw episodes have been associated with high mortality, and are likely to increase under certain global warming scenarios. Shrub and tree encroachment throughout the range is likely to further limit the potential for gene flow or recolonization of extirpated populations. Much of the knowledge of population trends is based on work conducted in the Ruby Range; more information is needed on the generality of population responses throughout the rest of the range of this species.

ACCOUNT AUTHOR: Hayley C. Lanier

Key References: Barrio and Hik 2013; Broadbooks 1965; COSEWIC 2011; Franken and Hik 2004; Galbreath and Hoberg 2012; Guthrie

1973; Harington 2011; Koh and Hik 2007; Lanier and Olson 2009, 2013; Lanier et al. 2015a, 2015b; Lynch et al. 2007; MacDonald and Jones 1987; Morrison and Hik 2007; Morrison et al. 2004, 2009; Rausch 1961; Slough and Jung 2007; Smith and Weston 1990; Trefry and Hik 2009; Tufts et al. 2015; Weston 1981; Zgurski and Hik 2012.

Ochotona coreana Allen and Andrews, 1913
Korean Pika

OTHER COMMON NAMES: Korean piping hare; Uneutokki, Jwitokki (Korean); Gaoli shutu (Chinese)

DESCRIPTION: In summer the dorsal pelage of the Korean pika is a dull to bright orange. The rostrum may have black stripes above. The ears have a blackish tone, but with a white border; they are paler internally and sport a tuft of long white hairs from the base. The winter pelage becomes more light brown to brownish orange, or assumes a grayish to yellowish hue. The head is gray, and the ventral surface whitish washed with a pale clay color.

The skull is relatively large with long nasals and a flat cranial portion. The anterior frontal is protuberant, but the posterior part and the anterior parietal are flat. The zygomatic arch is large, and sword-shaped jugals project posteriorly. The elliptical incisive foramina reach a line

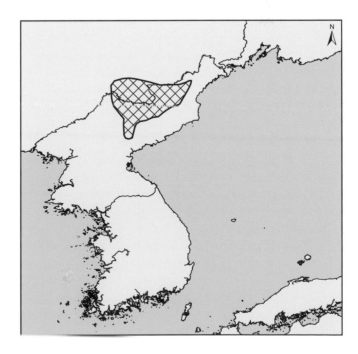

Ochotona coreana. Photo courtesy Tae-Young Choi

between the upper second molars. The tympanic bullae are large.

SIZE: Head and body 160–204 mm; Hind foot 27–33 mm; Ear 18–21 mm; Greatest length of skull 43–45 mm; Weight 132–194 g

CURRENT DISTRIBUTION: The Korean pika is confined to the higher mountains of the N Korean peninsula (such as Myohyangsan, Ganggye, Jagang-do, Musan, Hamgyeongbuk-do, and Bocheon, Ryanggang-do), and extends into the Baekdusan (Korea) / Changbaishan (China) regions.

TAXONOMY AND GEOGRAPHIC VARIATION: Subgenus *Pika.* No subspecies. *O. coreana* was initially named as an independent species, and later referred to as a subspecies of the northern pika (*O. hyperborea*). New molecular evidence, as well as its geographic isolation, indicates that it should be treated as an independent species that is most closely related to the northern pika.

ECOLOGY: This species is very poorly known. It is apparently common throughout its range.

HABITAT AND DIET: This is a talus-dwelling form of pika (as are all species in the subgenus *Pika*), although in some parts of its range it lives among the root mass of montane shrubs within the forest zone. In this respect it is similar to the closely related northern pika, which also can be found living in forests among tree-falls. It is most commonly found between 1,000 and 2,500 m in elevation.

BEHAVIOR: The Korean pika is very vocal. In other aspects, its behavior is assumed to be similar to that of the northern pika.

REPRODUCTION AND DEVELOPMENT: The mating season extends from May to July. Litter size is four to five, and two litters are produced per year.

CONSERVATION STATUS:

IUCN Red List Classification: Not Evaluated (NE) due to the recent recognition of its independent status

National-level Assessments: DPR Korea (Susceptible = Near Threatened)

MANAGEMENT: North Korea designated the Korean pika with its habitat at Sanyang District, Beagam-gun, Ryanggang-do, as a natural monument in 1980. There are no recent status reports on this species, although its range does overlap with several national protected areas in China (Changbaishan) and North Korea (Baekdusan and Myohyangsan).

ACCOUNT AUTHORS: Andrew T. Smith and Yeong-Seok Jo

Key References: Allen and Andrews 1913; Kishida and Mori 1930; Lissovsky 2014; Lissovsky et al. 2008; Pak et al. 2002; H. Won 1968; P. Won 1967; Won and Smith 1999.

Ochotona curzoniae (Hodgson, 1858)
Plateau Pika

OTHER COMMON NAMES: Black-lipped pika; Gaoyuan shutu (Chinese)

DESCRIPTION: In summer the dorsal pelage of the plateau pika is sandy russet or sandy brown, and the ventral pelage ranges from a sandy yellow to a grayish white. The winter pelage is softer and longer than the summer pelage; the dorsal surface becomes a light sandy yellow to a yellowish white. The thin round ears have a distinct white margin; the dorsal side is rust-colored. The lips are ringed with black, extending up to the tip of the nose. The soles of the feet are hairy and both the forefeet and the hind feet have long black claws.

When viewed from the side, the skull is greatly arched and elevated in the frontal area. The incisive and the palatal foramina are combined into a single foramen. The auditory bullae are small.

SIZE: Head and body 140–192 mm; Hind foot 28–37 mm; Ear 18–26 mm; Greatest length of skull 39–44 mm; Weight 130–195 g

PALEONTOLOGY: The Qinghai-Tibetan Plateau, the highest plateau on Earth, began its uplift in the Pliocene, continuing through the Quaternary glaciations. The most significant phase of this uplift, the Kunlun–Yellow River Movement, occurred 1.0–0.60 mya and caused the

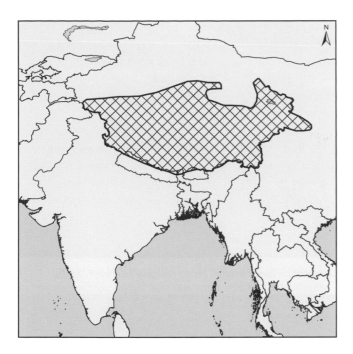

Ochotona curzoniae. Photo courtesy Andrew T. Smith

during the LGM were more spatially restricted, allowing some central populations to survive; thus population movements following the LGM tended to be bidirectional with an increase in gene flow between central and edge populations.

CURRENT DISTRIBUTION: The range of the plateau pika is roughly coincident with the Qinghai-Tibetan Plateau: S Xinjiang, S Gansu, Qinghai, Xizang, and W Sichuan, China; trans-Himalayan Nepal (the Mustang District); and N Sikkim, India. It occurs at elevations ranging from 3,000 to 5,000 m.

TAXONOMY AND GEOGRAPHIC VARIATION: Subgenus *Ochotona*. No subspecies. *O. curzoniae* has been included in the Daurian pika (*O. dauurica*), although all contemporary authors recognize it as a distinct species. Diploid chromosome number (2n = 46) differs from that of the Daurian pika (2n = 50), and recent molecular analyses clearly separate the two species. These analyses indicate that the plateau pika is most closely related to the Nubra pika (*O. nubrica*). The form *melanostoma* (Büchner, 1890) was apparently described without knowledge of *O. curzoniae*, as it was compared only to *O. dauurica*. Currently, *melanostoma* is considered a synonym of *O. curzoniae*. The form *seiana* was initially described as possessing characters very much like those of *curzoniae*. Today multiple lines of evidence indicate that *seiana* is a synonym of the Afghan pika (*O. rufescens*).

ECOLOGY: Plateau pikas live in family-controlled and -defended burrow system territories, which can be composed of a monogamous pair, a single male and multiple females (polygyny), or multiple males with a female or females (polyandry or polygynandry). The average diameter of a family burrow system on productive meadows is approximately 25 m. Adults co-exist with the young from sequential litters born during the breeding season, and young do not disperse away from their family burrow system in their summer of birth. The population density is low in the spring, following high overwinter mortality. Catastrophic mortality of pika populations can occur during winters with uncharacteristically heavy snowfall. During the breeding season populations can approach 300 individuals/ha. High populations are often seen where there is low cover height of vegetation (presumably due to intensive grazing of livestock), and under these conditions pikas may further impact rangelands. The numerous burrows dug by plateau pikas have been linked to the destruction of alpine grassland. Due to the dietary overlap with livestock and the implicated effects of burrows on the alpine grassland, the Chinese government has labeled

altitude of the plateau to increase by up to 3,000 m. This uplift increasingly exposed the endemic fauna, such as ochotonids, to glaciation events. Plateau pikas, whose current geographic distribution coincides with the Qinghai-Tibetan Plateau, expanded and contracted throughout the extensive glacial period (EGP), which began 0.5 mya and continued until 0.17 mya, and were also influenced by the Last Glacial Maximum (LGM), which occurred 0.021–0.017 mya. Significant population expansion occurred during the interglacial period of 0.17–0.021 mya. Much of this movement was unidirectional, from the periphery of the range into more central regions, as extensive glacier systems during the EGP precluded occupancy. Glaciations

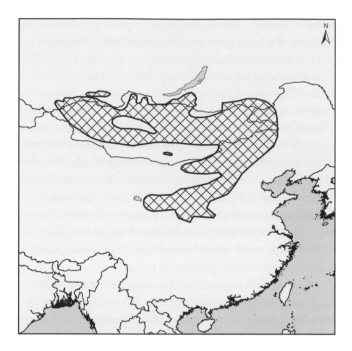

Ochotona dauurica. Photo courtesy Wendy Strahm

nously pale sandy yellow, grayish russet, or sandy brown. Thick hair conceals the pads at the tips of the toes. Vibrissae are 40–55 mm in length. The ears have a clear white margin, and the dorsal side of the ears is a blackish brown color; there is no rust-colored patch as is found in the plateau pika (*O. curzoniae*).

The skull is rather large and slightly arched. The long and narrow nasals are slightly widened anteriorly, and there is no constriction at the base of the widest part. The incisive foramen and the palatal foramen are combined into one foramen, and the frontal is slightly convex. The bullae are rounded and slightly flattened laterally.

SIZE: Head and body 150–220 mm; Hind foot 25–33 mm; Ear 16–26 mm; Greatest length of skull 39–45 mm; Weight 130–260 g

CURRENT DISTRIBUTION: The Daurian pika is distributed widely across Mongolia, extending into N China as far south as Qinghai and across the border into neighboring regions of Russia. It occurs at elevations between 400 and 4,000 m. It is largely a Gobi Desert species, although it extends up onto xeric northeastern reaches on the Tibetan Plateau.

TAXONOMY AND GEOGRAPHIC VARIATION: Subgenus *Ochotona*. Three subspecies: *O. d. dauurica* (*altaina* and *mursaevi* are synonyms; the primary form found across Mongolia, N China, and Russia); *O. d. latibullata* (Ubsu-Nur Basin, Mongolia, and into Tuva, Russia); and *O. d. huangensis* (*annectens*, *bedfordi*, and *shaanxiensis* are synonyms; Loess Plateau and into E Qinghai, China). The systematics of this species has been confusing, largely due to several misidentifications of specimens and inappropriate attribution of names (not following the law of priority). Some earlier treatments included *O. curzoniae* within *O. dauurica*, as *O. d melanostoma*. These two ecologically similar species are distinct and independent based on morphological and molecular evidence.

ECOLOGY: The Daurian pika is ecologically and behaviorally similar to the plateau pika. It is a burrow-dwelling pika that typically lives in large colonies. These burrow systems are quite complex, with multiple branches often used as lavatories, a single nest chamber (usually), and multiple entrances (sometimes as many as 40). Burrow tunnels normally lie at a depth of 30–40 cm, although they may extend to a depth of 1.5 m. The area encompassed by a single burrow varies greatly, but may encompass up to 1,000 m². Most burrows have a single nest chamber (roughly 30 × 40 cm), which is approached by two tunnels. Other tunnels are used as a lavatory or a storage area for hay. The burrowing activity of this species has been shown to have positive impacts on the surrounding landscape. Digging loosens the soil; burrows increase the concentration of certain soil aspects like nitrogen, humus, calcium, phosphorus, and humidity; and burrows also promote the growth of specific flora and increase the overall vegetation cover.

Populations can become quite numerous (densities of up to 300/ha have been observed), while crashes can occur yielding a single active family burrow in a hectare. Population density is subject to habitat type, and fluctuates both seasonally and annually. Causes range from flooding to food competition, and snowless winters to summer drought. Population explosions of Brandt's voles (*Lasiopodomys brandtii*) that consume almost all local vegetation have been known to cause crashes in populations of Daurian pikas.

Predators include eagle owls (*Bubo bubo*), steppe eagles (*Aquila nipalensis*), upland buzzards (*Buteo hemilasius*), saker falcons (*Falco cherrug*), corsac foxes (*Vulpes corsac*), wolves (*Canis lupus*), steppe polecats (*Mustela eversmanii*), and Pallas's cats (*Otocolobus manul*). For example, up to 80% of the diet of steppe eagles is Daurian pikas, and pikas are the most frequently consumed prey of Pallas's cats. Daurian pikas are the main food of these predators, especially during winter.

HABITAT AND DIET: The Daurian pika occupies semi-desert and desert-steppe habitat, preferring low-lying desert grassland habitat, although they may also occupy slopes of mountains and foothills. They tend to occupy lowland depressions with green vegetation, including wet and swampy bottomlands. In Tuva, Daurian pikas live in wormwood (*Artemisia*) and cinquefoil (*Potentilla*) or peashrub (*Caragana*) thickets. In parts of their range they are found in birch or larch forests, or on the edge of deciduous forest groves.

Like many pikas, the Daurian pika is a generalized herbivore that stores a haypile at the entrance of or within its burrow. These haypiles can be large (up to 6 kg) and are often targeted by wild and domestic ungulates. Species composition in any haypile is dependent on the plants surrounding the burrows. A wide variety of plants have been reported in the haypiles, generally up to 30 plant species in any pika colony, but with only 9 species (and sometimes only 1 or 2) in any particular haypile. Forbs represent the predominant plant growth form in haypiles (up to 90%), including such species as pygmy pea shrub (*Caragana pygmaea*), pygmy cinquefoil (*Potentilla* spp.), cypress (*Kochia prostrata*), fringed sagebrush (*Artemisia frigida*), and Lena alyssum (*Alyssum lenense*).

BEHAVIOR: Daurian pikas are active from sunrise to sunset, although in summer when it is warm the maximum activity is observed in the morning or early evening. They hide in their burrows during the hottest part of the day, and they resist venturing out of their burrows on windy days.

Daurian pikas reside in territorial family groups consisting of breeding adults and the young of the year. It is a highly social species, displaying allogrooming, boxing (rare and brief), huddling, and sitting in contact. They appear reticent to engage in intraspecific aggressive behaviors, but they do exhibit interspecific aggression when they encounter other small mammals such as sousliks (*Spermophilus*).

They utter three vocalizations: a short high whistle for an alarm, a short trill, and a long call usually given by males, which often precedes the short trills. Frequency of calls increases in the late afternoon to early evening hours, and some calling may be heard at night.

PHYSIOLOGY AND GENETICS: Diploid chromosome number = 50

REPRODUCTION AND DEVELOPMENT: Two or more litters are produced a year, with a litter size between 1 and 11. Reproductive periodicity varies according to location, beginning in April and extending to August in the Trans-Baikal region and April to September for much of Mongolia. Females from the first litter can obtain sexual maturity and reproduce in their summer of birth. Young Daurian pikas are born naked, but their growth and development have not been studied. Most populations in summer are composed primarily of young of the year (83%), with only 14% being overwintered adults and 3% consisting of individuals surviving two winters. These data demonstrate the rapid turnover in populations of Daurian pikas.

PARASITES AND DISEASES: Forty species of fleas are known to infect Daurian pikas, and of these two are known only from this species (*Ctenophyllus hirticrus*, *Amphalius runatus*), each of which are particularly abundant on pikas during winter. In summer, *Frontopsylla luculenta* and *Neopsylla bidentatiformis* are most numerous on Daurian pikas. Among ticks, *Dermacentor nuttalli*, *Haemogamasus kitanoi*, *Hirstoinyssus ochotonae*, *Laelaps nilaris*, *L. cleithronomydis*, and *Eulaelaps cricetuli* are found on Daurian pikas. Several warble flies penetrate the skin of Daurian pikas, one of which (*Ferrisella ochotonae*) may impose an infection rate of 98.5%. Additionally, 17 species of parasite have been found in the intestines of Daurian pikas.

Daurian pikas may also be subject to plague, but only as a secondary host, as plague epizootics primarily course through rodents.

CONSERVATION STATUS:

IUCN Red List Classification: Least Concern (LC)

MANAGEMENT: The management of Daurian pikas is very similar to that of the plateau pika. Daurian pikas have been considered to be pests and poisoned indiscriminately, while careful analysis of their role in the ecosystem demonstrates that they are a keystone species for biodiversity and serve as ecosystem engineers. Certain steppe shrubs only grow in areas occupied by Daurian pikas, and for these and other native plants the biomass of roots is greater, plants are taller, and plant cover is greater than in areas without pikas. The important aspect of seasonal phenology also favors plants growing in pika colonies, in that growth, maturation, and plant size are greater than in nearby areas without pikas. As highlighted above, Daurian pikas constitute a primary and valuable food source for the majority of avian and mammalian carnivores throughout their range. Thus poisoning of Daurian pikas runs counter to their importance in maintaining the functioning of the local ecosystem and maintenance of its native biodiversity.

ACCOUNT AUTHOR: Andrew T. Smith

Key References: Allen 1938; Batsaikhan et al. 2010; Borisova et al. 2008; Kawamichi and Dawanyam 1997; Komonen et al. 2003; Liao et al. 2007; Lissovsky 2014; Loukashkin 1940; Ognev 1966; Proskurina 1991; Ross et al. 2010; Smith and Xie 2008; Smith et al. 1990; Zhang et al. 2001; Zhong et al. 2008.

Ochotona erythrotis (Büchner, 1890)
Chinese Red Pika

OTHER COMMON NAMES: Hong'er shutu (Chinese)

DESCRIPTION: In summer the pelage of the Chinese red pika is a uniform bright rusty red covering the entire dorsal surface of the body, including the head. Ventrally the pelage including the legs and feet is a contrasting grayish white, although in some specimens there is a light brush of yellow. The large ears are also red, but lightly covered with hair, and there is a tuft of long white hairs at the anterior base of the ear. In contrast to its striking summer pelage, the winter coat is almost a uniformly drab gray. The ears are the only exception, which maintain their ferruginous coloration.

The incisive and the palatal foramina are separated, with the palatal foramen tapering forward to be positioned just behind the incisive foramen. The frontal bone is broad, and there are two vacuities, one in the anterior dorsal surface of each frontal. The overall profile of the skull is slightly arched. The sagittal crest is conspicuous, and the auditory bullae are large.

SIZE: Head and body 181–185 mm; Hind foot 32–42 mm; Ear 32–39 mm; Greatest length of skull 48–51 mm; Weight 184–352 g

CURRENT DISTRIBUTION: The Chinese red pika is endemic to China, being distributed from W Gansu and across N Qinghai at elevations between 2,000 and 4,000 m.

TAXONOMY AND GEOGRAPHIC VARIATION: Subgenus *Conothoa*. No subspecies (*vulpina* is a synonym). Previous treatments have included *O. erythrotis* in the Turkestan red pika (*O. rutila*), but these forms are genetically and morphologically distinct, as well as having widely separated geographical ranges. *O. gloveri* is closely related to and parapatric with *O. erythrotis*, and some have consid-

Ochotona erythrotis. Photo courtesy Wenjing Li

ered *gloveri* as a subspecies of *erythrotis*; the two forms appear to be sister species.

ECOLOGY: The Chinese red pika is a rock-dwelling species and is apparently individually territorial. They are not uncommon, although they have never been studied in depth. Most aspects of their natural history probably mirror those of other rock-dwelling pikas.

HABITAT AND DIET: Rock faces and crags that lie adjacent to alpine shrubland or meadows are the typical habitat types for this species. They are also known to live commensally with people, occupying mud walls of houses or adjoining agricultural fields. They create simple burrows in rocks or soil faces of 1–2 m in length. The Chinese red pika is a generalized herbivore that caches food during summer for winter survival.

BEHAVIOR: In winter, Chinese red pikas have been seen sunbathing. They will sit motionlessly when alarmed. It is unknown if this species utilizes alarm calls or songs.

REPRODUCTION AND DEVELOPMENT: Reproductive periodicity extends from May to August. Two litters are produced per year, and each litter can have between three and seven young.

CONSERVATION STATUS:

IUCN Red List Classification: Least Concern (LC)

MANAGEMENT: The Chinese red pika has been poisoned in areas where it lives commensally with people, but in general it has not been negatively impacted by humans. Its low density and remote habitat have contributed to the lack of information about this species.

ACCOUNT AUTHOR: Andrew T. Smith

Key References: Allen 1938; Howell 1928; Li 1989; Lissovsky 2014; Smith and Xie 2008.

Ochotona forresti Thomas, 1923
Forrest's Pika

OTHER COMMON NAMES: Huijing shutu (Chinese)

DESCRIPTION: The dorsal and the ventral pelage of Forrest's pika is blackish brown to dark reddish brown in summer. The dorsal sides of the ears are a light chestnut and have white edges. There are dark gray spots behind each ear that occasionally form a collar that may extend to the face, but the forehead remains brown. In winter the dorsal pelage becomes a grayish brown, while the ventral pelage is only slightly lighter. The feet are a dull white

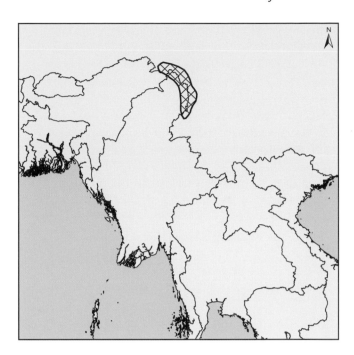

color, and the claws are considerably longer than those of the Moupin pika (*O. thibetana*). There is color variation throughout the range of Forrest's pika, and it is not uncommon to find dark or melanistic forms.

The skull is more arched (convex) than in the Moupin pika. The incisive foramen and the palatal foramen combine into one, mostly with a wavelike margin, so its profile is violin-shaped. When viewed from the front, the nasals are narrow and rectangular, and the interorbital region broad. The auditory bullae are of medium size.

SIZE: Head and body 155–185 mm; Hind foot 27–30 mm; Ear 18–23 mm; Greatest length of skull 37–41 mm; Weight 110–148 g

CURRENT DISTRIBUTION: Forrest's pika appears confined to the Gaolingong ranges in E Yunnan, bordering Myanmar and far NE India.

TAXONOMY AND GEOGRAPHIC VARIATION: Subgenus *Conothoa*. In the confusing world of *Ochotona* systematics, *O. forresti* has had a particularly checkered existence. Following its original description, it has at one time or another been assigned to *O. thibetana*, Royle's pika (*O. roylei*), or the steppe pika (*O. pusilla*). More recently two forms from within the core range of *forresti*, the Gaoligong pika (*O. gaoligongensis*) and the black pika (*O. nigritia*), were carved off as independent species, and an additional form, *O. f. duoxionglaensis*, was described from a region outside its core distribution. Detailed morphometric and genetic studies have shown that *O. forresti* is an independent species and not aligned with *thibetana*, *roylei*, or *pusilla*, and that it is confined to the Gaolingong ranges. The form *duoxionglaensis* from Xizang may be synonymous with *thibetana*, but this situation has not been clarified. The forms *gaoligongensis* and *nigritia* are synonymous with *forresti*, being based on melanistic individuals that are otherwise typical *forresti*.

ECOLOGY: Little is known about the ecology of Forrest's pika, but it is believed to be a burrowing form of pika.

HABITAT AND DIET: This species inhabits mixed conifer and broad-leafed forest, conifer forest, and shrub thickets at high altitudes (2,600–4,400 m). Its diet has not been studied, but it is probably a generalized herbivore like most pikas.

PHYSIOLOGY AND GENETICS: Diploid chromosome number = 54

CONSERVATION STATUS:

IUCN Red List Classification: Least Concern (LC)

National-level Assessments: China Red List (Near Threatened—NT)

MANAGEMENT: This species occurs in high and very remote areas, and there are no known threats. Obviously there is a need for studies on Forrest's pika.

ACCOUNT AUTHOR: Andrew T. Smith

Key References: Ge et al. 2012; Jiang et al. 2016; Lissovsky 2014; Molur et al. 2005; Smith and Xie 2008; Smith et al. 1990; Thomas 1923.

Ochotona gloveri Thomas, 1922
Glover's Pika

OTHER COMMON NAMES: Chuanxi shutu (Chinese)

DESCRIPTION: During summer the muzzle and the forehead of Glover's pika are tinged with yellow-brown, rufous yellow, or dull chestnut, while the cheeks are gray. Ears are yellow-brown, and long white hairs form at the base of the ears in front. The palms and soles are covered with gray-black short hairs. The rest of the dorsal surface is grayish brown or pale brown. Thus the summer pelage is distinctively different from the all-red dorsal pelage of the otherwise similar Chinese red pika (*Ochotona erythrotis*). The vibrissae are long, up to 60 mm. There are slight differences in the pelage of the various subspecies; *O. g. gloveri* appears to be darker than the other two subspecies. The ventral pelage is a uniform dull white, with the tips of the hairs white or buffy and their bases a slate

gray color. In winter the muzzle is orange-yellow, and the dorsal surface of the ears is orange-yellow. The remainder of the pelage, with the exception of the palms and soles, is lighter than it is during the summer. *O. g. brookei* has the most reddish color, *O. g. muliensis* is grayer, and *O. g. calloceps* is intermediate between these two forms.

The incisive foramen and the palatal foramen are distinct from each other, and there is no bony division between the two foramina. The incisive foramen is very constricted at its posterior end, and the palatal foramen ends anteriorly in a distinct notch in each premaxilla. There is a small oval foramen at the anterior dorsal surface of each frontal, although this may be missing in some individuals. The rostrum is narrow and long, and the nasals are very elongate, averaging 36% of the occipito-nasal length. Auditory bullae are small (smaller than in the Chinese red pika).

SIZE: Head and body 165–250 mm; Hind foot 31–41 mm; Ear 31–39 mm; Greatest length of skull 44–53 mm; Weight 140–270 g

CURRENT DISTRIBUTION: Glover's pika is endemic to China. Its range includes SW Qinghai, NE Xizang, NW Yunnan, and W Sichuan Provinces. It is typically found at elevations ranging from 3,500 to 4,200 m, although it may extend down to 1,700 m in Sichuan.

TAXONOMY AND GEOGRAHIC VARIATION: Subgenus *Conothoa*. Four subspecies: *O. g. brookei* (SW Qinghai, NE Xizang); *O. g. calloceps* (NW Yunnan, E Xizang); *O. g. glo-*

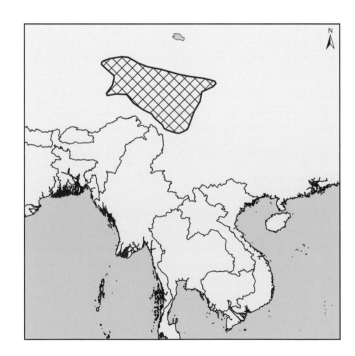

Ochotona gloveri. Photo courtesy Shaoying Liu

1–3 kg of vegetation) during summer that are typical of most pika species. Some species that have been noted in their haypiles include *Setaria viridis, Incarvillea sinensis, Artemisia argyi, Eragrostis nigra, Sophora davidii,* and *Bauhinia purpurea.* In areas where crops have intruded into areas occupied by Glover's pikas, they have been recorded storing apples and castanea.

BEHAVIOR: Glover's pika is primarily diurnal, although its peak activity periods are in the morning and late afternoon. They are seldom seen active during winter.

REPRODUCTION AND DEVELOPMENT: The reproductive season extends from May to July. Litter size averages three young.

CONSERVATION STATUS:

IUCN Red List Classification: Least Concern (LC)

MANAGEMENT: Glover's pikas are sometimes poisoned in situations where they live commensally with people and are considered to be pests. They also may be controlled in areas where plantations of apple, plum, or cherry trees have encroached on the habitat of Glover's pika.

ACCOUNT AUTHORS: Shaoying Liu and Andrew T. Smith

Key References: Allen 1938; Feng et al. 1986; Li 1989; Lissovsky 2014; Pen et al. 1962; Peng and Zhong 2005; Smith and Xie 2008; Smith et al. 1990; Thomas 1922; Yu et al. 1996.

veri (includes *kamensis,* W Sichuan); and *O. g. muliensis* (Muli County, Sichuan).

O. gloveri has previously been included in the Turkestan red pika (*O. rutila*) and *O. erythrotis.* Morphological and genetic evidence clearly separate *gloveri* and *erythrotis* from *rutila,* but these data also indicate that *gloveri* and *erythrotis* are very closely related. The Muli pika (*O. muliensis*) was originally described as a subspecies of *gloveri,* but it has been considered a separate species in many recent treatments. Its noteworthy characteristic is that it does not possess oval foramens in the dorsal surface of the frontals, and its color is much duller than that of *O. g. gloveri.* However, molecular systematics show that *muliensis* is clearly an independent subspecies within *O. gloveri.*

ECOLOGY: Glover's pikas are common throughout their range, but have never been studied in detail. They are a rock-dwelling pika, and thus much of their natural history is likely to be similar to that of other rock-dwelling pikas.

HABITAT AND DIET: Glover's pikas live in rocky cliffs, in talus, and even commensally in towns and lamaseries where they occupy lacunae in adobe walls. In some parts of their range they have been observed living among stones in sparse coniferous forests or in brushy areas. They have also been found living in caves where there are no large stones.

Like most pikas, Glover's pikas appear to be generalist herbivores, primarily eating forbs, sedges, and grasses. They construct haypiles (some containing as much as

Ochotona hoffmanni Formozov et al., 1996
Hoffmann's Pika

OTHER COMMON NAMES: Pischuha hoffmanna (Russian)

DESCRIPTION: The dorsal fur of Hoffmann's pika is ochraceous brown. Ventrally the fur is grayish ochraceous or ochraceous, and the chest is rufous. In winter the pelage is longer, softer, and grayer. Hairs above the neck gland are chestnut. The ears are rounded with white margins. The skull is large and stout, and the incisive and the palatal foramina are separated. The auditory bullae are of medium size.

Hoffmann's pika is one of the largest pikas in the *Pika* subgenus. Only representatives of several local peripheral populations of the alpine pika (*O. alpina*) may be larger than this species. All talus-dwelling Palearctic pikas are very similar in skull appearance, being different mainly in size and, consequently, in allometric features. Thus

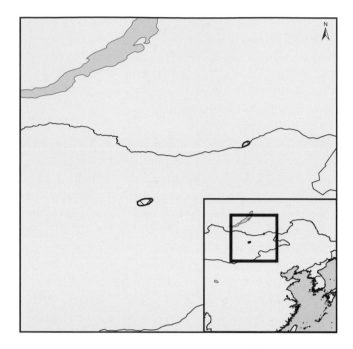

Ochotona hoffmanni. Photo courtesy Andrey Lissovsky

morphological identification of Hoffmann's pika without a large series of comparative material is difficult.

SIZE: Head and body 194–208 mm; Hind foot 30–32 mm; Ear 20–22 mm; Condylobasal length of skull 46–50 mm; Weight 260–310 g

CURRENT DISTRIBUTION: The range of Hoffmann's pika is restricted to the Bayan-Ulan Mountains of Mongolia and the Erman range of the Russian Federation. These two parts of the distribution are divided by a wide gap.

TAXONOMY AND GEOGRAPHIC VARIATION: Subgenus *Pika*. *O. hoffmanni* was initially described as a subspecies of *O. alpina*, and subsequently raised to species status based on differences in morphology, karyology, and behavior. Molecular analyses demonstrate that this species is not closely related to *O. alpina*, but that it is closely related to the species group that includes the northern pika (*O. hyperborea*) and Manchurian pika (*O. mantchurica*). Geographic variation is absent.

ECOLOGY: This is a rock-dwelling pika, and its natural history is similar to that of other rock-dwelling pikas. Hoffmann's pika inhabits talus composed of different size rocks as well as cavities under large rocks in the taiga. The distributional range is situated in larch taiga and light birch forests.

HABITAT AND DIET: Hoffmann's pika lives at relatively low altitude, like its close relative and neighbor the Manchurian pika.

BEHAVIOR: Hoffmann's pika is diurnal. It has a characteristic alarm call that is different from that of related species (alpine pika, northern pika, and Manchurian pika). Hoffmann's pika is very secretive, and although it is active during the daytime, it spends most time under the cover of stones.

PHYSIOLOGY AND GENETICS: Diploid chromosome number = 38

REPRODUCTION AND DEVELOPMENT: Breeding starts in May and lasts until August. Juveniles first appear in mid-June.

CONSERVATION STATUS:
   IUCN Red List Classification: Endangered (EN)—B1ab(iii)

MANAGEMENT: This is a very isolated form of pika, thus one that is likely to be highly sensitive to modifications of its habitat and climate change. Hoffmann's pika is common in the southern part of the Erman range in Russia; its status in Mongolia is unknown.

ACCOUNT AUTHOR: Andrey Lissovsky

Key References: Formozov and Baklushinskaya 1999; Formozov et al. 1996, 2006; Lissovsky 2014; Lissovsky et al. 2007.

Ochotona hyperborea (Pallas, 1811)
Northern Pika

OTHER COMMON NAMES: Siberian pika; Dongbei shutu (Chinese); Ezo naki-usagi (Japanese); Severnaya pischuha (Russian)

DESCRIPTION: The dorsal pelage of the northern pika varies according to subspecies. In summer it is generally light grayish brown to ochraceous brown. The sides are lighter and yellower, often tinged with gray. The ventral pelage is also lighter, usually white or gray with a yellow tinge. The ears are gray and edged in white. The feet are covered with whitish gray, brownish, or black hairs. The winter pelage is variable according to location, but it is typically lighter and grayer than the summer pelage.

In comparison with the skull of the alpine pika (*O. alpina*), the skull of the northern pika is shorter and more rounded, the orbits are not as elongated, and the rostrum is comparatively shorter. The frontal bones lack foramina. The small auditory bullae are relatively wider and flatter than those of *alpina*, and the cranium is less flexed posteriorly.

SIZE: Head and body 132–190 mm; Hind foot 21–27 mm; Ear 13–20 mm; Condylobasal length of skull 32–41 mm; Weight 52–165 g

CURRENT DISTRIBUTION: The northern pika has the largest geographic range of all the pikas, extending from the Ural Mountains through NE Eurasia to Sakhalin. It also occurs on Hokkaido, Japan, and several islands in the Bering Sea and the Sea of Okhotsk. It occurs at elevations from sea level to 2,200 m.

TAXONOMY AND GEOGRAPHIC VARIATION: Subgenus *Pika*. Eleven subspecies: *O. h. cinereofusca* (Transbaikalia, Russia); *O. h. davanica* (N Baikal, Russia); *O. h. ferruginea* (includes *cinereoflava, kamtschaticus, kolymensis, normalis*; from Kamchatka uplands to W Sayan Mountains, Russia); *O. h. hyperborea* (includes *litoralis, svatoshi*; Chukchi Peninsula, Russia); *O. h. minima* (Anadyr River basin, Russia); *O. h. naumovi* (Putorana Plateau, Russia); *O. h. shamani* (Indigirka River basin, Russia); *O. h. stenorostrae* (Tuva, Russia); *O. h. uralensis* (N Ural Mountains, Russia); *O. h. yesoensis* (Hokkaido Island, Japan); *O. h. yoshikurai* (Sakhalin Island, Russian far east). *O. hyperborea* is closely affiliated with *O. alpina*, and historically has been synonomized with that form—and also at times with the collared pika (*O. collaris*) and the American pika (*O. princeps*).

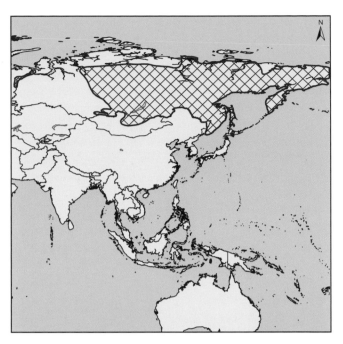

Ochotona hyperborea. Photo courtesy Makoto Shinohara

However, *hyperborea* and *alpina* (with its closely related Turuchan pika *O. turuchanensis*) have different chromosome numbers, and these species can occur in sympatry (in which case the northern pika exhibits character displacement and becomes smaller than in areas where it is the only pika species). Attesting to the confusion in placement of various forms of *hyperborea*, the forms *cinereofusca* and *svatoshi* have been linked with *O. alpina* in earlier treatments. The Korean pika (*O. coreana*) and the Manchurian pika (*O. mantchurica*; *scorodumovi* a synonym although sometimes treated as a subspecies of *O. alpina*) have recently been determined to be independent forms based on their morphology, ecology, and genetics.

ECOLOGY: The northern pika is primarily a rock-dwelling species of pika, but it has been recorded inhabiting driftwood mounds along riverbanks. In some circumstances it will occupy areas under fallen trees or in moss banks, where it will create burrows. It lives in large colonies with individual family territories consisting of an adult pair and their young. Individuals within each family have their own territory, although they may contribute to a joint haypile.

Population density tends to be stable over time throughout its distribution, but density may vary according to location. Typical density is around 20 individuals/ha.

The main enemies of northern pikas are small mustelids—weasels (*Mustela* spp.), steppe polecats (*M. eversmanii*), and sables (*Martes zibellina*)—although wolves (*Canis lupus*), foxes (*Vulpes* spp.), lynx (*Lynx lynx*), and other mammalian carnivores are known to prey on pikas. A variety of avian raptors, e.g., Eurasian eagle owls (*Bubo bubo*), golden eagles (*Aquila chrysaetos*), rough-legged buzzards (*Buteo lagopus*), and merlins (*Falco columbarius*) have been known to hunt pikas.

HABITAT AND DIET: Northern pikas are typically found in the montane taiga or tundra of boreal Asia. Rock piles made up of medium to large size stones with areas of vegetation are their preferred habitat, but they have also been known to inhabit lava flows. Showing great flexibility, the northern pika can also be found in the subalpine belt and zone of mountain tundra if there is suitable habitat, even occupying hollows between tree roots. They normally do not burrow, relying on tunnels within the talus or broken rock piles for shelter. In instances where there is a shortage of natural shelters, they have been known to excavate a complex system of tunnels (sometimes along steep banks or under tree roots).

The diet of the northern pika includes herbaceous plants, shrubs, forbs, pileate fungi (in many regions), and berries and seeds (in summer and fall). Montane-dwelling pikas will also consume lichens. They construct haypiles from the seasonal plant selection within the vicinity of their territory. Onset of food caching during the summer growing season is determined by location and altitude. The total amount of food cached depends on the size and number of places in which the animals can store haypiles, and at any locality they appear to be very selective of the species chosen. Generally the number of plant species in a haypile ranges from 5 to 10, while some haypiles consist of a single species and others contain more than 10 species.

Often the most common plant growing on a territory may be absent from the haypile. Northern pikas are not strictly central-pace foragers; they may scatter-hoard a number of smaller haypiles at distances of 0.5–1 m from each other. In the Urals there are two to six haypiles per family territory and the stored weight varies from 100 g to 8 kg. In other parts of the range single pikas make haypiles up to 6 kg. These haypiles are often browsed on by other fauna. Coprophagy occurs in this species, and the fecal matter is not always ingested immediately after excretion, as in hares, but may be stored.

BEHAVIOR: Northern pikas are diurnally active, with the duration of activity increasing during the summer months when daylight hours become extended. Most activity occurs during the brighter part of the day, during which they forage and store food.

Colonies are composed of individual families that consist of adults and their young. These family territories do not overlap and are normally centered less than 120 m apart. Spacing of family territories within a colony remains constant, even following the replacement of occupants. Young of the year are generally philopatric and stake out individual territories near the periphery of the family territory; dispersal of young is rare. Most social interactions among family members are amicable (friendly), although body contact between male and female "partners" is rare. Aggressive behavior between individuals (typically adult males and females) includes chasing, attacking, and boxing, and adult males are more likely to cross territorial boundaries and interact aggressively with neighbors.

Vocalization is well developed in this species. Calls are loud and sharp. The long call (song) is used by adult males, while both sexes use the short call. The short call may be used as a warning about predators or to announce territory occupancy. There is a noticeable increase in calls during the caching period and the mating season. During the mating season the increase is accompanied by greater diversity in call types; at this time males utter a short, dull trill ("krrrr") in addition to the long call. Trills may also be given when pikas become frightened. Northern pikas also mark their family territories by rubbing the apocrine gland on their cheek on rocks or by scent marking with their urine.

PHYSIOLOGY AND GENETICS: Diploid chromosome number = 40

REPRODUCTION AND DEVELOPMENT: Reproductive periodicity and fecundity are variable across the range. The breeding season is generally from mid-April to mid-

Ochotona hyperborea calling. Photo courtesy Yukito Toba

lurus andrejevi, Dermatoxys schumakovitchi). Of particular interest is the presence of several species of coccidia (Apicomplexa: Eimeriidae) that infect not only *O. hyperborea*, but also the North American *O. collaris*: *Eimeria banffensis, E. calentinei, E. circumborealis, E. cryptobarretti, E. klondikensis,* and *Isospora marquardti.*

CONSERVATION STATUS:
  IUCN Red List Classification: Least Concern (LC)
  National-level Assessments: Japan (Near Threatened [NT]; Ministry of the Environment Japan)
MANAGEMENT: The northern pika was harvested for its pelt until the 1950s (700–6,000 hides being processed annually). There currently is no active management throughout most of its wide distributional range, which is largely wild inaccessible habitat. It is known to occur in protected areas within its range. Population fragmentation of *O. h. yesoensis* on Hokkaido Island is a potential threat. Development near the Yubari-Ashibetsu Mountains, Japan, may pose a threat to the isolated population.

ACCOUNT AUTHOR: Andrew T. Smith

Key References: Formozov and Yakhontov 2003; Gliwicz et al. 2005, 2006; Haga 1960; Hobbs and Samuel 1974; Kawamichi 1969, 1970, 1971a; Kishida 1930, 1932; Lissovsky 2003, 2014; Lissovsky et al. 2007, 2008; Lynch et al. 2007; Ognev 1966; Rausch 1963; Smith et al. 1990; Sokolov et al. 2009.

August, but may be shorter for some regions. There can be up to three litters per year, but two are more typical and for pikas in the northern part of the range only a single litter may be produced. Usually, only a single litter is successfully weaned. Litter size is similarly variable, ranging from 1 to 11 young, with an average size that ranges from 2.0 to 5.9. Gestation lasts 28 days. Newborn animals are altricial, being born with their eyes and auditory meatuses closed. Young grow rapidly (doubling their weight during their first month following weaning) and may become sexually mature in their year of birth (particularly those from early litters).

PARASITES AND DISEASES: Northern pikas can be infected with a variety of ectoparasites (10 species of fleas, six of gamasids, two of trombiculids, one ixodid mite species, and three of lice). Similarly, they can be heavily infected with warble flies (*Oestromya falax, O. leporina, Oestroderma schubini*) in different parts of the species range; in some populations 60% of pikas have harbored warble flies. Endoparasites include roundworms (family Oxyuridae; *Lubiostomum vesicularis*), cestodes (*Schizorchis altaica, S. yamashitai, Taenia tenuicollis*), and nematodes (*Cepha-*

Ochotona iliensis Li and Ma, 1986
Ili Pika

OTHER COMMON NAMES: Yili shutu (Chinese)
DESCRIPTION: Among *Ochotona*, the Ili pika is large and very thickly furred. A dark rusty collar surrounds the neck, and light fulvous borders the mouth and extends up the face between the orbits. Otherwise, the face is gray, as is much of the dorsal surface. The white vibrissae are long, some extending half the length of the body. Patches of fulvous are found on the hips and along the sides. It has large ears that are also thickly furred frontally and without a distinct margin; the ears do not appear to be as rounded as in many pikas. The hind feet are large, and there is a light yellowish thick wool pad below the foot.

  The incisive and the palatal foramina are confluent and form a single pear-shaped opening. The frontals possess no foramen.

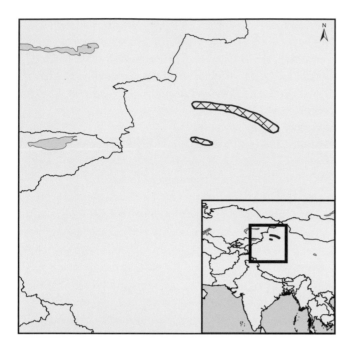

Ochotona iliensis. Photo courtesy Weidong Li

SIZE: Head and body 203–204 mm; Hind foot 42–43 mm; Ear 36–37 mm; Greatest length of skull 45–48 mm; Weight 217–250 g

CURRENT DISTRIBUTION: The Ili pika has a restricted distribution in two northeastern branches of the Tian Shan west of Ürümuqi, Xinjiang-Uyghur Autonomous Region, China. The southern spur is situated north of the Taklimakan Desert, and the northern spur is situated south of the Gurbantünggüt Desert. The species

ranges between 2,800 and 4,100 m in elevation. Due to the remote nature of the habitat in which they live, Ili pikas have not been censused throughout their distributional range. The species may not occur throughout the portrayed distribution, and may be extirpated from the southern spur.

TAXONOMY AND GEOGRAPHIC VARIATION: Subgenus *Conothoa*. No subspecies. This recently described and geographically isolated pika appears most closely related to Koslov's pika (*O. koslowi*) and the Ladak pika (*O. ladacensis*).

ECOLOGY: Typical of rock-dwelling pikas, Ili pikas are territorial and live at low density within their occupied habitat. They appear to occupy individual territories.

Predators of Ili pikas include steppe polecats (*Mustela eversmanii*) and beech martens (*Martes foina*). Red foxes (*Vulpes vulpes*) and snow leopards (*Panthera uncia*) also inhabit the area occupied by Ili pikas and may prey on them.

HABITAT AND DIET: The Ili pika is a rock-dwelling form, but unlike most pika species in this eco-type that inhabit talus or piles of broken rock, the Ili pika prefers to occupy slightly sloping large rock walls or cliff faces punctuated with gaps and holes that serve as their dens. Ili pikas construct haypiles, primarily composed of *Dracocephalum*, *Rhodiola*, and snow lotus herb (*Saussurea involucrata*), but also containing the poisonous Kusnezoff monkshood (*Aconitum kusnezoffii*).

BEHAVIOR: Unlike many pika species, the Ili pika is less likely to utter vocalizations. It is asocial and territorial. Recent data from infrared cameras show that Ili pikas mainly use feces and urine to mark their territory, as well as to communicate with other individuals. Ili pikas appear to be cathemeral, having activity periods throughout both day and night. It also appears that they seasonally vary their percent time active during the day and the night. Ili pikas become more active during the daylight hours than at night during the winter, but during spring and fall their nocturnal activity exceeds that during the day.

REPRODUCTION AND DEVELOPMENT: Little is known about the reproductive activity of the Ili pika; one observation of two young in a nest indicates that their litter size is small. Apparently, one to two litters are born during each summer reproductive season.

PARASITES AND DISEASES: The flea *Amphalius clarus* has been found on Ili pikas; endoparasites include intestinal nematodes and subcutaneous tapeworm larvae cysticerci. There are no data linking the Ili pika with any diseases that can be spread to humans, such as plague or tularemia.

Ochotona iliensis habitat. Photo courtesy Weidong Li

CONSERVATION STATUS:

IUCN Red List Classification: Endangered (EN)—A2abc; C2a(i)

National-level Assessments: China (Endangered (EN)—A2c)

MANAGEMENT: More than 71% of the known populations of the Ili pika have disappeared within the past 20 years, and it is likely that the conservation status of the species is approaching the Critically Endangered (CR) threshold.

Following the discovery of this species in 1983, the Ili pika was studied actively for about a decade, during which time a total of 27 confirmed sightings of the species were made (coupled with 10 sightings documented based on interviews with local pastoralists). This period was followed by a decade with no observations. Follow-up surveys in 2002 and 2003 found that pikas were gone from 8 of 14 known locations, including the type locality, and no living pikas were observed. All extant populations were classified on the basis of fresh signs (feces, urine deposits, footprints in the snow, and green active haypiles). Additional surveys in 2006, 2010, and 2014 continued to document this decline. Some new occupied sites were located and some populations became recolonized, but there continued to be a net loss of occupied sites. Throughout this period no living pikas were seen until two were observed in summer 2014, a gap of 20 years since the previous sighting.

Thus, our knowledge of the Ili pika is rudimentary, but the species is inescapably in decline. The most likely direct cause of decline appears to be an increase in the utilization of forage in high alpine meadows of the Tian Shan by pastoralists. Sheep grazing has intensified throughout the period during which the pikas have been investigated, leading to overgrazing of meadow on which the pikas rely for food; people now occupy the highest reaches of the Tian Shan and its pastures, which were unoccupied in the 1980s. It is also possible that climate change, as indicated by widespread melting of glaciers in the Tian Shan, may be a contributing factor to the decline of the Ili pika.

The Ili pika is not included on China's list of wildlife under special state protection under the country's 1988 Wildlife Protection Law, and its habitat is not under protection status.

ACCOUNT AUTHORS: Weidong Li and Andrew T. Smith

Key References: Ge et al. 2012; Jiang et al. 2016; Li 1997, 2003, 2004; Li and Ma 1986; Li and Smith 2005, 2015; Li and Zhao 1991; Li et al. 1991, 1993a, 1993b; Smith and Xie 2008; Smith et al. 1990.

Ochotona koslowi (Büchner, 1894)
Koslov's Pika

OTHER COMMON NAMES: Keshi shutu, Tulu shutu (Chinese)

DESCRIPTION: Koslov's pika is a relatively large, pale pika. The dorsal pelage is sandy yellowish in color, and the ears are a yellowish white or buffy with white margins. The

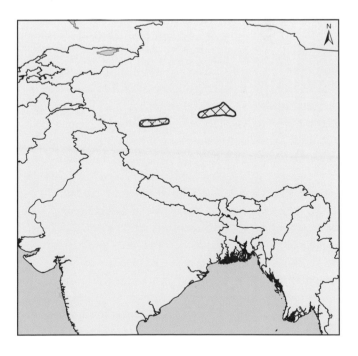

Ochotona koslowi. Photo courtesy Weidong Li

ventral surface is a grayish to a yellowish white. In winter the pelage becomes a light yellowish white. The lips are white, in contrast to the black lips of the plateau pika (*O. curzoniae*) and the Ladak pika (*O. ladacensis*).

The skull is large and strongly arched. The rostrum is short, and the nasals relatively short and wide. The orbit is well developed, and the interorbital region is narrow. The zygomatic arch extends outward. The auditory bullae are small.

SIZE: Head and body 220–240 mm; Hind foot 38–42 mm; Ear 16–20 mm; Greatest length of skull 40–47 mm; Weight 150–300 g

PALEONTOLOGY: Fossil specimens identified as being Koslov's pika have been found in strata dated to the Pleistocene at Zhoukoudian, China; thus it appears that the species may have one of the longest histories of all living ochotonids.

CURRENT DISTRIBUTION: Koslov's pika is endemic to China. The type locality is apparently at Fe Chen Ko, on the northern slope of the Arjinshan National Nature Reserve in Xinjiang. The range of Koslov's pika extends about 800 km east to west along the Kunlun Mountains from the juncture of Xizang and Xinjiang, and then along the border of Xizang and Xinjiang. This is a high-altitude species that is commonly found between 4,200 and 4,800 m in elevation, and may extend even higher.

TAXONOMY AND GEOGRAPHIC VARIATION: Subgenus *Conothoa*. No subspecies. *O. koslowi* has generated no taxonomic confusion within the sometimes contentious systematics of *Ochotona*. Craniometric measurements of *O. koslowi* are unique among the pikas, including the distinctiveness of *koslowi* from *O. ladacensis*, the species most closely linked to it based on molecular investigations. Other treatments also identify *O. koslowi* as a close relative of the Ili pika (*O. iliensis*). Some consider Koslov's pika to be primitive among the pikas based on its unique suite of morphological characters.

ECOLOGY: This is a meadow-dwelling burrowing form of pika. Burrow densities can range from 44 to 152/ha. Burrow tunnels are relatively simple, extending only about 2 m in length, with two or three sub-tunnels. Burrow depth is about 30–40 cm. This simple burrow structure apparently makes these pikas vulnerable to predators. Population density has been reported to be high in some studies, and as low as 3–4/ha in others.

HABITAT AND DIET: Koslov's pika inhabits smooth-sandy soils in typical alpine desert grassland habitat in high-elevation basins or mountain valleys. This environment is marked with a small number of plant species, including Moorcroft sedge (*Carex moorcroftii*), compact ceratoides (*Ceratoides compacta*), bifurcate cinquefoil (*Potentilla bifurca*), and various Cruciferae and Leguminosae species. Koslov's pika's favored food sources include falcate crazyweed (*Oxytropis falcata*), rock jasmine (*Androsace acrolasia*), and creeping false tamarisk (*Myricaria prostrata*).

BEHAVIOR: Koslov's pika is diurnally active. Typical of burrowing pikas, the social structure is composed of communal family groups.

PHYSIOLOGY AND GENETICS: Koslov's pika has a high haplotype diversity, but low nucleotide diversity, indicating past demographic expansion. One survey found a low level of genetic differentiation among local populations. Three positively selected amino acid sites have been identified, each of which might have biological significance, helping individuals adapt to the extremely high elevations the species occupies.

REPRODUCTION AND DEVELOPMENT: The reproductive period of Koslov's pika extends throughout mid-summer. Litter size ranges from four to eight young, and females have at least two litters during the reproductive season. Females have eight nipples, the first pair tucked in under the forelimbs.

PARASITES AND DISEASES: *Hirstionyssus citelli*, a gamasid mite, has been found on the surface of Koslov's pika, a range extension for this species.

CONSERVATION STATUS:

IUCN Red List Classification: Endangered (EN)—B1ab(iii)

National-level Assessments: China (Endangered (EN)—A2c)

MANAGEMENT: Following its discovery in 1884, 100 years passed before the species was rediscovered by scientists. Additionally, several dedicated expeditions following the rediscovery failed to locate any extant populations. Thus the species is poorly known, and may be in decline. There have been observations that the common ecologically similar plateau pika may be encroaching on the habitat of Koslov's pika.

ACCOUNT AUTHORS: Weidong Li and Andrew T. Smith

Key References: Ge et al. 2012; Jiang et al. 2016; Lanier and Olson 2009; Li 2013; Li et al. 2006; Lin et al. 2010; Lissovsky 2014; Smith 2008; Smith and Xie 2008; Smith et al. 1990; Ye et al. 2000; Yu et al. 2000; Zheng 1986.

Ochotona ladacensis (Günther, 1875)
Ladak Pika

OTHER COMMON NAMES: Ladake shutu (Chinese)

DESCRIPTION: Ladak pikas are comparatively large among the pikas. The woolly dorsal pelage is a pale buff, fulvous, with a slight rufous tint. The ventral surface is whitish, with a gray hue. The head is a dark rufous, trending to pale gray along the snout and pale white in front of and below the eyes. The rufous tint becomes stronger and extends around the neck and somewhat farther back on the upper body. The tip of the snout and upper and lower lips are dark blackish. The vibrissae are long,

up to 80 mm. The ears are well covered with hair, which is thickest on the outside, and are distinctly rufous. In winter the pelage is longer and its general color becomes a lighter fawn with a slightly grayish hue; only the ears remain distinctly rufous.

The skull is highly arched. The nasals are much narrower behind than in front and convex anteriorly. The incisive foramen and the palatal foramen are separated. The sides of the zygomatic arches are nearly parallel. The auditory bullae are small and widely spaced.

SIZE: Head and body 180–241 mm; Hind foot 32–40 mm; Ear 24–35 mm; Greatest length of skull 47–60 mm; Weight 190–320 g

CURRENT DISTRIBUTION: The geographic distribution of the Ladak pika extends from SW Xinjiang, through W Qinghai and across N Xizang, China. Its range also extends into the northwestern regions of Pakistan and Kashmir (Ladak), India. Although it is broadly sympatric with the plateau pika (*O. curzoniae*), it is not as widely distributed across the Tibetan Plateau. This species occurs at elevations from 4,200 to 5,790 m asl.

TAXONOMY AND GEOGRAPHIC VARIATION: Subgenus *Conothoa.* No subspecies (includes *auritus*). *O. ladacensis* was initially mistaken as a form of *O. curzoniae.* However, *ladacensis* is significantly larger than *curzoniae,* and molecular analyses place them in separate subgenera of *Ochotona. O. ladacensis* appears to be most closely related to Koslov's pika (*O. koslowi*) and the Ili pika (*O. iliensis*). The Ladak pika is very distinctive morphologically: its

Ochotona ladacensis, Photo courtesy James Eaton / Birdtour Asia

topsylla ornata, Hoplopleura ochotonae), gamasid mites (*Haemogamasus dauricus, Laelaps* spp., *Dermanyssus* spp.), and ixodid ticks (*Haemophysalum warburtoni, Ixodes per-sulcatus, Dermacentor marginatus*). Among endoparasites, several flatworms (*Schizorchis altaica, Dicrocoelium lance-atum, Hasstilesia ochotonae*) and roundworms (*Cephaluris andrejevi, Dermatoxis* spp., *Labiostomum vesicularis, Murie-lus tjanschaniensis*) have been found in large-eared pikas.

CONSERVATION STATUS:

IUCN Red List Classification: Least Concern (LC)

MANAGEMENT: From the 1930s to the 1950s, large-eared pikas were hunted for their fur, although the numbers of animals harvested cannot be determined as they were lumped together with gerbils. Otherwise, there have been no economic uses of large-eared pikas, and there are no current known threats to the species. The species is not subject to any current conservation or management initiatives.

ACCOUNT AUTHOR: Andrew T. Smith

Key References: Bernstein 1970; Blanford 1888; Feng et al. 1986; Kawamichi 1971b; Lissovsky 2014; Ognev 1966; Sludskii and Strautman 1980; Smith and Xie 2008; Smith et al. 1990; Sokolov et al. 2009; Thomas 1911a.

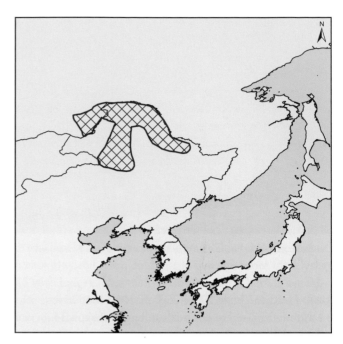

Ochotona mantchurica. Photo courtesy Andrey Lissovsky

Ochotona mantchurica Thomas, 1909
Manchurian Pika

OTHER COMMON NAMES: Manzhou shutu (Chinese); Manzhurskaya pischuha (Russian)

DESCRIPTION: In summer the dorsal pelage of the Manchurian pika is a cinnamon, fading to a pale cinnamon on the flanks. The ventral coloration is an ochraceous buff. In winter the upper surface is a light brown that becomes gray on the head and shoulder area and a clay color on the rump; ventrally the coloration is a dull white color with a wash of clay.

The mandible height in Manchurian pikas is greater than that of the closely related northern pika (*O. hyperborea*), and the length of the skull (average = 41.7 mm) is also larger than that in the northern pika (average = 37.7 mm).

SIZE: Head and body 135–220 mm; Hind foot 22–32 mm; Ear 17–26 mm; Condylobasal length of skull 37–46 mm; Weight 110–260 g

CURRENT DISTRIBUTION: The Manchurian pika is found from along the south bank of the Shilka and Amur Rivers through the Greater Khingan Range and extending south into the Lesser Khingan Range, NE China.

TAXONOMY AND GEOGRAPHIC VARIATION: Subgenus *Pika.* Three subspecies: *O. m. mantchurica, O. m. loukashkini,* and *O. m. scorodumovi.* This form was not recognized as an independent species until recently. Some parts of the distribution were included with *O. hyperborea*, and others in *O. alpina*. The nominative form was originally described as a subspecies of *O. hyperborea*, while Manchurian pikas from Russian Transbaikalia (*scorodumovi*) were described as and always referred to *O. alpina*. Meanwhile, the population in the Lesser Khingan Range (*loukashkini*) has at times been referred to as *O. alpina cinereofusca*, confounded by the similarity between *alpina* and *hyperborea*, thus leading these two forms to sometimes be synonymized within *alpina*. Recent molecular, bioacoustical, and morphological analyses have clarified that *O. mantchurica* combines the forms from the Greater and Lesser Khingan Ranges and Shilka-Argun Rivers interfluve in Russian Transbaikalia, and is indeed independent. The closely related *O. hyperborea* is found only north of the Amur River. Additional confusion has reigned with all these forms in this remote region of Asia; e.g., a large series of *O. hyperborea* collected close to Ulaanbaatar, Mongolia, which was used as a voucher series by many taxonomists, was originally designated as *mantchurica*. Thus *O. mantchurica* is closely related to *O. hyperborea*, but also to Hoffmann's pika (*O. hoffmanni*); these three species represent allospecies.

ECOLOGY: The Manchurian pika is a characteristic rock-dwelling species and expresses the typical ecology and behavior of these forms. Their territories are large, and thus the population density is low. Independent juveniles disperse locally to fill vacancies in nearby territories. Its elevational range is 400–1,300 m.

HABITAT AND DIET: Commonly occurring in rocky habitat, the Manchurian pika lives in a wide variety of environments from dry steppe and broad-leaved forest to taiga and mountain tundra. A generalized herbivore, it constructs very small haypiles compared with those of many other pika species.

BEHAVIOR: Interactions between males and females are few. The vocal repertoire of the Manchurian pika is distinctive from the closely related and geographically close northern pika. The characteristic deep modulated calls uttered by northern pikas are absent in the Manchurian pika. In contrast, Manchurian pikas have a complicated spectral structure of calls induced by biphonations. Short calls are given by both males and females to advertise their territory or to warn conspecifics of approaching predators. Unlike many pika species, the Manchurian pika does not produce a long call.

PHYSIOLOGY AND GENETICS: Diploid chromosome number = 39–40

REPRODUCTION AND DEVELOPMENT: The reproductive season starts in May and lasts until July, and a single litter of two to six young is produced.

CONSERVATION STATUS:

IUCN Red List Classification: Not Evaluated (NE) due to the recent recognition of its independent status

MANAGEMENT: The Manchurian pika is very common throughout its range and there are no known threats. No management initiatives are in place concerning the Manchurian pika.

ACCOUNT AUTHORS: Andrey Lissovsky and Andrew T. Smith

Key References: Formozov and Baklushinskaya 2011; He 1958; Lissovsky 2005, 2014, 2015; Lissovsky et al. 2008; Ognev 1966; Pan et al. 2007; Skalon 1934; Smith and Xie 2008; Thomas 1909.

Ochotona nubrica Thomas, 1922
Nubra Pika

OTHER COMMON NAMES: Nubula shutu (Chinese)

DESCRIPTION: In summer the Nubra pika tends to be a very pale form, and the degree of saturation in the pelage varies clinally from west (the palest forms) to east. In

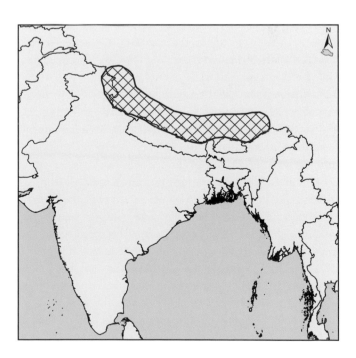

Ochotona nubrica. Photo courtesy Andrey Lissovsky

summer the dorsal pelage is gray to brownish red, with interspersed black hairs. The face, nape, and shoulders are gray with a slight fulvous tinge, and the flanks are a straw gray. The ventral surface is dull white to yellowish. Ears are blackish behind, with white edges; there is a distinct light patch on the outer surface of the ear. The pelage in winter is longer and a more uniform gray throughout.

The skull is noticeably narrow and flat. The palatal foramina are widely expanded in their posterior half. The auditory bullae are small and narrow.

SIZE: Head and body 140–184 mm; Hind foot 27–35 mm; Ear 20–27 mm; Greatest length of skull 37–43 mm; Weight 96–135 g

CURRENT DISTRIBUTION: The Nubra pika is essentially a trans-Himalayan species, ranging from E Xizang, China, west to Ladak, India. It is normally found between 3,000 and 4,500 m in elevation.

TAXONOMY AND GEOGRAPHIC VARIATION: Subgenus *Ochotona.* Two subspecies: *O. n. lhasaensis* (SE Xizang, China) and *O. n. nubrica* (*lama* and *aliensis* are synonyms; from the Mustang Region of N Nepal, across SE Xizang to Ladak). *O. nubrica* has generated significant taxonomic confusion. Following its description as an independent species, *O. nubrica* had been synonymized with the steppe pika (*O. pusilla*), Royle's pika (*O. roylei*), and the Moupin pika (*O. thibetana*), but molecular analyses clearly show its independence and separation from these forms. It is most closely related to the plateau pika (*O. curzoniae*), but detailed analyses demonstrate that these forms do not overlap in craniometric features. In contrast with these morphological data, molecular data from specimens of *O. nubrica* do not form a single clade; rather, they are aligned within various branches of *O. curzoniae* (see below).

ECOLOGY: The few observations that have been made on the Nubra pika comment that it is quite common.

HABITAT AND DIET: The Nubra pika occupies patches of dense vegetation, mainly common sea-buckthorn (*Hippophae rhamnoides*), saltcedar (*Tamarix* spp.), willows (*Salix* spp.), pea shrubs (*Caragana* spp.), and honeysuckle (*Lonicera* spp.), at various localities throughout its range. It is a burrowing pika and does not occupy talus habitat. It is believed to be a generalized herbivore like other pikas. Burrows tend to be simple and positioned in areas with thick thorny vegetation.

BEHAVIOR: Some reports indicate that the Nubra pika is not colonial, but most highlight that it lives in well-defined family groups typical of burrowing pikas.

REPRODUCTION AND DEVELOPMENT: Juvenile Nubra pikas have been reported from June through August. Nothing else is known about reproduction in this species.

PHYSIOLOGY AND GENETICS: Diploid chromosome number = 48. While the Nubra pika and the plateau pika are clearly independent species, given their distinct morphological differences and ecological preferences, they show very little separation in mitochondrial analyses. Apparently the Nubra pika displays haplotypes of plateau pikas due to past recurrent hybridization, followed by elimination of the mitochondrial DNA within the Nubra pika.

CONSERVATION STATUS:

IUCN Red List Classification: Least Concern (LC)

MANAGEMENT: Very little is known about population trends or even the natural history of the Nubra pika. Its range encompasses very remote habitats where it is unlikely to be jeopardized by anthropogenic disruption.

ACCOUNT AUTHOR: Andrew T. Smith

Key References: Feng and Kao 1974; Feng et al. 1986; Ge et al. 2013; Lanier and Olson 2009; Lissovsky 2014; Mallon 1991; Mitchell 1978; Smith and Xie 2008; Smith et al. 1990; Thapa et al. 2011; Thomas 1922; Yu and Zheng 1992.

Ochotona opaca Argyropulo, 1939
Kazakh Pika

OTHER COMMON NAMES: Kazakhskaya pischuha (Russian)

DESCRIPTION: In summer, the dorsal pelage of the Kazakh pika is a gray-brown color tinged with ochraceous or yellowish fur. The ventral pelage is sandy to white. The

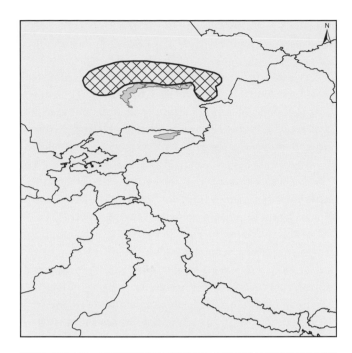

Ochotona opaca. Photo courtesy Andrey Lissovsky

TAXONOMY AND GEOGRAPHIC VARIATION: Subgenus *Pika*. No subspecies. *O. opaca* has long been treated as *O. pallasii pallasii*, and thus closely linked with *O. p. pricei* of Mongolia (herein: *O. pallasii pallasii*). Much of this confusion has stemmed from the original type locality of *pallasii* being placed within the range of *opaca*, in Kazakhstan, whereas it is now considered to have originated from within the range of Pallas's pika in Mongolia. Thus the name *O. pallasii* was incorrectly assigned to pikas from Kazakhstan. These two forms (*opaca* and *pallasii*) appear to be sister species, but sufficiently differentiated as to be considered independent.

ECOLOGY: Although in the subgenus *Pika*, along with most of the truly obligate rock-dwelling pika species, the Kazakh pika is best classified as an intermediate species—one that can occupy both rocky substrates and open meadow landscapes. And it constructs burrows in both situations. Its flexible habitat occupancy appears somewhat related to its density (but also geography). Kazakh pikas are more likely to occupy rocky outcrops in low-density years (termed "survival habitats"), and to settle in places devoid of rocks in high-population years. Population density of the Kazakh pika may vary greatly, but normally ranges from about 5 to 30/ha.

The burrows of Kazakh pikas, mostly occupied by individuals, may consist of many entrances that are usually associated with deep niches, voids, and crevices in rocks. They tend to avoid mesic environments. There are two characteristic types of burrows: temporary and permanent. Temporary burrows serve as a place to retreat when predators are identified. Permanent burrows may have from 5 to 12 entrances (average diameter 5–9 cm), and the burrow complex may cover an area of 10–15 m². Nearly all of the burrow system is connected by underground tunnels, although the system also includes lavatories at the end of blind alleys and two to three nesting chambers. These chambers are sizable (15–24 cm in diameter) and are lined with dry vegetation. In some open semi-desert regions, stones or cattle droppings are piled to construct a fence, up to 40 cm in height, that walls off the burrow entrance and appears castle-like. Kazakh pikas are also known to colonize abandoned burrows of marmots and ground squirrels.

The central core defended territory of Kazakh pikas (male and female, separately) is about 200 m². In contrast, the home ranges of the two sexes differ considerably in size. The home range of females may occupy 600–1,300 m², whereas that of males may encompass 4,200–5,200 m².

winter pelage is similar in tone, but lighter. The round ears have light margins.

The skull of the Kazakh pika is smaller than that of the closely related Pallas's pika (*O. pallasii*), but more strongly arched. In particular, the rostrum length is shorter (usually < 15 mm) than that found in Pallas's pika (usually > 18.5 mm). The auditory bullae of the Kazakh pika are moderately large, but still less inflated and smaller overall than those in Pallas's pika.

SIZE: Head and body 175–220 mm; Hind foot 26–32 mm; Ear 17–23 mm; Greatest length of skull 43–48 mm; Weight 100–173 g

CURRENT DISTRIBUTION: Kazakh pikas occur in E Kazakhstan, in an arc above Lake Balkhash. They occupy elevations that are generally lower than 1,000 m.

Kazakh pikas may co-occur with steppe pikas, although they occupy different habitats: the Kazakh pika dwells in rock crevices, while the steppe pika lives exclusively in thickets of pea shrub (*Caragana* spp.) or other shrubs. Strong interactions between these two species occur only in years when populations are very high.

HABITAT AND DIET: Kazakh pikas are most frequently found in rocky habitats formed by broken outcrops, yet they also occupy slopes characterized by shrubby vegetation where they construct burrows at the base of the shrubs. They are rarely found on the slopes of mountains containing soft soils. Occasionally they may occupy the rock walls of cattle enclosures.

The choice of vegetation utilized by the Kazakh pika varies geographically, among years, and among different local densities within a pika colony. In general, pikas feed on plants that grow close to their habitation. In places with a rich diversity of plants, pikas may show individual selectivity in those species that they prefer to harvest. Where local plant diversity is low, most pikas in a colony select similar plants for their haypiles. In years of high pika density, selectivity declines as almost all palatable plants are harvested. In early spring pikas are relegated to eating dry grass left behind from winter. As the season progresses they take advantage of young shoots of shrubs, herbs, and grasses as they become available. Haypiles are initiated earlier in regions that become desiccated earlier. At any site haying progresses over a period of one to two months. Generally young of the year are the first to initiate haypiles. Hay is stored near burrow entrances, in burrows, and on open areas.

Haypile size varies considerably across the range of the Kazakh pika. Where density is low, total haypile mass may reach 5 kg/ha. In some high-density situations with 28–30 haypiles per ha, total mass may range from 89 to 111 kg/ha. Individual haypile size ranges from 200 g to 20 kg. Some of the larger haypiles are anchored with a layer of rocks or ungulate dung to keep the vegetation from blowing away; when the wind subsides the rocks are removed and a new layer of vegetation is added. Rocks are also used to plug burrow entrances, particularly burrows containing hay, prior to winter.

BEHAVIOR: Timing of activity of the Kazakh pika appears related to weather conditions, season, and density. During summer they have been observed active at any hour of the day or night. They also appear to be more nocturnal during high-density years or during the period of dispersal when interactions among individuals are frequent. Overall activity levels are highest during the mating period when animals can be seen throughout the day. Feeding activity is highest early and late in the day, and often Kazakh pikas can be seen lying on rocks under the sun during midday for long periods. They tend to be inactive during midday when it is warm. Overall level of activity declines sharply in late summer to early fall after the reproductive and haypile-collecting season has been completed. With the onset of cooler weather, they become surface active only on warm sunny days.

Males cruise through their home range, and when they encounter females they utter soft, quiet trills. Males and females maintain separate core areas within their home ranges and they construct separate haypiles. Males protect their core areas from females, whereas females may be visited on their core areas by males. Males are even known to steal from the stores of females. The mating system appears to be polygynous. Males are highly aggressive toward other males, particularly individuals trespassing on the territories of neighbors. Aggressive interactions involve long chases, fights, and bouts of boxing. Males and females that share home ranges, as well as their young of the year, may be tolerated. Aggression becomes directed at juveniles as they become sexually mature, leading to their eventual dispersal. At this time also, males and females carve out individual territories. Kazakh pikas mark their entire home range with urine and scent marks (by rubbing their neck gland on rocks).

Young forced to disperse from their natal home range often relocate at considerable distances, up to 8–10 km. During dispersal young have been observed to swim across rivers 15 m wide.

Vocal behavior is well developed in the Kazakh pika. Both males and females utter a long call (song). The long call apparently has a reproductive function and is given primarily during the spring and early summer reproductive period. It consists of an introductory "pi-ik" followed by a prolonged vibrating trill.

PHYSIOLOGY AND GENETICS: Diploid chromosome number = 38

REPRODUCTION AND DEVELOPMENT: The high fecundity rate of the Kazakh pika is more reminiscent of characteristic meadow-burrowing species of pika than of rock-dwelling forms. Litter size ranges from 2 to 13 young, and several (2–3) litters are produced per year with a postpartum estrus. Some female juveniles may mature at an age of 1–1.5 months and reproduce in their summer of birth, whereas males first become reproductively active as yearlings. Gestation extends 25–27 days, and young are born naked and blind. Eyes open at day 10, and at this time young become highly mobile. They attain adult size 1–1.5 months following birth.

PARASITES AND DISEASES: Larvae of warble flies (*Oestromyia leporina*) have been reported to parasitize large numbers of the Kazakh pika. Additionally, many species of ectoparasites infect Kazakh pikas: ixodid mites (in particular *Dermacentor marginatus*, *Rhipicephalus rossicus*, *R. pumilio*); one gamasid mite (*Allodermanyssus sanguineus*); and 26 flea species (most numerous = *Ctenophyllus hirticrus*, *Amphalius runatus*, *Paramonopsyllus scalonae*). Seven species of helminth, four species of coccids, and blood parasites (*Grahamella*) are endoparasites of Kazakh pikas.

Kazakh pikas may be vectors of a specific form of trypanosomiasis caused by *Trypanosoma ochotona*. It is possible that some of the reported sharp declines in population density may have been driven by one of these epizootic diseases.

CONSERVATION STATUS:

IUCN Red List Classification: Least Concern (LC)
MANAGEMENT: Kazakh pikas could exert an adverse effect on rangeland vegetation; however, they are rarely found at densities that would trigger such an assessment. This species has also been trapped for the fur industry, but apparently this activity never jeopardized populations and it has now ceased. An isolated population of Kazakh pikas from the Dzhambul Mountains, Kazakhstan, has disappeared, and only restricted populations remain in the Kyzkach Mountains, Kazakhstan.

ACCOUNT AUTHOR: Andrew T. Smith

Key References: Formozov and Proskurina 1980; Lissovsky 2014; Lissovsky et al. 2016; Ognev 1966; Sludskii and Strautman 1980; Smirnov 1974, 1986; Smith et al. 1990; Sokolov et al. 2009.

Ochotona pallasii (Gray, 1867)

Pallas's Pika

OTHER COMMON NAMES: Mongolian pika; Menggu shutu (Chinese); Mongol ogdoi (Mongolian); Mongolskaya pischuha (Russian)
DESCRIPTION: The dorsal pelage of Pallas's pika is generally a dull grayish brown with slight yellow tinges becoming a rusty brown toward the rump. The fur on the top of the head is the same color and somewhat brighter as on the back, while the flanks tend to be lighter and have

more of a rusty tone. The overall pigmentation is paler than in the similar Kazakh pika (*O. opaca*). The ventral surface is whitish, although a yellow-brown transverse stripe appears on the chest and extends between the front legs. The inner ears are a whitish gray, and the margins are white (sometimes indistinct). In winter the longer and denser dorsal pelage becomes a straw gray with a pinkish rust tinge, while the top of the head is grayer. The vibrissae are long (60 mm).

The skull is slightly arched (less so than in the Kazakh pika). There are no fenestrae present in the frontals, and the incisive and the anterior palatine foramina are com-

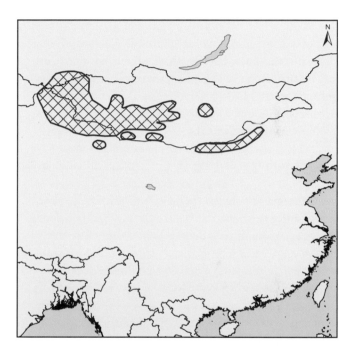

Ochotona pallasii. Photo courtesy Richard Reading

pletely separated. The orbit is well developed, and the interorbital region is narrow. The length of the rostrum (> 18.5 mm) is significantly longer than in the closely related Kazakh pika (usually < 15 mm). The auditory bullae are larger than in the Kazakh pika.

SIZE: Head and body 190–224 mm; Hind foot 26–36 mm; Ear 18–23 mm; Greatest length of skull 43–48 mm; Weight 160–280 g

PALEONTOLOGY: No fossil record is known, but Recent remains document the loss of Pallas's pika populations from isolated mountains throughout its range (see "Management").

CURRENT DISTRIBUTION: The range of Pallas's pika is highly disjunct, as the three subspecies have widely separated ranges. *O. p. pallasii* occurs from SE Altai, Russia, across W Mongolia, extending into NE Xinjiang, China; *O. p. hamica* is found in the Karlik Shan (mountains) of E Xinjiang, China; and *O. p. sunidica* is sporadically distributed across N Nei Mongol, China. They occupy elevations from 1,700 m to 3,200 m asl.

TAXONOMY AND GEOGRAPHIC VARIATION: Subgenus *Pika*. Three subspecies (their respective ranges given above): *O. p. hamica*, *O. p. pallasii* (*pricei* a synonym), and *O. p. sunidica*. The Helan Shan pika (*O. argentata*—including the synonym *helanshanensis*) is closely related. Historically, this form has been referred to as *O. pricei*, a junior synonym of *pallasii*, in much of the Russian literature and has included pikas ranging across S Kazakhstan. However, *O. opaca* is a clearly differentiated form (see that account), and the type locality, which was formerly believed to occur in the range of *opaca*, is now known to occur farther east within the distribution of what had been called *O. pallasii pricei*—which herein is considered *O. p. pallasii*. The isolated forms *hamica* and *sunidica* have not been thoroughly compared with *O. p. pallasii*, but they may represent recently isolated populations and actually be synonyms.

ECOLOGY: The ecology of Pallas's pika resembles that of the Kazakh pika, and it can also be best classified as an intermediate species—one that occupies both rocky substrates and open landscapes, such as semi-desert. Compared with the Kazakh pika, Pallas's pika is more likely to be found in open habitat. Population density of Pallas's pika may vary greatly and reach levels as high as 39/ha, although survival habitat densities are normally smaller. Population density varies greatly among years due to changes in climate and in some years due to a plague epizootic.

Burrows, mostly occupied by single pikas, may have many entrances. Most burrows are occupied in succession by individuals, and they persist for many years. In meadow/steppe habitat the density of burrow entrances ranges from 5 to 42. A burrow may have 4 to 12 feeding chambers, each with a diameter of up to 40 cm, and contain one nesting chamber situated roughly 70 cm deep. A single occupied burrow system may cover an area from 3 to 60 m². Often the pika burrow system is associated with the abandoned burrows of the Tarbagan marmot (*Marmota sibirica*).

Pallas's pikas may co-occur with Alashan ground squirrels (*Spermophilus alashanicus*), and there have been observations that overgrazing by the squirrels led to a significant reduction in the numbers of pikas.

HABITAT AND DIET: It is difficult to generalize about the preferred habitat of Pallas's pika, as it is so variable across the range of the species. In the Altai they appear confined to mountain steppes with occasional stones or piles of rocks, but also extend to desert steppes. The same range of habitats is occupied by Pallas's pika across Mongolia—from rocky outcrops, to semi-desert where they burrow under shrubs such as wormwood (*Artemisia* spp.) or other characteristic vegetation, to meadow-steppe characterized by *Stipa krylovii*, *S. gobica*, *Agropyron cristatum*, and wild onion (*Allium polyrhizum*). *O. p. sunidica* appears to preferentially occupy cracks in rocky sites.

The choice of vegetation utilized by Pallas's pikas varies widely. A dominant behavior is the construction during summer of a haypile from which animals feed during the following winter when snow blankets their world. Most vegetation is gathered selectively from the area immediately surrounding the burrow system. In SW Tuva, Russia, certain plants (*Euphorbia altaica*, *Artemisia* spp., and some crucifers) reach a high abundance and grow only above pika burrows. Like the Kazakh pikas, Pallas's pikas make plugs of stones at the base of burrow entrances where hay is stored.

BEHAVIOR: Compared with the Kazakh pika, Pallas's pika is considered to be exclusively diurnal. Overall activity levels are highest during the mating period when animals can be seen throughout the day. Feeding activity is highest early and late in the day, and often Pallas's pikas can be seen lying on rocks under the sun during midday for long periods. They tend to be inactive during midday when it is warm. Overall level of activity declines sharply in late summer to early fall after the reproductive and haypile-collecting season has been completed. With the onset of cooler weather, they become surface active only on warm sunny days.

The mating system appears to be polygynous. Most males remain close to their burrows, and males demonstrate high levels of intrasexual aggression, even if their burrows are in close proximity. Whereas aggressive interactions in most pika species are short (normally lasting about 1 second), those of Pallas's pika are long and extremely violent. Aggressive interactions include chases, boxing, and prolonged fights; individuals trespassing on the territories of neighbors may be killed by bites inflicted by the occupant. Even females tend to be aggressive toward each other. Overall, these interactions are about 10 times more intense in Pallas's pika than in the Kazakh pika. Males and females that share home ranges, as well as their young of the year, may be tolerated. Aggression becomes directed at juveniles as they become sexually mature, leading to their eventual dispersal.

Vocalizations in Pallas's pikas and Kazakh pikas differ significantly. Pallas's pikas do not utter the long call, and the dynamic spectrum of alarm calls differs between these species. Additionally, both male and female Pallas's pikas utter a unique territorial trill, and juveniles give this call only after they have dispersed and claimed a territory.

Another mechanism utilized by Pallas's pikas to demarcate their territories is scent marking. Rocks or sticks are rubbed with secretions from glands located on the sides of their neck, and often they urinate on these and pyramids of their characteristic round fecal pellets.

PHYSIOLOGY AND GENETICS: Diploid chromosome number = 38

REPRODUCTION AND DEVELOPMENT: The high fecundity rate of Pallas's pika is more reminiscent of characteristic meadow-burrowing species of pika than of rock-dwelling forms. Litter size ranges from 3 to 10 young, and several (2–3) litters are produced per year with a postpartum estrus. Some female juveniles may mature and reproduce in their summer of birth.

PARASITES AND DISEASES: Pallas's pikas are known vectors of plague, although extensive epizootics appear rare. Some observed population declines in the species may have been due to a plague epizootic.

CONSERVATION STATUS:
IUCN Red List Classification: Least Concern (LC)
MANAGEMENT: The grazing activities of *O. p. pallasii* could exert a negative effect on rangeland vegetation; however, they are rarely found at densities that would trigger such an assessment.

Of particular interest, however, are the isolated populations of Pallas's pika. No information exists on any of the disjunct populations of *O. p. sunidica*. An isolated population of *O. p. pallasii* from the Choyr Mountains (southeast of Ulaanbaatar) has declined greatly and should be considered threatened, and other southerly populations

of Pallas's pika have similarly declined over the past centuries. This decline has been documented based on the percent of pika bones in owl pellets on the Dzhinst-Ula Ridge, Mongolia: pikas constituted 38% of owl prey 11 centuries ago, 28% 10 centuries ago, and 21% six centuries ago. Only one individual was seen in a 1980 survey, and it is likely none occur there today.

Similarly, *O. p. hamica* is poorly known. The type series of six specimens was collected in 1911 in the Karlik Shan (Xinjiang, China); there have been no records of this animal at that locality since. Apparently the range of this subspecies extends into S Mongolia, where a record of its gradual decline has been documented. In the Tsagaan Bogd Uul (mountains) pikas made up 23.6% of the small mammal bones identified between 3,000 and 4,500 years ago. This declined to 7.8% between 2,000 and 3,000 years ago, 7.3% between 1,500 and 2,000 years ago, and only 0.4% in the period encompassing the past 200 years. Pallas's pika has not been collected there in historical time, although in 1979 fresh scats were located and pikas' calls were heard. Presence of Pallas's pika in another isolated range, the Atas Bogd Uul (mountains), is also enigmatic. One animal was captured there in 1951, and it was reported that the top of the mountains contained a healthy population of pikas. However, no pikas were recorded there in a survey conducted in 1988. These results collectively indicate that due to their initial small isolated populations, potential habitat degradation, and possible effects of climate change, these populations of *O. p. hamica* should be considered endangered.

ACCOUNT AUTHOR: Andrew T. Smith

Key References: Denisman et al. 1989; Kniazev and Savinetski 1988; Lissovsky 2014; Lissovsky et al. 2016; Ognev 1966; Sludskii and Strautman 1980; Smith and Xie 2008; Smith et al. 1990; Sokolov et al. 2009; Thomas 1912; Wesche et al. 2007; Yong et al. 1987.

Ochotona princeps (Richardson, 1828)
American Pika

OTHER COMMON NAMES: Cony, Rock rabbit, Hay-maker, Mouse hare, Piping hare, Whistling hare, Little chief hare, Southern pika
DESCRIPTION: American pikas are typical in shape and intermediate in size among the ochotonids. They undergo two molts annually; the gray winter pelage is nearly twice as long as the summer pelage and is worn for most of the

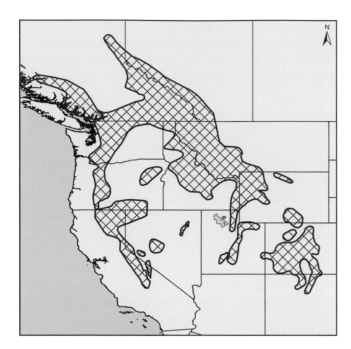

Ochotona princeps. Photo courtesy Andrew T. Smith

year. The dorsal pelage in summer varies geographically from a smoke gray to cinnamon brown to tawny, often with ochraceous overtones; populations living on dark (e.g., lava) substrate are often a darker hue. The ventral pelage is typically a buffy white. The fine dense underfur is predominantly slate gray. Juveniles remain gray throughout most of their summer of birth, gradually transitioning into adult coloration by late summer. American pikas lack the grayish collar that is present in the collared pika (*O. collaris*). The moderately large ears are haired on both sides with darker pelage, yet fringed with distinctive white margins. The hind limbs are digitigrade and similar

in length to the forelimbs. The hind feet have four toes, the forefeet have five, and the soles of the feet are densely furred. There is little to no sexual dimorphism in size; where it does occur, males are only slightly larger than females. Females have three pairs of mammae: pectoral, abdominal, and inguinal. Both sexes have a well-developed apocrine gland in the lower cheek, which is used to deposit scent marks.

The skull is slightly rounded (not inflated) in profile, with a broad and flat interorbital region. The rostrum is slender, and the nasals are widest anteriorly. The maxilla has one large fenestra, and the elongated jugals project far posterior to the zygomatic arm of the squamosal. The frontals lack postorbital processes. The length of tympanic bullae is typically less than 25% of the total skull length. **SIZE:** Head and body 157–216 mm; Hind foot 25–35 mm; Ear 20–27 mm; Greatest length of skull 39–47 mm; Weight 121–176 g

PALEONTOLOGY: Small and large forms of *Ochotona* have lived in North America since the Late Pliocene, and the genus had a far greater range across the continent during the Pleistocene than its current distribution. Fossil remains of animals putatively assigned to *O. princeps* have been found in numerous Pleistocene cave deposits in NE North America, where they persisted until the Early Holocene. By the Wisconsinian glacial period, pikas were confined to W North America, including throughout the current landscape of the Great Basin. The Wisconsinian glaciation led to the geographic separation of the American pika in the south from the collared pika in the north. Over time, as temperatures increased into the Recent, American pikas became isolated on mountain landforms. Frequent extinctions on mountain ranges and isolated habitats in the Great Basin have led to general impoverishment of pikas in this region between the Sierra Nevada and the Rocky Mountains, including a significant upward shift in the distribution of the species. Today the Great Basin remains a natural laboratory for the effects of a changing climate, as numerous isolated populations have been documented to become extirpated over the past century and several populations are today found at higher elevations than in the recent past.

CURRENT DISTRIBUTION: American pikas are distributed discontinuously, primarily in mountainous habitats and mountain chains, across W North America from C California to S New Mexico, northward to S British Columbia and Alberta. The Sierra Nevada, Cascade, and Rocky Mountains are current strongholds. In general, pikas are found at higher elevations in the southern extension of their range (generally > 2,500 m) and at lower elevations as latitude increases. There are some notable exceptions to this trend, as the American pika can be found at nearly sea level along the Columbia River Gorge, and numerous populations have been found recently at relatively low-elevation sites at lower latitudes.

TAXONOMY AND GEOGRAPHIC VARIATION: Subgenus *Pika*. Until recently, 36 subspecies of the American pika were recognized, largely based on the descriptions of isolated populations throughout the discontinuous distribution of the species. Currently, informed by a thorough analysis of all subspecies using modern genetic and morphometric techniques, five discrete lineages have been identified and the number of subspecies has been reduced to five: *O. p. princeps* (N Rocky Mountains); *O. p. fenisex* (Coast Mountains and Cascade Range); *O. p. saxatilis* (S Rocky Mountains); *O. p. schisticeps* (Sierra Nevada and Great Basin); and *O. p. uinta* (Uinta Mountains and Wasatch Range of C Utah).

The only congener in North America is *O. collaris*, whose southern limit is ~ 800 km north of the current northern extent of *O. princeps*. Some authorities have considered *O. collaris* and *O. princeps* to be one species, but all contemporary studies consider them to be separate species. Within the range of the American pika, its appearance is unlike that of any other small mammal.

ECOLOGY: The ecology of American pikas centers on their dependence on rocky environments, which in nature generally have a patchy distribution. Thus, how pikas are viewed ecologically is very scale-dependent. At a very fine scale, where there are numerous small patches of habitat, pikas exhibit one of the best mammalian models for metapopulation dynamics. A regional population of pikas may be spatially structured into assemblages of local populations on patches of habitat, which in turn are linked by dispersal. Persistence of such metapopulations is contingent on both the extinction rate of populations on patches and the probability that these patches may subsequently be recolonized. At a biogeographic scale, the occupancy pattern of pikas on mountaintop islands has informed hypotheses concerning island biogeography theory for several decades, and forms the background for examining how the modern-day American pika distribution may be impacted by extinction events caused by climate change or land-use patterns.

The American pika is sympatric across its range with various species of marmots (*Marmota* spp.), ground squirrels (various genera), chipmunks (*Neotamias*), and woodrats (typically the bushy-tailed woodrat *Neotoma cinerea*).

Pikas are preyed on by numerous avian and terrestrial predator species, especially weasels (*Mustela frenata*, *M. erminea*).

HABITAT AND DIET: American pikas typically inhabit talus or patches of broken rock in mountainous areas with cool, moist microclimates, particularly those containing rocks with diameters between 0.2 and 1.0 m. However, pikas also occur in talus-like formations such as mine tailings, lava-tube collapses, and lava-flow margins that provide thermal refugia and shelter, and occasionally they are found associated with human structures (e.g., old house foundations, riprap, roadcuts, rock quarries, piles of lumber or scrap metal). Sufficient vegetation must occur adjacent to or within these rocky habitats to allow localized foraging by American pikas.

Like all pikas, the American pika is a generalist herbivore. Individuals of this species exhibit two distinctly different foraging behaviors. One involves the direct consumption of food, a behavior that most often occurs very close (within 2 m) to the talus-forefield interface. Here, there is often a grazing halo that reflects the high intensity of foraging by pikas. The second behavior involves the gathering of plants to store in a cache (haypile) that serves as a source of food over winter for the non-hibernating pika. In general, plants selected for the haypile are collected farther from the talus-meadow interface than where direct feeding is concentrated. Because American pikas are individually territorial, the construction of a haypile (often placed relatively close to the talus-meadow border) makes them central-place foragers. Foraging follows a similar pattern throughout the species' range. Differences in species composition of plants harvested by pikas occur frequently and reflect variation in composi-

Ochotona princeps habitat. Photo courtesy Andrew T. Smith

tion of plant communities among sites; plants available to an individual pika on its territory, which may differ from those of neighbors; and individual variability in foraging tendencies. On any given territory occupied by an individual pika, some plants are selected over others and some are not harvested at all. For example, in typical alpine habitat, sedges, grasses, and forbs comprise the plants selected by pikas. At the xeric edge of their range, pikas may subsist entirely on rough forage from Great Basin shrubs (e.g., bitterbrush *Purshia tridentata*, big sagebrush *Artemisia tridentata*, rubber rabbitbrush *Ericameria nauseosa*). In the mesic Columbia River Gorge, moss comprises the majority of all food eaten, a higher percentage than for any free-living mammalian herbivore. Additionally, some of the plants harvested for the haypile are selected in a predictable sequence. Selected plants in any area generally are found to contain significantly higher caloric, protein, lipid, and water content than non-selected plants. Plant chemistry may also be an important aspect of diet selectivity. In some areas plants known to contain high levels of anti-herbivore chemicals (plant secondary compounds) are actively avoided, whereas in other regions plants with these characteristics are preferentially chosen, as they appear to exhibit superior preservation qualities (and as the poisons degrade, become available for direct consumption).

The construction of a haypile is the dominant behavior exhibited by American pikas during the summer, though this behavior is less pervasive where snow cover is infrequent. Pikas appear to gather as much food for their haypile as possible, with males (freed from parental care) harvesting the most, followed by adult females (who begin haying later than males), followed by young of the year who must mature and secure a territory before beginning the process of constructing their cache. While pikas may forage from snow tunnels during the winter when their environment is blanketed by snow, most analyses indicate that the haypile serves as an important source of food for pikas during the winter.

BEHAVIOR: American pikas hold individual territories, the size of which is equivalent for males and females. Much of their ecology and behavior revolves around aspects of their territoriality. In most regions territories are located close to the talus-forefield interface, and few territories are established in areas of barren talus where residents either would have no access to vegetation or would have to weave through a matrix of territory-holding pikas to access food. Territory size varies according to the level of productivity of adjoining vegetation or the con-

figuration of the talus habitat. The most reliable metric for determining density of American pikas is the nearest-neighbor distance between central-place haypiles, which function as the figurative behavioral center of each pika's territory. These distances range from 15 to 30 m. Adjacent home ranges are typically occupied by individuals of the opposite sex. This condition often persists because territories are occupied sequentially by individuals of the same sex as the previous occupant. American pikas, like most talus-dwelling pikas, are relatively long-lived for a small mammal, with territory holders often living to 3–4 years of age; the current longevity record is 8 years of age. Thus, vacancies on the talus matrix occur rarely, leading to the deterministic settlement pattern by gender. Ultimately, the density of pikas is relatively stable across years (partly due to their strong territoriality), and densities range from 2.2 to 9.9 individuals/ha.

American pika individuals are surface active ~ 30% of daylight hours, but times of peak activity vary widely across the species' range. At hotter, drier sites, pikas may become crepuscular or even exhibit some nocturnal activity. The most commonly observed behaviors of the American pika include surveillance, usually sitting on a prominent rock; haying and feeding; vocalizing; and activities related to territoriality, including vocalizations, scent marking, aggression, and social tolerance. Territory defense occurs primarily via aggressive acts of chases and fights between conspecifics; however, aggressive acts occur rarely and most commonly occur when unfamiliar individuals of the same gender (e.g., a resident and an immigrant) come into contact.

Vocalizations serve many functions in the behavioral repertoire of American pikas. They utter two basic types of call: short calls and long calls. Long calls (songs) are given by adult males; these calls are most common early in the summer season, and thus appear related to mating behavior. Short calls appear to be given primarily to advertise territory occupancy. Pikas often give short calls just prior to leaving the talus to forage and, if haying, give a follow-up call as soon as they return, after depositing vegetation into their cache. Nearest-neighbor male and female pikas also frequently give antiphonal calls, whereby a series of alternate short calls are given. When an animal disappears or perishes, the absence of a vocalization may lead a pika without a territory to quickly recognize the vacancy and claim the site, leading to territory establishment and, hence, survivorship. When pikas become aware of predators (avian or mammalian), they may utter a long series of short calls functioning as an alarm call to warn conspecifics of the threat. Interestingly, the alarm call is delayed when the predator is a weasel, as weasels can apprehend pikas in the talus interstices. The delay allows the weasel to move to another territory; thus the caller would not jeopardize its own fitness. The short call structure varies considerably across the geographic range of the American pika, and is distinctly different in different subspecies (e.g., an "eh-eh" in *O. p. schisticeps* and an "eeeeh" in *O. p. princeps*); also, the number of notes in the short call may vary. The palatal bridge in these different subspecies varies accordingly, just as a different reed in a woodwind instrument would yield a different sound.

Cheek rubbing, the sideward rubbing of the lower cheek apocrine gland on rocks, has been observed in most populations. Individuals can discriminate conspecifics on the basis of the odors emitted from scent marks, and they appear to serve to demarcate and advertise territorial limits. They may also act as a sexual advertisement prior to mating. Cheek rubbing occurs more frequently when a new pika claims a vacant territory on the talus, and more marking occurs near the periphery of territories.

A key characteristic of the American pika is its general inability to disperse, particularly over long distances. In fully marked populations, most juveniles that successfully claim a territory occupy sites neighboring or very close to their natal site. Animals that trespass on the territories of conspecifics are normally chased, and those that leave the talus face a foreign and hostile environment without any assurances of locating suitable vacant talus to occupy. During the summer of their birth, young pikas tend to occupy space on the talus near their putative mother and father, and they also are primarily active at times when adults are inactive—almost a time-sharing system. Nevertheless, there is evidence that marked pikas (or animals with identifiable genetic markers) may disperse up to 3 km. Dispersal distance appears related to elevation and aspect (temperature), with movements less restricted in cooler, more mesic settings and highly constrained in low, hot environments, such as near the edge of the species' geographic range.

PHYSIOLOGY AND GENETICS: Diploid chromosome number = 68. Commensurate with their reduced ability to disperse, American pikas show weak to strong levels of inbreeding and genetic structuring.

The American pika has a high resting body temperature (mean = 40.1°C) and a low upper lethal temperature (mean = 43.1°C). These values appear to be determined by adaptation for low thermal conductance to reduce energy consumption at low ambient temperatures. When temperatures are hot, pikas primarily thermoregulate behaviorally by moving between the talus surface and cooler microrefugia that exist deeper within the rocky habitat.

REPRODUCTION AND DEVELOPMENT: American pikas typically become reproductively active in the summer after birth, and apparently all females in a population breed and initiate two litters. It is common for only one of the two litters to be weaned, and throughout the range of American pikas most young produced are initiated from first litters. When, due to adverse climate or a predation event, the first litter is lost, females retain their physiological fitness and wean the second litter successfully. Average litter size ranges from 2.3 to 3.7 (based on embryo counts during late gestation). Most populations have an average litter size of three (range one to five), of which two young are normally weaned successfully. Litter size does not vary with age, with habitat productivity, or between first and second litters. Gestation time is 30 days. Resorption of entire litters may occur under extremely adverse weather conditions, and resorption of single embryos is common and normally occurs in larger litters. Typically, parturition begins in May and peaks in June. At lower elevations, however, the breeding season can begin as early as March. Phenology of reproduction is more synchronous where snowmelt is predictable than at comparatively unpredictable sites. Timing of reproduction at any locality is highly seasonal, and initiation of the first litter represents a compromise between two conflicting demands. First, early weaning of litters increases the likelihood that offspring will mature sufficiently early so as to successfully compete for vacant territories created by overwinter mortality. However, should a female initiate the first litter too early when snow still blankets her territory, she may be forced to abandon that breeding attempt because the high energetic cost of lactation cannot be accommodated.

American pikas exhibit a facultatively monogamous mating system, as the wide spacing of territories makes it difficult for males to control access to multiple females, particularly when most successful mating occurs when snow still blankets the slope. Nearest-neighbor distances between haypiles of alternating male-female pairs are shorter than the average separation of adult territories, thus giving a male close access to a particular female for mating. However, genetic markers indicate that some males successfully mate with non-neighboring females.

Young pikas are born altricial and slightly haired; their eyes are closed and their teeth are fully erupted. Mothers exhibit an absentee maternal care system characteristic of most lagomorphs; they visit the nest only once every 2 hours, and nurse for an average of 10 minutes during each visit. The growth rate of neonate pikas is among the highest of all lagomorph species. Young may become surface active and independent as early as 3–4 weeks of age.

PARASITES AND DISEASES: American pikas are known to harbor at least 37 genera and 66 species of ectoparasites and a wide variety of endoparasites. The coccidian and helminth faunas are shared with those of the collared pika from Alaska and N Canada and the northern pika (*O. hyperborea*) from Asia, an indication of the relatedness of these forms. A detailed analysis of the phylogenies of shared parasites of American and collared pikas has demonstrated that the collared pika was derived from the American pika before continental ice sheets isolated the southern and northern populations.

CONSERVATION STATUS:

IUCN Red List Classification: Least Concern (LC); declining

MANAGEMENT: Rangewide, direct utilization of American pikas is rare; most state and provincial regulations throughout its range protect the species as non-game wildlife. Several states (California, Nevada, New Mexico, Wyoming) list it as a species of special concern. A significant portion of the geographic range of the American pika includes designated protected areas, such as wilderness areas, national parks, and monuments. Outside strictly protected areas, grazing of cattle in areas immediately adjacent to pika-occupied talus can lead to changes in foraging behavior, which ultimately may jeopardize these populations.

The greatest threat to the American pika is climate change. Several life history aspects of the species make it inherently vulnerable to climate change. Pikas exhibit a poor ability to dissipate heat and possess a narrow difference between their normal (relatively high) body temperature and their upper lethal temperature; thus they are sensitive to warm and hot temperatures. As they do not hibernate, during winter they rely on snow cover for thermal insulation; diminishing snowpack may expose them to acute cold temperatures. Due to their strict territoriality, local population densities are low, and the distribution of talus patches results in low densities at larger scales as well. Should local populations become extirpated, the potential for recolonization is reduced due to the American pika's poor dispersal ability. Reproductive capacity of American pikas is low. Each of these factors may interact to make pikas vulnerable to population extirpation and,

potentially, regional extinction. These factors have seemingly played a role in the documentation of extinction of several isolated populations and upward shifts in minimum elevation of populations in the Great Basin. Many of these extirpations appear correlated with increasing temperatures or loss of winter snowpack. However, recent discovery of several low-elevation populations indicates that much remains to be learned about the resilience of pika populations and the potential threat from climate change. One life history feature works to bolster the persistence of pikas in the face of climate change: they can be relatively long-lived for a small mammal.

ACCOUNT AUTHORS: Andrew T. Smith, Erik A. Beever, and Chris Ray

Key References: Beever et al. 2003, 2011; Broadbooks 1965; Conner 1985; Dearing 1997a, 1997b; Erb et al. 2014; Galbreath et al. 2009, 2010; Hafner 1994; Hafner and Sullivan 1995; Henry and Russello 2013; Huntly et al. 1986; Ivins and Smith 1983; Jeffress et al. 2013; MacArthur and Wang 1974; Meaney 1987; C. Millar et al. 2013; J. Millar 1974; J. Millar and Zwickel 1972; Peacock and Smith 1997; Ray et al. 2012; A. Smith 1974, 1978, 1987; A. Smith and Ivins 1983a, 1983b, 1984; A. Smith and Nagy 2015; A. Smith and Weston 1990; J. Smith and Erb 2013; Whitworth 1984.

Ochotona pusilla (Pallas, 1769)
Steppe Pika

OTHER COMMON NAMES: Little pika, Small pika; Chekushka, Stepnaya pischuha (Russian)

DESCRIPTION: While sometimes sporting the common name "small pika," the steppe pika is actually a medium-sized species within the genus. The dorsal pelage is a dark grayish brown dappled with straw-colored hairs. The ventral pelage is grayish white. The ear has lighter hairs at the margin. There is ochraceous-yellow or ochraceous-brown on the upper lip and chin. Dense brown hair covers the feet and the pads of the digits. Vibrissae extend to 40 mm in length. In winter the coat is denser, longer, and a bit paler than it is in summer.

The skull appears massive, relatively broad, and bulged in the cranial section. The cranium is rounded and without a sagittal crest. The nasals are parallel and weakly expanded anteriorly. Laterally the maxilla is laced with delicate fenestration. The auditory bullae are large.

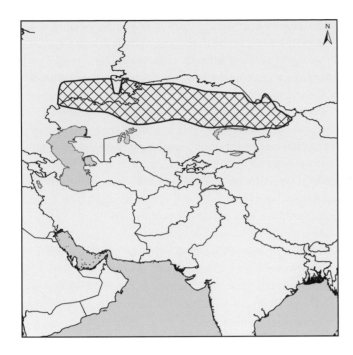

SIZE: Head and body 153–210 mm; Hind foot 25–36 mm; Ear 17–22 mm; Greatest length of skull 33–39 mm; Weight 95–277 g

PALEONTOLOGY: It appears that the steppe pika represents a terminal taxon of the earliest lineage of the genus *Ochotona*. As a living relict, it originated not later than in the Late Pliocene, and at the time was the most widespread species of pika throughout Eurasia.

Among contemporary mammals, the steppe pika has one of the longest histories of documented extinction. The steppe pika is known to have spread into W Europe on several occasions during Pleistocene glaciations, and most recently and extensively during the Younger Dryas period (12,800–11,500 ybp). Its distribution extended across Europe and as far west as the British Isles by the end of the last ice age, and since that time its range has steadily decreased. At the beginning of the Holocene, it was still found in Hungary, in the tenth century in Ukraine, and in the eighteenth century between the Don and the Volga Rivers. By the nineteenth century, the range of the steppe pika had contracted to east of the Volga River, approaching its contemporary range.

CURRENT DISTRIBUTION: The steppe pika has a large geographic distribution that extends from east of the Volga River in Russia (Saratov Region; thus the only European *Ochotona*) and across N Kazakhstan to the Altai Region of Russia. Recently the known range has been extended into NW Xinjiang, China.

TAXONOMY AND GEOGRAPHIC VARIATION: Subgenus *Lagotona*. Two subspecies: *O. p. pusilla* (west of the Ural River); *O. p. angustifrons* (east of the Ural River). Based on both morphological and molecular studies, the primitive *O. pusilla* appears to be an outlier among all pika species; most treatments list it alone in the subgenus *Lagotona*. In spite of this, some earlier treatments included the forms *nubrica*, *forresti*, and *osgoodi* in *O. pusilla*. Of these, the Nubra pika (*O. nubrica*) and Forrest's pika (*O. forresti*) are considered independent species, and *osgoodi* is a subspecies of the Moupin pika (*O. thibetana*).

ECOLOGY: The steppe pika is a burrow-dwelling species. It constructs two types of burrows. Temporary burrows are simple with only one to two entrances, have no brooding chamber, and are dug by females as emergency hideouts. Permanent burrows are constructed by adults and reproductive females. The number of entrances to permanent burrows varies by region but can range from 1 up to 14. Neighboring burrows may be as close as 2–4 m, and in most regions there are on average 12–24 burrows/ha. Population density can become quite high, but may be subject to substantial variation spatially and temporally. Reported densities range from a low of less than 1/ha to more than 100/ha. High densities in late summer following the reproductive season can be followed by severe crashes, the magnitude being dependent on the conditions during winter. Overwinter decreases in population are most severe in years with a shallow snow cover.

HABITAT AND DIET: The steppe pika is typically found in open steppe and semi-desert regions with moist soil that has substantial vegetative growth. Where it is found in arid regions, it is generally associated with wet areas and extremely dry conditions are avoided. That said, it also avoids settling on very wet soils. In various regions of their distribution steppe pikas favor thickets of Russian almond (*Prunus tenella*), ground cherry (*Physalis* spp.), willow (*Salix* spp.), Iberian spirea (*Spiraea hypericifolia*), wormwood (*Artemisia procera*), and groves of tamarisk (*Tamarix* spp.).

Steppe pikas feed exclusively on the aerial parts of plants, and selection of plants varies with the season. In early spring they concentrate on the previous year's dry vegetation, but switch to early season grasses when they appear in spring. Later during the growth season they may switch their feeding preferences to wormwood, giant fennel, and other plants. In winter they feed from their haypiles or on branches of shrubs and grasses under the snow.

Collection of vegetation during summer for their hay-piles varies according to age and reproductive status as well as local conditions. Young non-reproductive pikas tend to initiate haying earliest, and may delay doing so until after the reproductive season. Some young dispersing animals that fail to find a nest for themselves may not store food at all. Steppe pikas living in more xeric conditions tend to begin food storage collection before those living in more mesic areas. Haypile size varies from 300 g to 10.5 kg, and haypile height normally does not exceed 50 cm. Haypiles are normally placed either above the entrances of burrows or, more often, in shrubs located 11–15 m from the burrow entrance. Grasses are rarely found in the haypiles of steppe pikas; instead, they concentrate on woody forbs: Iberian spirea, sophora root (*Sophora alopecuroides*), wormwood, common licorice (*Glycyrrhiza glabra*), oriental dodartia (*Dodartia orientalis*), Russian almond (*Prunus tenella*), and giant fennel (*Ferula communis*). While 56 species of plant have been found in their haypiles, most individual stores are composed of one (or less often two) species of plant.

BEHAVIOR: The steppe pika exhibits both diurnal and nocturnal behavior. The species is highly vocal, and calls are usually heard at late dusk and throughout the night. These calls, composed of a series of long trills with rich, low sounds, can be heard up to 3 km away. Both males and females make the long call, or song. During the reproductive season steppe pikas will utter a mating song that differs from the usual call in that it is more sonorous, sharp, and polyphonic. Vocal activity during this period becomes more intense. Steppe pikas hide when a predator approaches; they do not utter an alarm call.

Social behavior of the steppe pika is poorly known, most likely because the shrubby habitat that they normally occupy makes observation difficult. They appear to live in family groups composed of an adult pair with their young, although they may live alone. During the reproductive season males become very active and patrol large areas; at this time strong fights between males are common and they accumulate many scars. Females are also known to pursue and chase other adult females at this time. The mating system appears to be promiscuous.

PHYSIOLOGY AND GENETICS: Diploid chromosome number = 68

REPRODUCTION AND DEVELOPMENT: The reproductive season lasts from as early as April through July, depending on weather conditions, and from one to five litters may be produced. Female young of the year, which become sexually mature at the age of 4–5 weeks, may produce from 1–3 litters. Males first become reproductively active in the spring following birth. Litter size ranges from 1 to 13 young, and varies with age (young of the year generally having smaller litters than overwintered adults). Gestation lasts 22–24 days. Young are born naked and blind, and first open their eyes on the eighth day following birth. Young become independent and begin eating grass at the age of 20 days.

PARASITES AND DISEASES: No infectious diseases or epizootics have been observed in steppe pikas. They commonly are host to larvae of subcutaneous bot flies (*Oestromyia leporina*), ixodid ticks (*Ixodes laguri, Dermacentor marginatus, Rhipicephalus pumilio*), and gamasid mites (*Haemolaelaps glasgovi, Laelaps stabularis*). Steppe pikas serve as host to several flea species, most commonly *Frontopsylla elata*. Eight species of helminths have been found in steppe pikas.

CONSERVATION STATUS:

IUCN Red List Classification: Least Concern (LC)

MANAGEMENT: In the past steppe pikas were trapped for the fur trade to make high-quality felt, but this activity has ceased. They are important food animals for many furbearers, such as foxes, weasels, and polecats. The species does not pose a risk of damaging pastures, largely because they primarily live among shrub-thickets; they do not apparently compete with livestock. In contrast, both domestic sheep and Pamir argali (*Ovis ammon polii*) may feed on hay stored by steppe pikas.

The contraction of the range of the steppe pika is one of the best-documented histories of extinction. And while the steppe pika does not now seem to be threatened, the processes that led to this retraction in range are still with us: naturally occurring climate change from the Pleistocene to the present (and current anthropogenic-caused climate change) and overgrazing and clearing (including fragmentation) of steppe habitat for conversion to agriculture.

The nominate subspecies, found at the leading edge of the species' contraction, may be threatened. The Red Book of the Bashkir Autonomous Republic lists *O. p. pusilla* as "rare."

ACCOUNT AUTHOR: Andrew T. Smith

Key References: Fisher and Yalden 2004; Fostowicz-Frelik et al. 2010; Lissovsky 2014; Ognev 1966; Shayilawu et al. 2009; Sludskii and Strautman 1980; Smith 1994; Smith et al. 1990; Sokokov et al. 2009.

Ochotona roylei (Ogilby, 1839)
Royle's Pika

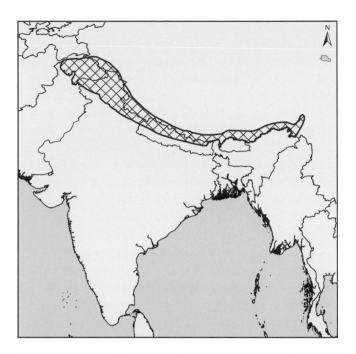

OTHER COMMON NAMES: Himalayan mouse hare; Hui shutu (Chinese); Runda (Garhwali-Indian hill dialect); Meeth (Garhwali-Indian hill dialect); Rongcha (Kumauni-Indian hill dialect)

DESCRIPTION: Royle's pika has reddish brown or bright chestnut-colored summer fur on the dorsal side, particularly on the fore parts of the body, and dark rufous gray on the remainder of the dorsal surface. The ventral parts of its body are whitish gray to dark gray. A pale band crosses the nape. The hair is exceedingly fine, straight, and glossy. The winter coat lacks rufous coloration. The coat color varies considerably among the different subspecies. Compared with the large-eared pika (*O. macrotis*), Royle's pika is more rufous, whereas the large-eared pika is more distinctly brown. Additionally, the border between the dorsal and the ventral pelage is less distinct in the large-eared pika than in Royle's pika. The ears are round and of a moderate size, with sparse hair.

The skull of Royle's pika is slightly arched with small bullae, on average 23% or less of the occipito-nasal length, and with a comparatively shorter rostrum than found in the large-eared pika. The palate length is usually shorter than 17 mm. Frontal fenestrae disappear in adults.

SIZE: Head and body 136–204 mm; Hind foot 25–34 mm; Ear 23–26 mm; Greatest length of skull 41–46 mm; Weight 127–186 g

CURRENT DISTRIBUTION: Royle's pika occurs across the Himalayan region through Pakistan, Kashmir, NW India, Nepal, and Xizang from 2,400 to 5,200 m in elevation.

TAXONOMY AND GEOGRAPHIC VARIATION: Subgenus *Conothoa*. Four subspecies: *O. r. roylei*, *O. r. wardi*, *O. r. nepalensis* (includes *angdawai* and *mitchelli*), and *O. r. himalayana*. There has been considerable controversy regarding the taxonomy of Royle's pika. It has been aligned at one time or another with the forms *macrotis*, *cansus*, *forresti*, *himalayana*, *lama*, and *nubrica*. Detailed genetic, morphological, ecological, and zoogeographical studies reveal that *O. macrotis*, the Gansu pika (*O. cansus*), Forrest's pika (*O. forresti*), and the Nubra pika (*O. nubrica*) are each distinct species. *O. lama* is now considered a synonym of *O. nubrica*. Initially named as an independent species, the Himalayan pika (*O. himalayana*) has been treated as a synonym of *O. roylei*, then it was determined to be distinct,

Ochotona roylei. Photo courtesy Katherine Solari

and now, following detailed genetic analyses, it is considered a subspecies of *O. roylei*. Many treatments consider *O. roylei* to be very closely related to *O. macrotis*, but this similarity is false and based on misidentification of specimens attributable to *O. macrotis chinensis* from C China as *O. roylei chinensis*. Currently *O. roylei* and *O. macrotis* are not considered to be sister species as was once thought.

ECOLOGY: Royle's pikas are synanthropic and solitary in nature. Like other talus-dwelling species their density is low and ranges from 12.5/ha in Nepal Himalaya to 16.2/ha in Garhwal Himalaya. Adults co-exist with the young from sequential litters born during the breeding season, and juveniles initiate dispersal by the end of summer. The spatio-temporal variation in adult pika abundance is positively influenced by duration of the snow cover period, and this influence is more prominent at lower altitudes compared with higher sites. Spring snow depth, food availability, and rock cover influence juvenile pika abundance. The amount of monsoon precipitation determines its global distribution pattern. The scaly-breasted wren-babbler (*Pnoepyga albiventer*) has a symbiotic relationship with Royle's pika in the Nepal Himalaya. Royle's pika forms the major prey base for several meso-carnivores like the red fox (*Vulpes vulpes*), the yellow-throated marten (*Martes flavigula*), and the Himalayan weasel (*Mustela sibirica*), as well as large carnivores like the common leopard (*Panthera pardus*). The home range size approximates 83 m^2.

HABITAT AND DIET: Royle's pika is a non-burrowing talus-dwelling pika species. They prefer open habitat, and their presence is significantly influenced by small size crevices (< 15 cm). The talus habitat occupied by Royle's pikas is also cooler and more humid than the outside environment; hence it acts as a thermal refuge for the species. Apart from rocky talus in forest openings and alpine meadows, they also actively use crevices in man-made structures.

Royle's pikas primarily feed on forbs, and their food selection is positively influenced by leaf area. In addition, they avoid plants high in fiber and those containing secondary metabolites.

BEHAVIOR: Royle's pikas show a bimodal activity pattern with a high level of activity during early morning and late evening hours. Their activity is negatively influenced by ambient temperature and positively by ambient moisture. Cessation of activity due to high temperature is more pronounced in the subalpine region than in the alpine region. Royle's pikas engage in distinct behavioral activities, such as foraging, locomotion, long musing, short musing, grooming, vigilance, and vocalization. Time spent foraging contributes the greatest amount in the overall time activity budget of adult pikas. Long musing behavior is observed most frequently during early morning and late evening hours, whereas short musing occurs primarily during midday and the late afternoon hours. Royle's pikas have fixed musing points close to, but above, their nests on prominent rocks. They spend more time in vigilance in parts of their home range that contain fewer rocks,

as rocks facilitate their ability to escape from predation. These pikas also spend more time searching for food in situations when there is less vegetation available to eat. Royle's pikas have a very weak vocalization; their calls are barely audible from a distance of a few meters.

Unlike other talus-dwelling pikas, Royle's pikas do not engage actively in food-hoarding behavior. They may construct temporary haypiles in habitats with sparse rock cover, possibly to avoid predation risk. In winter, they excavate tunnels (5–7 cm diameter) in snow close to their talus opening to facilitate locomotion across different parts of their talus habitat. Multiple individuals may use these snow tunnels. Royle's pikas appear to be less territorial than other pikas, but they may actively defend their nest range. During the period of dispersal, adult pikas have been observed to chase juveniles away from their territory.

PHYSIOLOGY AND GENETICS: Diploid chromosome number = 62. Royle's pika has three pairs of large metacentric, a submetacentric, and a series of smaller sub-telocentric and acrocentric chromosomes. This arrangement is very similar to that of other pika species such as the Turkestan red pika (*O. rutila*) and the large-eared pika. Additionally, a large metacentric X and significantly small Y form the sex chromosomes in Royle's pika.

Royle's pikas produce two different types of excrement, dry globular fecal pellets and soft elongated caecotrophs, which are enriched with essential nutrients and later re-ingested. Like other pika species, Royle's pika does not hibernate and remains active throughout the winter.

REPRODUCTION AND DEVELOPMENT: Royle's pikas become reproductively mature at the age of 1 year and have a low fecundity rate. They produce one litter per year. The average litter size is two to three young. The breeding season typically is initiated in late spring, but occasionally may extend into late summer. Juvenile pikas are generally encountered earlier at low-altitude sites and later at high-altitude sites. Years when there is a short period of snow cover in winter apparently delays the emergence date of juveniles.

PARASITES AND DISEASES: Different anoplocephaline cestodes such as Anoplocephalinae, *Schizorchis* cf. *altaica*, and *Ectopocephalium abei* have been reported from the small intestine of Royle's pika. Ectoparasites such as fleas (*Ctenophyllus orientalis*, *Chaetopsyll* spp.) and ticks (*Haemaphysalis danieli*, *Ixodes hyatti*, *Ixodes shahi*) have also been reported on Royle's pikas.

CONSERVATION STATUS:

IUCN Red List Classification: Least Concern (LC)

National-level Assessments: China Red List (Near Threatened—NT); India (Schedule IV—Wildlife Protection Act, 1972)

MANAGEMENT: The Api people of W Nepal often consume Royle's pikas, which they believe cures weakness and colds. Royle's pika occurs in several protected areas across its geographical range in Pakistan, China, India, and Nepal. At the local level minor disturbances due to livestock grazing and other activities (fuel wood collection, small-scale logging) are possible threats to the species, but the magnitude of these threats is not severe at present. In some protected areas (e.g., Kedarnath Wildlife Sanctuary) in W Himalaya, low-scale rock extraction from Royle's pika talus habitat has been carried out for various developmental activities such as building roads or houses. As this species does not make burrows and is totally dependent on naturally available rock talus, such large-scale rock extraction might significantly reduce habitat availability for the species. Like other talus-dwelling pikas, such as the American pika (*O. princeps*), Royle's pika is also sensitive to climatic fluctuations (high temperature, thin snow cover). Therefore, to ascertain the possible impact of environmental change on the species, long-term ecological studies across its distribution range are necessary.

ACCOUNT AUTHORS: Sabuj Bhattacharyya and Andrew T. Smith

Key References: Abe 1971; Bhattacharyya 2013; Bhattacharyya et al. 2009, 2013, 2014a, 2014b, 2015; Capanna et al. 1991; Černý and Hoogstraal 1977; Feng 1973; Feng and Zheng 1985; Feng et al. 1986; Jiang et al. 2016; Kawamichi 1968, 1969, 1971b; Khanal 2007; Lewis 1971; Mitchell 1978, 1981; Molur et al. 2005; Rausch and Ohbayaski 1974; Roberts 1977; Smith et al. 1990.

Ochotona rufescens (Gray, 1842)
Afghan Pika

OTHER COMMON NAMES: Collared pika, Rufescent pika; Ryzhevataya pischuha (Russian); Gurukses-sichan (Turkmenian)

DESCRIPTION: The dorsal surface and flanks of the Afghan pika are a pale gray to gray-brown. During the summer it has a distinctive cream-colored collar that is edged with russet hairs. It is gray to dirty gray-white and usually tinged with yellow ventrally. The winter pelage is brown and the collar becomes less apparent or unnotice-

able. Most vibrissae are dark brown (the longest ones are white) and relatively short (45–55 mm). The soles of the feet are whitish, and the ears are without distinct light margins.

The skull is large and shows varying degrees of being arched. Older specimens have a well-developed sagittal crest. The intraorbital region is narrow with well-developed lateral crests present on both sides. The nasals widen anteriorly and become narrow posteriorly after a slight constriction. The bullae are moderately large and closely spaced.

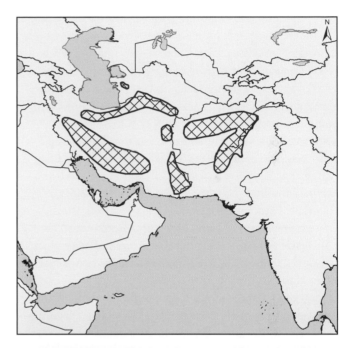

Ochotona rufescens. Photo courtesy Behzad Farahanchi

SIZE: Head and body 150–235 mm; Hind foot 31–37 mm; Ear 20–27 mm; Greatest length of skull 43–53 mm; Weight 155–240 g

PALEONTOLOGY: The fossil record of the Afghan pika is poor. A number of claims have been made for the presence of Afghan pika remains from archaeological sites farther north of its current range into Kyrgyzstan, Armenia, and environs, but these observations are complicated by the general similarity of morphology among ochotonids. These materials do demonstrate that *O. rufescens*, the steppe pika (*O. pusilla*), and a variety of other large-sized fossil ochotonids share archaic dental features and collectively are much less differentiated than all other extant *Ochotona*.

CURRENT DISTRIBUTION: The Afghan pika occurs in the mountainous regions of S Turkmenistan, Iran, northern C Pakistan, and Afghanistan. It occurs sporadically throughout its distribution due to habitat preference. It occupies elevations from 200 to 3,600 m.

TAXONOMY AND GEOGRAPHIC VARIATION: Subgenus *Conothoa*. Four subspecies: *O. r. regina* (Kopet Dagh Mountains between Turkmenistan and Iran); *O. r. rufescens* (includes *seiana, vulturna*; E Afghanistan); *O. r. shukurovi* (SW Turkmenistan in the Greater Balkans); *O. r. vizier* (Khorud Range, Iran). At one time *seiana* was treated with *O. curzoniae*, but these forms are clearly independent. The form *vulturna* was described on the basis of an aberrant specimen. Within *Ochotona*, *O. rufescens* is clearly separated from other species based on molecular and morphological (craniometric) data; it is most closely related to forms in the *macrotis* group, such as the Ladak pika (*O. ladacensis*), Koslov's pika (*O. koslowi*), the Turkestan red pika (*O. rutila*), and Royle's pika (*O. roylei*).

ECOLOGY: Unlike most pikas that can be classified as either a talus-dwelling or a steppe burrow-dwelling species, the Afghan pika is known to have adapted to both habitat types. Though it is most commonly associated with rocky desert habitats, where it occupies natural hollows, crevices, and niches, it will create burrow systems in rock-less fields. This species is known to create burrows in adobe walls when living within the vicinity of human habitations. Burrows are generally shallow (0.5 m in depth) and short (1–2 m). Population density may reach as high as 70 individuals/ha, although density may vary markedly from year to year depending on ambient conditions. The Afghan pika tends to live in isolated colonies throughout its range, and these occupy areas ranging from 0.25 to 6.1 ha. The density of Afghan pikas is normally much greater than typically found in talus-dwelling species, and most

life history features parallel those of burrowing pikas. Many other aspects of its behavior and life history are distinctly different from those of all other pika species.

Longevity of Afghan pikas appears to be 3 years, and females outnumber males in many populations.

Typical predators are little owls (*Athene noctua*), eagle owls (*Bubo bubo*), kestrels (*Falco tinnunculus*), black kites (*Milvus migrans*), northern goshawks (*Accipiter gentilis*), crows (*Corvus* spp.), martens (*Martes* spp.), and weasels (*Mustela* spp.). It is also preyed on by the Levantine viper (*Macrovipera lebetina*) in southwestern Turkmenia.

HABITAT AND DIET: The Afghan pika inhabits arid montane slopes of ravines and juniper forests.

Vegetation cover ranging between 30% and 60%, and no greater, is preferred. Afghan pikas may subsist on a diversity of native xeric plants such as thistles (*Cousinia* spp.), ephedra (*Ephedra* spp.), and wormwood (*Artemisia* spp.). Up to 58 species of plants have been recorded to be stored by the Afghan pika, and the resulting haypiles may weigh as much as 5 kg. Most haypiles are made by individual pikas, although occasionally collective haypiles have been observed. Whereas most pikas, living in northern or high-elevation seasonal environments, construct a single haypile each year, the Afghan pika collects stores of food twice per year. The first haypile is constructed in spring before local vegetation becomes desiccated, and the second one is gathered in fall when the vegetation has once again been restored by rain. Afghan pikas feed from their haypiles during cold winters and also warm summer periods. When green vegetation is available, such as during warm winters, the haypiles are not utilized.

Where cultivated crops, such as wheat or apple orchards, have encroached on the habitat of Afghan pikas, they are reported to be pests—dragging wheat into their haypiles and gnawing the bark off apple trees.

BEHAVIOR: The Afghan pika is diurnally active, and its daily activity appears to be regulated by the ambient temperature. During summer, it avoids midday heat by retreating to its hollow or burrow. Conversely, winter activity peaks during the warmest part of the day.

Complex family groups are typical, where a single male, two to three mating partners, and one to two generations of young may live in close proximity to one another in a communal territory. Most activity is confined to an area approximately 30 m in diameter. Little aggression is expressed among members within a family burrow system. Young are typically seen to engage in play behavior, such as boxing. Although males are generally solitary, they actively help in burrow digging and food storage with those females with whom they mate. Males also temporarily reside in those shelters during much of the reproductive season, from the first copulation and the birth of the first brood to the timing of the second copulation. Males display aggression only during the period of juvenile dispersal in mid-August.

Vocalizations are quite different from those of other pika species and are not well developed. The Afghan pika has a weak squeak and a prolonged whistle, but no long call, and its utterances are difficult to hear unless quite close to the animal. It typically does not use an alarm call in the presence of a predator, except for avian predators. Unlike other pikas, Afghan pikas are known to drum the ground rapidly with their hind feet to signal danger. Another form of communication is territorial marking by rubbing their neck glands on stones or against sticks.

PHYSIOLOGY AND GENETICS: Diploid chromosome number = 60. A detailed examination of the population genetic structure of the Afghan pika indicates that most genetic variance occurs within populations, although there is also significant genetic structure among populations. In comparison with other *Ochotona* species that have been studied, the Afghan pika appears to have far less nucleotide diversity.

REPRODUCTION AND DEVELOPMENT: The Afghan pika is highly fecund, producing up to five litters during the breeding season. The reproductive season is thought to last from mid-March to late September. Reported average litter sizes range from 5.2 to 7.1. Litters as large as 10–11 have been reported, and females have 4 pairs of teats. Gestation lasts 26 days. Females feed their young 3–4 times per day, for a duration of 6–12 minutes each time. Young begin to leave the nest 12 days following birth, and independent life begins 18 days after birth. By their 53rd day of life, young Afghan pikas average 145 g, demonstrating their rapid rate of growth. Young females may become reproductive actively during the first summer of their birth, attaining sexual maturity at the age of 5–6 weeks. Young males, however, do not reproduce during their summer of birth.

PARASITES AND DISEASES: Afghan pikas are host to a number of ectoparasites: four species of fleas, five species of ixodid ticks, and four species of red mites. It is unknown if any of these leads to transmission of disease.

Their helminth fauna is unique and species-specific. Afghan pikas harbor 19 species of helminths from 13 genera, and 2 of these (*Pikaeuris* and *Fastigiuris*) are only found in the Afghan pika. Members from *Cephaluris* (two species) and *Labiostomum* (three species) are known also from other species of pika.

CONSERVATION STATUS:

IUCN Red List Classification: Least Concern (LC)

MANAGEMENT: Because the Afghan pika is considered a pest in some areas where it damages agricultural crops, such as apples and walnuts, it has been subject to control. The spread of agriculture also constitutes a threat, as habitat is converted to farmland, after which the pikas feed on crops, particularly by debarking trees in orchards. However, as most populations of the Afghan pika occur in remote areas, this threat is minimal.

A population of the Afghan pika on the Small Balkan Ridge may be endangered due to isolation.

The Afghan pika is the only pika that has been domesticated for laboratory research, where it was once used in France and Japan; both of these captive colonies have died out.

At one time the Afghan pika was trapped for its fur, but due to its poor quality this practice has ceased.

ACCOUNT AUTHOR: Andrew T. Smith

Key References: Blanford 1888; Čermák et al. 2006; Fulk and Khokhar 1980; Hassinger 1973; Khalilipour et al. 2014; Lay 1967; Lissovsky 2014; Ognev 1966; Roberts 1997; Sapargeldyev 1987; Shafi et al. 1992; Smith et al. 1990; Sokolov et al. 2009.

Ochotona rutila (Severtzov, 1873)
Turkestan Red Pika

OTHER COMMON NAMES: Hong shutu (Chinese); Krasnaya pischuha (Russian)

DESCRIPTION: The summer pelage of the Turkestan red pika is bright rust reddish on the back and upper part of the head. There is sometimes a yellowish white collar behind the ears. The back of the ear is grayish black, while the inner hair is grayish white; the ears do not have distinctive light borders. At the base of the anterior inner side of the ear is a yellowish or white tuft of hair. The vibrissae are long (80–95 mm). The ventral surface is whitish with yellowish tones, and a rust-colored transverse stripe appears on the chest between the forelegs. In winter the pelage turns an ash gray, punctuated with blackish brown speckling, dorsally, while the upper part of the head remains rufous. The ventral pelage becomes lighter without any yellowish tones, and no rust-red stripe appears on the chest.

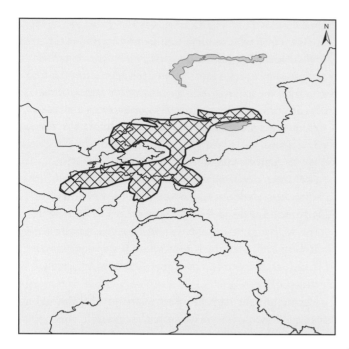

Ochotona rutila. Photo courtesy Marc Foggin

The skull is large and moderately arched. The frontal bones lack crests and are without fenestrae. The nasal bones become broadened anteriorly, and the zygomatic arches are parallel to each other.

SIZE: Head and body 180–260 mm; Hind foot 36–39 mm; Ear 27–29 mm; Greatest length of skull 48–53 mm; Weight 220–320 g

CURRENT DISTRIBUTION: The Turkestan red pika sporadically populates the western portions of the Tian Shan and the northern edge of the Pamirs in Central Asia. Its range extends through Uzbekistan, Tajikistan, Kyrgyzstan, Kazakhstan, and extreme W China (Xinjiang). Its range may include N Afghanistan. The lower elevational limit

is not known, and it rarely occurs above 3,700 m. The greatest populations are known from the vicinity of the Great Alma-Ata Lake, Kazakhstan, and Islander-Kul Lake, Tajikistan.

TAXONOMY AND GEOGRAPHIC VARIATION: Subgenus *Conothoa*. No subspecies. Historically, the forms *erythrotis*, *vulpina*, *gloveri*, and *brookei* have been included in *O. rutila*. However, the Chinese red pika (*O. erythrotis*, including *vulpina*) and Glover's pika (*O. gloveri*, including *brookei*) are clearly distinct and independent from *O. rutila* based on morphological and molecular analyses.

ECOLOGY: The Turkestan red pika is a typical rock-dwelling species, and most aspects of its natural history parallel those of other obligate rock-dwelling pikas. Families consisting of an adult pair and their young live on large territories at low density. Young remain with the parents during the summer of their birth and some juveniles may stay with them overwinter. The population density is low, roughly 12–20 individuals/ha, and there is little variation in population levels from year to year. The number of families/ha ranges from a low of 0.3–3.3/ha in different types of habitat throughout the range of the Turkestan red pika. Family territories are often separated by 50–100 m, measured from the center of each territory. Large territories are required for finding sufficient vegetation on which Turkestan red pikas persist.

The population dynamics of Turkestan red pikas are marked by their stable fecundity and relatively long life span, thus yielding a stable level of population across much of their geographic range. Attaining an age of 3 years is characteristic of this species.

The ermine (*Mustela erminea*) often preys on this species. The Tian Shan red-backed vole (*Myodes centralis*) and the silver mountain vole (*Alticola argentatus*) opportunistically feed on their haypiles.

HABITAT AND DIET: Talus and scree zones with large stones for cover are the preferred habitat of Turkestan red pikas; they do not dig burrows. They are strictly herbivorous, and they concentrate their feeding on plants that grow near their shelters, rather than on adjacent meadows as is typical of some other rock-dwelling pikas. Turkestan red pikas are known to forage on a wide variety of plants. In Tajikistan, more than 100 different species of plants belonging to 19 plant families have been recorded in their diet. This list includes 35 species of grasses, many forbs, and 10 species of trees and shrubs. In the Trans-Ili Alatau (Kazakhstan and Kyrgyzstan), 89 species from 37 families were included in the pika's diet, including 10 species of trees and shrubs, 2 species of mosses, and 2

Ochotona rutila habitat. Photo courtesy Andrew T. Smith

species of lichens. The diet of Turkestan red pikas changes throughout the summer foraging season, with a higher frequency of occurrence of grasses early in the summer and declining throughout the summer, a trend of increase in the frequency of herbs through mid-summer and then a decline, and an increase in trees and shrubs toward the end of summer. At the end of winter, when the pikas have completely used up their haypile stores, feeding is concentrated on spruce (*Picea*), juniper (*Juniperus* spp.), and mountain ash (*Sorbus aucuparia*) trees. At this time the bark of trunks of trees may be entirely stripped to a height of 0.5–1 m.

Turkestan red pikas construct large haypiles during summer, although the size varies considerably. One haypile occupied an area of 4–5 m². The mass of most haypiles ranges from 300 g to 2 kg; one large haypile weighed 8 kg. The selection of plants for caching in the haypiles tends to differ from the plants consumed directly. Grasses constitute only a very small percentage of stored plants. Some of the forbs included in haypiles, such as Przewalski's golden ray (*Ligularia przewalskii*), Jerusalem sage (*Phlomis fruticosa*), and rhubarb (*Rheum rhabarbarum*), are rejected as food items when offered to captive Turkestan red pikas.

BEHAVIOR: Turkestan red pikas are diurnal, with most activity occurring at dawn and dusk. Although male and female partners inhabit the same territory, they rarely interact with one another during the fall season. Young remain on their natal territory throughout their summer of birth, and they feed on vegetation stored by their parents. Some, but not all, juveniles overwinter on their natal home range. Social encounters among family members are rare, but may consist of boxing sessions between partners.

Unlike most pika species, the Turkestan red pika has none of the typical vocalizations for alarm, communication between other members of its familial unit, or a song. When potential predators threaten, they hide under rocks and utter chattering noises, and conspecifics react to this signal and become vigilant. Long chirping sounds may be uttered when members of a pair encounter each other.

PHYSIOLOGY AND GENETICS: Diploid chromosome number = 62

REPRODUCTION AND DEVELOPMENT: The mating season of the Turkestan red pika extends from April until July. It has a relatively low fecundity rate, averaging 4.2 young per litter (range 2–6), and typically 2 litters, with a postpartum estrus, are produced per year. In some parts of their range young females become sexually mature within 3–3.5 months and participate in breeding, although this is a rare occurrence.

Gestation lasts approximately one month. Following birth, eyes open on the 13th-14th day. Young become surface active and independent roughly on day 20, at which time their body weight has reached 60–70 g. Young attain adult size in approximately three months following birth, and most do not become reproductively active in their year of birth.

PARASITES AND DISEASES: Turkestan red pikas do not appear to be heavily parasitized. Among endoparasites, eight species of helminths have been recorded. Eight species of flea have been detected, of which *Paraneopsylla jatti* and *Amphalius clarus* are the dominant species. One pika from the Gissar Range (Tajikistan) was infected with 12 warble fly (*Oestromyia leporina*) larvae. Any possible transmission of infectious diseases by this species has not been investigated.

CONSERVATION STATUS:

IUCN Red List Classification: Least Concern (LC)

National-level Assessments: China Red List (Near Threatened—NT)

MANAGEMENT: There are no known threats to the Turkestan red pika, although its sporadic distribution and low density may make it vulnerable in the future. The species was hunted for its fur from 1936 until the 1950s, but this practice has ceased. The behavior of inflicting damage to trees by their girdling behavior may make them a target for control.

ACCOUNT AUTHOR: Andrew T. Smith

Key References: Bernstein 1963; Jiang et al. 2016; Lissovsky 2014; Ognev 1966; Sludskii and Strautman 1980; Smith et al. 1990; Sokolov et al. 2009.

Ochotona syrinx Thomas, 1911
Qinling Pika

OTHER COMMON NAMES: Qinling shutu (Chinese)

DESCRIPTION: In summer the dorsal fur of the Qinling pika is brown or rufous brown mixed with darker hair tips, and the throat and chest are rufous. The ventral pelage is dull brown or ochraceous. In winter the fur is longer, softer, and grayer. The ears are small and rounded, gray inside at the base and brown closer to the edge with a narrow white margin. The general variation in pelage color of the Qinling pika overlaps that of the Gansu pika (*O. cansus*) and the Moupin pika (*O. thibetana*). Additionally, this small pika is of the same approximate size as the Gansu pika and the Moupin pika, and thus cannot be differentiated from them by size.

The general stout construction of the skull, its flattened aspect, and its broader zygomatic arches allow the Qinling pika to be easily distinguished from that of the Gansu pika.

SIZE: Head and body 130–175 mm; Hind foot 23–32 mm; Ear 15–26 mm; Condylobasal length of skull 32–37 mm; Weight 64–110 g

CURRENT DISTRIBUTION: The Qinling pika is an endemic Chinese species. Its range consists of four isolated areas: the Qinling Mountain range, Shaanxi Province; Xunhua County, Qinghai Province; the peripheral Heng-

duan Mountains, Sichuan Province; and the Daba Mountains, Hubei and Hunan Provinces.

TAXONOMY AND GEOGRAPHIC VARIATION: Provisionally *O. syrinx* can be listed in the subgenus *Ochotona*, although recent mitochondrial DNA studies indicate that it may represent an isolated position within the matrix of *Ochotona* species. *O. syrinx* was originally described as an independent species closely related to *O. cansus*. The species was then "lost" in numerous taxonomic arrangements that split and aggregated the superficially similar *O. cansus* and *O. thibetana*. The first attempt to revive it as an independent species listed it as *O. huangensis*. *O. huangensis*, however, is a senior synonym of the forms *annectens* and *bedfordi*, each of which are properly assigned to the Daurian pika (*O. dauurica*) based on solid morphological analyses. Intraspecific variation among the isolated forms of *O. syrinx* is very poorly studied.

ECOLOGY: No ecological studies have been conducted specifically on the Qinling pika, but as a burrowing species it most likely shares most of its life history traits with other burrowing pika species. It inhabits elevations of 1,800–3,100 m asl.

HABITAT AND DIET: This species occupies mixed coniferous and broadleaf forests, as well as shrublands and occasionally grassland habitat. It is a generalized herbivore.

PHYSIOLOGY AND GENETICS: Diploid chromosome number = 42

REPRODUCTION AND DEVELOPMENT: Little is known about reproduction in the Qinling pika, although young pikas in natural history museums were collected from May to September and subadults from June to December.

CONSERVATION STATUS:

IUCN Red List Classification: Least Concern (LC), as *O. huangensis*; Not Evaluated (NE) as *O. syrinx*, due to the recent recognition of its independent status

MANAGEMENT: There are no known conservation threats or attempts to manage the Qinling pika.

ACCOUNT AUTHOR: Andrey Lissovsky

Key References: Allen 1938; Lissovsky 2014; Lu et al. 1997; Smith and Xie 2008; Smith et al. 1990; Thomas 1911b; Vakurin et al. 2012; Yu and Zheng 1992.

Ochotona thibetana (Milne-Edwards, 1871)
Moupin Pika

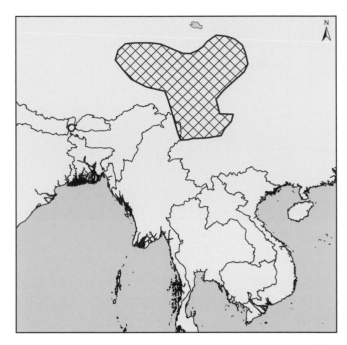

OTHER COMMON NAMES: Zang shutu (Chinese)

DESCRIPTION: In summer the dorsal surface of the Moupin pika is dull dark reddish brown to grayish brown, shading to a pale buffy brown on the flanks. The ventral pelage is a dirty gray and a well-marked buffy collar extends down the middle line of the throat to the belly. The marginal surface of the inner ear is brown to gray, and the extreme edge is white. While a small pika, the Moupin pika is slightly larger than the similar Gansu pika (*O. cansus*).

The skull is slightly convex in profile. It is longer, and the zygomatic arch and interorbital wider, than in the similar Gansu pika. The incisive foramina are triangular in shape, and their margins tend to flare out gradually. The auditory bullae are small.

SIZE: Head and body 140–180 mm; Hind foot 24–32 mm; Ear 17–23 mm; Greatest length of skull 37–42 mm; Weight 72–136 g

CURRENT DISTRIBUTION: The Moupin pika occurs in the mountains of the E Tibetan Plateau and along the Himalayan massif in China (SE Qinghai, Yunnan, Sichuan, Gansu, and SE Xizang), India (Sikkim), Bhutan, and N Myanmar. The Moupin pika is broadly parapatric with the Gansu pika across much of its range and with the Qinling pika in the Hengduan Mountains, Sichuan.

TAXONOMY AND GEOGRAPHIC VARIATION: Subgenus *Ochotona*. Six subspecies: *O. t. nangqenica* (S Xizang); *O. t. osgoodi* (NE Myanmar); *O. t. sacraria* (W Sichuan); *O. t. sikimaria* (Sikkim); *O. t. thibetana* (*hodgsoni* and *zappeyi* are synonyms; W Sichuan, S Qinghai, SE Xizang, NW Yunnan); and *O. t. xunhuaensis* (E Qinghai). *O. thibetana* has had a checkered systematic history, and it is safe to say that not all controversies have been sorted out. The forms *lama*, *aliensis*, and *lhasaensis* have been included in *thibetana*, whereas now they represent *O. nubrica*, and the form *morosa* herein is assigned to *O. cansus* rather than *thibetana*. *O. osgoodi* was at one time included as a subspecies of *O. pusilla*, and several treatments have included *O. forresti* in *O. thibetana*. The form *sikimaria* is geographically separate from the major range of *O. thibetana*, and recent analyses indicate that it may be an independent species and even more closely related to *O. curzoniae*. Similarly, the forms *sacraria* and *xunhuaensis* appear to be genetically distinct and may also represent independent species.

Ochotona thibetana. Photo courtesy Shaoying Liu

ECOLOGY: The Moupin pika is a burrow-dwelling form that occurs at high density.

HABITAT AND DIET: The Moupin pika lives in bamboo and rhododendron forests in the low elevations of its range and in subalpine forest in the high elevations. In elevations between 2,100 and 4,100 m, it can inhabit rocky areas under the forest canopy. Moupin pikas dwell in alpine pine forests, mixed cypress and birch forests, sparse needle-leaved forests, thickets, or thick growths of grass. They tend to avoid open grassland/meadow habitat.

The Moupin pika feeds on roots, stems, leaves, buds,

Ochotona thibetana habitat. Photo courtesy Shaoying Liu

and seeds of plants. It is a generalized herbivore that creates haypiles to store vegetation during winter months.

BEHAVIOR: The Moupin pika is a social species.

REPRODUCTION AND DEVELOPMENT: The reproductive season for the Moupin pika extends from at least April to July, and it generally has a litter of one to five young.

CONSERVATION STATUS:

IUCN Red List Classification: Least Concern (LC)

MANAGEMENT: Throughout S China, the Moupin pika may be threatened by deforestation, although this has not been thoroughly investigated. This species is also targeted as a pest and poisoned in S Gansu and N Sichuan. No conservation or management plans are in place for the Moupin pika, although it is present in several protected areas across S China.

ACCOUNT AUTHORS: Andrew T. Smith and Shaoying Liu

Key References: Allen 1938; Anthony 1941; Bonhote 1904; Chen et al. 1982; Feng et al. 1986; Lissovsky 2014; Pen et al. 1962; Peng and Zhong 2005; Shou and Feng 1984; Smith and Xie 2008; Smith et al. 1990; Thomas 1922.

Ochotona thomasi Argyropulo, 1948
Thomas's Pika

OTHER COMMON NAMES: Xialu shutu (Chinese)

DESCRIPTION: Thomas's pika is a small pika. The coat (summer and winter) is dorsally sandy-colored and the hairs are coarse. The underparts are usually pale white or tinged with yellow in the middle of the chest region.

Because the digits are covered with dense hairs, the claws are hidden in the hair. There is some regional variation in pelage, as specimens from SE Qinghai (Banma and Jiuzhi Counties) possess a black-brown coat that closely resembles that of the Gansu pika (*O. cansus*); the claws of these specimens are not covered with hair.

The skull of Thomas's pika is the narrowest, smallest, and flattest of the ochotonids'. The zygomatic width averages only 14 mm (range 13.5–14.8 mm), and the interorbital breadth is very narrow, averaging only 3.5 mm (range 3.1–3.8 mm). The height of the skull is very small, averaging 11.44 mm (range 11.1–11.8 mm). Each of these measurements do not overlap with those of the similar Gansu pika with which it is widely sympatric (skull height of the Gansu pika: 14.6–17.9 mm). The zygomatic width of Thomas's pika averages 40.4% (range 39.3%–41.1%) of the occipito-nasal length, whereas in the Gansu pika this value ranges from 43% to 50%. The profile of the braincase of Thomas's pika is elliptical in shape. The bullae are convex and elongated.

SIZE: Head and body 130–152 mm; Hind foot 24–27 mm; Ear 17–20 mm; Greatest length of skull 33–36.3 mm; Weight 45–76 g

CURRENT DISTRIBUTION: Thomas's pika is endemic to China, occurring on isolated mountains of the E Qilian mountain range in Qinghai Province. The localities include Qilian, Menyuan, Tianjun, Jiuzhi, Banma, and Kuze Counties. It occurs between elevations of 3,600 and 4,020 m. There have been reports of Thomas's pika

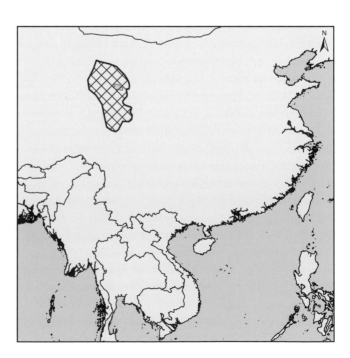

in Sichuan, but on close inspection and analysis utilizing molecular systematics, it was determined that these specimens belonged to the Gansu pika. Apparently Thomas's pika tends to converge morphologically on the Gansu pika in some regions where they are sympatric. This species is poorly known, and it is possible that its range in the E Qilian Mountains extends to the northern side of this range, and thus into Gansu Province.

TAXONOMY AND GEOGRAPHIC VARIATION: Subgenus *Ochotona*. No subspecies (*ciliana* is a synonym). *O. thomasi* differs morphologically (craniometric measurements) and molecularly from all other known pikas and thus has not generated any systematic confusion, with the exception of the putative convergence with *O. cansus* in part of its range. Molecular systematics based on mtDNA showed that this species is located at the root of the subgenus *Ochotona*.

ECOLOGY: The ecology of Thomas's pika is not well known, but it is assumed to be similar to that of the Gansu pika. This species mainly dwells in cracks of rocks. It also excavates burrows for dwelling below the moss layer of thickets and the soil layer. The burrows have complex passageways, and many entrances are connected in an underground maze.

HABITAT AND DIET: In the E Qilian Mountains, Thomas's pika is the dominant small mammal species in the high-elevation shrub environment composed of willows (*Salix* spp.), the shag-spine (*Caragana jubata*), and shrubby cinquefoil (*Potentilla fruticosa*). In this way it is separated from the Gansu pika, which is the dominant species in the low-elevation shrubby cinquefoil environment. The habitat occupied by Thomas's pikas is also abundant in rocks, and they mainly occupy cracks in the rocky substrate, surrounded by shrubs.

Thomas's pika is a generalized herbivore that primarily feeds on plants of the sedge and grass families. Because the habitat of Thomas's pika grows very few species of grass, it is speculated that its diet includes the sedge *Kobresia* as well as alpine knotweed (*Polygonum alpinum*).

BEHAVIOR: Thomas's pika is diurnally active, and it may live in small family groups.

PHYSIOLOGY AND GENETICS: Although Thomas's pika is distributed in a very narrow area, it possesses a very high genetic diversity. The species contains two genetically defined subclades that parallel the differences in morphology between forms found north of the Amnye Machen Mountains (Tianjun, Kuze, Molie, Menyuan Counties, Qinghai Province) and those found south of this range (see "Description").

CONSERVATION STATUS:
    IUCN Red List Classification: Least Concern (LC)
    National-level Assessments: China Red List (Near Threatened—NT)

MANAGEMENT: Thomas's pika is a very poorly known species, and to the best of our knowledge is not utilized at all. It is found in high and remote settings, and the habitat zone it occupies, that of shrubs, makes it an unlikely competitor with any human use of the environment. However, research is needed to determine whether the effect of pika pest control, primarily directed at the plateau pika (*O. curzoniae*), may collaterally threaten Thomas's pika. More information is needed on its ecology and natural history.

ACCOUNT AUTHORS: Shaoying Liu and Andrew T. Smith

Key References: Jiang et al. 2016; Li 1989; Lissovsky 2014; Peng and Zhong 2005; Smith and Xie 2008; Smith et al. 1990.

Ochotona turuchanensis Naumov, 1934
Turuchan Pika

OTHER COMMON NAMES: Turuhanskaya pischuha (Russian)

DESCRIPTION: The dorsal summer pelage of the Turuchan pika is ochraceous red with a dark blurred band on the back and nape formed by hairs with black tips. The

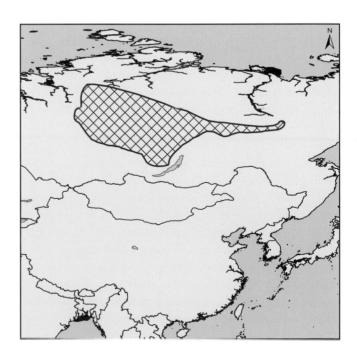

sides and ventral surface are a conspicuous ochraceous red. There are small chestnut-red spots on the sides of the neck, which are barely distinguishable on the common red background. The feet are dark brown. The ears are fringed with a narrow light stripe. In winter the red coloration changes into a yellow-brown or a yellow-gray. The principal color is produced by the graphite-gray undercoat visible through the long hairs. On the gray background the chestnut-red spots on the neck are well emphasized.

The shape of the skull closely resembles that of the northern pika (*O. hyperborea*), and it is shorter and more rounded than that of the alpine pika (*O. alpina*). The cranium is less flexed posteriorly, and the auditory bullae are relatively wide and flat. In comparison with sympatric northern pikas, however, the fronto-parietal suture assumes a different shape: the suture in the Turuchan pika is U-shaped, compared to the V-shaped appearance of this suture in the northern pika. Additionally, in the Turuchan pika the incisive bones do not form a noticeable protrusion on the lateral side of the palatine foramen and the palatal opening is spear-shaped, whereas in all northern pikas the incisive bones form a pair of flat jags that cover the front edge of the palatine foramen and the palatal opening assumes a trapezoidal or rectangular outline.

SIZE: Head and body 158–220 mm; Hind foot 25–30 mm; Ear 16–24 mm; Condylobasal length of skull 37–43 mm; Weight 113–171 g

CURRENT DISTRIBUTION: The Turuchan pika is a Russian endemic, and its range extends from the middle part of the Yenisei basin of the Angara River to the north along the east bank of the Yenisei River up to the northern border of the Putorana Plateau and then east to the Lena River.

TAXONOMY AND GEOGRAPHIC VARIATION: Subgenus *Pika*. *O. turuchanensis* was formerly included in *O. hyperborea* and more recently treated as a subspecies of *O. alpina*. There is uncertainty surrounding the identity of this species. Genetically *O. turuchanensis* appears very similar to *O. alpina*, while several aspects of its morphology (skull size and shape) are nearly identical to those of *O. hyperborea*. The call structure of *turuchanensis* differs from that of *alpina*. Where *turuchanensis* occurs in sympatry with *hyperborea*, it is the larger form of pika.

ECOLOGY: The Turuchan pika is not a typical rock-dwelling pika throughout its range. It does assume the ecology of rock-dwelling pikas in areas with significant talus features, such as on the Putorana Plateau and several other localities. However, throughout the major part of its range it lives in flat taiga where it occupies cavities between buried stones, preferring hilltops with spring sources. In these cases it constructs runways in moss and behaves very secretly. It is preyed on by ermines (*Mustela erminea*), rough-legged hawks (*Buteo lagopus*), and northern hawk owls (*Surnia ulula*).

HABITAT AND DIET: Throughout the major part of its range the Turuchan pika occupies springheads at the tops of hills, as these are the only places in the flat taiga landscape where stones can be found near the surface. In taiga they can also occupy piles of fallen logs. Otherwise, they occupy talus in regions where this habitat occurs.

The Turuchan pika is a selective herbivore, and occasionally eats mushrooms. Like most other rock-dwelling pikas, it subsists on dry vegetation stored in a haypile during the long winter. In any locality, its choice of vegetation to cache is highly selective.

BEHAVIOR: Diurnally active, the Turuchan pika avoids high temperatures and does not tolerate heavy winds like other pikas. It abstains from nocturnal activity due to low ambient temperatures.

Like many pika species the Turuchan pika is vocal, and the structure of the short call very closely resembles that of the northern pika, although being a larger pika than the northern pika the basic frequency of its call is higher (in kHz). The long call (song) is distinct, most closely resembling that of the alpine pika.

REPRODUCTION AND DEVELOPMENT: The period of reproduction extends from mid-June to late August. Juveniles first appear mid-July. An average of 3.4 embryos are produced per litter.

CONSERVATION STATUS:

IUCN Red List Classification: Least Concern (LC)

MANAGEMENT: The Turuchan pika is known to occur in Putoranski State Nature Reserve (UNESCO World Heritage Site), as well as in the Central Siberian Reserve; it may occur in other protected areas, but it has not been investigated across its full geographic range. Given its remote distribution, there are no known active threats to this species, nor are there any active conservation measures in place.

ACCOUNT AUTHOR: Andrey Lissovsky

Key References: Formozov and Baklushinskaya 2011; Formozov et al. 1999; Lissovsky 2003, 2005, 2014; Lissovsky and Lissovskaya 2002; Lissovsky et al. 2007.

Family Leporidae Fischer, 1817

The leporids comprise the rabbits and the hares (genus *Lepus*), totaling 63 species. These animals tend to be well adapted for foraging in open habitats. Their long hind limbs allow them to run speedily over open ground. Tails are short (in contrast to those of most rodents). Compared with the pikas, they have long ears and an elongated nasal region. These species also have large eyes that are adapted to their crepuscular or nocturnal activity patterns. The distribution of the leporids is now worldwide, given the introduction, primarily of the European rabbit (*Oryctolagus cuniculus*) and European hare (*Lepus europaeus*), to Australia and remote islands.

One important distinction between the rabbits and the hares concerns their reproductive mode. Rabbit kittens are born naked (altricial) following a relatively short period of gestation. Their eyes open about 4 to 10 days following birth. In contrast, hare leverets are generally well developed (precocial) at birth. Gestation in hares is longer, and leverets are born fully furred with their eyes open.

One has to be careful when employing common names for some species within the Leporidae. For example, the hispid hare (*Caprolagus hispidus*) is a rabbit, and jackrabbits are hares (genus *Lepus*).

The Rabbits

There are 10 genera and 31 species of lagomorph that are colloquially called rabbits—leporids not in the genus *Lepus*. Of these one genus, *Sylvilagus* (the cottontails), includes 18 species, and the remaining 13 species demonstrate a remarkable generic diversity across the globe. Seven of the remaining nine genera are monotypic, i.e., containing a single species: *Brachylagus* Miller, 1900;

Bunolagus Thomas, 1929; *Caprolagus* Blyth, 1845; *Oryctolagus* Lilljeborg, 1873; *Pentalagus* Lyon, 1904; *Poelagus* St. Leger, 1932; and *Romerolagus* Merriam, 1896. One genus contains two species (*Nesolagus* Forsyth-Major, 1899), and another has four (*Pronolagus* Lyon, 1904). The species accounts that follow highlight this morphological and biogeographical diversity, including a black rabbit (*Pentalagus furnessi*), two striped rabbits (*Nesolagus netscheri* and *N. timminsi*), and a dwarf form (*Brachylagus idahoensis*). Many of these monotypic genera are not only evolutionarily unique, but are also range-restricted and threatened with extinction. On the other hand, the European rabbit (*Oryctolagus cuniculus*) is probably the most common and widely distributed of all lagomorphs, due to its widespread introductions around the world. Even so, this species is threatened within its native range.

Brachylagus idahoensis (Merriam, 1891)
Pygmy Rabbit

DESCRIPTION: The pygmy rabbit is the smallest leporid species. The ears are shorter and more rounded than in sympatric cottontails, and the legs are shorter. The pygmy rabbit has a soft, fine pelage of tan gray along the dorsal surface and white to buff on the ventral surfaces with a rufous color at the nape of the neck and on the feet. The

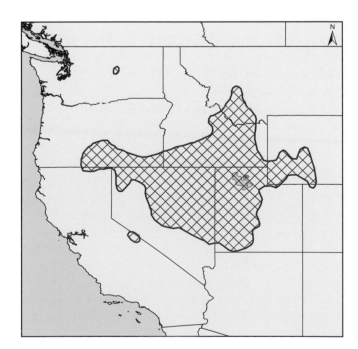

Brachylagus idahoensis. Photo courtesy Betsy L. DeMay

tail is inconspicuous, being the same color as the back, and is not visible even when the animal flees. Females are, on average, larger than males.

Skull shape and dentition are similar to that of the cottontail rabbits (*Sylvilagus*), but distinguished by a pronounced supraorbital process with anterior and posterior projections on the frontal bone; the anterior extension is greater than half the length of the posterior extension. Also, the anterior surface of the first upper cheek tooth has only one reentrant angle, in contrast with two or more in cottontails.

SIZE: Head and body 250–290 mm; Tail 20–30 mm; Hind foot 44–85 mm; Ear 39–57 mm; Greatest length of skull 48–53 mm; Weight 314–690 g

PALEONTOLOGY: Paleoenvironmental information suggests that the range of pygmy rabbits has expanded and contracted with the distribution of sagebrush (*Artemisia* spp.) vegetation in the Great Basin and Columbia Basin. During the Middle Holocene (8,000–5,000 years ago), climates in the Great Basin became warmer and drier, sagebrush declined in abundance in the lower elevations, and a concomitant decrease in abundance of pygmy rabbits is noted in the fossil record. In contrast, this arid period coincided with an expansion of sagebrush vegetation in the Columbia Basin, which encompasses the northwestern portion of the species' range. Remains support the scenario that the now disjunct Washington population became isolated from the rest of the species' range during the Late Pleistocene. This population likely occupied much of the Columbia Basin during the Middle Holocene, but its range contracted subsequently, as increased moisture resulted in a contraction of sagebrush-dominated vegetation.

CURRENT DISTRIBUTION: The pygmy rabbit occupies the Great Basin of the W United States, historically spanning eight states (Washington, Oregon, Idaho, Utah, Wyoming, Montana, California, and Nevada), and was recently documented in Colorado. Occurrence is highly patchy throughout the species' range. The population in Washington is isolated and separated from the rest of the species range by more than 200 km. This population was extirpated in 2004, and ongoing reintroduction efforts began in E Washington during 2007.

TAXONOMY AND GEOGRAPHIC VARIATION: No subspecies. *Brachylagus* was historically considered a subgenus of *Sylvilagus*, but evidence based on dental and cranial characters and on genetic analyses distinguishes *Brachylagus* as a separate genus with only one extant member. Although pygmy rabbits are superficially similar to cottontails, more recent genetic evidence suggests that the species might be a surviving representative of a primitive lineage, and that it is not closely related to *Sylvilagus*. Although no subspecies are described for the pygmy rabbit based on either genetic data or morphological characteristics, the population in Washington has been isolated from other populations for an estimated 10,000 years. Genetic analyses support the identification of this population as distinct, and in the United States it is listed as a federally endangered distinct population segment (DPS), the Columbia Basin pygmy rabbit. Ongoing semi-captive breeding and reintroduction efforts have been managed to conserve the genetic signature of the Columbian Basin ancestry while also enhancing genetic diversity. Pygmy rabbits from Idaho were introduced into the initial breeding program, and individuals from Oregon, Nevada, Utah, and Wyoming have been translocated to Washington for both breeding and augmentation of the reintroduced population.

ECOLOGY: Pygmy rabbits are obligate burrowers that excavate residential burrows of 1–2 m in depth that typically have two to eight entrances located under shrubs. Individuals occupy multiple burrows throughout the year and frequently switch among them. Burrows can remain occupied for more than 10 years by generations of rabbits. Foraging, digging, and deposition of fecal pellets around burrows influence soil nutrients and sagebrush reproduction, and alter plant species composition. Extensive subnivean tunnels also are created during winter in regions with snow cover.

Reported home range sizes vary among studies and estimation methods; however, consistent results indicate that males use larger areas than females during the breeding season, and both sexes use smaller, similar sized areas during the non-breeding season. Mortality rates are high and variable for both juveniles and adults. Predation by avian and mammalian species is an important cause of mortality, and like most lagomorphs, predation risk shapes many aspects of the species' ecology. Selection of microsites appears to be influenced by both concealment of rabbits from predators and visibility of the surrounding habitat that facilitates visual detection of potential predators. Most individuals live to reproduce during only one season.

HABITAT AND DIET: The two habitat features consistently associated with pygmy rabbits are shrub-steppe vegetation dominated by sagebrush and relatively deep soils that are suitable for construction of burrows. Often these attributes are associated with microtopography, including drainages, alluvial outwashes, and mima mounds. The species is considered a habitat specialist because it does not occur in areas without sagebrush.

Winter diet is highly specialized and consists almost exclusively of sagebrush. During summer, grasses and forbs are important forage, but sagebrush still composes approximately half of the diet. Pygmy rabbits exhibit strong preferences for types of sagebrush and even individual shrubs, and evidence suggests that variation in plant chemistry and nutrients influences foraging behav-

Brachylagus idahoensis habitat. Photo courtesy Stephanie M. DeMay

ior. Because pygmy rabbits forage heavily around burrows, cumulative effects of browsing on slow-growing sagebrush shrubs potentially alter habitat quality around long-occupied burrows. Pygmy rabbits typically forage from the ground, but have been observed climbing and feeding in the sagebrush canopy.

BEHAVIOR: Individuals often rest at burrow entrances or in forms under shrubs. Pygmy rabbits are most active above ground around dawn and dusk, although they can be active at any time, and patterns shift seasonally with increased nocturnal activity occurring during summer. Sexual behavior also influences activity, with males exhibiting elevated levels during the early breeding season. Pygmy rabbits typically rely on cryptic coloration and behaviors to avoid predation, but will retreat to dense vegetation or burrows if disturbed. Gait differs from the typical leporid bounding, likely due to their relatively shorter hind legs, but movement is rapid and can be erratic when they are pursued.

Pygmy rabbits are considered non-gregarious and most often are observed alone; however, relatively little is known about their social interactions. The tendency to give warning calls when disturbed has led some to speculate that the species might be more social than previously believed. In Idaho, about half of juveniles fitted with radio transmitters dispersed at 8–12 weeks of age and settled 1–12 km from their natal areas. Some adults also exhibited movements of 3–7 km before and during the breeding season. Both sexes range more widely during the summer and restrict movements around burrow systems during winter.

PHYSIOLOGY AND GENETICS: Diploid chromosome number = 44. The pygmy rabbit deals with environmental conditions that present significant physiological challenges. Pygmy rabbits forage primarily on sagebrush, which is relatively high in energy, but heavily defended with plant chemicals that are toxic to most herbivores. Pygmy rabbits have the ability to detoxify sagebrush and tolerate terpenes and other plant chemicals that occur in sagebrush to a greater degree than sympatric cottontails. The thermal environment also is likely challenging for this small-bodied lagomorph. Throughout much of the species range, temperatures can exceed 39°C during summer and drop below −20°C during winter. Evidence suggests that the species is thermally stressed during winter and likely also during summer; however, individuals select microsites, including burrows, where the thermal environment is modified.

REPRODUCTION AND DEVELOPMENT: Most detailed information about reproduction is known from the captive breeding and reintroduction program for the endangered Columbia Basin pygmy rabbit. Testes of males become scrotal during February, which coincides with greater movements documented by free-ranging males. Mating behavior consists of chases and short copulations. Most breeding occurs from February to June. Gestation lasts 22–25 days, and pygmy rabbits have 3 litters per year, although some in captivity had 4. Litter size is two to seven young (average four). Although births typically occur at 1 year of age, genetic analyses have provided evidence of juveniles breeding during their first year in a semi-captive population of Columbia Basin pygmy rabbits. Females construct natal burrows that consist of a single tunnel ending in a nest chamber 29–74 cm from the ground surface. Nest chambers are lined with grass, sagebrush bark, and hair, and soil is backfilled into the burrow entrance. Females return one to two times per day (typically during the night) to nurse the young, and entrances to natal burrows are backfilled with soil and concealed by females following nursing. Young emerge from natal burrows 15–16 days after birth and typically weigh 60–70 g. The species exhibits a promiscuous mating system, and multiple paternity has been documented via genetic analyses in free-ranging and semi-captive populations.

PARASITES AND DISEASES: Free-ranging pygmy rabbits support a suite of ectoparasites including fleas, ticks, lice, mites, and warble flies. Prevalence appears to vary markedly among populations and seasons. Disease ecology of free-ranging pygmy rabbits is largely unstudied. Anecdotal evidence of rapid collapse of some populations suggests that disease might be responsible. Mycobacteriosis due to *Mycobacterium avium* infection and coccidiosis caused by a newly described species (*Eimeria brachylagia*) in pygmy rabbits has been documented in the captive and semi-captive populations of the Columbia Basin pygmy rabbit.

CONSERVATION STATUS:

IUCN Red List Classification: Least Concern (LC)

National-level Assessments: USA Endangered Species Act (Endangered—Columbia Basin Distinct Population Segment)

MANAGEMENT: Currently, there is no utilization of pygmy rabbits. Previously they were considered a game animal, in part because of the difficulty in distinguishing the species from juvenile cottontails. Abundant remains of pygmy rabbits in archaeological sites support their prior use by Native Americans, likely for food and hides.

Loss and alteration of sagebrush habitats, especially those associated with deep soils, as a result of agriculture, wild fires, energy development, and urbanization have reduced and fragmented available sagebrush habitats. Invasion of exotic plants like cheatgrass (*Bromus tectorum*) following fires continues to degrade sagebrush habitats and alter historic fire cycles, resulting in state transitions to exotic grasslands that are difficult to restore to sagebrush-steppe systems. Research is needed to address questions related to fire and sagebrush restoration as well as land and population management, and to fill in gaps about the life history, disease ecology, and distribution of the species.

ACCOUNT AUTHORS: Janet L. Rachlow, Penny A. Becker, and Lisa A. Shipley

Key References: Adams et al. 2011; Camp et al. 2014; Crawford et al. 2010; DeMay et al. 2016; Elias et al. 2006, 2013; Estes-Zumpf and Rachlow 2009; Estes-Zumpf et al. 2010; Green and Flinders 1980a, 1980b; Katzner and Parker 1997; Katzner et al. 1997; Larrucea and Brussard 2008, 2009; Lawes et al. 2012, 2013; Lee et al. 2010; Lyman 1991, 2004; Pierce et al. 2011; Price and Rachlow 2011; Price et al. 2010; Rachlow et al. 2005; Sanchez and Rachlow 2008; Schmalz et al. 2014; Shipley et al. 2006, 2012; Thimmayya and Buskirk 2012; Thines et al. 2004; Wilson et al. 2011; Woods et al. 2013.

Bunolagus monticularis (Thomas, 1903)
Riverine Rabbit

OTHER COMMON NAMES: Bushman hare; Vleihaas, Pondhaas, Oewerkonyn (Afrikaans)

DESCRIPTION: A characteristic feature of riverine rabbits is the dark brown band that runs along the sides of the lower jaw and tails out upward toward the base of the ears. There are distinct white rings around the eyes. The limbs are short and heavily furred. The pelage is variable, but usually a reddish brown shade grizzled with black, and it is silkier than that of hares (*Lepus*). The underparts and throat are cream-colored, and the riverine rabbit has a uniformly rufous-gray woolly tail. The wide, long ears have fringes of white hair on their inner margins.

The skull is comparatively short for a leporid, including a shortened hard palate. Each upper incisor has a single groove that is not filled with cement. In general, the cranial characters of the riverine rabbit are distinct from those of red rock hares (*Pronolagus*), although not clearly distinct from those of *Lepus*. While having similar

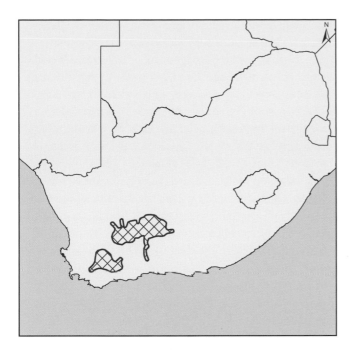

Bunolagus monticularis. Photo courtesy Paul Carter

morphology, on average riverine rabbit skulls are smaller than those of *Lepus*. The premolar foramen is absent in the riverine rabbit.

SIZE: Head and body 337–470 mm; Tail 70–108 mm; Hind foot 90–120 mm; Ear 107–124 mm; Greatest length of skull 78.9–85 mm; Weight 1,400–1,900 g

PALEONTOLOGY: The ancestral Leporidae was widespread throughout North America and Asia, and their complex evolutionary history can best be explained by several intercontinental exchanges and vicariance events between 14 and 8 mya. The first divergence in the group separated an Asian lineage (currently represented by *Nesolagus*, *Pronolagus*, and *Poelagus*) from the remainder of North American leporids, including the ancestor of *Bunolagus*. Between 11.8 mya (± 1.24 mya) and 10.28 mya (± 1.34 mya), a dispersal occurred from North America to Asia, and this exchange resulted in the establishment of ancestral *Pentalagus*, *Caprolagus*, *Bunolagus*, and *Oryctolagus* in Asia. Shortly hereafter, approximately at 9.44 mya (± 1.15 mya), *Bunolagus* dispersed into Africa and subsequently gave rise to the present form isolated at the southern tip of the continent.

CURRENT DISTRIBUTION: The riverine rabbit is endemic to the semi-arid Central, Upper, and Little Karoo of South Africa (Nama Karoo, Succulent Karoo, and Fynbos Biomes). The total extent of occurrence is 40,493 km², and the area of occupancy is estimated to be 3,011 km². The distribution of the total population of the riverine rabbit comprises two separate regions, one northern population on top of the central plateau demarcated by the Great Escarpment and a second southern population occurring at lower altitudes. The northern population is roughly bounded by the Karoo towns of Carnarvon and Williston (north), Victoria West (east), Beaufort West (southeast), and Calvinia (west). Within this distribution area the riverine rabbit occupies a very restricted and specialized niche: the discontinuous and dense vegetation on soft and nutrient-rich alluvial soils associated with the seasonal rivers of the Karoo. This area falls within the Central Karoo, which is characterized by a summer rainfall regime with most of the rainfall occurring between October and March, and is subject to extended periods of drought. The southern population is within the predominantly winter-rainfall regions of the Little Karoo and Fynbos Biomes of South Africa, and is roughly bounded by Touwsrivier in the north, Barrydale in the south, and Montagu in the west. The southern population is distributed within three catchments: the Breede, Gouritz, and Olifants river systems.

TAXONOMY AND GEOGRAPHIC VARIATION: *Bunolagus monticularis* was originally described as *Lepus monticularis* in 1903, but separated generically in 1929. Although the species superficially resembles the Cape hare (*L. capensis*) in external morphology, the presence of a black stripe along the lower jaw and a uniformly brown tail distinguish it from *Lepus* at both the species and genus levels. The diploid chromosome number (2n = 44) differs from that of the Cape hare (2n = 48). *Bunolagus* is a monotypic genus. No subspecies have been recognized, but population genetic analyses provide some evidence that the *Bunolagus* populations within the northern distribution (on the central plateau above the Great Escarpment) are genetically distinct from the populations in the southern distribution (below the Great Escarpment).

ECOLOGY: Riverine rabbits are solitary, and they are primarily active at night and rest during the day in shallow depressions (forms) that are scraped out under Karoo shrubs. Density estimates range from 0.06 to 0.17 rabbits/ha. Females have a home range of about 12 ha, and males have home ranges of about 20 ha in optimal habitat. Home ranges overlap relatively little intra-sexually. Adult male-female couples and adult female-juvenile couples form for short periods during mating and during rearing of the young, respectively. The riverine rabbit is a habitat specialist, occupying a river zone landscape that is also of economic importance to landowners in terms of cultivation and small-stock grazing. It thus functions as a key indicator species for these critical river zones, as its regional extinction in many areas of its former natural distribution range is indicative of the degradation, fragmentation, and loss of riverine vegetation caused by overutilization, cultivation, and transformation of river channels.

HABITAT AND DIET: The riverine rabbit is strictly herbivorous. Within the Nama-Karoo, it browses predominantly on the sunflower (*Pteronia erythrochaeta*), *Bassia salsoloides*, *Salsola glabrescens*, and members of the Mesembryanthemaceae during winter, and on available grasses during the vegetation growth season (summer and fall). The riverine rabbit occurs in the following vegetation types starting with the most widely occupied: E Upper

Bunolagus monticularis habitat. Photo courtesy John E. C. Flux

Karoo, W Upper Karoo, Bushmanland Vloere, W Little Karoo, Roggeveld Karoo, Montagu Shale Renosterveld, Tanqua Karoo, S Karoo Riviere, Upper Karoo Hardeveld, Gamka Karoo, N Upper Karoo, Matjiesfontein Shale Fynbos, Swartberg Shale Renosterveld, and Robertson Karoo. It should be noted that these are broad descriptors of habitat type, and rabbit populations in the northern part of the distribution are strongly associated with alluvial floodplains and narrow belts of azonal riparian vegetation. The southern population of the riverine rabbit is not so strongly associated with riparian vegetation and has been frequently observed on re-vegetated fallow lands in Renosterveld vegetation not associated with drainage areas.

BEHAVIOR: Although the riverine rabbit is mostly active at night, there is evidence of cathemeral activity patterns, with extensive diurnal and crepuscular activity apparent across all seasons. This could be a consequence of a number of factors, such as fluctuations in food availability, predator avoidance behavior, or thermoregulatory stress.

PHYSIOLOGY AND GENETICS: Diploid chromosome number = 44. Based on analyses of mtDNA and microsatellite data, analyses of molecular variance suggest that there are two genetically distinct geographic clades (a northern population on top of the central plateau demarcated by the Great Escarpment and a second southern population occurring at lower altitudes). These two populations should be regarded as distinct evolutionary significant units (ESUs), but the exact boundaries of the two populations are not secure. Additional sampling is still needed to adequately delineate the exact boundaries of the genetic provinces and to determine the extent of potential gene flow among the clades.

REPRODUCTION AND DEVELOPMENT: The riverine rabbit displays a polygamous mating system. It is the only indigenous burrowing rabbit in Africa and depends on the soft alluvial soils in the river floodplains to construct breeding stops (burrows). One or two altricial young, weighing 40 g at birth, are born from August through May and reared in the fur- and grass-lined breeding stop. The entrance is plugged with soil and twigs to camouflage it from predators. The burrow has an entrance 90–105 mm wide, is approximately 200 mm long, and widens into a chamber 120–170 mm broad. The youngest juvenile noted in the wild weighed 500–600 g, indicating that juveniles remain in the breeding stop for relatively long periods before leaving to forage independently. Population growth is slow as females produce only one litter a year, with one to two young per litter. Generation length for this species is two years. Limited observations indicate that gestation time is 35–36 days. Riverine rabbits display a postpartum estrus and breeding is synchronized.

PARASITES AND DISEASES: Very little is known about riverine rabbit parasites or diseases. Circumstantial evidence suggests that the ability of the myxoma virus to circumvent the immune responses of the European rabbit (*Oryctolagus cuniculus*) is due to a specific gene conversion, which is shared by the riverine rabbit, and this suggests that accidental exposure to the virus could exterminate the species.

CONSERVATION STATUS:

IUCN Red List Classification: Critically Endangered (CR)—C2a(i)

National Red List: South Africa (Critically Endangered (CR)—C2a(i))

MANAGEMENT: Loss and degradation of habitat are the main threats to the riverine rabbit. Over the past century, 50% to 80% of habitat has been lost as a result of cultivation and livestock farming within the riparian areas of the Karoo. Other threats to the species include hunting (the rabbit is hunted for sport and for bushmeat by farm workers), accidental mortality in traps set for pest animals on farmlands, and vehicle collisions. Furthermore, projected climate models illustrate that, with predicted climate change impacts, the majority of current habitat in the northern population could become unsuitable.

Unconventional shale gas development ("fracking") in the Nama-Karoo is an emerging threat that could have significant impacts on rabbit populations in the Northern Cape as a result of ancillary activities associated with fracking, including increased roadkill mortalities, habitat fragmentation, and altered hydrology of Karoo river systems. The rapid increase in renewable energy developments, such as wind farms, also pose a risk to the species' habitat.

At present, very little of the riverine rabbit habitat is protected within formal protected areas or within private reserves, and thus the conservation of the species is largely under the goodwill of private landowners or Karoo farmers. The Riverine Rabbit Conservation Project is a collaboration between the Endangered Wildlife Trust's Drylands Conservation Programme, Provincial Nature Conservation authorities, and research institutions. This project coordinates the biodiversity of the Karoo region, and promotes integrated land management practices that can sustain the riverine rabbit, its habitat, and many other species while ensuring socioeconomic benefits to landowners and communities. Thus the riverine rabbit functions as the flagship species of the Karoo.

The complex genetic structure evident within and between populations sampled thus far emphasizes the fact that the riverine rabbit will only survive through habitat conservation. Connectivity among populations and the conservation of habitat throughout the range are of critical importance to maintain the evolutionary potential of the species.

ACCOUNT AUTHORS: Christy J. Bragg, Conrad A. Matthee, and Kai Collins

Key References: Abrantes et al. 2011; Ahlmann et al. 2000; Coetzee 1994; Collins 2001, 2005; Collins and du Toit 2016; Collins et al. 2004; Duthie 1989; Duthie and Robinson 1990; Duthie et al. 1989; EWT 2014; Fostowicz-Frelik and Meng 2013; Happold 2013; Hughes et al. 2008; Matthee et al. 2004e; Mucina and Rutherford 2006; Nowak 1991; Robinson 1981a, 1981b; Robinson and Dippenaar 1987; Robinson and Skinner 1983; South African Mammal CAMP Workshop 2013; Thomas 1903, 1929.

Caprolagus hispidus (Pearson, 1839)
Hispid Hare

OTHER COMMON NAMES: Assam rabbit, Bristly rabbit; Khargorkata, Ha ha pahoo (Assamese); Jungli Kharayo (Nepali); Lapin de L'Assam (French); Conejo de Assam (Spanish)

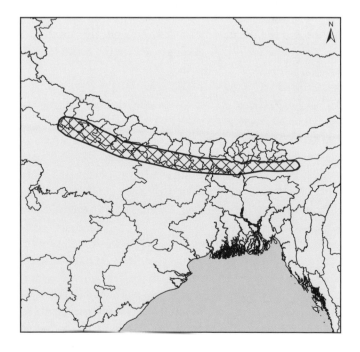

Caprolagus hispidus. Photo courtesy Bed Bahadur Khadka

DESCRIPTION: The hispid hare is also called the "bristly rabbit" because of its coarse dark brown bristly coat on the dorsal surface, due to a mixture of black and brown hairs. The chest is brown and the abdomen whitish. The short tail is brown throughout and the ears are shorter in the hispid hare in contrast to the longer ears and white underside to the tail found in the sympatric rufous-tailed hare (*L. nigricollis ruficaudatus*). Hispid hares have a conspicuously large nose compared with those of other lagomorphs.

The top of the skull is very solid and flat. Compared with those of other rabbits the frontals are longer and have a shorter extension. The palate is as long as it is broad, and the incisive foramina are small. The postorbital processes are small and united with the frontals anteriorly.

SIZE: Head and body 380–500 mm; Tail 25–38 mm; Hind foot 79–86 mm; Ear 54–61 mm; Greatest length of skull 76 mm; Weight 1,810–3,210 g

PALEONTOLOGY: Lagomorph fossils from the Late Miocene found in the Siwalik group of sediments on the Potwar Plateau in NE Pakistan may be assignable to *Caprolagus*. There is fossil evidence that in the Pleistocene the distribution of *Caprolagus* extended as far as Central Java (Indonesia).

CURRENT DISTRIBUTION: Historically, the hispid hare has been recorded from tracts along the S Himalayan foothills from Uttar Pradesh (India) through Nepal and West Bengal to Assam (India), extending southward as far as Decca (Bangladesh) and to Kanha National Park, Madhya Pradesh (India). The Kanha record, however, was based exclusively on fecal pellet analysis and has not been subsequently confirmed by the sighting of any hispid hares. Currently the hispid hare is believed to occur only in N India, S Nepal, and E Bhutan.

TAXONOMY AND GEOGRAPHIC VARIATION: No subspecies. *C. hispidus* was originally described as a *Lepus*, but its unusual morphological characteristics led to it being considered a distinct genus shortly thereafter. Recent molecular studies have confirmed the generic independence of *Caprolagus*.

ECOLOGY: The hispid hare does not construct burrows, but rather takes shelter in surface vegetation or burrows made by other animals. Nests are circular in shape and composed of narenga (*Narenga porphyrocoma*) grass leaves. Home ranges of males and females overlap, with the ranges of males being larger than those of females.

HABITAT AND DIET: The hispid hare primarily occupies tracts of early successional tall grasslands, locally termed "elephant grass." They preferentially occupy habitats with tall grass species, although the height of preferred grasses varies with season. This habitat provides both cover and food for the hispid hare. With increased burning of these grasslands, their distribution is increasingly confined to riverine regions that are not burned. These habitats, however, are the first to be flooded during the monsoon, which may negatively impact their survival.

The diet of the hispid hare is known to include at least 23 plant species. Their preferred diet is the inner part of stems and also leaves of narenga, kans (*Saccharum spontaneum*), and siru (*Imperata cylindrica*). Their diet may also include halfa grass (*Desmostachya bipinnata*) and Bermuda grass (*Cynodon dactylon*).

BEHAVIOR: Hispid hares mostly prefer to stay in a certain area and eat grass species present in that area instead of foraging outside the safety of their home range. Compared with most lagomorphs, they are described as "slow-moving."

REPRODUCTION AND DEVELOPMENT: Litter size is small, normally two to three young. Females have only four teats, which also reflects the small litter size. Breeding normally occurs from January to March. The small size of scrotal testes in males is indicative of a monogamous mating system.

CONSERVATION STATUS:

IUCN Red List Classification: Endangered (EN)—B2ab(ii,iii,v); decreasing

CITES: Appendix I

National-level Assessments: India (Endangered (EN)—B2ab(ii,iii,iv); Schedule 1 Indian Wildlife Protection Act);

Nepal (Critically Endangered (CR)—B1ab(ii,iii,iv) + 2ab(ii,iii,iv); Schedule 1 National Parks and Wildlife Measures Act)

MANAGEMENT: Existing utilization of hispid hares includes subsistence harvesting. The major threats to the species, however, are fire, gathering of thatch, grazing, invasion of grasslands by broadleaved trees, floods, and predation. Of these the burning of the grasslands and thatch collection are the most critical because they simultaneously remove both cover and food supply for the hispid hare. The timing of these activities is also important, as they normally occur during the dry season, which coincides with the reproductive season of the hispid hare. While much of the currently occupied range is in protected areas such as Jaldapara National Park (India), Chitwan and Bardia National Parks and Suklaphanta Wildlife Reserve (Nepal), Manas National Park (Bhutan), and Manas National Park (India), maintenance, conservation, and restoration of habitat are needed. Similarly, there is a need for greater awareness of the species and its status at the local level and by park/reserve staff. Research in the areas of population biology and ecology, threats, and monitoring of population trends is needed.

ACCOUNT AUTHORS: Andrew T. Smith and Charlotte H. Johnston

Key References: Aryal et al. 2012; Bell 1986; Bell et al. 1990; Blanford 1888; Maheswaren 2002; Molur et al. 2005; Nath and Machary 2015; Oliver 1979, 1980; Tandan et al. 2013; Winkler et al. 2011; Yadhav et al. 2008.

Nesolagus netscheri (Schlegel, 1880)
Sumatran Striped Rabbit

OTHER COMMON NAMES: Sumatran short-eared rabbit; Kelinci hutan (Indonesian)

DESCRIPTION: The Sumatran striped rabbit is characterized by its unique striped pattern. A thin brown stripe runs dorsally from the nose to the tail. The sides of the neck and shoulders are brown. The foreneck is dark brown. There is a tapering, broad brown stripe that originates at the median dorsal stripe and curves downward to the shoulders. Another tapering, broad brown stripe runs down the sides at a right angle to the median line, but does not extend onto the abdomen. The ears are short

and brown on the exterior. A narrow brown line runs from the ear toward the lower mandible, and surrounds the eye. The background color is light brown to gray on the dorsum, and it lightens to a buff color ventrally and on the legs. The digits are dark brown. The rump and tail are a bright rust color. The pelage consists of soft, dense underfur and longer guard fur. The skin is apparently very thin and delicate. The tail is short and barely visible. The body and legs are compact.

On first examination, the Sumatran striped rabbit is almost identical in appearance to the Annamite striped rabbit (*N. timminsi*) that occurs in the Annamite Mountains

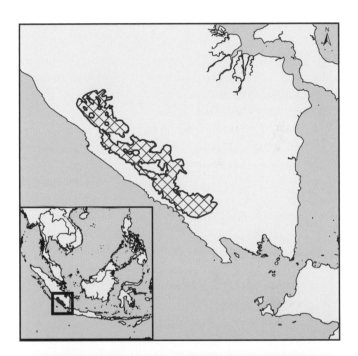

Nesolagus netscheri. Photo courtesy Jeremy Holden

along the Laos-Vietnam border. However, the Sumatran striped rabbit has a smaller skull length, a larger foramen lacerum, and a shorter second upper premolar.

SIZE: Head and body 368–417 mm; Tail 17 mm; Hind foot 67–87 mm; Ear 34–45 mm; Greatest length of skull 67–74 mm; Weight 1,500 g

PALEONTOLOGY: The Sumatran striped rabbit is one of two extant members of the genus *Nesolagus*. The earliest known *Nesolagus* fossil is *N. sinensis*, which was found in SW China. *N. sinensis* retains a simplified paedomorphic pattern during the ontogenetic process, a trait also present in the two extant *Nesolagus* species. This indicates that *N. sinensis* is more primitive than the extant *Nesolagus* species, and is likely a direct ancestor of both species. However, despite having similar morphology, there is a large degree of genetic divergence between the two extant species, and it is estimated that they have been separated for more than eight million years.

CURRENT DISTRIBUTION: The Sumatran striped rabbit is endemic to the montane forest of the Barisan Mountains in W Sumatra (Indonesia). The historic distribution is thought to have stretched from South Sumatra to Aceh Provinces (Indonesia) along the Barisan Mountains, although most specimens have been collected from W Sumatra Province. The species was described in 1880 from a specimen collected by E. Netscher in the Padang Highlands, a mountainous region east of the city of Padang. In 1895, a sailor brought a live specimen to the Zoological Gardens at Amsterdam that had apparently also been collected from that area. J. A. Piepers reported that the species was seen occasionally in the Padang Highlands on Gunung Talang as forest was cleared for coffee plantations. Maarseveen, the owner of a coffee plantation in the eastern portion of the Padang Highlands, also reported finding three or four Sumatran striped rabbits as the forest was cleared for coffee in 1884, with additional later sightings as nearby plantations were cleared. At this time several live specimens were maintained for up to a year in captivity. However, after 1909 no more individuals were observed for a significant time, and Piepers believed that cultivation on the mountain may have extirpated the species. E. Jacobson collected several museum specimens in West Sumatra in 1914 and 1915: one from the eastern region of the Padang Highlands, near the coffee plantation owned by Maarseveen, and two from Sungai Kumbang on Gunung Kerinci. In 1916, Stolz also collected an individual from the eastern region of the Padang Highlands in Alahan Pandjang District. Historically, the species was also documented in the South Sumatra and

Bengkulu Provinces. In 1916, E. Jacobson collected several museum specimens from that region: one from Rimbo Pengadang and three from Gunung Dempo.

The Sumatran rabbit was not officially documented again until 1972, when M. Borner reported a sighting of the species in Gunung Leuser National Park in Aceh Province. In 1978, it was once again sighted in West Sumatra, near Gunung Kerinci, by J. Seidensticker, and the first photograph of the species was a camera trap photograph recorded in 1997 by J. Holden (Flora and Fauna International) in Kerinci Seblat National Park. Since 1997, there have been additional sightings and photographs from the general area in, and around, both Kerinci Seblat National Park and Bukit Barisan Selatan National Park. However, there have been no recent reports of the species outside these areas, despite an increasing number of surveys and camera trap studies being conducted across Sumatra. Therefore, it is possible that the species distribution is now restricted to these two protected forests.

The Sumatran striped rabbit is thought to be most common at elevations above 600 m. However, it has occasionally been recorded from lower elevations. In 1997, it was sighted in the lowland forest outside Pemerihan, and it was recorded again in a lowland forest (544 m) in 2011.

TAXONOMY AND GEOGRAPHIC VARIATION: The genus *Nesolagus* was thought to be monospecific until the discovery of the Annamite striped rabbit in the 1990s. The two *Nesolagus* species are very similar in general appearance, but genetic analyses indicate that they diverged nearly eight million years ago.

There is some variation in the coloration and marking of Sumatran striped rabbit specimens, but there is no indication of the presence of subspecies on the island of Sumatra.

ECOLOGY: Little is known about the Sumatran striped rabbit throughout its range. As it is not commonly seen, it is thought to be rare. It is not generally known to local people, although some refer to it as a "marmot." The paucity of sightings may also be attributed in part to its presumed nocturnal nature. All specimens were trapped at night, and captive rabbits are most active at night. All recent camera trap photographs of the species have also been recorded at night. Yet, there is some evidence that Sumatran striped rabbits are at least occasionally active during the day, as they have been seen feeding on fallen fruit during the day. In some diurnal observations, the rabbit has quickly run away, whereas in others the rabbit froze for a long time before moving off.

The species is thought to seek shelter in burrows or hollows at the base of trees. There are several reports of Sumatran striped rabbits being discovered in burrows or tree hollows as land was cleared for coffee plantations. As a captive individual did not show any predisposition toward digging, it has been presumed that the species might inhabit natural holes or burrows dug by other animals.

There is no information on the social structure of the species. There have been some historical and unconfirmed reports of the species occurring in small groups. However, all camera trap photographs and sightings are of a single animal, so it appears to be largely solitary.

HABITAT AND DIET: Although it has been occasionally reported from newly cleared areas or degraded forests, the Sumatran striped rabbit appears to be forest dependent, and has not been recorded from open areas or cultivated fields. It is most commonly found at higher elevations in montane and submontane primary habitat, but there are uncommon reports from lowland forests.

The Sumatran striped rabbit does not feed in clearings or open areas, foraging only in forested areas. A captive Sumatran striped rabbit preferred different species of *Cyrtandra*, a common genus of understory vegetation in tropical forests. Plants in the Araceae have also been accepted, but not preferred. Several other species of plant were reluctantly eaten, but cultivated vegetables and fruit were avoided. Some local people report that the species has been seen eating fruit on the ground, but odorous guavas do not seem to attract them to camera traps. There have been no analyses of food habits in the wild, but a camera trap video from Kerinci Seblat National Park showed an individual slowly moving along the forest floor, feeding on several different plants in the understory.

BEHAVIOR: The Sumatran striped rabbit is relatively slow-moving compared to other rabbit species. It was reported to freeze upon discovery during several observations, although after some time it moved quickly for cover. An individual that was brought to the Zoological Gardens at Amsterdam was described as "shy and timorous" while in captivity.

CONSERVATION STATUS:

IUCN Red List Classification: Vulnerable (VU)—B1ab(i,ii,iii,iv,v)

National-level Assessments: Indonesia (Protected)

MANAGEMENT: There is no evidence that hunting or capture for the pet trade poses a significant threat to the Sumatran striped rabbit, likely owing to its rarity and difficulty to catch. Local people also report that the meat has a bad taste, which may be another reason that it is not targeted. However, individuals may be vulnerable to capture in snares intended for other species.

As the Sumatran striped rabbit appears to be forest dependent, the most significant threat is apparently the considerable level of habitat loss and degradation across the island of Sumatra. From 1990 to 2010, Sumatra lost 7.54 million ha of primary forest and roughly 70% of its total forest cover. This loss may devastate species such as the Sumatran striped rabbit that do not appear to tolerate human-dominated landscapes.

In the International Union for Conservation of Nature (IUCN)/Species Survival Commission Action Plan, a study to elucidate the status and ecology of the Sumatran striped rabbit was recommended. However, to date there has been no focused study of the species, and current information is derived from anecdotal camera trap photographs and incidental sightings. All recent observations of the species are concentrated in and around two national parks, Kerinci Seblat and Bukit Barisan Selatan, which may reflect a contraction of their range. This lack of knowledge highlights the urgency of ascertaining information on the species' current status and ecology in order to implement effective conservation strategies.

ACCOUNT AUTHORS: Jennifer L. McCarthy, Jeremy Holden, and Debbie Martyr

Key References: Can et al. 2001; Chapman et al. 1990; Dinets 2010; Flux 1990b; Jacobson 1921; Jacobson and Kloss 1919; Jin et al. 2010; Margono et al. 2012; McCarthy et al. 2012; Surridge et al. 1999.

Nesolagus timminsi Averianov, Abramov, and Tikhonov, 2000

Annamite Striped Rabbit

OTHER COMMON NAMES: Tho van (Vietnamese); Ka tai lai seua (Lao)

DESCRIPTION: The Annamite striped rabbit is similar in appearance to the Sumatran striped rabbit (*N. netscheri*). It sports black and brown dorsal stripes against a pale buff-colored background and a rust-colored rump. A dark stripe runs from the tip of the nose along the back and ends near the base of the tail. On the face two stripes extend from an area just below the eyes and upward along

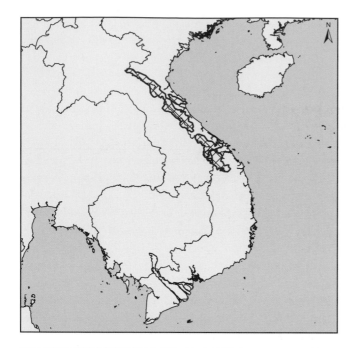

Nesolagus timminsi. Photo courtesy Andrew Tilker

the ears. There is a fork at the base of this stripe near the eye. There appears to be some degree of distinct frontal and dorsal striping patterns among individuals. The underside is white. The tail and ears are short.

The skull is slightly larger (12%) than that of the Sumatran striped rabbit. The Annamite striped rabbit has a relatively small frontal supraorbital process, a short posterior projection, and no anterior projection. Its relatively long posterior zygomatic process extends posteriorly beyond the zygomatic process of the squamosal.

SIZE: The type specimen of the Annamite striped rabbit consists only of a skull and partial skeleton; no full specimens exist, therefore typical measurements do not exist. Greatest length of skull 78.9 mm.

PALEONTOLOGY: Fossil evidence of a related species, *N. sinensis*, has been found in Chongzuo, Guangxi, and S China. *N. sinensis* is believed to be more primitive than and probably directly ancestral to *N. timminsi*.

CURRENT DISTRIBUTION: The Annamite striped rabbit occurs in the N and C Annamite Mountains along the Vietnam and Lao PDR border. In Vietnam it is believed to occur in six provinces: (from N to S) Nghe An, Ha Tinh, Quang Binh, Quang Tri, Thua Thien Hue, and Quang Nam. In Laos it is believed to occur in six provinces: (from N to S) Xiangkhouang, Bolikhamxay, Khammouane, Savannakhet, Saravane, and Xekong. There are no records from the S Annamites, but its existence in this region cannot be ruled out based on purely ecological factors. The exact westward extent of occurrence in Laos is uncertain, although the species appears to be closely associated with wet evergreen forest restricted to areas in close proximity to the border with Vietnam. Most records come from relatively low altitudes (below 1,000 m), and the species probably could utilize habitats down to sea level (although few such areas now exist). The Annamite striped rabbit has been recorded at elevations as low at 200 m asl and as high as 1,400 m asl.

TAXONOMY AND GEOGRAPHIC VARIATION: No subspecies have been described, and insufficient material is available to judge whether or not there is any substantial geographic variation.

ECOLOGY: Few, if any, studies have focused on the Annamite striped rabbit, and little is known about the ecology of this forest-dwelling lagomorph. It appears to be uncommon in some parts of its range and at low to moderate densities in other areas; this pattern probably reflects the influence of both natural and anthropogenic factors.

HABITAT AND DIET: The Annamite striped rabbit is predominantly found in wet evergreen forests with little to no dry season. Most records come from low to midelevation broadleaf forest. However, there is one record in Vietnam from montane forest and one record in Laos from hill evergreen forest with a significant Dalat pine (*Pinus dalatensis*) component.

Like all lagomorphs the Annamite striped rabbit is herbivorous; the specific diet is unknown.

BEHAVIOR: The Annamite striped rabbit appears to be nocturnal; camera trap records (n = 104 independent events) from three protected areas in the C Annamites occur exclusively at night. It also appears to be a solitary species; camera trap records from the same study recorded lone individuals, with a single exception in which a pair was recorded.

PHYSIOLOGY AND GENETICS: Assuming a constant rate of divergence, *N. timminsi* and *N. netscheri* would have diverged approximately eight million years ago and separated in the Pliocene epoch. Currently, much of the range of the Annamite striped rabbit is being determined by extracting DNA from blood obtained from leeches.

REPRODUCTION AND DEVELOPMENT: Reproductive information is lacking.

PARASITES AND DISEASES: Unknown.

CONSERVATION STATUS:

IUCN Red List Classification: Data Deficient (DD)

National-level Assessments: Not listed in the Red Book of Vietnam

MANAGEMENT: The Annamite striped rabbit is threatened by hunting, primarily by the setting of indiscriminant snares. Snaring operations range from low-level subsistence hunting to well-organized poaching gangs focused on the subnational and international trade. Wildlife products from these snaring operations end up in both the bushmeat and medicinal markets; a wide range of species are involved in both trades. Hunters also use dogs to exploit a range of mammal species, and it is likely that the Annamite striped rabbit is impacted. Given the significant levels of anthropogenic habitat alteration in Vietnam and Laos, it is probable that habitat loss and degradation have negatively affected the species from its historic population levels and distribution. However, in terms of threat severity, habitat loss and degradation are almost certainly second to hunting. The fact that there are records from heavily logged areas (not clear felled), may suggest that habitat degradation per se may not be a major threat. Although numerous protected areas have been established throughout the range of the species, many of them lack effective enforcement, and poaching is severe and widespread. There are currently no protection efforts solely focused on this species.

ACCOUNT AUTHORS: Andrew Tilker, Nguyen The Truong An, and Thomas Gray

Key References: Abramov et al. 2008; Averianov et al. 2000; Can et al. 2001; Jin et al. 2010; Schnell et al. 2012; Surridge et al. 1999.

Oryctolagus cuniculus (Linnaeus, 1758)
European Rabbit

OTHER COMMON NAMES: Wild rabbit, Rabbit; Conejo (Spanish); Coelho (Portuguese); Lapin de garenne (French)

DESCRIPTION: The European rabbit is intermediate in size and similar in shape to other leporids; generally, there is no prominent sexual dimorphism. As in other lagomorphs, European rabbits have strong hind limbs, adapted to running, and relatively long ears, although shorter than

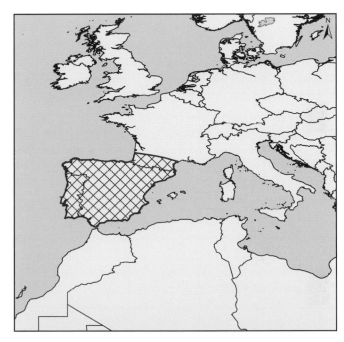

Oryctolagus cuniculus. Photo courtesy Daniel Buron / © danielburon.net

those of hares. The feet are well furred beneath and have large claws. The hind feet have four toes and the forefeet have five. The coat is normally grayish brown, with rufous fur on the neck and light gray belly fur. The ears are the same color as the coat, but lack the black tip typical of the hares that occur in Europe. The short tail is white below and brownish black above.

The European rabbit is the only lagomorph species that has been domesticated, and the only mammal domesticated in Europe. Domestication occurred during the Middle Ages, when the European rabbit was used as a source of meat, skin, and fur. More recently domestic rabbits have been widely used for commercial meat and fur production but have also been important for scientific research and as pets. Domesticated forms bear little resemblance to the original wild stock; some are small, and those bred for meat production can be very large and heavy. Domesticated forms come in a variety of coat colors.

The skull length is short, and antero-external shoulders are present on the zygoma. A single groove on each principal upper incisor tooth is not filled with cement. The skull differs from that of hares in having a wider bridge across the palate between the molar teeth.

SIZE: Head and body 340–450 mm; Tail 40–60 mm; Hind foot 70–85 mm; Ear 65–80 mm; Greatest length of skull 89.7 mm (108.4 mm in domestic forms); Weight 900–1,400 g (Iberian Peninsula). Rabbits introduced elsewhere can be larger; wild rabbits in Australia commonly approach 1,800 g adult body weight. Domestic forms can weigh more than 9,000 g.

PALEONTOLOGY: The genus *Oryctolagus* appeared in the fossil record before any other modern leporid genus. *O. layensis* was detected in the Middle Pliocene and is the first known species of the genus. To date, this ancient rabbit has only been recorded from 3.5 to 2.5 mya in Spain. No transitional forms are known between this species and the modern *O. cuniculus*. However, they appear closer to each other than to *O. lacosti*, a larger-sized rabbit species recorded between 2.5 and 0.6 mya in SW Europe. Current forms of *O. cuniculus* were first recorded in S Spain in the Middle Pleistocene (about 0.6 mya) associated with a relict, warm fauna. In the Late Pleistocene *O. cuniculus* ranged across the peri-Mediterranean area and C Europe, but during the maximum glacial period and Early Holocene was again confined to the Iberian Peninsula and probably S France.

CURRENT DISTRIBUTION: European rabbits originated in the Iberian Peninsula, where they mostly occur in Mediterranean ecosystems, and probably expanded naturally across S France. In general, they are found at low-intermediate altitudes, with some exceptions. Rabbits have been introduced in many areas worldwide, where they often have acquired a feral status, behaving in general as natural wildlife. The high plasticity of this species promoted the successful establishment on all continents, except Antarctica, as well as on more than 800 islands worldwide. European wild rabbits are present in most of Europe, where they may reach high numbers in southern countries (Spain, Portugal, and France) and Great Britain. Nevertheless, they also occur in central and northern regions within Europe (Austria, Belgium, Denmark, the Netherlands, Germany, Sweden, Poland, Hungary, and even the Ukraine). Spaniards transported European rabbits to America from the fifteenth century on, resulting in frequent introductions in wild ecosystems in both central and southern regions. Some introductions were successful, although only a few wild populations persisted in the long term in Chile and Argentina. European rabbits were also successfully introduced into Australia and New Zealand in the mid-nineteenth century, where they became major pests. Rabbits also occur in North Africa and on Mediterranean, Atlantic, Pacific, and Indian islands.

TAXONOMY AND GEOGRAPHIC VARIATION: Two subspecies: *O. c. cuniculus* and *O. c. algirus*. The divergence between the two subspecies, detected at multiple loci and at the genomic level, suggests that they have evolved independently beginning approximately 2 mya during the Quaternary glaciations. Additional data on parasitology, behavior, reproduction, and morphology support the genetic data. *O. c. algirus* is lighter, has shorter ears and hind feet, and produces a smaller litter size. The subspecies *algirus* only occurs in SW Iberia (Portugal and S Spain), in some parts of N Africa, and as introductions in the Azores, Madeira, and Canary islands, whereas *O. c. cuniculus* is present in NE Iberia and in most regions where the species has been introduced, including the rest of Europe, America, Australia, and most islands. *O. c. cuniculus* is the subspecies that has been domesticated. Currently, the two subspecies overlap in the central part of the Iberian Peninsula, where they also hybridize. The noticeable differences between both subspecies, reinforced by the recent evidence of genomic incompatibilities and partial reproductive isolation, have raised the question of whether *O. c. algirus* and *O. c cuniculus* are only nascent or already well-separated species.

ECOLOGY: In Spain, Portugal, and S France, European rabbits are strongly linked to the Mediterranean shrub land ecosystem (mosaic habitat), which is sometimes termed "the rabbit's ecosystem." Here, the rabbit is the prey for more than 30 predator species, including some of conservation concern. Among these, the highly endangered Iberian lynx (*Lynx pardinus*) and the Spanish imperial eagle (*Aquila adalberti*) rely heavily on rabbits and are considered to be specialist rabbit predators.

European rabbits also conspicuously alter plant species composition and vegetation structure through grazing, promoting the maintenance of open scrubland, which is their preferred habitat. They disperse seeds of dozens of plant species, contributing to plant population recruitment and facilitating long-distance dispersal of seeds into habitats not previously colonized. Moreover, rabbit latrines have a demonstrable effect on soil fertility and plant growth, and provide feeding resources for many invertebrate species. Rabbit warrens also provide shelter for a number of vertebrates and invertebrates.

In summary, the European rabbit acts as multifunctional keystone species in the Iberian Mediterranean ecosystem, notably affecting communities and ecosystem processes, and thus can truly be classified as an ecosystem engineer.

Beyond their native range in SW Europe, European rabbits have been proven to be a highly successful introduced species, with many and varied impacts on native communities. In those situations, their capacity as ecosystem engineers has had important negative consequences for the composition and distribution of vegetation, contributing to the decline of many endemic plant species. Introduced rabbits affect native fauna, both directly and indirectly via a range of mechanisms that may interact synergistically and result in population crashes or extinction of native species. In Australia, for example, competition with rabbits for warrens was said to be a decisive factor in the extinction of the Nalpa bilby (*Macrotis lagotis grandis*). But more important, the severe impact rabbits have on vegetation has had negative effects for those native animals dependent on native plants. Equally importantly, numerous rabbits have enabled predators such as dingoes (*Canis lupus dingo*), red foxes (*Vulpes vulpes*), and feral cats to reach high numbers and in turn prey more heavily on native animals. For all these reasons, European rabbits are considered one of the most harmful pest species in many of the areas where they have been introduced.

HABITAT AND DIET: The European rabbit shows remarkable environmental tolerance, and therefore it has an enormous ability to subsist in a wide range of habitats and conditions. Nevertheless, it generally reaches high numbers only in areas containing both cover, which can be used as refuge against predators and climatic extremes, and pastures and crops, which provide high-quality food. In the Iberian Peninsula, rabbits prefer a mixed habitat of Mediterranean oak savanna or scrub-forest and open feeding areas. Rabbits therefore benefited from traditional Iberian agriculture, in which farmers cultivated small patches of crops or pastures interspersed with areas dominated by scrubland or forest. Rabbit abundance and distribution are also strongly influenced by soil type. Soil consistency determines the ability of rabbits to dig warrens, which are vital for rabbit breeding and protection when vegetation cover is poor.

In Spain and Portugal, rabbits mostly occupy areas of typically Mediterranean climate, characterized by warm to hot dry summers and mild winters, especially where rainfall between 400 and 600 mm annually favors the winter growth of the pasture species that provide their main food. Although they may be found at almost any altitude, rabbits rarely occur above 1,500 m.

Outside their native range, rabbits can occur almost anywhere, from arid to sub-arctic regions, owing to their high environmental plasticity. In Australia, for example, European rabbits can be found in many different habitat types, ranging from deserts to coastal plains, yet they are especially abundant in arid and "Mediterranean-like" climatic regions. Moreover, in some areas, particularly in C Europe, rabbits co-exist with humans in cities, making their homes in parks and cemeteries, as well as in gardens and lawns.

Like all lagomorphs, European rabbits are generalist herbivores, as they consume a large spectrum of plant species to obtain the quantity and quality of food they require. Rabbits need ~ 100 g dry matter daily and lactating females require ~ 50% more. High energy and protein intake is achieved by highly selective grazing and browsing, making palatable or preferred plants less abundant where rabbits reach high numbers. Grasses are usually the preferred food source for the European rabbit. A feature of rabbit digestion is the importance of hind-gut fermentation of fibrous plant materials by micro-organisms. These bacteria also recycle urea into new protein, which in turn is utilized by the rabbit through the process of coprophagy, or re-ingestion of soft fecal pellets. Undigested fiber is voided as hard pellets. Up to 15% of the rabbits' protein requirements are obtained in this way, helping to maintain their high reproductive capacity.

BEHAVIOR: Unlike most of the lagomorphs, European rabbits are considered highly social animals. This is reflected in their behavior within the warren, where dominant males use chin-gland secretions to mark objects and other group members with a distinctive scent, or leap high to spray urine over other individuals. Special anal glands are also used to provide a distinctive individual odor when scent marking latrines on territorial boundaries. This behavior allows all individuals in the social group to identify each other and drive away intruders from other nearby groups.

There are important benefits from living in groups, including cooperative maintenance of the warren, diminished risk of predation through the dilution effect, and the reduction in individual vigilance for the detection of approaching intruders or predators among social group members. Social groups often hold up to four females (often genetically related) with a couple of males, one dominant, and their progeny. However, this may vary according to population density.

Males behave very aggressively toward other males, dominants being the ones that breed, while the subordinates will either live in surrounding areas or be expelled from the group. Dominant females nest in the safest part of the warren, and nest limitation will even cause subordinate females to nest peripherally or even to abandon the warren. Infanticide by females is frequent. Adults drive away their youngsters as they approach sexual maturity.

Home ranges vary in size (average ~ 0.5 ha; range 0.2–10.0 ha) depending on the distance between food patches and differences in sheltering vegetation or warren size, with smaller home ranges being found in areas with more vegetation.

European rabbits may be active throughout the day, but their maximum activity occurs during the night, and especially at dawn and dusk. Rabbit behavior is also affected by the type and structure of habitat, hunting pressure, or abundance of predators.

PHYSIOLOGY AND GENETICS: Diploid chromosome number = 44. European rabbits were one of the first mammal species to be genetically sequenced, and complete genome sequences are now available for both rabbit subspecies. Taking advantage of these genomic resources, several studies have been conducted in order to understand the genomic architecture of speciation, domestication, host-parasite co-evolution, and other processes and

mechanisms of general interest in evolutionary biology. Genetic data strongly support the separation of the two subspecies, indicating some level of reproductive isolation that results in a marked and narrow hybrid zone, bisecting the Iberian Peninsula (NW to SE). Some portions of the genome seem to cross the subspecies barrier; nevertheless, there are highly divergent regions that are exclusive for each subspecies and likely associated with low hybrid fitness. Genetic studies also have shown that rabbits have one of the most diverse mammal immune systems, in which the genetic diversity of the antibody repertoire reaches a peak two months after birth. Such information may prove important in understanding age-related immune responses to major diseases such as rabbit hemorrhagic disease (RHD).

REPRODUCTION AND DEVELOPMENT: The interaction of climatic variables, nutritional state, and light regimes (latitude) are main factors influencing the breeding season of rabbits. The annual cycle of male fertility is determined by day length, with testes size increasing with increasing day length, reaching a peak just as the main spring breeding period begins and correspondingly decreasing in the fall with decreasing day length. The female reproductive cycle appears influenced by maternal age and nutritional changes in pastures, thus factors that depend largely on rainfall and temperature. Thus, in their native range, rabbits start the main breeding period soon after the first rainfall in spring or early summer, finishing in late summer or early fall as annual pastures mature and protein intake falls. At these latitudes, females may breed at any time of the year if there is sufficient rich food available. Ovulation is induced by the stimuli associated with coitus and occurs 10–12 hours after copulation. The gestation period lasts approximately 1 month (28–31 days), but there is a postpartum estrus, in which females can be fertilized immediately after giving birth. This means that rabbits can produce 5 or more litters over as many months, with 4 or 5 young per litter (average litter size *O. c. algirus*: 3.5; *O. c. cuniculus*: 4.7; range 2–6).

Pregnant females build a nest either in a short offshoot of an established warren or in a separate breeding "stop" usually less than 1 m long, so called because the doe stops up the entrance with spoil. The female lines the nest with vegetation and her own belly fur loosened under hormonal influence several days before parturition. Newborns (~ 38 g) are blind, deaf, and hairless, and open their eyes after 7–10 days; hearing becomes functional during their second week. While fully dependent on the mother, rabbits gain weight at an approximate rate of 9 g per day.

During this time the female feeds the young only once daily for a few minutes and usually at night. Males are not involved in caring for young. The young emerge from the nest weaned at about 19 days and leave the nest permanently at day 23–25. At this age, young *O. c. algirus* weigh approximately 90 g and young *O. c. cuniculus* weigh about 125 g. In captive rabbits, and in areas where they have been introduced, young may weigh more than 200 g at weaning. Until rabbits reach sexual maturity, usually four to eight months later (depending on the date of birth), young survive poorly because of predation, diseases, territoriality, and aggression, making the survival of young one of the population-limiting factors. *O. c. algirus* females and males may reach sexual maturity at only 900 and 750 g, respectively, half the weight at maturity compared with that of introduced rabbits in some regions. Mortality rates in the first year of life frequently exceed 90%. Adult wild rabbits generally survive little more than 2 years, and very few survive beyond 6 years of age.

PARASITES AND DISEASES: The European rabbit is known to host a vast diversity of parasites. Ectoparasites include *Pulex irritans, Spilopsyllus cuniculi, Xenopsylla cunicularis, Caenopsylla laptevi, Echidnophaga iberica, Rhipicephalus pusillus, Ixodes bivari, I. frontalis, I. ventalloi, Haemaphysalis hispanica,* and *Hyalomma lusitanicum;* endoparasites include *Dicrocoelium dendriticum, Taenia pisiformis, Andrya cuniculi, Leporidotaenia wimerosa, L. pseudowimerosa, Neoctenotaenia ctenoides, Cittotaenia denticulata, Trichuris leporis, Graphidium strigosum, Nematodiroides zembrae, Protostrogylus cuniculorum, Dermatoxys hispaniensis, D. veligera, Gongylonema neoplasticum, Passalurus ambiguus, Trichostrongylus retortaeformis, T. axei, T. vitrinus,* and *Eimeria* spp. Several parasite species are found only on the European rabbit, but others are shared with other leporids or even other species.

Two viral diseases, myxomatosis and rabbit hemorrhagic disease (RHD), have had a huge impact on European rabbit populations. The myxoma virus originated from the South American tapetí (*Sylvilagus brasiliensis*), and was first observed in laboratory European rabbits in the late nineteenth century. This virus causes skin tumors, and in some cases blindness, followed by fatigue and fever. It was introduced to Australia in 1950 to control the rabbit population, and later to France, from where it became established in rabbits throughout Europe. It is now endemic in European rabbit populations, generally producing annual outbreaks.

RHD was first described in China, and rapidly spread in both wild and domestic rabbits in Europe. It was also deliberately introduced to Australia and New Zealand for controlling wild rabbit populations. A non-pathogenic virus strain and a new emerging RHD virus strain have also been detected. Recombinant viruses among the classic virus, the new strain, and the non-pathogenic form have been recently reported.

CONSERVATION STATUS:
IUCN Red List Classification: Near Threatened (NT)
National-level Assessments: Spain (Vulnerable (VU)); Portugal (Near Threatened (NT))

MANAGEMENT: In its native range, the European rabbit is an ecosystem engineer and keystone species. Additionally, rabbits are a very important game species in Spain and Portugal. During the past 50 years, rabbits have declined sharply in Iberia, mainly as a consequence of habitat loss and the incidence of two viral diseases, RHD and myxomatosis, which has had important negative consequences both ecologically and economically. This decline is continuing following the emergence of yet another RHD variant in 2012. This has resulted in a general rise in management actions aimed to increase numbers of rabbits, both for the conservation of rabbit-dependent predators and for hunting activity. These management actions include habitat management, predator control, rabbit translocations, and vaccinations, among others. Habitat management consists generally of creating feeding habitats (by providing crops or pastures or supplementary feeders in dense scrubland areas), or providing shelter through scrubland management, or the construction of artificial warrens where refuge availability for rabbits is scarce. Legal predator control aimed to increase rabbit numbers is usually carried out by hunters through shooting, cage trapping, or snaring, and these activities are mostly focused on the red fox. The illegal killing of other carnivores and raptors also sometimes takes place. Nevertheless, despite these conservation concerns there are also parts of Spain where rabbits are relatively numerous. Rabbits are largely managed by hunting in some farmland areas where the rabbit is regarded as an agricultural pest.

Elsewhere in the world, where European rabbits have been introduced, they have perturbed communities in varying manners by altering the composition and local abundance of both native flora and fauna, and have caused devastating problems for agriculture by eating crops, damaging pastures used for livestock production, and making the land more vulnerable to soil erosion. Consequently, a considerable array of control methods is used to eradicate invasive rabbits. For example, poison baits are commonly used in both Australia and New Zealand, while fumigation and physical destruction of warrens using bulldozers (warren ripping) have also been particularly effective at reducing rabbit numbers in Australia. Physical exclusion of rabbits from crops through fencing was also widely used to reduce rabbit damage areas, although less so since the introduction of biological controls such as myxomatosis and RHD.

Being an important prey species, a game animal, but simultaneously considered a pest, the European wild rabbit constitutes an intriguing conservation paradox, where the continuous monitoring and management of natural populations are mandatory.

ACCOUNT AUTHORS: Miguel Delibes-Mateos, Rafael Villafuerte, Brian D. Cooke, and Paulo C. Alves

Key References: Abrantes et al. 2012; Alves et al. 2008; Branco et al. 2002; Callou 2003; Calvete 2006; Capucci et al. 1998; Carneiro et al. 2014a, 2014b; Cooke 2012b, 2014; Dalton et al. 2012; Delibes and Delibes-Mateos 2015; Delibes-Mateos et al. 2007, 2008, 2009, 2014; Eldrige and Simpson 2002; Esteves et al. 2015; Ferrand 2008; Ferreira and Alves 2009; Ferreira et al. 2014, 2015; Flux and Fullagar 1992; Geraldes et al. 2006; Gibb 1990; Gonçalves et al. 2002; Lees and Bell 2008; Lombardi et al. 2007; López-Martínez 2008; Martins et al. 2002; Rouco et al. 2011; Schwensow et al. 2012; Silva et al. 2015; Smith and Boyer 2008a; Tablado et al. 2009; Thompson and King 1994; Villafuerte and Moreno 1997; Villafuerte et al. 1997.

Pentalagus furnessi (Stone, 1900)
Amami Rabbit

OTHER COMMON NAMES: Amami-no kuro-usagi (Japanese)

DESCRIPTION: The hair of the Amami rabbit is very dark, almost black. The soft underfur is plumbeous, and the dominant pelage is coarse and hispid, brownish black, much with buff annulations or tips, becoming mahogany on the rump. The feet are a brighter yellowish brown, except at the base of the claws. The heavy, curved claws are unusually long for rabbits (10–20 mm). This species has the most primitive characters in the Leporidae.

The skull as a whole is low and flat, not as arched as in most Leporidae. It is broad between the orbits, and the rostrum is shorter and heavier than in other Leporidae. The nasals are short and broad, and as wide in front as behind. The postorbital process consists of the posterior limb only, as is the case with red rock hares (*Pronolagus* spp.), the hispid hare (*Caprolagus hispidus*), and the

volcano rabbit (*Romerolagus diazi*). The sutures of the interparietal are obliterated. The bony plate is long. The incisive foramina are narrow, their sides approximately parallel, resembling in shape those of red rock hares, but much smaller. The zygoma is moderately heavy, its posterior free extremity is moderately long, and its antero-inferior angle is slightly enlarged but considerably flared outward. The auditory bullae are small, being even more reduced in size than they are in red rock hares. The mandible has a large, rounded angular process, which is separated from the condyle by a small, shallow notch.

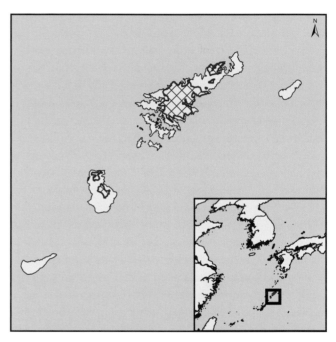

Pentalagus furnessi. Photo courtesy Hiromitsu Katsu

The bones of the upper extremity, like those of the lower extremity, are relatively shorter and stouter than in the other genera. The structure of the teeth, the relative size of the radius and ulna, and the short tarsus and metatarsus are peculiar and unlike anything in the Leporidae.

SIZE: Head and body 398–530 mm; Tail 20–35 mm; Hind foot 80–92 mm; Ear 40–50 mm; Greatest length of skull 76–85 mm; Weight 2,000–2,800 g

PALEONTOLOGY: Two upper molariform teeth (right first upper molar and left third upper premolar) reported from Tokuno-shima Island constitute the first fossil record of the Amami rabbit and date to the Late Pleistocene. Recently, fossils of *Pentalagus* spp. have been reported from Okinawa Island and date to the Early Pleistocene (1.7–1.5 mya). Fossils of *Pliopentalagus*, which is ancestral to the current Amami rabbit, are found in Huainan, Anhui Province, China, and date from the Late Miocene (approximately 6 mya) to the Late Pliocene (approximately 3 mya). Fossils of *Pliopentalagus* have also been recorded from Moldavia and Slovakia in Europe.

CURRENT DISTRIBUTION: The Amami rabbit occurs only on Amami-ohshima Island and Tokuno-shima Island in the Ryukyu archipelago in SW Japan. It occurs from near sea level (coastal cliff areas) to mountaintops (Mount Yuwan, elevation 694 m, on Amami-ohshima Island; and Mount Inogawa, elevation 645 m, on Tokuno-shima Island). The size of its distributional range, as estimated by counting fecal pellets during 2012–2014, is 364.15 km² (51% of the island) on Amami-ohshima Island, and 45.28 km² (18%) on Tokuno-shima Island. The size of its range on Amami-ohshima Island in 2012–2014 was 0.9% and 17% smaller than the estimate in 1992–1994 and in 1974, respectively. The size of its range on Tokuno-shima Island in 2012–2014 was 5% larger than its estimate in 1992–1994. It is distributed in three small fragmented populations (from N to SW: 34, 6, and 2 km²) in the north of Amami-ohshima Island and two fragmented populations (from N to S: 19 and 26 km²) in northern and southern areas on Tokuno-shima Island.

TAXONOMY AND GEOGRAPHIC VARIATION: The Amami rabbit is a monotypic species without any subspecies. It is a distinct species of rabbit based on morphological and molecular phylogenetic aspects.

ECOLOGY: The average home range size is 1.3 ha for males and 1.0 ha for females as revealed by radio-telemetry surveys. Female home ranges do not overlap, whereas male home ranges overlap with those of both other males and females. The only native predator is the habu pitviper (*Protobothrops flavoviridis*), and originally there were no other predators on both islands. However,

invasive predators such as feral dogs and cats now occur on both islands. In addition, the small Indian mongoose (*Herpestes auropunctatus*) was introduced in 1979 to Amami-ohshima Island to control the habu pitviper and non-native rats (*Rattus rattus*). However, mongooses did not prey on the habu pitvipers, but instead became major predators on the rabbits and other native species.

HABITAT AND DIET: When the islands were originally covered by dense primary forests, the Amami rabbit lived mainly in primary forests. After deforestation in the 1970s–1980s resulted in losses of 70% to 90% of primary forests, the rabbit now inhabits not only primary forests but also cut-over areas and forest edges covered by Chinese silver grass (*Miscanthus sinensis*). Rabbits are active mainly at night, moving for feeding and dropping their pellets in open places, such as forest roads where food plants are common. Most active ranges are 100–200 m away from their burrows, which are usually located in small valleys covered by dense forests.

Amami rabbits eat mainly the sprouts and young parts of plants, including cambium and mast crops from a wide range of plant species. The Amami rabbit feeds on more than 29 species of plants, including 12 species of herbaceous plants, such as common medicine plant (*Adenostemma lavenia*), *Carex* spp., *Peucedanum japonicum*, and miniature beefsteakplant (*Mosla dianthera*), and 17 species of shrubs, such as Itajii chinkapin (*Castanopsis sieboldii*), Asian melastome (*Melastoma candidum*), Molucca raspberry (*Rubus sieboldii*), Japanese snowbell (*Styrax japonicus*), and crow prickly ash (*Zanthoxylum ailanthoides*). They also feed on acorns of the Itajii chinkapin that fall to the ground during fall and winter.

BEHAVIOR: The Amami rabbit has a unique communication system that includes vocalizations similar to those

Pentalagus furnessi burrow. Photo courtesy Fumio Yamada

found in pikas (*Ochotona*) and beating the ground with their hind feet. At dusk, rabbits appear at the entrances of their burrows before they become active, and make calls that can be heard loudly and clearly in small valleys. A mother also vocalizes to attract her offspring when she approaches her nursing burrow.

PHYSIOLOGY AND GENETICS: Diploid chromosome number = 46. FN (including the X chromosome) = 80. No other leporid has the same diploid number.

REPRODUCTION AND DEVELOPMENT: Litter size is generally one, and parturition takes place during late March to May and September to December. Thus, females may give birth two or three times a year. One week prior to parturition, mothers dig burrows (entrance = 15 cm diameter, depth = 150 cm) that contain a chamber (30 cm diameter) full of leaves in the back of the burrow where the juvenile resides. Mothers visit the burrow after 20–21 hours and commence suckling the juvenile. After suckling, the entrance is quickly (within 30 seconds) covered by soil and camouflaged with twigs and leaves by the mother. Females have three pairs of nipples.

A neonate of the Amami rabbit 2 days after birth in captivity weighs 100 g, and has short brown hair on the body, closed eyes and ears, erupted incisors, and claws with white tips. By the age of 3–4 months, juveniles are 25–35 cm long. At this time they are actively rejected by their mother from both the nursing burrow and the mother's burrow, leading to their independence and ultimately to their dispersal. The gestation period and timing of sexual maturation of the Amami rabbit are unidentified.

PARASITES AND DISEASES: Amami rabbits host 13 species of eight genera of trombiculid mites and five species of three genera of ticks. *Cordiseta nakayamai* and *Walchia pentalagi* are host-specific to the Amami rabbit and can be used to distinguish active burrows from inactive ones. The trematode *Ogmocotyle* spp. and a kind of cestode (Anoplocephalidae gen. sp.) are reported as endoparasites. Amami rabbits are known to have systemic protozoal infection (probably toxoplasmosis), purulent bronchopneumonia due to gram-negative bacilli infection and fibrinous pericarditis, accumulations of foamy macrophages (suspected endogenous lipid pneumonia), focal fungal pneumonia, focal pyogranulomatous pneumonia, pulmonary abscesses, and renal abscesses.

CONSERVATION STATUS:

IUCN Red List Classification: Endangered (EN)—B1ab(ii,iii,v)+2ab(ii,iii,v); declining

National-level Assessments: Japan, Ministry of the Environment (Endangered (EN)); Nationally Endangered

Species of Wild Fauna and Flora by the Act on Conservation of Endangered Species of Wild Fauna and Flora (Ministry of the Environment, Japan); Special Natural Monument (Ministry of Education, Culture, Sports, Science, and Technology, Japan). Amami-ohshima Island and Tokuno-shima Island are candidate sites of the Amami and Ryukyu Islands UNESCO World Heritage Sites, respectively.

MANAGEMENT: Impact of habitat loss is severe for the Amami rabbit because extensive logging operations on the two islands have resulted in the area of old forests being reduced to less than 10% to 30% of their extent in 1980. Furthermore, the impact of invasive predators such as feral dogs, feral cats, and mongooses is severe, and road kills often occur. A new eradication program of invasive mongoose was restarted in 2005 by the Ministry of the Environment of Japan to protect native species, including the Amami rabbit, and the ecosystem on Amami-ohshima Island. With the eradication program, the population of mongooses shrunk (from 10,000 in 2005 to 130 mongooses in 2013) and their distribution was more limited (from 500 km^2 in 2005 to 50 km^2 in 2013). Endangered species including the Amami rabbit and rodents and amphibians have recovered dramatically. A program to complete the elimination of the mongoose on the island began in 2014.

Control of feral cats and dogs and prevention of road kills are urgent measures on both islands. However, feral cat control in habitats occupied by Amami rabbits was interrupted in October 2013 because of opposition by animal rights activists. Since then, the number of feral cats, the number of rabbits killed by feral cats, and the frequency of rabbits in feral cat scats (as well as native rodents) have all increased. If left in this state, the Amami rabbit and other endangered rodents will become extinct, especially on Tokuno-shima Island. Therefore, governments, authorities, and nongovernmental organizations are considering resuming feral cat control while gathering scientific data.

In addition, prevention of habitat loss due to deforestation is also needed. Current secondary forests on both islands will soon attain an appropriate cutting age (40 years old) following previous deforestation in the 1970s–1980s. Should these forests be harvested, connection of separated habitats by forest corridors and expansion of buffer zones of small separated habitats are necessary.

The population of the Amami rabbit on Amami-ohshima Island seems to have recovered due to the decrease of the mongoose population compared to its high level in 2002–2003. But, the current population of the Amami rabbit on Tokuno-shima Island has decreased due to the increase of feral cat predation compared to the population in 2003–2004. Population size estimated by fecal pellet counts ranged from 2,500 to 6,100 rabbits in 1993–1994 and from 2,000 to 4,800 rabbits in 2002–2003 on Amami-ohshima Island, and from 120 to 300 in 1998 and from 100 to 200 rabbits in 2003–2004 on Tokuno-shima Island.

ACCOUNT AUTHOR: Fumio Yamada

Key References: Corbet 1983; Daxner and Fejfar 1967; Fukasawa et al. 2013; Harada et al. 1985; Kubo et al. 2013; Matsuzaki et al. 1989; Matthee et al. 2004e; Nagata et al. 2009; Robinson and Matthee 2005; Sako et al. 1991; Stone 1900; Sugimura and Yamada 2004; Sugimura et al. 2000, 2014; Tomida and Jin 2002; Tomida and Otsuka 1993; Watari et al. 2013; Yamada 2002, 2008, 2015b; Yamada and Cervantes 2005; Yamada et al. 2000, 2002.

Poelagus marjorita (St. Leger, 1929)
Bunyoro Rabbit

OTHER COMMON NAMES: Uganda grass hare
DESCRIPTION: The Bunyoro rabbit is a medium-sized lagomorph with comparatively short ears. The dorsal pelage is a buffy brown, grizzled (agouti) with black hairs; hairs are whitish gray at the base, with a wide black subterminal band, a pale brown to buff terminal band, and a black tip. The underfur is grayish white. The flanks are paler, mainly buffy brown; most hairs are without black tips. The ventral pelage is yellowish buff; hairs are whitish gray on the basal half and yellowish buff on the terminal half. A white mid-ventral stripe runs from the chest (20–30 mm wide) to the lower abdomen (40–50 mm wide), extending posteriorly to the inner surface of the hind limbs; hairs are pure white. The ventral underfur is pure white. The head is similar in color to the dorsal pelage; the chin and throat are white. The ears are relatively short (~ 77% of GLS) compared with those of sympatric lagomorphs, and are similar in color to the dorsal pelage, usually without a fringe of white hair on the ear margins and with brown hairs on the inner surface; there is no black on the tips. Ears are shorter than the hind feet. The nuchal patch is rufous and does not extend to the sides of the neck. The forelimbs and hind limbs are brownish-buff; the soles of the feet have dense rufous or blackish hairs.

Pages 107–109 remade.

The tail is short and the same color as the dorsal pelage above and on the sides, while paler (often with some white hairs) ventrally. Both sexes have glandular slits on either side of the genitalia. Juveniles have a deep rufous nuchal patch; the hairs on the soles of the feet are whitish or gray.

The skull is of medium length; the minimum length of the hard palate is longer than in sympatric *Lepus*; the ratio of the mean width of the mesopterygoid space to the mean minimum length of the hard palate is low (~ 84%) and less than in *Lepus*. Antero-external shoulders are present on the zygoma. A single groove is present on each

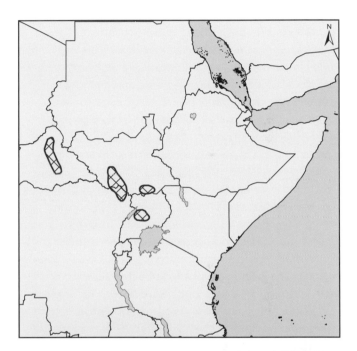

Poelagus marjorita. Photo courtesy Thierry Aebischer/Chinko Project

principal upper incisor tooth; they are not filled with cement.

SIZE: Head and body 400–605 mm; Tail 38–70 mm; Hind foot 65–108 mm; Ear 61–70 mm; Greatest length of skull 78.0–89.6 mm; Weight 2,260–3,170 g

CURRENT DISTRIBUTION: The Bunyoro rabbit is recorded from restricted areas of C and W Uganda, South Sudan, NE Democratic Republic of Congo, and NE Central African Republic. There is no evidence that the species occurs in Rwanda, Burundi, Kenya, S Chad, S Democratic Republic of Congo, and N Angola as reported in earlier publications.

TAXONOMY AND GEOGRAPHIC VARIATION: *Poelagus* is a monotypic genus, and combines features of both hares and rabbits. No subspecies. Phylogenetically, *Poelagus* is most closely related to the African red rock hares (*Pronolagus*). One hypothesis suggests that the *Poelagus/Pronolagus* clade is descended from an Asian leporid ancestor (close to *Nesolagus*) that entered Africa approximately 11.3 mya and gave rise to the extant *Poelagus/Pronolagus* assemblage.

ECOLOGY: The Bunyoro rabbit is terrestrial and primarily nocturnal. During the day, these rabbits rest alone in a form in thick vegetation. Locomotion is more similar to that of a rabbit than of a hare; this is probably because the skeleton is rabbit-like, e.g., the scapula is long and narrow (broad and "shovel-like" in hares), the ulna is sturdy (reduced in hares), the transverse processes of the lumbar vertebrae are narrow (expanded in hares), and the cervical vertebrae are short (elongated in hares). In contrast, one authority suggests that several aspects of the skull and skeleton suggest a saltatorial (rather than a cursorial) lifestyle. Predators are likely to include servals (*Leptailurus serval*), genets (*Genetta* spp.), hawks, and owls.

HABITAT AND DIET: The Bunyoro rabbit primarily inhabits woodland savanna, but also stony habitats and hills with short grass. It may also occur in forests (e.g., in South Sudan). It forages at night on flowers and sprouting grasses. It tends to prefer pastures that have been heavily grazed by larger mammals, and newly mown fields and burned areas where the grasses are sprouting. Quantitative data on diet are not available.

BEHAVIOR: The Bunyoro rabbit is probably solitary when resting in a form. It feeds at night in small groups (pairs or females with young), and may be found on rocky habitats with rock hyraxes (*Procavia* spp.).

REPRODUCTION AND DEVELOPMENT: Newborn young have been recorded in January, February, March, June, August, and October in Garamba National Park, NE DR Congo, and juveniles (weighing 185–200 g) have been recorded in January, February, May, and August. These data suggest that reproduction occurs in most (if not all) months of the year. Gestation is thought to be about five weeks. Litter size is small (one to two young). Young are born in a short burrow whose entrance is concealed by grass and soil. At birth, young are blind and helpless, with a sparse covering of short hair, as in the European rabbit (*Oryctolagus cuniculus*).

CONSERVATION STATUS:

IUCN Red List Classification: Least Concern (LC)

MANAGEMENT: In Uganda, the Bunyoro rabbit is hunted with nets and dogs. There are no specific measures to conserve this species. They occur in some conservation areas (e.g., Garamba National Park), but elsewhere changing land-use patterns have reduced the suitability of many habitats for the species. The restricted geographical distribution and rarity of sightings in recent years (compared with the early and mid-twentieth century) are cause for concern.

ACCOUNT AUTHOR: David C. D. Happold

Key References: Happold 2013; Happold and Wendelen 2006; Hatt 1940; Kingdon 1974; Kraatz et al. 2015; Matthee et al. 2004e; St. Leger 1929, 1932; Verheyen and Verschuren 1966.

Pronolagus crassicaudatus
(I. Geoffroy, 1832)
Natal Red Rock Hare

OTHER COMMON NAMES: Greater red rockhare, Natal red rockhare; Natal se Rooi Klipkonyn (Afrikaans); Lièvre Roux de Natal (French); Natal-Rotkaninchen (German)

DESCRIPTION: The Natal red rock hare is the largest of the rock hares. It has a thick, coarse pelage. The back and sides are grayish, rufous brown with black flecks. The rump is bright rufous compared to the back, and the underfur is gray. The chest, belly, and anal region are light rufous. The ears are short and lightly furred, whitish gray posteriorly and gray anteriorly. The crown and sides of the head are grayish brown. It has a gray to brown nuchal patch, with a whitish stripe that extends along the jawline to the chin. The lower cheeks, chin, and throat are grayish white to gray. The legs are rufous with the forelimbs mar-

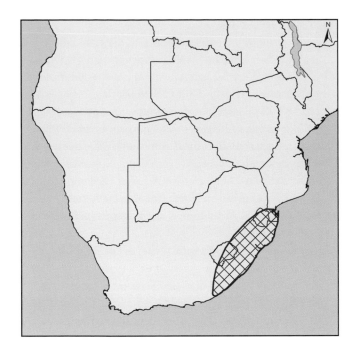

ginally shorter than the hind limbs. The tail is short and bright rufous, and lighter than that of other rock hares. The nape patch color is variable by region, ranging from gray to brown.

SIZE: Head and body 460–560 mm; Tail 35–110 mm; Hind foot 100–125 mm; Ear 60–80 mm; Greatest length of skull 85.3–94.8 mm; Weight 2,400–3,050 g

CURRENT DISTRIBUTION: The Natal red rock hare has the most limited distribution in South Africa of the rock hares. It is found along the eastern coast of South Africa (Eastern Cape and KwaZulu-Natal Provinces) extending north into the southernmost portion of Maputo Province, Mozambique. Its eastern range encompasses almost the entirety of Swaziland and the eastern half of Lesotho, and there is a minor distribution in SE Mpumalanga Province, South Africa. It inhabits areas from sea level to 1,550 m.

TAXONOMY AND GEOGRAPHIC VARIATION: *Pronolagus crassicaudatus* was originally described as being in the genus *Lepus*, and later treated as a subspecies of *P. randensis*. Previous treatments of the Natal red rock hare list five subspecies, but there is a general consensus that the validity and geographic distribution are uncertain and require revision: *P. c. crassicaudatus*, *P. c. bowkeri* (considered a subspecies of *P. rupestris* by some), *P. c. kariegae*, *P. c. lebombo*, and *P. c. ruddi*.

ECOLOGY: Known predators of rock hares in general are Verreaux's eagle (*Aquila verreauxii*), Cape eagle-owls (*Bubo capensis*), and leopards (*Panthera pardus*).

HABITAT AND DIET: The Natal red rock hare is an herbivore with a preference for sprouting grasses. It can be found inhabiting rocky kopjes and hills, as well as ravines, where grasses and shrubs grow among and at the bases of rocks. Preference for these rocky outcroppings creates a discontinuous patchwork of suitable habitat surrounded by isolating open areas.

BEHAVIOR: The Natal red rock hare will form colonies that are typically made up of a small number of individuals. Primarily nocturnal, it will seek out rock crevices during the day, occasionally resting in dense grass. It has been observed foraging for food in the late afternoon, but will rarely move far from the safety of its cover to forage. It deposits feces some distance from its daytime resting area. Rock hares are generally considered more vocal than other African lagomorphs.

PHYSIOLOGY AND GENETICS: Diploid chromosome number = 42

REPRODUCTION AND DEVELOPMENT: The reproductive season likely lasts throughout most of the year. Pregnant females have been collected in June and August, and lactating females in August, October, and February in KwaZulu-Natal. Four females had one to two fetuses.

CONSERVATION STATUS:

IUCN Red List Classification: Least Concern (LC); decreasing

National-level Assessments: South Africa (Least Concern (LC))

MANAGEMENT: The Natal red rock hare is considered to be locally abundant, especially in KwaZulu-Natal. It is hunted for subsistence and has protection as a game species under provincial nature conservation agencies. There are no known conservation measures in place for this species. Though it occurs in protected areas within its distribution, the isolated nature of its preferred habitat should be considered with regard to future management and conservation efforts. Encroaching agricultural practices and expanding urbanization are also potential concerns for the Natal red rock hare, as there has been a 21% to 50% loss of habitat since the 1900s.

ACCOUNT AUTHOR: Charlotte H. Johnston

Key References: Duthie and Robinson 1990; Happold 2013; Hoffmann and Smith 2005; Matthee et al. 2004a; Pringle 1974; Robinson 1981c; Skinner and Chimimba 2005; Taylor 1998.

Pronolagus randensis Jameson, 1907
Jameson's Red Rock Hare

OTHER COMMON NAMES: Jameson's Rooi Klipkonyn (Afrikaans); Lièvre Rouge de Jameson (French); Jamesons Rotkaninchen (German)

DESCRIPTION: Jameson's red rock hare is marginally smaller than the Natal red rock hare (*P. crassicaudatus*), the largest of the rock hares. The pelage is dense and silky; it is grizzled, rufous-brown dorsally, fading to a light rufous rump. Fur on the sides and limbs are paler than the dorsal fur. Jameson's red rock hare has a pinkish buff venter that may have white patches. The ear fur is sparse and brownish gray. The head is a silvery, brownish gray. The chin, lower cheeks, and throat are whitish. There is a rufous nuchal patch. The gular patch coloration is similar to that of the dorsal pelage. The tail is ochraceous brown with a black tip.

SIZE: Head and body 420–500 mm; Tail 60–135 mm; Hind foot 87–110 mm; Ear 80–100 mm; Greatest length of skull 85.5–96.3 mm; Weight 1,820–2,950 g

CURRENT DISTRIBUTION: Jameson's red rock hare occurs in two disjunct expanses in S Africa. The western population has a north-south distribution in W Angola, south to W and C Namibia. The eastern population occurs through C Zimbabwe, extending east with minor distribution in

Pronolagus randensis front. Photo courtesy Paul Carter

Pronolagus randensis hind. Photo courtesy Paul Carter

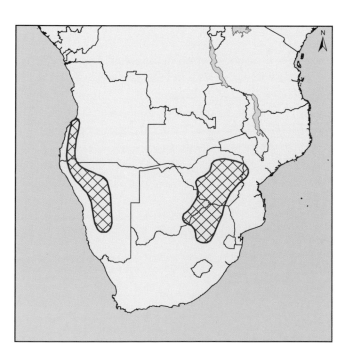

Mozambique, south into NE South Africa, and west into E Botswana.

TAXONOMY AND GEOGRAPHIC VARIATION: *Pronolagus randensis* formerly included *P. crassicaudatus* as a subspecies. No subspecies; but four subspecies were previously recognized (*P. r. randensis*, *P. r. capricornis*, *P. r. makapani*, and *P. r. powelli*), but a mtDNA analysis showed no variation among six S African locations spanning these forms.

ECOLOGY: Known predators of rock hares in general are Verreaux's eagle (*Aquila verreauxii*), Cape eagle-owls (*Bubo capensis*), and leopards (*Panthera pardus*).

HABITAT AND DIET: Jameson's red rock hare occurs on rocky hillsides, kloofs (canyons), kopjes, cliffs, and gorges. As for other rock hares, these rocky environments can be isolating owing to inhospitable landscapes between suitable habitat patches, though Jameson's red rock hare appears capable of dispersing up to 22 km to adjacent habitat. It grazes on grasses and is partial to fresh shoots after a fire. Where its distribution is sympatric with Hewitt's red rock hare (*P. saundersiae*), it occupies lower elevations.

BEHAVIOR: Rocky crevices under boulders and grass patches are used for daytime shelter by Jameson's red rock hare. Predominantly nocturnal, it will sun on rocks in the early morning and forage for food in the late afternoon. If grass cannot be found among the rocky outcroppings they occupy, Jameson's red rock hares will venture into flat surrounding areas. They create fecal middens up to 1 m in diameter. Feces are identifiable by their round, but flattened disc-like shape. Typically solitary, they will congregate in small groups of an estrous female with one to two males or a female with young. When startled they retreat up the rocky outcropping, leaping from one rocky surface to another to reach shelter.

PHYSIOLOGY AND GENETICS: Diploid chromosome number = 42

REPRODUCTION AND DEVELOPMENT: Available data from Zimbabwe suggest that Jameson's red rock hare breeds year-round. Pregnant females have been documented in July, August, and January. Average litter size is 1.1, with the number of fetuses ranging from 1 to 2.

CONSERVATION STATUS:
IUCN Red List Classification: Least Concern (LC)

MANAGEMENT: Jameson's red rock hare can be found in many national parks (like Matobo National Park, Zimbabwe) and protected areas throughout its range. In South Africa, it is afforded seasonal protection under provincial nature conservation agencies as a species hunted for both game and sport. Commercial plantations of pine and eucalyptus have contributed to the loss of habitat quality. Total habitat loss since the 1900s ranges from 21% to 50%.

ACCOUNT AUTHOR: Charlotte H. Johnston

Key References: Duthie and Robinson 1990; Happold 2013; Hoffmann and Smith 2005; Matthee 1993; Matthee et al. 2004b; Robinson 1981c; Skinner and Chimimba 2005.

Pronolagus rupestris (A. Smith, 1834)
Smith's Red Rock Hare

OTHER COMMON NAMES: Smith se Rooi Klipkonyn (Afrikaans); Lapin Roux de Smith (French); Smiths Rotkaninchen (German); Sungura ya Mawe (Swahili)

DESCRIPTION: Smith's red rock hare is a medium sized rock hare. It is rufous-brown dorsally with black grizzling, becoming bright rufous on the rump. The sides are lighter than the dorsal pelage. The limbs are bright rufous like the rump, with the hind limbs slightly lighter than the forelimbs. The ventral fur is light in contrast to that of the upper body, ranging from light rufous to whitish rufous. The tail is typically black or rusty brown with a black tip. The head is primarily gray, and the cheeks are tinged whitish.

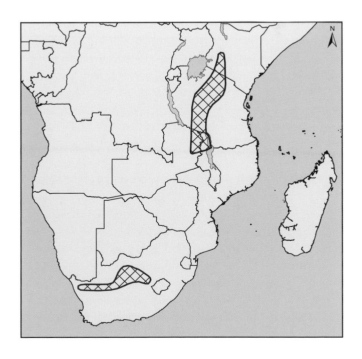

Pronolagus rupestris. Photo courtesy Ann and Steve Toon Wildlife Photography

The muzzle, forehead, and ears are tinged with brown. Gular patch coloration is comparable to that of the dorsal pelage. Smith's red rock hare has a rufous nuchal patch.

SIZE: Head and body 380–535 mm; Tail 50–115 mm; Hind foot 85–100 mm; Ear 80–105 mm; Greatest length of skull 75.1–85.3 mm; Weight 1,350–2,050 g

CURRENT DISTRIBUTION: Smith's red rock hare has two disjunct populations approximately 1,200 km apart, one found in South Africa and the other in Kenya, Tanzania, Malawi, and Zambia. In South Africa, its distribution lies in a primarily east-west orientation from C Free State Province, north to SW North West Province, and southwest into C Northern Cape Province. The range of this population previously extended into Namibia, but Smith's red rock hare is no longer thought to occur there. The northeastern population has a primarily north-south orientation from SW Kenya, south through C Tanzania, NE Zambia, and N Malawi.

TAXONOMY AND GEOGRAPHIC VARIATION: *Pronolagus rupestris* was originally described in the genus *Lepus*, and subsequently included in *P. crassicaudatus*. No subspecies. Previous treatments listed six subspecies, but the validity is considered uncertain and they are listed as synonyms: *P. r. curryi*, *P. r. fitzsimonsi* (treated as a subspecies of *randensis* by some), *P. r. melanurus*, *P. r. mülleri*, *P. r. nyikae*, and *P. r. vallicola*. Coloration varies among locations. A thorough review of the taxonomic status of the East African population is recommended.

ECOLOGY: Known predators of rock hares in general are Verreaux's eagle (*Aquila verreauxii*), Cape eagle-owls (*Bubo capensis*), and leopards (*Panthera pardus*).

HABITAT AND DIET: Smith's red rock hare is dependent on rocky environments (krantzes, hillsides, kopjes, and ravines), like other rock hares. Herbivorous, it is partial to fire-initiated, sprouting grasses. It commonly occupies lower elevations than Hewitt's red rock hare (*P. saundersiae*).

BEHAVIOR: Primarily a nocturnal species, Smith's red rock hare will forage at sunset on its rocky outcropping. Daytime shelters are sought in rock crevices. When disturbed at night adults will emit a "tu . . . tu" sound; when disturbed before sunrise they will grunt. Juveniles make a churring sound when distressed and when handled young (< 4 months) will scream. When escaping or flushed from a hiding spot, they retreat up the rocky outcropping while jumping in a zigzag manner to another hiding spot. Nests are created at the base of shrubs in open habitat.

PHYSIOLOGY AND GENETICS: Diploid chromosome number = 42

REPRODUCTION AND DEVELOPMENT: Breeding occurs from September to February, and 3–4 litters per year with 1–2 young (40–50 g each) are produced. Gestation is 35–40 days. Young are born altricial; they are nearly furless, blind (9–11 days before opening), with plugged ears at birth, and will stay in a fur-lined nest prepared by the female.

CONSERVATION STATUS:

IUCN Red List Classification: Least Concern (LC)

MANAGEMENT: Hunting for subsistence poses a threat to Smith's red rock hare, but it is seasonally protected as a game species by provisional nature conservation agencies. Habitat loss due to human encroachment could potentially become a threat to the species. It has been reported that 20% of its habitat has been lost since the 1900s. Like other rock hares, it occurs in national and provincial parks and wildlife refuges. Studies concerning its status and life history are recommended. The paucity of data for the East African population should prompt studies regarding population and habitat status, as well as potential threats.

ACCOUNT AUTHOR: Charlotte H. Johnston

Key References: Bronner et al. 2003; Duthie 1997; Duthie and Robinson 1990; Happold 2013; Hoffmann and Smith 2005; Matthee and Robinson 1996; Matthee et al. 2004c; Skinner and Chimimba 2005; Whiteford 1995.

Pronolagus saundersiae (Hewitt, 1927)
Hewitt's Red Rock Hare

OTHER COMMON NAMES: Hewitt's Rooi Klipkonyn (Afrikaans); Lièvre Roux de Hewitt (French); Hewitts Rotkaninchen (German)

DESCRIPTION: Hewitt's red rock hare is a medium sized rock hare. The pelage is thick, dense, and woolly. Dorsal coloration is grizzled brown. The sides are lighter than the back. The limbs are bright rufous, with the hind limbs

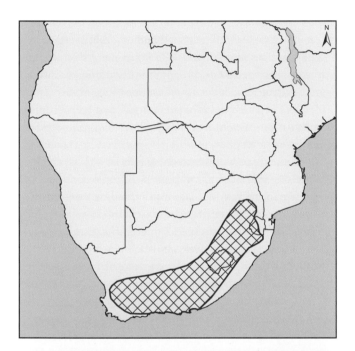

Pronolagus saundersiae. Photo courtesy Paul Carter

marginally paler than the forelimbs. The underside is light to whitish rufous. Tail color varies from sandy to red. Hewitt's red rock hare has a grayish-brown head, with cheeks that are grayish white. Nuchal patch color is rufous and the gular patch is brownish.

SIZE: Head and body 380–535 mm; Tail 50–115 mm; Hind foot 85–100 mm; Ear 80–105 mm; Greatest length of skull 70.6–91.9 mm; Weight 1,350–2,050 g

CURRENT DISTRIBUTION: Hewitt's red rock hare occurs in South Africa, Lesotho, and S Swaziland. Specifically, its distribution in South Africa encompasses C Western Cape Province, east to Eastern Cape Province excluding the coastal region, S Free State Province, W KwaZulu-Natal Province, and south C Mbabane Province in Swaziland.

TAXONOMY AND GEOGRAPHIC VARIATION: *Pronolagus saundersiae* was initially assigned as a subspecies of *P. crassicaudatus* and later included in *P. rupestris*. *P. saundersiae* and *P. barretti* had been identified as distinct groups; however, mtDNA and morphological data show only *P. saundersiae* to be a true species, with *barretti* considered a junior synonym. There are no subspecies.

ECOLOGY: Known predators of rock hares in general are Verreaux's eagle (*Aquila verreauxii*), Cape eagle-owls (*Bubo capensis*), and leopards (*Panthera pardus*).

HABITAT AND DIET: The habitat of Hewitt's red rock hare is similar to that of other rock hares: isolated rocky kopjes and hillsides. There are no data specifically regarding the diet of Hewitt's red rock hare. It is presumed to have preferences similar to those of Smith's red rock hare (*P. rupestris*), being partial to consuming grasses found among its rocky habitat. It typically occupies higher elevations than Smith's red rock hare.

BEHAVIOR: Hewitt's red rock hare is presumed to be a nocturnal forager that does not stray far from its daytime shelter, like Smith's red rock hare. Boulders and rock crevices are utilized for sheltering.

PHYSIOLOGY AND GENETICS: Diploid chromosome number = 42

REPRODUCTION AND DEVELOPMENT: Reproduction is likely to mirror that of *P. rupestris*, breeding from September to February with 3–4 litters per year, litter size 1–2, and gestation lasting 35–40 days.

CONSERVATION STATUS:

IUCN Red List Classification: Least Concern (LC); decreasing

MANAGEMENT: Hewitt's red rock hare was only recently separated from Smith's red rock hare. As such, there are no studies specific to this species. Research regarding all aspects of its natural history, current population status, and identification of potential threats is needed. While not considered a threat to the species as of yet, habitat loss due to commercial plantations has been ongoing and is expected to continue in the future.

ACCOUNT AUTHOR: Charlotte H. Johnston

Key References: Duthie 1997; Happold 2013; Matthee and Robinson 1996; Matthee et al. 2004d; Skinner and Chimimba 2005; Whiteford 1995.

Romerolagus diazi (Ferrari-Pérez, 1893)
Volcano Rabbit

OTHER COMMON NAMES: Zacatuche, Teporingo, Burrito, Conejo de los volcanes (Spanish); Lapin des volcans (French)

DESCRIPTION: The volcano rabbit has short hind feet, unusually small and rounded ears, and a tail so short as to be externally invisible. The dorsal pelage is a yellowish brown, and individual hairs are black at both the tip and the base, which results in a grizzled appearance. The ventral pelage is soft, very short, and dense, and there is no contrast in color with the dorsal pelage. The volcano rabbit does not have a distinctive winter and summer pelage.

The fur is molted throughout the year in four stages: hair loss in random areas, disappearance of melanin, melanin deposition in the bare areas, and regrowth of the hair. The footprint leaves an impression of four fingers.

The first lower premolar is divided into two sections by two reentrant folds. The external auditory meatus of the volcano rabbit is relatively larger than in other lagomorphs, it has a smaller than average sternum, and the remainder of the pre-sternum is long and narrow.

SIZE: Head and body 270–315 mm; Tail 18–31 mm; Hind foot 42–55 mm; Ear 40–45 mm; Greatest length of skull 45–47 mm; Weight 386–602 g

PALEONTOLOGY: The volcano rabbit is thought to have diverged during the Late Eocene, 35–38 mya. Recent molecular data indicate that most rabbit and hare genera arose from a single rapid diversification event during the Miocene (between 12 and 16 mya). Although no fossil data are available to indicate when *Romerolagus* diverged from other leporids, it is considered by many scientists to be the most primitive of living rabbits and hares. Its chromosome evolution must have ceased only at the end of the Pleistocene. Unlike most lagomorphs, the volcano rabbit retains an ancestral genetic polymorphism. This set of atypical morphological features, including the rudimentary fusion of vertebrae, peculiar dental pattern, and ancestral genetic characteristics, highlights the primitive status of the volcano rabbit.

CURRENT DISTRIBUTION: The volcano rabbit is endemic to Mexico. The likely ancestral geographic range was ap-

Romerolagus diazi. Photo courtesy Andrew T. Smith

parently limited to two ranges, namely, the Sierra Nevada and the Sierra Chichinautzin. Currently it is restricted to the Transverse Neovolcanic Belt, where its distribution is discontinuous and limited to four volcanoes (Popocatepetl, Iztaccihuatl, El Pelado, and Tlaloc) spanning approximately 386 km². The volcano rabbit occurs between 2,800 and 4,250 m in elevation, but is found at the highest density between 3,150 and 3,400 m. The species has apparently disappeared from some of its historical range in the C Transverse Neovolcanic Belt, including the eastern slopes of Iztaccihuatl. Our understanding of the current distribution is incomplete. Recent landscape surveys have allowed a detailed cartographic representation of 12 occupied distribution patches covering 18,410 ha. Two of these distribution areas have recently become reconnected: the Malacatepec-Quepil-Pelado volcanoes and the Iztaccihuatl-Popocatepetl volcanoes. In other patches, populations have either expanded or contracted.

TAXONOMY AND GEOGRAPHIC VARIATION: In spite of the disrupted distribution pattern, there is no evidence of recognized subspecies of *R. diazi*. However, rigorous research to document the current genetic population status has not yet been conducted. Paleoecological studies have suggested that the Sierra Nevada and Sierra Chichinautzin have been unconnected for at least the past 300 years, and this emerging evidence may explain the disruption and likely absence of the volcano rabbit from Volcano Xinantecatl (also known as Nevado de Toluca).

ECOLOGY: The abundance of volcano rabbits varies significantly among vegetation communities. Subalpine bunchgrass (*Festuca tolucensis*) and Hartweg's pine (*Pinus hartwegii*) communities are the habitats with the highest rabbit densities. Volcano rabbits give birth in the grass, whereas in rocky terrain, where bunchgrasses are less abundant, they are less abundant as they must use cracks as burrows. The close relationship between this rabbit and bunchgrass has been appreciated from the time of the Aztecs, as evidenced by its Aztec name "zacatochtle"—from *zacatl* = *zacate* (bunchgrass) and *tochtle* = *conejo* (rabbit). This Aztec name, slightly modified to "zacatuche," is still widely used.

Groups of three to five volcano rabbits are commonly recorded within this subalpine bunchgrass, often regarded as a habitat patch encompassing 5 ha. Density varies from patch to patch, but on average a high-quality patch contains a maximum of five animals, whereas poorer quality habitat may support only two animals in a group.

There are no up-to-date population estimates of volcano rabbits for the entire distribution area, but it is believed that the total population ranges from 11,000 to 25,000 individuals. The volcano rabbit's ecology differs from that of many other lagomorphs. Volcano rabbits are highly selective in their habitat choice, whereas most widely distributed rabbit genera such as the European rabbit (*Oryctolagus cuniculus*) and cottontails (*Sylvilagus*) occupy a greater variety of habitats.

Rattlesnakes (*Crotalus* spp.), red-tailed hawks (*Buteo jamaicensis*), weasels (*Mustela* spp.), bobcats (*Lynx rufus*), and coyotes (*Canis latrans*) are regarded as the main predators of the volcano rabbit. It has been observed that predation rates vary significantly among habitat types. Around villages, feral dogs also prey extensively on the volcano rabbit.

HABITAT AND DIET: The volcano rabbit is a habitat specialist. Its diet largely depends on three species of native grasses: *Festuca amplissima*, *Muhlenbergia macroura*, and *Stipa ichu*. The volcano rabbit feeds on the young leaves of grasses and some spiny herbs. The nutritional intake of these species has been analyzed and the results show that these sources of food are inadequate to provide the nutrition necessary to maintain the stable body condition observed in most wild populations. Thus, other species must complement the diet, and detailed studies must be conducted to fully understand the volcano rabbit's nutritional requirements and forage selectivity.

BEHAVIOR: Volcano rabbits gather in colonies of two to five animals that live together in burrows which are small depressions hidden in the bunchgrass or cracks. Burrows are up to 5 m long and extend up to 40 cm under the surface. The species' social organization in the wild has yet to be studied. It is reported that social dominance hierarchies form among groups of six individuals (two males, four females) housed in semi-natural enclosures, with only one male and one or two females breeding in each group. Captive females are clearly aggressive toward both sexes, whereas males were never observed to initiate aggression toward a female. Female-female aggression is much more frequent and violent than female-male aggression. Dominant individuals are always females. Parental care is still largely unknown. Newborn volcano rabbits are curious explorers and dare to approach open habitats in order to get fresh young grass leaves. In nature, groups of young volcano rabbits play joyfully in groups far from thick bunchgrass habitats where predation risks are higher.

The volcano rabbit's peak hours of activity are during the morning and late afternoon. Foreign individuals are never allowed into the group burrow system. Like pikas, volcano rabbits use acute and strident calls to alert other members in their burrow of a potential danger.

PHYSIOLOGY AND GENETICS: Diploid chromosome number = 48 (FN = 78); in this respect the volcano rabbit resembles hares and differs from rabbits.

The speediness and hind limb development relative to body size in volcano rabbits correlate with their need for evasive action. They are relatively slow and vulnerable in open habitats; therefore, they take comfort in thick vegetation. They also have difficulty breeding in small enclosures. Volcano rabbits have been bred in captivity, but there is evidence that the species loses a significant amount of genetic diversity when it reproduces in such conditions. A comparative study done on wild and captive volcano rabbits found that the latter lost a substantial amount of DNA loci, and some specimens lost 88% of their genetic variability. There was, however, one locus whose variability was higher than that of the wild population.

REPRODUCTION AND DEVELOPMENT: Volcano rabbits breed throughout the year, although they are more sexually active during the summer. The nests, hidden small depressions in the brush and covered with skin and bits of plants, are built only from April to September. The period of gestation lasts 39 days, and a female can have up to 3 pups per litter at most. Young are born with hair and eyes closed and they cannot move or feed themselves for the first three weeks of life.

PARASITES AND DISEASES: Volcano rabbits have a great variety of specific ecto- and endoparasites such as nematodes (*Boreostrongylus romerolagi*, *Dermatoxys romerolagi*, and *Lamothiella romerolagi*), cestodes (*Anoplocephaloides romerolagi*), fleas (*Cediopsylla tepolita* and *Hoplopsyllus pectinatus*), and mites (*Cheyletiella mexicana* and *C. parasitovorax*). *C. mexicana* is host-specific, and the occurrence of these two mites on one host species is unique because otherwise species of *Cheyletiella* exclude each other. The volcano rabbit is a relict species that is characterized by its ancestral morphological features and primitive parasites.

CONSERVATION STATUS:

IUCN Red List Classification: Endangered (EN)—B2ab(i,ii,iii,v)

CITES: Appendix I

National-level Assessments: Mexico (Endangered—Mexican Official Norm NOM-059-ECOL-2010 [SEMARNAT])

MANAGEMENT: It is illegal to hunt volcano rabbits according to Mexican law, but this legislation is poorly en-

forced. The volcano rabbit occurs within several protected areas: Izta-Popo, Zoquiapan, and Ajusco-Chichinautzin National Parks; but hunting, grazing, and grass burning persist within the park boundaries. Captive breeding programs have been established with some success, but infant mortality in captivity is very high and the species appears highly prone to inbreeding depression. It has been recommended that conservation measures focus on habitat management, particularly the control of burning and overgrazing of the bunchgrass "zacaton" habitat, and enforcement of the existing laws prohibiting the hunting and trade of volcano rabbits. Captive colonies, especially those in the zoos of Mexico, D.F., should be used to educate the public about the protected status of the volcano rabbit. Conversely, during the past decade, local inhabitants of Milpalta, a village within the range of the volcano rabbit, have taken the lead in enforcing conservation measurements to protect their lands. This includes protecting bunchgrass and pine forest habitats. Environmental concern was gained largely because a number of workshops were conducted, aimed at empowering local communities with scientific evidence of the large number of endemic and endangered species harbored on their lands. The contribution of local communities in protecting Mexican lagomorphs has been documented as a new opportunity rather than a threat. The volcano rabbit is recognized as a conservation icon for the significant number of endemic, sympatric, and endangered species sharing its habitat.

Another concern is the lack of law enforcement in the region inhabited by the volcano rabbit, such that criminal actions have increased significantly during the past 20 years. Large areas originally devoted for leisure by inhabitants of people from towns and cities such as Cuernavaca, Puebla, Cholula, and Mexico City have been abandoned. This has also changed the land-use patterns since even sheepherders are reticent to continue doing what they used to for at least two centuries. Research activities have also diminished significantly.

On a positive note, some native habitat conditions have improved. This is well illustrated by the reconnected set of volcanoes Malacatepec and Pelado. In the Sierra Nevada, the reconnection is largely explained by the ongoing volcanic activity of Popocatepetl. Visitors, climbers, and herders are no longer allowed in the neighborhood areas. As a result, native habitat conditions have improved.

The most striking management practice significantly impacting the quality of the habitat of the volcano rabbit concerns the digging of ditches presumably for enlarging water infiltration. This practice was implemented by the National Commission of Protected Areas (Comisión Nacional de Áreas Naturales Protegidas, or CONANP) on most volcanoes around Mexico City and adjacent areas. Indisputable evidence has shown that these ditches have not increased water infiltration. Similarly, CONANP has launched a large reforestation program in native alpine grasslands where pines have not grown in the past 30,000 years. These ditches and reforestation practices have neglected existing scientific knowledge and discourage conservation actions launched by local communities. These communities have become the most promising allies in long-term conservation when they engage in sustainable land-use practices. In this way, protected areas are foreseen as machines of local empowerment and resources as a heritage of local stakeholders.

Taking into account the foreseen natural events such as current volcanic activity of Popocatepetl and man-made threats such as global climate change, urban expansion, and land tenure conflicts, the population size of volcano rabbits may be reduced to half within the next 20 years. This scenario would represent the shrinking of the species' distribution areas to 7 from the current 15. It is indisputable that population trends will decrease in the coming years and therefore sustainable conservation and reliable monitoring actions are critical.

ACCOUNT AUTHORS: Alejandro Velázquez and Fernando Gopar-Merino

Key References: Bell et al. 1985; Bray and Velázquez 2009; Cabrera-García et al. 2006; Cervantes et al. 1990; Fa and Bell 1990; Hoth et al. 1987; Lozano-Garcial and Vazquez-Selem 2005; Ong 1998; Pérez-Amador et al. 1985; SEMARNAT 2010; Velázquez 1993, 1994, 2012; Velázquez and Romero 1999; Velázquez et al. 1993, 1996, 2001, 2003, 2011.

Genus *Sylvilagus* Gray, 1867

In contrast with other rabbits, cottontail rabbits (genus *Sylvilagus*) are confined to the Americas. The 18 species are remarkably similar in general morphology: small and compact, long ears (longer than those of pikas and shorter than those of hares), and generally sporting a round fluffy tail. Their fur is usually soft and pliable. The many species occupy a variety of habitats, as indicated in the accounts that follow.

Sylvilagus aquaticus (Bachman, 1837)
Swamp Rabbit

OTHER COMMON NAMES: Swamper, Cane cutter, Cane jake

DESCRIPTION: The swamp rabbit is the largest form of the genus *Sylvilagus* in North America and is not sexually dimorphic. The dorsal pelage is grayish to rusty brown and is heavily grizzled with black. The sides are slightly

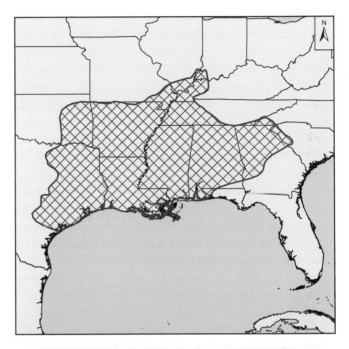

Sylvilagus aquaticus. Photo courtesy Greg W. Lasley

paler than the back, and the belly and the underside of the tail are white. The inside of the ear is naked and pinkish, between the ears there is a black line or spot, and the nape patch is orange in color. Although highly similar in appearance to the eastern cottontail (*S. floridanus*), the swamp rabbit has a larger and more robust skull with relatively shorter and rounder ears. The dorsal portion of the hind legs and feet is distinctly orange rather than tan or white like the eastern cottontail's.

The skull is larger than those of most congeners and has a wide rostrum. The auditory bullae are smaller than the foremen magnum. Projections of the supraorbital process are entirely fused to the skull or lacking. The premolar foramina are present in adults and symmetrically or slightly asymmetrically paired at the third premolar or third-fourth premolar level. The inter-parietal is distinctly separated with sutures.

SIZE: Head and body 450–550 mm; Tail 50–74 mm; Hind foot 90–113 mm; Ear 60–80 mm; Greatest length of skull 84–100 mm; Weight 1,600–2,700 g

PALEONTOLOGY: The Laurentide Ice Sheet extended as far south as the northernmost portion of the range of the swamp rabbit during the Illinoian Stage 0.191–0.130 mya. During this time and subsequent glaciations, the Lower Mississippi Valley may have served as a refugium for displaced taxa including the swamp rabbit. Swamp rabbit fossil teeth found in Missouri date to the Late Pleistocene (0.126–0.012 mya). Following Pleistocene glaciations, sea levels rose, drainage along the Gulf of Mexico decreased, and a zone of secondary contact occurred among mostly aquatic and semi-aquatic organisms that inhabited the Florida peninsula and those inhabiting the greater part of the continent. The approximate location of this contact zone among closely related taxa is in W Georgia, which coincides with the divide between the current range of the swamp rabbit (continental, west of the Appalachian Mountains) and its closest relative, the marsh rabbit (*S. palustris*; peninsular, east of the Appalachian Mountains).

CURRENT DISTRIBUTION: Swamp rabbits are found in the lowlands of the SE United States; their western border is E Texas and E Oklahoma. They are found in most of Arkansas, Louisiana, Alabama, and Mississippi. The northern border of their range is southernmost Missouri, Illinois, and Indiana. In the east they occur in westernmost Kentucky and W Tennessee, and extend farther southeast into small portions of W South Carolina and W North Carolina and occupy the northern half of Georgia. The southeastern edge of their range extends as far as the westernmost portion of the panhandle of Florida.

TAXONOMY AND GEOGRAPHIC VARIATION: Two subspecies: *S. a. littoralis* (which occupies a narrow band of marshes along the coast of the Gulf of Mexico from Mississippi to Texas) and *S. a. aquaticus* (which inhabits the rest of the range). Cranial characteristics, cytogenetics, and molecular phylogenies support the close taxonomic relationship between *S. aquaticus* and *S. palustris*. Molecular phylogenies group the swamp rabbit and the marsh rabbit with cottontail rabbits in the monophyletic genus *Sylvilagus*, though higher-order relationships within this genus have not yet been resolved.

ECOLOGY: The annual home range size of the swamp rabbit is from 1 to 12 ha, depending on habitat. Home ranges are largest during the winter-spring period. Swamp rabbit density in timber habitat was estimated at two rabbits/ha. Predation by avian and mammalian predators is the primary cause of mortality; annual survival rate is 42%. Seasonal survival is lowest during the winter-spring period. Swamp rabbits are especially vulnerable to predators during periods of deep snow in the northern portion of their range. They will rarely cross large open spaces, even during flooding events. Swamp rabbits co-exist with eastern cottontails and other leporids in portions of their range, but habitat use usually differs among species when in sympatry.

HABITAT AND DIET: The swamp rabbit lives in bottomland hardwood forests, swamps, wetlands, river bottoms, and lowland areas. It is always associated with water as this is its primary route of escape from predators. In the northern portion of its range it is found in mature forests, regenerating forest tracts that are at least 15 years old, and canebreak communities. In the southern portion it is found in swamp forest communities that include coastal and riparian areas that are about 24°C isotherm. Within their preferred landscape, habitat use by swamp rabbits is determined primarily by site-level factors. Swamp rabbits prefer dense ground-level vegetation generally found in early successional forests; canopy gaps are increasingly important as forests mature. The species consumes a variety of plants and herbs, including sedges (*Carex*), grasses, cane (*Arundinaria*), blackberry (*Rubus*), hazelnut (*Corylus avellana*), poison ivy (*Toxicodendron radicans*), deciduous holly (*Ilex decidua*), and in some cases tree seedlings. It actively feeds at dusk and is coprophagous.

BEHAVIOR: Once thought to be territorial, swamp rabbits appear to tolerate moderate levels of home range overlap with conspecifics. The swamp rabbit's social structure consists of a linear dominance hierarchy among the males established in late winter and early spring. Fighting is rare while dominance displays, such as charging and paw raking, and submissive displays, such as submissive posture and retreat, are common. There is a direct relationship between the male's social status and the frequency of dominance displays, with the majority of dominance displays directed at the next lower-ranking male. Males practice "chinning," which is the marking of a territory through the spreading of pheromones by rubbing glands that are located on the bottom of the chin on objects. Higher-ranking males move more among female ranges, and the alpha male is responsible for the most copulations.

During spring, swamp rabbits form definite breeding groups whose territories are defended by a dominant male. Subordinate males are tolerated in the breeding group's range when females are not present. Females are generally mutually tolerant of one another within and among breeding groups and do not exclude one another from occupied areas. When confronted by a male, females often adopt a threat posture: crouched with ears pinned back and chin raised. If rushed by the male, females will stand on their hind feet and box with their forefeet in an aggressive response. Females also engage in a variable and repetitive jump sequence in which the male rushes at the female and she jumps over the male sometimes urinating at the height of her jump. The length of these sequences appears to increase with male arousal, but do not result in copulation. Female swamp rabbits exhibit synchronous estrus behavior on a 12-day cycle.

Swamp rabbits are quite vocal and produce squeaks, chirps, squeals, and alert calls during social interactions. When handled, they often grunt and may make loud distress cries. Live-trapped swamp rabbits are more difficult to handle than similar eastern cottontails and often strike out with their hind feet. Swamp rabbits use downed logs and similar raised objects for repeated defecation. It has been suggested that this behavior communicates social information or provides a better view of their surroundings. Swamp rabbits are active at dawn and dusk during spring and summer, but tend to be mostly nocturnal during the winter months. During the day they rest in and on forms such as dense tangles of vines, heaps of brush, or logs. Swamp rabbits take to the water when pursued. They dive and swim very well, sometimes with little more than their eyes and nose above the water's surface.

PHYSIOLOGY AND GENETICS: Diploid chromosome number = 38. The swamp rabbit exhibits relatively low genetic diversity at the northernmost portion of its range potentially due to small effective population sizes and fragmented habitat. Genetic data indicated no evidence of hybridization in the wild or genetic introgression with the more common eastern cottontail. Dramatic flooding events may cause embryo resorption due to increased adrenal stress hormones. Otherwise, little information currently exists regarding the physiology of swamp rabbits.

REPRODUCTION AND DEVELOPMENT: In general the swamp rabbit breeding season lasts from late January to August, although there is slight variation throughout its range. Southern areas may experience a year-round breeding season. The gestation period is around 37 days. Litter sizes vary from one to six and there are two to five litters per year. Young are born nearly hairless with their eyes tightly closed. Females will create nests near or under fences, at the bases of trees (often in hollows), in brush and lumber piles, or in abandoned buildings. Females nurse the young in the nest during dawn and dusk. It has been noted that orphaned rabbits from other nests are often adopted by other females. Little is known about juvenile recruitment in wild populations.

PARASITES AND DISEASES: Several parasites have been reported in swamp rabbits, including trematodes (*Hasstilesia texensis*, *H. tricolor*), cestodes (*Cittotaenia ctenoides*, *C. variabilis*, *Multiceps serialis*, and *Raillietina stilesiella*), and nematodes (*Graphidium strigosum*, *Nematodirus leporis*, *Obeliscoides cuniculi*, *Passalurus ambiguus*, *Trichostrongylus calcaratus*, and *Trichuris leporis*). Tularemia is a major disease contracted by the swamp rabbit.

CONSERVATION STATUS:

IUCN Red List Classification: Least Concern (LC)

MANAGEMENT: The swamp rabbit is hunted for sport, meat, and fur, and is of interest to state conservation agencies. Hunting does not appear to significantly affect swamp rabbit populations. Populations declined considerably during the twentieth century because of habitat loss and fragmentation due to agricultural development. The alteration of habitat resulting from hydrologic projects also constitutes a threat to the species as these changes can increase the frequency and severity of flooding within suitable habitat for these rabbits. Swamp rabbits are classified as State Endangered in Indiana and a Species of Special Concern in Missouri and South Carolina.

ACCOUNT AUTHORS: Clayton K. Nielsen and Leah K. Berkman

Key References: Allen 1985; Berkman et al. 2009, 2015; Bunch et al. 2012; Chapman and Feldhamer 1981; Chapman and Flux 2008; Chapman and Litvaitis 2003; Conaway et al. 1960; Crawford 2014; Elbroch 2006; Fostowicz-Frelik and Meng 2013; Fowler and Kissell 2007; Hunt 1959; Kjolhaug and Woolf 1988; Lowe 1958; Marsden and Holler 1964; Matthee et al. 2004e; Platt and Bunch 2000; Robinson et al. 1983b, 1984; Roy Nielsen et al. 2008; Scharine et al. 2009, 2011; Schauber et al. 2008; Scheibe and Henson 2003; Schwartz and Schwartz 2001; Soltis et al. 2006; Vale and Kissell 2010; Watland et al. 2007; Wilson and Ruff 1999; Zollner et al. 1996, 2000a, 2000b.

Sylvilagus audubonii (Baird, 1857)
Desert Cottontail

OTHER COMMON NAMES: Audubon's cottontail

DESCRIPTION: Desert cottontail rabbits are mid-sized relative to other species of the genus, but are substantially smaller and weigh less than their cousins the eastern cottontail (*S. floridanus*) and the New England cottontail (*S. transitionalis*). Their hair color is typical of the genus with salt and pepper gray mixed with brown pelage across their backs and sides and a whiter underbelly. The ears are long relative to head size, but not exceptionally so as in some other species of lagomorphs, e.g., the antelope jackrabbit (*Lepus alleni*). The backs of the ears are covered with short hair, while the insides only have a sparse short covering of hair. There are no strong color markings on the face except for a small amount of white hair around the nares and a lighter brown ring around the eyes. Typical of this genus, the tail is bicolored with brown-gray

above and white below. The hind feet are large compared to the body, as is common for most lagomorphs.

The rostrum is long, stout, and triangular in shape, and the skull has a prominent (upturned) supraorbital process. The nasals are rectangular and widest posteriorly. The tympanic bullae are much inflated.

SIZE: Total length 372–397 mm; Tail 51–56 mm; Hind foot 88–90 mm; Ear 70–73 mm; Greatest length of skull 53 mm; Weight 755–1,250 g

PALEONTOLOGY: Fossil specimens of *S. audubonii* are well documented from Pleistocene deposits from Rancho La Brea in California and likely from Texas. More recent excavations in Colorado have identified abundant remains of possible desert cottontails from 1,000-year-old archaeological sites. The abundance of the remains found indicates that likely through the history of humans in the SW United States this genus was an important resource.

CURRENT DISTRIBUTION: On the eastern edge of its range, the desert cottontail is found across the western Great Plains from just south of the Canadian border in the north to south of Mexico City in Mexico. On the western edge, it extends from C California, Nevada, and Utah south through the Baja Peninsula and along the Sonoran coast of Mexico. The species is found throughout the Chihuahuan and Sonoran Deserts. Desert cottontail rabbits can be found from below sea level in Death Valley up to elevations of at least 1,800 m above sea level.

TAXONOMY AND GEOGRAPHIC VARIATION: Eleven subspecies: *S. a. audubonii*, *S. a. arizonae*, *S. a. baileyi*,

Sylvilagus audubonii. Photo courtesy David E. Brown

S. a. cedrophilus, S. a. confinis, S. a. goldmani, S. a. minor, S. a. neomexicanus, S. a. parvulus, S. a. sanctidiegi, and *S. a. vallicola.* The subspecies *S. a. parvulus* and *minor* range throughout most of the Chihuahuan Desert, and *S. a. arizonae* is found throughout most of the Sonoran Desert north to inland S California, W Arizona, and S Nevada. *S. a. confinis* occupies most of the Baja Peninsula, and *S. a. neomexicanus* and *baileyi* occupy the southern and north-eastern edges, respectively, of the species within the United States. The other subspecies have more limited distribution along the Pacific coast (*S. a. audubonii, sanctidiegi,* and *vallicola*), the S Sonoran Desert (*S. a. goldmani*), and the four corners region (*S. a. cedrophilus*).

ECOLOGY: Desert cottontails can be consumed by a wide variety of avian and mammalian carnivores, including coyotes (*Canis latrans*), bobcats (*Lynx rufus*), badgers (*Taxidea taxus*), red-tailed hawks (*Buteo jamaicensis*), golden eagles (*Aquila chrysaetos*), and horned owls (*Bubo virginianus*). Reptilian predators include rattlesnakes (*Crotalus* spp.) and gopher snakes (*Pituophis catenifer*). However, because of their relative rarity on the landscape, the proportion desert cottontails make of individual carnivore species diets is often low, generally less than 5%. Bobcats, however, appear to rely more on cottontails, possibly because bobcats hunt more often in areas of dense cover, where cottontails are most likely to be found. As desert cottontails are not normally important food items in their predators' diets, top-down influences on their abundance are not clear. At least one study in Texas reported that the removal of coyotes did not affect rabbit densities.

Desert cottontail densities have been found to be as low as 0.02 individual/ha in the Chihuahuan Desert to 4.7/ha in C California. Densities however, can fluctuate widely, with some estimated densities as high as 24/ha at a C Chihuahuan Desert study site. Data from New Mexico and Arizona indicate a general decline in abundance in that region starting around the 1980s. However, factors causing the decline are unclear.

Annual changes in rabbit densities appear to be influenced by precipitation patterns, which in turn affect levels of plant productivity. However, the strength of that bottom-up relationship is uncertain. In S New Mexico, the relationship between total above-ground primary productivity and rabbit abundance appeared relatively weak. In contrast, in the central part of the Chihuahuan Desert in Mexico, rabbit abundance was highly related to annual above-ground forb productivity of the previous year.

Because of the uncertainty of cause and effect in an-

nual density changes, reasons for the long-term large-scale declines being noted in rabbit abundance are difficult to identify. Desert cottontail rabbits appear to not be affected by low to moderate cattle grazing. Intensive housing developments in land use across the Southwest very likely have impacted the amount of habitat available for desert cottontails. However, studies show that they prosper in low-density exurban housing developments. In contrast, several studies report that drought conditions can depress numbers significantly. As herbaceous species abundance appears to be highly sensitive to precipitation amounts and is correlated with rabbit abundance, reductions in the average annual precipitation over the past 10 years in Arizona and New Mexico could be contributing factors to the declines seen there. If this is the case, then global climate change, if it results in reduced precipitation, could have long-term impacts on regional desert cottontail populations within its range.

Home range sizes of desert cottontails can vary greatly depending on habitat variables such as shrub cover. Reported home range sizes of males have been up to 6.1 ha. Females inhabit much smaller areas of around 0.5 ha.

HABITAT AND DIET: Because of their limited running capability compared to that of jackrabbits (*Lepus* spp.), desert cottontails prefer habitat with a higher density of cover, rarely venturing out into more open areas. Within this dense cover, they use mid-story plants such as triangle-leaf bursage (*Ambrosia deltoidea*) and prickly pear cactus (*Opuntia* spp.) for their forms (depressions). It appears that these mid-story plants provide more cover than other species such as creosote (*Larrea tridentata*), which they rarely use. The selection of denser habitats is related to predator avoidance in terms of both reduced detection and escape, as this species tends to rely on cover rather than speed to escape predators. Such denser vegetation would also provide some protection from adverse weather conditions.

The diet of desert cottontail rabbits consists of a mixture of grass, forb, and shrub species. In Arizona, 43 plant species were identified in their diet, with forbs (41.3%) being dominant over grass (37.9%) and shrubs (20.8%) relative to levels of use. There is a high amount of seasonal variability in use. Common grass species used included fluff grass (*Tridens pulchellus*) and red brome (*Bromus rubens*). Important forb species were woolly plantain (*Plantago purshii*) and Coulter lupine (*Lupinus sparsiflorus*). The main shrub species used were cactus (*Opuntia* spp.). Changes in seasonal use patterns appear to be influenced by plant moisture content, availability, phenological growth state, palatability, and relative abundance. The resulting general seasonal use patterns are higher use of succulent forbs in spring and wetter summers with a shift to grasses and shrubs, primarily cactus, during drier seasons and years.

BEHAVIOR: Though not as social as some leporids, desert cottontails appear to tolerate the presence of others. Both sexes will maintain home ranges as areas of familiarity, but there is little evidence of territorial behavior (defended areas). Females, at least, have been observed feeding in close proximity to one another. Males, however, especially during the breeding season, may not be as tolerant of other males and have been observed chasing each other. As population densities are often fairly low over much of their arid habitat, the lack of interactions between individuals may be due more the low incidence of individuals meeting than any type of avoidance behaviors. When they are in groups, some degree of intercommunication relative to danger, e.g., position of the tail, appears to exist. Apart from this, little is known about the level and type of communication that exists in this species.

Desert cottontails can be found active at any time of the day, but they are most active from around sunset to sunrise with a peak around early morning. There is some evidence that the timing of their activity is influenced by the activity patterns of their predators. When active, they will forage either under shrubs or in open areas, but most times not very far from nearby shrubs. When startled, they quickly return to dense shrub areas for protection.

PHYSIOLOGY AND GENETICS: Diploid chromosome number = 42. Like most desert-dwelling mammals, desert cottontails are physiologically adapted to the temperature and precipitation extremes of arid environments. This species is capable of shifting its thermoneutral zone upward in the summer, which is likely related to its ability to reduce its basal metabolism by 18%. These factors result in less heat load and thus lower evaporative water loss during hot summer temperatures. Overall, a higher lethal body temperature of 44.8°C and high evaporative cooling capacity add to this species' ability to survive hot desert temperatures. Additionally, behaviorally they seek out microenvironments, e.g., shade at the base of shrubs, to further reduce thermoregulation demands on their systems.

REPRODUCTION AND DEVELOPMENT: The breeding season for the desert cottontail is quite variable over its range. Discrete seasons of reproduction have been reported for California (December–June), Arizona (January–August), and Texas (February–August). However, there is at least one report of year-round breeding from California. The gestation period is 28 days. Mean litter sizes reported from various areas of its range were between 2.6 and 3.6 young. It is estimated that females can have up to five litters/season. Young are born essentially hairless and measure approximately 90 mm in length. Sexual maturity in females appears to occur around 80 days after birth.

PARASITES AND DISEASES: Desert cottontails are known to be infected with various species of cestodes, including *Raillietina* spp., *Cittotaenia variabilis*, and *Taenia pisiformis*. Nematode parasites include *Dermatoxys veligera*, *Nematodirus leporis*, *Obeliscoides cuniculi*, and *Passalurus ambiguus*. At least two intestinal protozoans have been reported (*Trichomonas* and *Chilomastix*). Serological studies of desert cottontails from Texas have found low rates of incidence of *Babesia* spp., *Borrelia* spp., and *Anaplasma* spp. External parasites include fleas (Pulicidae and Ceratophyllidae), ticks (Ixodidae), and bots (*Cuterebra* spp.).

CONSERVATION STATUS:

IUCN Red List Classification: Least Concern (LC)

MANAGEMENT: The desert cottontail is considered a game species over much of its range. However, it can be considered a low-priority species because hunting of desert cottontails is not as prevalent as it is for its relative, the eastern cottontail. In general, populations are considered stable. Recent declines in abundance in several regions of its range have raised concerns regarding its long-term survival and may indicate a sensitivity to climate change.

ACCOUNT AUTHOR: John W. Laundré

Key References: Arias-Del Razo et al. 2011; Bock et al. 2006; Brown and Krausman 2003; Burgess and Windberg 1989; Chapman and Willner 1978; Grajales-Tam and González-Romero 2014; Henke and Bryant 1999; Hinds 1973; Ingles 1941; Joseph et al. 2003; Lightfoot et al. 2010; López-Vidal et al. 2014; Nelson et al. 1997; Pfaffenberger and Valencia 1988; Richerson et al. 1992; Smith et al. 1996; Stout 1970; Turkowski 1975; Vargas Cuenca and Cervantes 2005; Yang et al. 2005.

Sylvilagus bachmani (Waterhouse, 1839)
Brush Rabbit

OTHER COMMON NAMES: Riparian brush rabbit; Conejo matorralero (Spanish)

DESCRIPTION: More than most other members of the Leporidae, the brush rabbit exhibits a high degree of geographic variation across its range. Nevertheless, it is one of the smaller cottontails, generally weighing less than

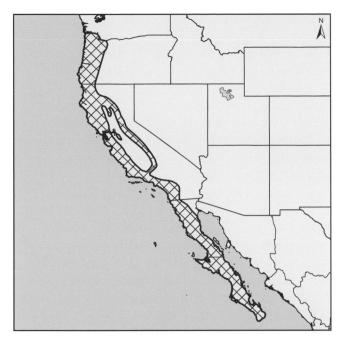

Sylvilagus bachmani. Photo courtesy Moose Peterson / WRP

900 g. Much of its range overlaps with that of the desert cottontail (*S. audubonii*), so the two species are sometimes confused. However, the brush rabbit is smaller, has a more rounded forehead and a shorter rostrum, has smaller ears that do not have dark tips, and has a distinctive cottontail that is smaller and more hidden on the underside of the tail. The back is a speckled brown to gray brown, which grades to white on the belly and underside of the tail. The legs are relatively short and are rusty brown or tan-colored. The feet are small, sparsely haired, and typically pale gray or pale orange. The nape patch is a pale orange, but not as distinctive as in the desert cottontail. The eye-rings are cream-colored. Females are slightly larger than males.

When the skull is viewed from the side, the rostrum appears relatively short (by comparison with that of the desert cottontail). The braincase is slender. The palatine foramina vary geographically. The auditory bullae vary from small to large.

SIZE: Head and body 303–369 mm; Tail 10–30 mm; Hind foot 70–86 mm; Ear 45–63 mm; Greatest length of skull 49–55 mm. Weight: 511–917 g

PALEONTOLOGY: There is little published information on the brush rabbit fossil record. However, it has been found at two well-studied sites in California: LaBrea (Los Angeles area) and Pacheco (580 km north of LaBrea in the San Francisco Bay area), which are both Pleistocene assemblages.

CURRENT DISTRIBUTION: Brush rabbits are confined to the Pacific coast, occurring from the Columbia River in the north to the tip of Baja California in the south. They do not occur east of the Cascade-Sierra Nevada mountain ranges. They occur at elevations ranging from sea level to at least 2,070 m.

TAXONOMY AND GEOGRAPHIC VARIATION: As well as the desert cottontail, the mountain cottontail (*S. nuttallii*) is also somewhat similar to *S. bachmani*, but the mountain cottontail also is larger and differs in other ways as well, including in distribution and habitat preference. Thirteen subspecies: *S. b. bachmani* (along the C California coast between the Salinas River and Morro Bay); *S. b. cerrosensis* (Cedros Island, Baja California, Mexico); *S. b. cinerascens* (SW California from around Bakersfield to N Baja California); *S. b. exiguus* (C Baja California from El Crucero north to San Vincente); *S. b. howelli* (Baja California from the California-Mexico border south along the Sierra de Juarez Mountains to about San Vincente); *S. b. macrorhinus* (San Mateo and Santa Cruz Counties, California); *S. b. mariposae* (foothills of the Sierra Nevada Mountains in C California from Sacramento to Bakersfield); *S. b. peninsularis* (S Baja California from El Crucero south); *S. b. riparius* (confined to some riparian areas in N San Joaquin Valley, Stanislaus and San Joaquin Counties, California); *S. b. rosaphagus* (Baja California along the coastal plain between Ensenada and Rosario, Mexico); *S. b. tehamae* (north of Klamath Falls, Oregon, southward along the east side of the coast range to Sacramento, California, and from Redding, California, south along the Sierra Nevada range to Placerville); *S. b. ubericolor* (W Oregon and California between the Columbia River in the north to San Francisco Bay in the south, and east to the summit of the Cascade-Sierra Nevada Mountains); *S. b. virgulti* (C California coast range between Alameda County [Berkeley] and Kern County [Maricopa]). Recent genetic and morphological analyses indicate that *S. mansuetus* is likely not an independent form and may be considered a subspecies of *bachmani* (herein we treat *mansuetus* as independent).

ECOLOGY: Brush rabbits create networks of trails and runways through their brushy habitat and seasonally through tall stands of dense herbaceous cover adjacent to patches of shrubby or woody cover. They do not dig burrows, but they will use burrows dug by other animals or cavities in tree trunks and downed logs. Their home ranges are small, less than 2,000 m², and tend to depend on the distribution and uniformity of the habitat. For *S. b. riparius*, home range size varies by season; home ranges are larger in the breeding season than in the nonbreeding season and they do not vary by sex. Also for *S. b. riparius*, home range overlap is high (80% to 90%) and does not differ with respect to dyad types (FF, FM, MM), but overlap is greater during the nonbreeding season than during the breeding season. Brush rabbits appear to be mainly nocturnal, but diurnal movements and behavior have also been observed (e.g., sunbathing).

Brush rabbits are important prey for many predators, from rattlesnakes to raptors, throughout their range. In California, predators of adult and young brush rabbits include but are not limited to barn owls (*Tyto alba*), great-horned owls (*Bubo virginianus*), red-tailed hawks (*Buteo jamaicensis*), Cooper's hawks (*Accipiter cooperii*), coyotes (*Canis latrans*), gray foxes (*Urocyon cinereoargenteus*), mink (*Neovison vison*), bobcats (*Lynx rufus*), and long-tailed weasels (*Mustela frenata*). Data on survivorship of brush rabbits are scarce. However, in the early phases of a *S. b. riparius* captive breeding and reintroduction program, more than 50% of released rabbits were dead within 12 weeks. This rate was subsequently reduced with a modified release strategy and some rabbits lived for 800 to 1,500 days post-release.

HABITAT AND DIET: The brush rabbit lives up to its name through its obligate need for habitat that provides dense cover. Although this cover is often provided by dense thickets of bramble, rose, willow, chaparral, or other woody brush, it also inhabits forest edges, old clear-cuts, or burns in forests. Furthermore, *S. b. riparius* readily uses tall grass and herbaceous cover during spring and summer. Although brush rabbits live in dense cover, they often forage within a few meters of cover, eating grasses, thistle, clover, berries, and a wide variety of other plants. In winter they rely more on woody plants.

BEHAVIOR: Brush rabbits are very cautious when emerging from their dense brushy cover to feed in adjacent herbaceous cover. When interacting with conspecifics, they generally maintain a separation of 0.3–8 m, without a "chase" resulting, yet they will often touch noses and sniff each other preceding a chase. This species also exhibits the usual grooming behavior associated with cottontails. They have been known to climb low shrubs as well as trees when threatened or to escape flooding. Introduced eastern cottontails (*S. floridanus*) exhibited aggressive behavior toward brush rabbits when they were penned together, but brush rabbits also attack each other if live-trapped together.

PHYSIOLOGY AND GENETICS: Diploid chromosome number = 48

REPRODUCTION AND DEVELOPMENT: Brush rabbits typically breed from midwinter to early summer, but the breeding season can be longer or shorter depending on local environmental conditions. Although litters can have one to seven young, they are typically smaller with two to three young. An adult female can produce five to six litters a year, and subadult *S. b. riparius* females can reproduce. There is some variation by region for both the breeding season (California: December–May; Oregon: February–August) and litter size. The gestation period is 24–30 days. The young are raised in a concealed, fur-lined nest chamber, which the female covers with grass; she only returns to at night to feed the nestlings. For *S. b. riparius*, nest chambers are in the form of a fist-sized cavity dug into the ground and well hidden in tall grass or other herbaceous cover a few to several meters away from brushy cover. The young only spend about two weeks in the nest before moving into the brushy cover.

PARASITES AND DISEASES: Brush rabbits are hosts to a variety of ecto- and endoparasites. They can serve as hosts of the raccoon and skunk roundworm (*Baylisascaris procyonis* and *B. columnaris*, respectively) infection, which can cause neurologic disease. Brush rabbits are also hosts to fleas and ticks, some of which carry diseases that are a threat to humans. In research on *S. b. riparius*, the rabbit tick (*Haemaphysalis leporispalustris*) was the only tick found, and it occurred on 47 of 48 rabbits (0 to > 20/ rabbit). A number of diseases are known to occur in brush rabbits and related *Sylvilagus* species. Tularemia, plague, myxomatosis, silverwater, California encephalitis, equine encephalitis, listeriosis, Q-fever, and brucellosis have been documented in California *Sylvilagus* populations.

CONSERVATION STATUS:

IUCN Red List Classification: Least Concern (LC)

National-level Assessments: USA ESA for *S. b. riparius* (Endangered); Mexico for *S. b. cerrosensis* (Endangered Official Norm NOM-059-ECOL-2010)

MANAGEMENT: The riparian brush rabbit (*S. b. riparius*) is listed as endangered by the state of California and by the U.S. federal government. A number of significant conservation and recovery actions have been implemented since 2001. These include establishment of new populations, restoration and creation of new habitat, and the provision of high ground with suitable cover for shelter and protection when rivers are at flood stage and go over their banks (twice since 2004). Although the subspecies is still on the endangered species lists, its status is significantly improved.

Recently, *S. b. exiguus* has been proposed for consideration as threatened. Additionally, *S. b. peninsularis* needs to be assessed as it might even be extinct. Little is known about the status of other subspecies, but in the context of a changing climate and its impact on ecosystems, a range-wide assessment is warranted.

Brush rabbits are considered a game species throughout much of their range. However, because the habitat that they occupy is brushy and difficult to hunt in, they are not hunted to the same extent as leporids that use more open habitats. Since *S. b. riparius* is a listed species, hunting for rabbits in or near its habitat areas is banned. Hunting is not a major threat to brush rabbits anywhere in their range. Wildfire and flooding are major threats in most areas. Climate change may well be a significant threat also.

ACCOUNT AUTHOR: Patrick A. Kelly

Key References: Black et al. 2009; Chapman 1974; Hamilton et al. 2010; Kelt et al. 2014; Lorenzo et al. 2013; Mossman 1955; Orr 1940; Schmitz et al. 2014; SEMARNAT 2010; Tomiya et al. 2011; Williams et al. 2008.

Sylvilagus brasiliensis sensu stricto

Contemporary treatments of *Sylvilagus* (such as the 1990 IUCN / Species Survival Commission Lagomorph Action Plan and the IUCN Red List—http://www.iucnredlist .org/) show that *S. brasiliensis* has a huge distributional range extending throughout most of N South America and into Central America and as far as S Mexico. Indeed, the original treatment of the species by Linnaeus in 1758 gave the type locality as being "South America." As such, *S. brasiliensis* has been portrayed as having one of the broadest geographic distributions of any small mammal, extending

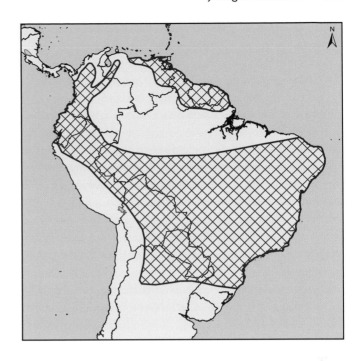

from Veracruz, Mexico, in the north, to Argentina in the south, and from the Atlantic coast of Brazil in the east, to the Pacific coast of South America in Peru and Ecuador in the west. Within this range it inhabits elevations from sea level to 5,000 m asl in the Andes, and occupies a total area of ~ 10,900,000 km². At least 37 taxa have been described as species or subspecies and subsumed into synonymy with *S. brasiliensis*.

It has become apparent that the arrangement that considers most cottontail rabbits from South America and Central America, with the exception of the extension of the eastern cottontail (*S. floridanus*) into extreme N South America, as being subsumed in a single taxon is incorrect, and work has begun to examine and define the forms that previously made up this conglomerate of a species. Herein we treat the extreme northern forms as Gabb's cottontail (*S. gabbi*), and it is possible that the Venezuelan lowland rabbit (*S. varynaensis*) is also a variant of *brasiliensis* (although it could have a phylogenetic link to the eastern cottontail).

Within the range of the remaining "*brasiliensis*," only three forms have been clarified: the tapetí (*S. brasiliensis*), the Andean cottontail (*S. andinus*), and the Rio de Janeiro dwarf cottontail (*S. tapetillus*); other forms are currently under investigation. Little is known of the natural history of any of these species; thus here we distinguish each, identify the species' geographic distribution, and comment briefly on aspects of their biology known at the present time.

Sylvilagus brasiliensis (Linnaeus, 1758)
Tapetí

DESCRIPTION: The tapetí is a relatively small cottontail. The nuchal patch is orange-rufous to mahogany-red and short (~ 0.8 mm). Mid-dorsal fur is about 25 mm in length, with thick, fine, white to gray underfur approximately 10 mm in depth; emergent hair is coarse and agouti with a basal band above the underfur black, a buff yellow central band with some orange, and a black tip. Midventral hair is coarse, white, about 22 mm in length, with thick, primarily gray underfur constituting the basal 10–12 mm. Fur in the throat area is a pale horn in color, about 20 mm thick, with the basal portion consisting of thick, fine, gray underfur. The external pinnae are relatively densely furred externally for the basal 25 mm, then the remainder of the length is lightly furred, with skin visible; the internal aspect is lightly furred in orange-rufous transitioning to a mixture of buff and primarily white on the interior edge of the pinnae. The hind feet are primarily white dorsally, with some pale pinkish buff beneath, particularly between the phalanges.

The skull is medium to large compared with those of other *Sylvilagus*. It is generally longer and narrower than that of other species in the genus: longer than that of the eastern cottontail, the desert cottontail (*S. audubonii*), the brush rabbit (*S. bachmani*), the San José brush rabbit (*S. mansuetus*), the mountain cottontail (*S. nuttallii*), and the New England cottontail (*S. transitionalis*), overlapping that of the Omiltemi rabbit (*S. insonus*), and smaller than that of the swamp rabbit (*S. aquaticus*) and the marsh rabbit (*S. palustris*). The zygomatic breadth is narrower than that of most remaining *Sylvilagus* species, being larger than that of the brush rabbit and the mountain cottontail, equal to that of the desert cottontail, and smaller than that of the swamp rabbit, the eastern cottontail, the New England cottontail, and the Tres Marías cottontail (*S. graysoni*). The lacrimal tubercles are prominent, and the maxillary is heavily fenestrated. The postorbital process is tightly fused to the frontal. The frontal bones are generally smooth anteriorly, but increasingly pitted in their posterior half. The parietal bones are heavily pitted. The incisive foramina are long and broad. Premolar foramina are absent, a diagnostic in contrast to other species long considered conspecific (e.g., the Andean cottontail).

SIZE: Head and body 375 mm; Tail 20 mm; Hind foot 83 mm; Ear 58 mm; Greatest length of skull 69–72 mm; Weight 950 g

CURRENT DISTRIBUTION: While the exact current distribution of the tapetí is unclear, an ecological niche modeling approach based on existing locality data from available specimens suggests a restricted range of 500–700 km^2 within a narrow band along Atlantic coastal Brazil. This range is congruent with the Pernambuco Endemism Center, in the northern portion of the Atlantic Forest Biome, a particularly endangered ecosystem. Only a careful revision of *S. brasiliensis* as broadly understood, including further, more detailed morphological, chromosomal, and molecular assessments, will result in an accurate assessment of its distribution.

TAXONOMY AND GEOGRAPHIC VARIATION: The problem with the original species description is that Linnaeus did not designate a holotype. Only recently was a neotype designated, permitting delimitation of this taxon, resulting in the restricted geographic range outlined above. In time this species may be found to also occur in more southern regions of the Atlantic Forest Biome.

CONSERVATION STATUS:
IUCN Red List Classification: Not Evaluated (NE)

Sylvilagus andinus (Thomas, 1897)
Andean Cottontail

OTHER COMMON NAMES: Páramo cottontail

DESCRIPTION: The Andean cottontail is a medium-sized cottontail. The dorsal pelage is agouti, with a long gray undercoat and hairs tipped in buff then black, giving an overall brown appearance, but not as dark brown as the lowland forest cottontails. Agouti banding appears narrower than on lowland cottontails. The dorsum blends imperceptibly into a brownish orange rump patch (absent in juveniles). The ventral surface is gray mixed with white (white more dominant in juveniles). The hind feet are orange-cream dorsally, dark gray-brown, and thickly furred on the plantar surface. The forefeet are similar in coloration with an orange tinge running from the ankle to the body. The ankles show some brighter orange. The sides are slightly paler than the dorsum. The nuchal patch is bright orange and extends dorsally to double the distance of the ears. Some individuals show a buffy circummorbital band; others have only a postorbital patch. The gular patch is buff-gray in juveniles; those of adults vary from gray with some orange-brown to brighter orange-brown with black underfur. The crown is darker than the sides, but lightens at the base of the ears. The tail is imperceptibly short on specimens. Females are larger than males.

The skull is medium sized in comparison to those of other *Sylvilagus*, and differs from those of typical South American lowland forest *Sylvilagus* in being smooth, rather than pitted, on the parietal and frontal bones. The postorbital processes taper to a point and are separated from the frontal bone (very short in some individuals), versus ending bluntly and fixed to the frontal as in the tapetí. The frontal bones taper to a point between the nasal bones, versus ending in a blunt, rounded terminus, as in the tapetí. Premolar foramina are present in the maxillary bone, lingual to premolar three, versus being absent in the tapetí. The dentition, in particular lower premolar three, is quite distinctive: extremely complex with highly convoluted rostral entrants in the Andean cottontail, versus relatively simple, with one or few shallow invaginations as in the tapetí. The hypoflexid separating the anterior from posterior lobes of that tooth shows a similar difference in complexity.

SIZE: Head and body 338–355 mm; Tail 20–31 mm; Hind foot 64–73 mm; Ear 52–61 mm; Greatest length of skull 64.9–70.5 mm; Weight 680–1,052 g

PALEONTOLOGY: Although currently restricted to Ecuador (as *S. a. andinus*; but see below), there is a highland fossil record of "*S. brasiliensis*" from Peru (Departamento Ancash) potentially attributable to this species found at Guitarrero Cave, ~ 2,590 m asl, in the Callejón de Huaylas, just above the Río Santa floodplain, between the Cordillera Negra and Cordillera Blanca. Remains of *Sylvilagus* from Guitarrero are potentially as old as 12,560 ±360 ybp to as recent as 2,315 ±125 ybp.

CURRENT DISTRIBUTION: The Andean cottontail is restricted to the treeless Páramo zone of the Andes, ranging from 3,000 to 4,800 m asl. *S. andinus sensu stricto* is restricted to Peru. It has been suggested that they could descend to 2,500 m or less in inter-Andean valleys, but competition from lowland *Sylvilagus* would restrict them at the tree line on the eastern ("*S. defilippi*") and western ("*S. kelloggi*") slopes of the Andes.

TAXONOMY AND GEOGRAPHIC VARIATION: *Sylvilagus andinus* was described from the western slope of Volcán Cayambe, Province of Pichincha, Cantón Cayambe, eastern cordillera of Ecuador, altitude 4,000 m asl. Described originally as a species, it has gone back and forth between species and subspecies of *S. brasiliensis*, and only recently has been removed from synonymy with that species. As a result, the taxonomic and geographic boundaries remain somewhat nebulous. Some have defined the *S. andinus* group as being constituted by the forms *S. capsalis*, *S. kelloggi*, *S. chillae*, *S. andinus*, *S. a. canarius*, *S. a. chimbanus*, *S. nivicola*, *S. salentus*, *S. fulvescens*, *S. apollinaris*, *S. purgatus*, *S. nicefori*, and *S. meridensis*. Thus, if the "*S. andinus* group" is translated into "a polytypic *S. andinus*," then *S. andinus* would include 13 subspecies. It is unlikely that *S. a. purgatus* should be included in that list, while *S. a. capsalis*, a highland form from Peru, should probably be included in this arrangement. Subsequently *S. a. carchensis* and *S. a. chotanus* were named, and an alternative grouping of *S. andinus* was suggested that included *S. a. carchensis*, *S. a. chotanus*, *S. a. andinus*, *S. a. nivicola*, *S. a. chimbanus*, and *S. a. canarius*. Either hypothesis of taxonomy results in a distribution in the Andean Páramos from Venezuela (*S. a. meridensis*) to Peru (*S. a. capsalis*), although the latter grouping is somewhat more circumscribed than the former. A molecular study has suggested deep divergences even across small distances in the fractured Páramo habitat. The question of species limits in the cottontails of this group should be addressed in order to effect coherent conservation actions for these animals.

ECOLOGY: The sex ratio among adult Andean cottontails is roughly even at some localities, but almost a 2:1 ratio of males to females in others; there is an even proportion of juveniles to adults. Population density averages 2.85/ha (range 2.2–3.9), among the lowest reported for *Sylvilagus* species. Western black-chested buzzard-eagles (*Geranoaetus melanoleucus australis*) have been observed preying on Andean cottontails.

HABITAT AND DIET: The Andean cottontail is restricted to Páramo habitat above the tree line (3,000–3,500 m) to the snowline (~ 4,700–4,800 m). It apparently avoids forested habitats. The diet is largely unknown. Some observations indicate that it has a very restricted diet, feeding only on grasses (*Bromus catharticus*, two species of *Poa*) and one sedge (*Carex* sp.). During drier months (December–April) when food becomes scarce, mortality increases.

BEHAVIOR: The Andean cottontail has been reported to be diurnal, in contrast to *Sylvilagus* occupying the forested areas immediately below the tree line, which are nocturnal. However, there are also indications that Andean cottontails may exhibit some nocturnal activity. They build nests characterized by a central chamber ~ 27 cm in diameter by ~ 21 cm in height. The chamber is surrounded by three to five smaller satellite chambers connected to the central chamber by tunnels ~ 95 cm in length by 11 cm high and 12 cm wide. Females construct a nursery chamber on the side of the central chamber (~ 22 cm in diameter by 19 cm in height) that is lined with grasses and various species of Asteraceae, all mixed with their own fur. Nest temperatures are typically 5°C–10°C warmer than nighttime minima and 5°C–10°C cooler than daytime maxima. They also make underground nests ("warrens") that may be occupied by two to five rabbits. The daily temperature in some localities ranges from –17.8°C to 40.0°C, suggesting that the Andean cottontail is able to adapt, whether physiologically or behaviorally, to a variety of temperatures.

PHYSIOLOGY AND GENETICS: Sequences derived from cytochrome *b* and 12S mitochondrial genes definitively place *S. andinus* and *S. brasiliensis* in distinct clades from one another, reinforcing their status as valid species.

REPRODUCTION AND DEVELOPMENT: Litter size of *S. a. meridensis* has been reported to average 1.18 neonates.

CONSERVATION STATUS:

IUCN Red List Classification: Not Evaluated (NE)

MANAGEMENT: As noted above, the Andean cottontail is limited to the Andean zones above the tree line. The Páramo ecosystem lies within the Tropical Andes biodiversity hotspot, which has lost 75% of its original area of remaining primary vegetation. These are areas that are under intense pressure from humans. The species is under pressure from uncontrolled hunting, agriculture, and cattle operations, and apparently it does not survive in human-altered habitat. These data, along with the "sky island" nature of its distribution (disparate patches in the higher elevations of the Andes), suggest that strong conservation measures should be enacted in order for the species to prosper or at the very least to maintain its current distribution and numbers. This biodiversity hotspot, notwithstanding its loss in area, is nevertheless hypothesized to represent at the present time ~ 15% of primary vegetation for all biodiversity hotspots and to hold 16% of global endemic vertebrate biodiversity. Conserving this cottontail would therefore result in conservation of many additional habitat-restricted species.

Sylvilagus tapetillus Thomas, 1913
Rio de Janeiro Dwarf Cottontail

OTHER COMMON NAMES: Small tapetí, Dwarf tapetí

DESCRIPTION: The Rio de Janeiro dwarf cottontail is similar to the already small tapetí in general appearance, but even somewhat smaller in general; its ears are approximately 1 cm shorter than in the tapetí. The pelage coloration is similar to that of the tapetí, although generally not as dark brown / black as in the latter. An orangish nuchal patch is present; the venter is primarily whitish. Females weigh slightly more than males.

The skull of the holotype (only accessible skull) is relatively small. The postorbital process is strongly fused to the frontal bone for over half of its length (in the tapetí, the most caudal quarter is lightly fused to the frontal). The posterodorsal process of the premaxilla barely extends to the caudal end of the nasal bones, whereas in the tapetí it extends to or beyond the caudal end of the nasal bones. A process of the frontal bone extends between the caudal ends of the posterodorsal process of the premaxillae and the nasal bones. This process is definitively absent in the tapetí. Pitting on the dorsal surface of the cranium is present, and not as conspicuous as that of the tapetí, but extending farther rostrally than in the latter, to the interorbital region, where pitting ends abruptly, rather than gradually, as in the tapetí. No premolar foramina are present, as in the tapetí, but in contrast with the Andean cottontail.

SIZE: Head and body 295 mm; Tail, inconspicuous; Hind foot 70 mm; Ear 46 mm; Greatest length of skull 61 mm (unreliably measured on a broken cranium); Weight 476–1,127 g

CURRENT DISTRIBUTION: Only three specimens definitively identified as *S. tapetillus* are known, and all are from a single locality in Brazil: Rio de Janeiro, Porto Real, Rio Paraíba, near Rezende (estimated coordinates: 22°24'40.43"S, 44°19'14.72"W, elevation ~ 390 m). This location is in an upland region ("Vale do Paraíba") varying between ~ 375 and 525 m, bounded to the northwest by the Serra da Mantiqueira, and dropping precipitously in the southeast at the Serra das Araras, to the lowland coastal plain in the vicinity of metropolitan Rio de Janeiro. Given our hypothesis of South American cottontails as a collection of ecologically restricted taxa, it is unclear at present how far beyond the confines of this small area the Rio de Janeiro dwarf cottontail could be distributed.

TAXONOMY AND GEOGRAPHIC VARIATION: No subspecies. The taxonomic status of *S. tapetillus* has reflected varying opinions as to whether its small size, based on the holotype, was evidence of juvenile or subadult condition, or whether it actually represents an adult. We have examined the holotype of *S. tapetillus* and are confident that, based on ossification of cranial sutures, as well as on wear of cranial and dental features, it represents a specimen of an adult cottontail that is taxonomically distinct from other *Sylvilagus*.

HABITAT AND DIET: The Rio de Janeiro dwarf cottontail lives in hilly terrain, where grasses abound near small streams. Although the type locality is near a stream, these cottontails apparently are more common near hilltops within their habitat, in areas with sparse, almost denuded, vegetation, with isolated tussocks of the grass "barba de bode" (goat's beard *Aristida pallens*).

BEHAVIOR: In captivity, the Rio de Janeiro dwarf cottontail is reported to be nocturnal and to feed preferentially at dusk. During the day, this cottontail generally remains in a grass nest, not moving much, and when doing so, moving in small leaps. They are normally docile, but also extremely skittish in the presence of humans, and are difficult to trap in their native habitat.

PHYSIOLOGY AND GENETICS: Diploid chromosome number = 40 and FN = 72. This karyotype differs in FN from that of other forms that have been included in *S. brasiliensis*, and there also appear to be distinct differences in mitochondrial cytochrome *b* sequences between *S. tapetillus* and other *brasiliensis* forms. Body temperature (measured rectally) varies between 38°C and 38.6°C.

REPRODUCTION AND DEVELOPMENT: In captivity the litter size of the Rio de Janeiro dwarf cottontail averages 3 kittens, following a gestation period of 30–35 days. Two periods of parturition have been observed in captivity: in March and in September. Females become reproductively active starting 8–10 months after birth.

CONSERVATION STATUS:
IUCN Red List Classification: Not Evaluated (NE)

MANAGEMENT: As noted above, these animals are very skittish and appear to avoid contact with humans. Because of their restricted distribution in the uplands of the Paraíba River valley, it is likely that they are under relatively severe threat, if indeed they even remain in existence. In contrast with this perspective, however, is the suggestion that they prefer somewhat denuded habitats with sparse tussock grass. This ecological proclivity might suggest that some accommodation with humans is possible within their range. Further studies of this taxon are strongly recommended to ascertain their conservation status and morphological and geographical species limits.

Other South American *Sylvilagus brasiliensis sensu lato*

Comparisons, both morphological and molecular, of the neotype of *S. brasiliensis* with other named taxa in its synonymy suggest that many of the latter are distinct at the species level. Careful taxonomic assessments of the cottontails of South America will be an ongoing effort. For now, only the Central American Gabb's cottontail and Dice's cottontail, along with the South American Andean cottontail and the Rio de Janeiro dwarf cottontail, have been excised from *S. brasiliensis*. What to do, from a taxonomic perspective, with rabbits inhabiting the remaining 10,899,400 km² of the range of *S. brasiliensis sensu lato*? On the basis of priority, and barring *S. brasiliensis* and *S. tapetillus*, the oldest available name for lowland taxa south of the Amazon River appears to be *S. minensis* Thomas, 1901. Taxa north of the Amazon are somewhat more problematic. It might be argued that "*incitatus*" (Bangs, 1901) is available. However, as Bangs himself noted, this taxon likely is, if anything, allied with *S. gabbi*. Furthermore, that taxon is from San Miguel Island, in Panama's Archipiélago de las Perlas. That strengthens (based on geographic proximity) a possible relationship with *S. gabbi*, and also underscores the distinction of *incitatus* with any *Sylvilagus* west and south of the northern reaches of the Andes, particularly the Cordillera Central and Oriental. At present, therefore, and as unlikely as it is unrealistic, the only name available for the *S. brasiliensis* group taxa north of the Amazon is *S. sanctaemartae* Hershkovitz, 1950.

We note in passing, however, that *S. sanctaemartae* is readily distinguishable from other *S. brasiliensis* group taxa, except *S. gabbi*, in having two foramina in the basisphenoid: a craniopharyngeal foramen and another, innominate, foramen that may be a division of the craniopharyngeal or another, independent, foramen (material is lacking to test these alternative hypotheses). In addition, *S. sanctaemartae* has small but distinct antorbital processes, a feature largely absent from lowland *S. brasiliensis*. These and other differences between *S. sanctaemartae* and lowland *S. brasiliensis* suggest that *S. sanctaemartae* will eventually be found to have a restricted distribution in NE Colombia (lower skirts of the Sierra Nevada de Santa Marta), and that other lowland *S. brasiliensis* north of the Amazon will be found to be distinct from both *S. brasiliensis sensu stricto* and *S. sanctaemartae*.

ACCOUNT AUTHORS: Luis A. Ruedas and Andrew T. Smith

Key References: Bonvicino et al. 2015; Durant 1980; Hershkovitz 1938, 1950; Oliveira e Silva and Dellias 1973; Ruedas et al. 2017; Stone 1914; Tate 1933; Wing 1980.

Sylvilagus cognatus Nelson, 1907
Manzano Mountain Cottontail

DESCRIPTION: The pelage of the Manzano Mountain cottontail is overall light in coloration, particularly on the flanks, and appears similar to that of the desert cottontail (*S. audubonii*), from which it may be distinguished on the basis of size, preferred habitat, and elevation of habitat. The general appearance is of a predominantly gray animal, with some black-agouti bands, particularly concentrated in the mid-dorsal region. There is a relatively small (approximately the length of the ears), triangular nape patch that is orange suffused with gray. The head is darker than the body, appearing mostly brown flecked with gray and black, particularly dorsally. There is a patch of predominantly brown flecked with black at the haunches that is visible from a lateral perspective. The tail is relatively prominent, gray with brown dorsally and white ventrally. The hind feet are buff orange closer to the body, with white dorsally and more prominent laterally toward the distal end of the foot. The forefeet are bright orange-brown proximally, becoming buff with white distally. The prominent gular patch is yellowish buff with a gray underfur. There may be seasonal variation in pelage coloration, but all adults nevertheless still have more gray overall and are lighter than the eastern cottontail (*S. floridanus*). Juveniles have the same appearance, but with a less prominent nuchal nape and much narrower agouti banding, resulting in a more homogeneous appearance to the color; subadults have agouti banding intermediate between that of juveniles and adults. The coloration of the forefeet and the hind feet does not appear to vary seasonally.

The skull comparisons that follow are based on the

type series of *S. cognatus* and the holotype of *S. floridanus* (note that *S. floridanus* is quite variable throughout its range, and these characters may be variable). The skull of the Manzano Mountain cottontail is similar to that of the eastern cottontail, but may be distinguished from the latter by the more prominent dorsally flared supraorbital ridge. In addition, the antorbital processes in the Manzano Mountain cottontail are more prominent, revealing a distinct antorbital notch. The postorbital processes of the Manzano Mountain cottontail are triangular in shape, tapering to a terminus that is lightly fused to the skull,

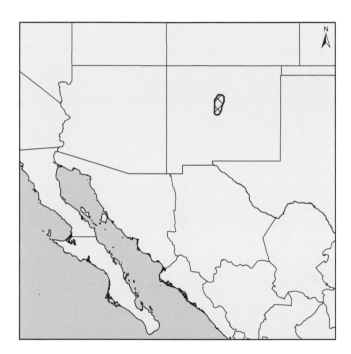

Sylvilagus cognatus. Photo courtesy Tony Godfrey @ artfullbirds.com

whereas the postorbital processes of the eastern cottontail are homogeneously thick and fused with the frontal bone throughout a greater proportion of their length (~ 50%). The dentition is characteristically distinct. The groove in the first upper incisor of the Manzano Mountain cottontail is more lingually located than that of the eastern cottontail; in the eastern cottontail, the ratio of labial to lingual lobes is between 50:50 and 60:40, whereas in the Manzano Mountain cottontail, the labial lobe is closer to two-thirds of the tooth.

SIZE: Head and body 384–462 mm; Tail 55–75 mm; Hind foot 100–105 mm; Ear 67–73 mm; Greatest length of skull 68–72 mm; Weight > 1,100 g

CURRENT DISTRIBUTION: The Manzano Mountain cottontail occurs only in the Sandia and Manzano Mountains of C New Mexico.

TAXONOMY AND GEOGRAPHIC VARIATION: No subspecies. The monotypic *S. cognatus* was recently separated from the similar *S. floridanus*.

ECOLOGY: Little is known about the natural history of the Manzano Mountain cottontail.

HABITAT AND DIET: The Manzano Mountain cottontail occurs at high elevations (2,300–3,100 m) in areas dominated by conifer forests, but may also include lower-altitude areas with montane scrub and subalpine-montane grassland. It is likely limited at lower elevations by Piñon-Juniper woodlands, Juniper savanna, and Plains-Mesa grassland, which are dominated by desert cottontails.

REPRODUCTION AND DEVELOPMENT: There are few data on reproduction in the Manzano Mountain cottontail. A juvenile individual and several subadults were collected in late July. A number of small adults have also been collected in January. It is likely that there is a prolonged development to adulthood, certainly of at least one year, possibly longer.

CONSERVATION STATUS:

IUCN Red List Classification: Data Deficient (DD)

MANAGEMENT: Although the geographic range of the species is depicted as being continuous across the Sandia and Manzano Mountains, this extent of range should be considered only as a conservative working hypothesis. The two mountain chains are separated by the Tijeras Canyon, which drops to an elevation of ~ 2,100 m. This is below the generally preferred elevational range of the species, although it may range that low or even lower. However, while suitable habitat may have existed during recent glaciations, the current habitat in the pass is not suitable for the Manzano Mountain cottontail, but rather more propitious to the desert cottontail. Furthermore, the Sandia and

Manzano Mountains are separated in the Tijeras Canyon by a large freeway, which constitutes an impassable barrier to the Manzano Mountain cottontail. Thus, the Sandia and Manzano populations are functionally discontinuous. Additionally, suitable habitat in the Sandia range only covers ~ 100–110 km², an area that is shrinking due to the effects of global warming. Furthermore, the Sandia Range is bisected by the heavily trafficked road from the town of Sandia Park on the east slope, to Sandia Peak and its ski area. The threats to the conservation and the well-being of this species are therefore somewhat severe.

Additional museum records identified as *S. cognatus* exist from the Capitan Mountains and the Mount Withington Range, also in New Mexico. These should be viewed askance, particularly the latter, as it is separated from the Sandia and Manzano Ranges not only by extensive areas of desert, but also by the Rio Grande River. The "mountain islands" geographical ecology of New Mexico has likely resulted in interesting evolutionary phenomena in *Sylvilagus*; e.g., the Guadalupe Mountains of S New Mexico are occupied by the Davis Mountains cottontail (*S. robustus*). The taxonomy of *Sylvilagus* in New Mexico could bear further scrutiny.

ACCOUNT AUTHOR: Luis A. Ruedas

Key References: Frey et al. 1997; Ruedas 1998.

Sylvilagus cunicularius
(Waterhouse, 1848)
Mexican Cottontail

OTHER COMMON NAMES: Conejo montés, Conejo de monte, Conejo serrano, Conejo (Spanish)

DESCRIPTION: The Mexican cottontail is the largest rabbit in Mexico, equaling medium-sized jackrabbits in biomass. It has coarse pelage, a massive skull, and soft fur. The dorsal color is dirty yellowish or grayish, without rufous, except on the nape, and the light sub-terminal rings on the hairs are uniformly pale cream-color. The ears are as long as the head, and their backs are thinly haired and gray with the extreme tips and outer edges darkening to black. The orbital area is clear, deep buff; the sides of the head are dark, dingy buff, and the nape is dull rusty rufous. The forelegs are similar in color to the nape, but

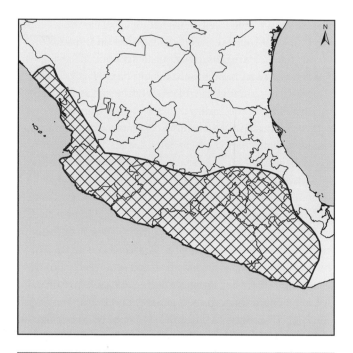

Sylvilagus cunicularius. Photo courtesy Jorge Vázquez

duller and less rufous; the hind legs and sides of the hind feet are duller, more rusty brownish than the forelegs; the tops of the hind feet are buffy whitish or pale, dull rusty. The tail is short and grayish brown above and pure white on the under surface.

The skull is large, heavy, and broad across the braincase. The rostrum is heavy with a massive base, flattened in the frontal region, and arched along the upper outline; the nasals are sharply compressed into a dorso-lateral, pit-like indentation about one-third of the distance from

the tip and expand again toward the tip. The anterior edge of the palatal bridge is level with the front of the anterior premolar, and its posterior edge is level with the division between the last premolar and the first molar. The jugals are proportionally light and slightly grooved with a deep pit anteriorly. The supraorbital process is light and narrow and slightly raised above the plane of the frontals; the postorbital process is usually joined to the skull posteriorly, enclosing a narrow flattened, oval foramen. The interparietal is triangular shaped, its antero-posterior length is nearly two-thirds its transverse diameter, and the occipital shelf is unusually broad. The ventral border of the ramus of the lower jaw usually sits on the posterior angle and the tip when resting on a plane, leaving the middle free; the middle ventral border outline of the skull of old individuals sometimes becomes convex, raising its tip.

SIZE: Head and body 485–515 mm; Tail 54–68 mm; Hind foot 108–111 mm; Ear 60–63 mm; Weight 1,800–2,355 g

PALEONTOLOGY: The few known fossils of the Mexican cottontail from the United States and Mexico represent the Pliocene and the Pleistocene. Pliocene fossils (upper premolars) have been collected in Graham County, SE Arizona. In Mexico, a broken skull and two right dentary bones (one without teeth) of Pleistocene age were collected at "Barranca Seca," Veracruz. It is thought that the remains were deposited when the environment was no more tropical than it is today. Other fossil records of the Mexican cottontail date from the Late Pleistocene in La Presita, San Luis Potosí, and Guilá Naquitz, Oaxaca. Additionally, a 12,000-year-old anklebone (calcaneum) of the Mexican cottontail was reported from Tlapacoya, in the state of Mexico.

CURRENT DISTRIBUTION: The Mexican cottontail is a Mexican endemic species that ranges widely from the Pacific Coastal Plain from Sinaloa south to Oaxaca, and encompassing the Transverse Volcanic Axis from the highlands of Michoacán to Veracruz. Marginal records in the northern part of its distribution include the states of Sinaloa, Nayarit, Jalisco, Michoacán, Mexico, Hidalgo, Puebla, and Veracruz. The Mexican cottontail occurs from sea level to 4,300 m in elevation.

TAXONOMY AND GEOGRAPHIC VARIATION: Three subspecies: *S. c. cunicularius* (the central portion of their distribution; Neovolcanic Axis in the states of Michoacán, Guerrero, and Oaxaca); *S. c. insolitus* (northern distribution from mid-Sinaloa along the Pacific Coastal Plain to south C Jalisco); *S. c. pacificus* (S Jalisco south to the Pacific Coastal Plain in Oaxaca). Mexican cottontails from the Sierra Madre of Michoacán, in the western part of their range, are slightly larger in both skin and skull dimensions than those from elsewhere, but there are no color differences. Considering the wide range of this species under varied climatic conditions, the amount of variation is surprisingly small.

A shared common ancestor with biogeographic and morphological affinities has been hypothesized between *S. cunicularius* and the Tres Marías cottontail (*S. graysoni*). Phylogenetic studies of *Sylvilagus* using data from mitochondrial sequences of the 16S gene show that *S. cunicularius* and the desert cottontail (*S. audubonii*) constitute a monophyletic group, and that the Mexican cottontail is closely related to the Omiltemi rabbit (*S. insonus*).

ECOLOGY: The population density of the Mexican cottontail averages 27 individuals/km² in C Mexico, but it varies according to the rainy and dry seasons. Individual home range size is ~ 3.6/ha, and home ranges overlap between the sexes. The Mexican cottontail is sympatric with other lagomorphs in portions of its range, including the volcano rabbit (*Romerolagus diazi*), the desert cottontail, the eastern cottontail (*S. floridanus*), the white-sided jackrabbit (*Lepus callotis*), the black-tailed jackrabbit (*L. californicus*), and the antelope jackrabbit (*L. alleni*). Mexican cottontails from the highlands of C Mexico have been collected in abandoned burrows of Merriam's pocket gophers (*Cratogeomys merriami*) and in natural rocky hollows and crevices on and underneath the ground.

Fecal pellets of the Mexican cottontail are regularly found next to the base of grasses, forming dunghills or latrines that lie at least 20 m from each other. Fecal pellets of adults are characteristic and rarely confused with those from other sympatric leporids, although their form and size approach those of jackrabbits. Fecal pellets of Mexican cottontails are brown, composed of regurgitated plant materials, and compact (flat and round at their widest part). Fecal pellets are 1.0–1.5 cm in diameter. Mexican cottontails eat seeds of alligator juniper (*Juniperus deppeana*) and deposit them in latrines away from the mother tree, and therefore are a dispersing agent for this tree species.

The most important predators of the Mexican cottontail are mammalian carnivores such as grey foxes (*Urocyon cinereoargenteus*), coyotes (*Canis latrans*), bobcats (*Lynx rufus*), and pumas (*Puma concolor*); some birds of prey, including barn owls (*Tyto alba*) and hawks (*Buteo* spp.); occasionally snakes; and American crocodiles (*Crocodylus acutus*).

HABITAT AND DIET: The Mexican cottontail occurs in tropical, semi-arid, temperate, open forest, dense shrub,

Sylvilagus cunicularius habitat. Photo courtesy Jorge Vázquez

and grassland. In C Mexico it is abundant in pine and pine-oak forests, and in W Mexico it occurs in pastures, dry deciduous forest, and areas of disturbed vegetation, where it selects habitats characterized by grasses and herbs. In the state of Guerrero, this cottontail is usually found in upland valleys in both tropical deciduous and pine-oak zones. In Guerrero *S. c. pacificus* occupies the narrow coastal area below 457 m asl. In the Mexican Transvolcanic Belt it inhabits forests of pine and oak-pine with understory of clumped grasses. It rarely has been seen in conifer forests of "oyamel" (*Abies religiosa*). In Ixtacuixtla, Tlaxcala, this cottontail is found in xeric scrub dominated by kidneywood tree (*Eysenhardtia polystachya*), shadbush (*Amelanchier* spp.), and sumac (*Rhus* spp.); alligator juniper forests; and oak woodlands (*Quercus* spp.). In La Malinche National Park it inhabits pine forest and grasslands of *Muhlenbergia macroura* and Peruvian feathergrass (*Stipa ichu*), being more abundant in grasslands and mixed sites of pine forest and grasslands than in pine forest alone.

Mexican cottontails from C Mexico feed on clumped grasses, including Peruvian feathergrass, *M. macroura*, and *Festuca amplissima*. They select tender shoots of grasses, young leaves of forbs, cortexes of shrubs, and cultivated plants such as oats (*Avena sativa*) and corn (*Zea mays*). In C Mexico the diet of the Mexican cottontail includes 23 species, such as *M. virletii*, Hall's panicgrass (*Panicum hallii*), speargrass (*Piptochaetium virescens*), and *Baccharis conferta*.

BEHAVIOR: Mexican cottontails are solitary and active mainly at dusk and dawn, although they may be active at night and during the day. The periods of greatest visible activity in the Mexican cottontail are 0700–0900 h and 1400–2000 h. They engage in a wide variety of individual behaviors: resting, grooming, foraging, coprophagy, and scent marking. These rabbits also participate in social behaviors such as group feeding, expulsion among individuals, and running in pairs. Strongly aggressive behaviors are displayed by Mexican cottontails. The maximum number of animals that can be maintained in a colony under semi-natural conditions without severe fighting and deaths is three to four females and two males in an enclosure of 530 m² and two females and one male in a 108 m² enclosure.

PHYSIOLOGY AND GENETICS: Diploid chromosome number = 42. The FN and morphology of the sexual chromosome are unknown since only females have been examined.

White blood cells of Mexican cottontail contain lymphocytes (35% to 60%), monocytes (1%), eosinophils (0—2%), basophils (0% to 2%), and neutrophils (37% to 62%).

REPRODUCTION AND DEVELOPMENT: The Mexican cottontail reproduces throughout the year. However, reproductive females have been recorded in greater numbers from March to October, even while in the same population males with descended testes and juveniles could be found throughout the year. The onset of breeding may be associated with an increase in day length and temperature, whereas the birth of young is associated with increased rainfall, hence with the availability of succulent vegetation. Litter size ranges from 1 to 6, and the gestation period is 28–31 days. Each female may reproduce several times a year.

Nursery burrows are extremely difficult to find because mothers close the entrance with soil and cover it with leaves and grass like other lagomorphs. Burrows dug under semi-natural conditions consist of a single tunnel with a simple entrance ranging from 15 to 53 cm in length. These tend to be dug beneath grass tussocks (*M. macroura*), shrubs such as willow ragwort (*Senecio salignus*), or herbs including *Eupatorium pazcuarense* and pride of the mountain (*Penstemon roseus*). The nursery burrow ends in a small spherical chamber containing a nest 17 cm (range 13–22 cm) beneath the surface. Nests consist of dry grass, fragments of woody plants, pine needles and hay, fur pulled from the mother's body, and fecal pellets probably also from the mother.

Births occur at the nursery burrow entrance, and then newborns crawl into the nest. Young at birth are covered with fine hair, and their eyes are closed. Mothers open the nursery burrow entrance to nurse their young, and fol-

lowing nursing they reclose the burrow. After the nursery burrow is opened, young come to the surface on their own to nurse. Litters are nursed once a day in the afternoon, and nursing takes place until about 12 days following parturition.

Males possess one scrotal sac, containing the testis and epididymis. The paired testes of adults are scrotal throughout the year. The penis is cylindrical and covered by the preputial skin structure. The male reproductive system possesses four accessory sexual glands: ampulla, prostate, bulbourethral, and preputial. Females have a lower level of testosterone than males in any condition of reproduction. In males, testosterone levels are low four months after the start of the breeding season, and then soar before initiating breeding.

PARASITES AND DISEASES: Ectoparasites on Mexican cottontails include fleas (*Cediopsylla inaequalis interrupta*, *Euhoplopsyllus glacialis affinis*, and *Pulex irritans*). Gut parasites identified in the Mexican cottontail include protozoans (*Eimeria*), cestodes (*Cittotaenia* and *Taenia*), and nematodes (*Strongyloides*, *Passalurus*, *Trichostrongylus*, and *Heterakis*).

CONSERVATION STATUS:

IUCN Red List Classification: Least Concern (LC)

MANAGEMENT: Mexican cottontails are hunted for food and sport. They are also killed for being considered competitors with livestock. This killing has caused a decline of some populations of the Mexican cottontail. Legal hunting does not account for the reproduction season or age structure of Mexican cottontails, which additionally contributes to the decrease of Mexican cottontail populations. Until a few years ago the Mexican cottontail was a common mammal near the coast in Jalisco, however, at present it is rare in the area; local residents claim the decline is due to hunting. Similarly, the Mexican cottontail is no longer abundant in the state of Morelos after intensive hunting by residents. In some parts of their range, fires and excessive logging have reduced drastically the rabbit's habitat (grasslands and pine forest). In C and W Mexico the Mexican cottontail faces the threat of habitat fragmentation. Additionally, it has been suggested that deforestation, expansion of cattle pastures and agriculture, and introduced feral fauna and grasses are negatively affecting conservation of this species.

ACCOUNT AUTHORS: Consuelo Lorenzo, Jorge Vázquez, Luisa Rodríguez-Martínez, Amando Bautista, Antonio García-Méndez, and Fernando A. Cervantes

Key References: Aguilar et al. 2014; Álvarez 1969; Álvarez et al. 1987; Armstrong and Jones 1971; Cervantes and Vázquez 2008; Cervantes et al. 1992; Chapman and Ceballos 1990; Dalquest 1961; Davis 1944; Davis and Lukens 1958; Davis and Russell 1954; Diersing and Wilson 1980; Gilcrease 2014; González et al. 2007; Hall 1981; Leopold 1959; Lorenzo et al. 1993, 2015; Morgan and White 2005; Nelson 1909; Pérez et al. 2008; Ramírez-Albores et al. 2014; Ramírez-Pulido et al. 1977; Rodríguez-Martínez 2015; Rodríguez-Martínez et al. 2014; Thomas 1890; Vázquez et al. 2007, 2013; White 1991.

Sylvilagus dicei Harris, 1932
Dice's Cottontail

OTHER COMMON NAMES: Conejo de montaña, Conejo de monte (Spanish)

DESCRIPTION: Dice's cottontail is a large-bodied cottontail with relatively short ears and a very short tail that, in the field, makes it appear tailless. The dorsum is agouti, with a prominent buffy orange band tipped distally with an equally prominent black band, and the sides are a blackish gray. The inconspicuous tail is blackish. The venter is white with gray underfur and there is a brown-orange throat patch. The legs and feet are orange-brown dorsally and laterally, with an abrupt transition to dark gray underneath. A buffy eye-ring circumscribes the upper and rear portion of the lower eyelids. The eye-shine is red and bright. The dorsal surface of the head, from the rhinarium to the ears, is darker than the remainder of the sides of the head, with a dark, mostly black patch on the uppermost surface of the head. The ears have a black patch, and the fur is tipped with white, giving a somewhat grizzled appearance. Laterally, from the ears to the mental area, a yellowish white band follows the caudal aspect of the head, separated from the buffy lateral region of the face by a thinner black band. Females are slightly larger than males. The forest-dwelling Gabb's cottontail (*S. gabbi*) is slightly smaller than Dice's cottontail, and has more lightly colored fur with a small orange patch on the nape. The eastern cottontail (*S. floridanus*) has a larger orange patch on the nape, longer ears, and a distinctly white tail.

The skull is large within *Sylvilagus*. It is unusual in displaying extensive pitting on the cranial surfaces, a condition typically associated with the volcano rabbit (*Romerolagus diazi*), but also seen in many of the South American species allied with *S. brasiliensis*. In the tapetí (*S. brasiliensis*), the pitting is mostly restricted to the parietal bones, with some pitting on the frontal bones near the fronto-parietal suture and some on the rostral end of the postorbital process. In the potentially parapatric Gabb's cottontail, pitting extends only into the dorsal interobital region. In contrast, Dice's cottontail has extensive and prominent pitting also on the frontal bones, almost to the fronto-nasal suture. In addition, the nasal bones extend well past the posterodorsal process of the premaxilla (in Gabb's cottontail the posterodorsal processes of the premaxillae extend well past the nasal bones), and end smoothly at the frontonasal suture, rather than jaggedly,

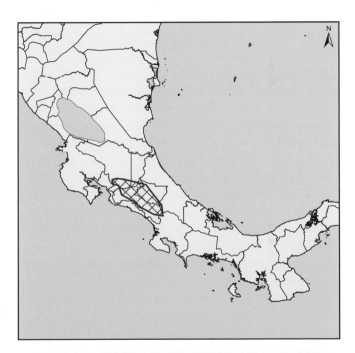

Sylvilagus dicei. Photo courtesy Jan Schipper

at an angle almost perpendicular to the long axis of the animal, rather than diagonally, as in Gabb's cottontail. The caudal end of the posterodorsal process of premaxillae is separated from the nasal bones by the short but prominent maxillary processes of the frontal bones; this process appears much longer in Gabb's cottontail, and tapers to a fine rostral point, but ends in a blunt point in Dice's cottontail. The major palatine foramina are extremely prominent. Porosities at the base of the upper premolars may correspond with premolar foramina, but a more extensive sample would be required to indubitably ascertain this. These foramina are absent in the tapetí and present in Gabb's cottontail.

SIZE: Head and body 340–448 mm; Tail 21–35 mm; Hind foot 83–103 mm; Ear 45–57 mm; Greatest length of skull 76–77 mm; Weight 700–1,500 g

PALEONTOLOGY: There is no fossil record for Dice's cottontail. The Cordillera de Talamanca Mountain Range, where Dice's cottontail is endemic, is the oldest mountain range in Costa Rica: the volcanoes constituting the range are hypothesized to have been present as early as the earliest Paleocene, ~ 66 mya. However, the uplift leading to the current elevations, due to subduction of the Cocos Ridge below SE Costa Rica, is more recent, dating to ~ 5 mya, although some controversy has arisen recently about this topic. The age of the genus *Sylvilagus* is more recent (~ 4.4 mya), supporting *in situ* speciation, presumably driven by ecological factors such as increasing oak and Páramo vegetation types. Páramo vegetation in Costa Rica is found mainly in the Talamanca Range.

CURRENT DISTRIBUTION: Dice's cottontail is generally found at higher elevations: from 1,640 m in Cervantes, Costa Rica, to 3,800 m in Cerro Chirripó, Cordillera de Talamanca. There is a record from an elevation of 1,180 m in W Panama (Rancho de Rio Jiménez). However, some lower-elevation records also exist from Costa Rica: Navarro, 1,235 m; Cervantes, 1,333 m; and Juan Viñas, 1,151 m; all three sites are in Cartago Province. One additional record from Heredia Province, from an elevation of 1,075 m, is from close to Braulio Carrillo National Park (see below).

TAXONOMY AND GEOGRAPHIC VARIATION: No subspecies. The species was described from El Copey de Dota (Talamanca, Costa Rica), but was synonymized with *S. brasiliensis*. Subsequent authors followed this taxonomy until it was demonstrated that *S. dicei* constitutes a distinct rabbit species. Some authors believe that cottontails observed in the central region of Costa Rica, particularly in the region of Braulio Carrillo National Park, may not correspond to *S. dicei*, a hypothesis that requires further testing.

ECOLOGY: Dice's cottontail is generally restricted to elevations above 1,100 m within its area of distribution, and is increasingly common above 1,600 m. At higher elevations it constitutes an important component of the diet of several carnivores, particularly coyotes (*Canis latrans*). This species seems to prefer forest edges and open areas. Dice's cottontail is a good swimmer.

HABITAT AND DIET: Dice's cottontail is especially common in Páramo habitat, e.g., in areas such as Cerro de la Muerte, Cerro Chirripó, and other high areas of the Talamanca Mountain Range. It is locally abundant along roads and forest edges at La Auxiliadora and areas of Páramo in Los Quetzales National Park (both at Cerro de la Muerte area). It is also found close to gardens and agriculture areas in the Coto Brus Valley and adjacent areas. In this region it is common to find rabbits along roads and trails.

Dice's rabbit feeds on short grasses and other green vegetation mainly at clearings near forest edges, roadsides, and open Páramo. Especially after fires, the young shoots of high-elevation bamboo (*Swallenochloa subtessellata*) are abundant, and Dice's cottontail is fond of them. The species' feeding areas in Páramo habitat are conspicuously marked with many groups of small, round fecal pellets typical of *Sylvilagus*.

BEHAVIOR: Dice's cottontail is mainly nocturnal, but is often seen active early in the morning and at dusk. It stays motionless if it is approached with a flashlight, but quickly escapes when approached. All individuals we have observed are solitary.

REPRODUCTION AND DEVELOPMENT: It is very probable that the species breeds year-round; however, kittens of Dice's cottontail have been observed only between September and April.

CONSERVATION STATUS:

IUCN Red List Classification: Data deficient (DD)

MANAGEMENT: Before 2012, Dice's cottontail was classified as a small game species and hunted; however, hunting is no longer legal in Costa Rica. Dice's cottontail is generally restricted to the protected areas of Chirripó and Los Quetzales National Parks (Costa Rica) and La Amistad National Park (Costa Rica–Panama), as well as adjacent montane areas. Another potentially important area for the species is Braulio Carrillo National Park; however, the identity of the species in this area remains to be definitively ascertained. The species is still hunted for meat and pest control in some areas, including Cañas Gordas and Coto Brus, Costa Rica, near the Panamanian border, and sites around Cerro de la Muerte. The species remains common and abundant in the protected areas mentioned above, where it appears unaffected by human activities. However, a status survey is required before adequate conservation measures can be recommended. Research remains to be conducted on population status, basic ecology, threats, and human use of Dice's cottontail.

It is occasionally common to find dead Dice's cottontails killed by cars in the area of Cerro de la Muerte, particularly on the Inter-American Highway. Additionally, the Páramos in the Cordillera de Talamanca burn periodically and completely over large areas. This disturbance likely has some short-term effect on cottontail populations by removing forage and a longer-term effect by removing cover. In addition, there has been an extensive reduction in forested habitat along the lower reaches of the species' distribution range. This has theoretically led to range contraction, as Dice's cottontail is restricted to the forested montane areas. Much of this type of forest is being logged and cleared for pasture right up to the border of protected areas. In addition, it has been hypothesized that there has been an increase of the number of mesocarnivores such as coyotes, tayras (*Eira barbara*), and other potential rabbit predators, due to human-mediated habitat changes, changes that in turn negatively impact Dice's cottontail populations. Our observations at Cerro de la Muerte and the Coto Brus Valley do not support this hypothesis; notwithstanding, research is needed on predation and habitat reduction in this cottontail.

ACCOUNT AUTHORS: José M. Mora and Luis A. Ruedas

Key References: Chapman and Ceballos 1990; Diersing 1981; Harris 1932; Hershkovitz 1950; Humphreys and Barraclough 2014; Janzen 1967; Madriñán et al. 2013; Montes et al. 2015; Mora 2000; Reid 1997; Ruedas et al. 2016; Smith and Boyer 2008b; Timm et al. 1989; Vaughan and Rodriguez 1986; Wainwright 2002; Zeilinga de Boer et al. 1995.

Sylvilagus floridanus (J. A. Allen, 1890)
Eastern Cottontail

OTHER COMMON NAMES: Florida cottontail
DESCRIPTION: The eastern cottontail is a medium to large *Sylvilagus*, and females are slightly larger than males. Throughout its large range, body size increases from south to north and from west to east. The dorsal pelage is long

and dense and varies from brown to gray. The underside of the body and tail is white, and the ears are generally longer in proportion to its head size than found in most cottontail rabbits. The eastern cottontail is similar in appearance to other members of *Sylvilagus* with which it is sympatric, and diagnostic characteristics vary by region. It is often lighter in color than congeners in E North America and Central America. A white spot on its forehead is usually present and can be diagnostic when compared to the New England cottontail (*S. transitionalis*) and the Appalachian cottontail (*S. obscurus*). The hind legs are tan

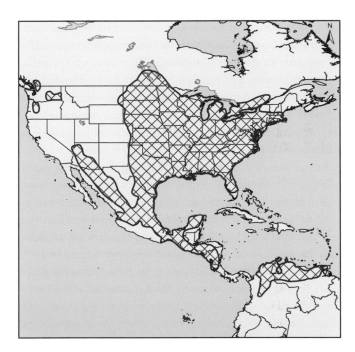

Sylvilagus floridanus. Photo courtesy Randall D. Babb

or white and not orange like in the swamp rabbit (*S. aquaticus*). The eastern cottontail's conspicuous white tail underside can distinguish it from the marsh rabbit (*S. palustris*), similar species in the W United States (brush rabbit *S. bachmani*, pygmy rabbit *Brachylagus idahoensis*) and Central American congeners (Gabb's cottontail *S. gabbi*, Dice's cottontail *S. dicei*).

The braincase is rounded with a broad and high interorbital region. Projections of the supraorbital process may be useful to discern the eastern cottontail from similar species. The posterior projections are transversely broad and generally fused to the braincase on the posterior ends. The anterior projections are slim, short, and nearly lacking or fused to the frontals. The presence of the premolar foramina is variable, and they can be single symmetrically or asymmetrically paired between the second and fourth upper premolars.

SIZE: Head and body 395–477 mm; Tail 25–61; Hind foot 90–105 mm; Ear 55–67 mm; Greatest length of skull 71–76 mm; Weight 800–1,500 g

PALEONTOLOGY: The earliest fossils identified as *S. floridanus* were reported from Florida and date to the Late Pliocene to Early Pleistocene (~ 2.6–1.6 mya). Later fossils from the Irvingtonian (1.8–0.3 mya) occurred in Florida and the C United States (Kansas, Texas, Arkansas, South Dakota). Scant evidence also places them as far east as Pennsylvania and outside their current range in S California during this time. Late Pleistocene fossils (< 0.1 mya) were found in much of the same area (Texas, Tennessee, Nebraska, Georgia) and also more of their contemporary range, including Virginia, Maryland, New Mexico, C Mexico, Colombia, Venezuela, and Alberta, Canada. The eastern cottontail may have remained in Florida and the C United States during glacial maxima, but its distribution likely ebbed and flowed with post-glacial climate changes. Such changes in the Holocene significantly altered the landscape for the eastern cottontail in the SW United States. Hotter and drier conditions created impassible lowlands and isolated populations in the highlands, leading to the divergence of subspecies and species in this region. The eastern cottontail may have occupied the Appalachian Mountains and surrounding foothills relatively late, well after post-glacial climate change (< 10 kya).

CURRENT DISTRIBUTION: The eastern cottontail is the most widely distributed member of *Sylvilagus* and occurs from S Canada into NW South America as far as Venezuela. Historically, it inhabited the E United States from the Rocky Mountains to the East Coast and as far north as New York. Since European settlement in the United States, it has been introduced to areas outside its historic range and N Italy where it is considered invasive.

TAXONOMY AND GEOGRAPHIC VARIATION: Molecular phylogenies suggest *S. floridanus* is a sister taxon to the Appalachian cottontail (*S. obscurus*) and the New England cottontail. More than 30 subspecies of *S. floridanus* have been described reflecting the species' broad geographic distribution and regional adaptations. Four subspecies were described from most of the United States, east of the Rocky Mountains (*S. f. alacer*, *S. f. mallurus*, *S. f. mearnsii*, *S. f. similis*); one from an island off Virginia (*S. f. hitchensi*); three in peninsular Florida (*S. f. ammophilus*, *S. f. floridanus*, *S. f. paulsoni*); three from W Texas, New Mexico, and NE Mexico (*S. f. chapmani*, *S. f. llanensis*, *S. f. nelsoni*); two inhabiting Arizona and Mexico (*S. f. hesperius*, *S. f. holzneri*); seven ranging from C Mexico through Guatemala (*S. f. aztecus*, *S. f. chiapensis*, *S. f. connectens*, *S. f. orizabae*, *S. f. restrictus*, *S. f. russatus*, *S. f. subcinctus*); one from the Yucatán Peninsula (*S. f. yucatanicus*); one occupying Honduras and Nicaragua (*S. f. hondurensis*); one in Costa Rica (*S. f. costaricensis*); six from mainland Colombia and Venezuela (*S. f. continentis*, *S. f. cumanicus*, *S. f. orinoci*, *S. f. purgatus*, *S. f. superciliaris*, *S. f. valenciae*); and three from islands off Venezuela (*S. f. avius*, *S. f. margaritae*, *S. f. nigronuchalis*). The Davis Mountains cottontail (*S. robustus*) and the Manzano Mountains cottontail (*S. cognatus*), previously considered *S. floridanus* subspecies, have been recognized as distinct species based on cranial and dental characteristics. *S. f. holzneri* also bears distinct cranial characteristics and species status has been suggested. The parapatric *S. f. chapmani*, *S. f. mearnsii*, and *S. f. alacer* can be distinguished by blood serum proteins. Within the easternmost United States (the range of *S. f. mallurus*), translocations of multiple subspecies for hunting and subsequent hybridization have influenced the population and no concordance of geographic and genetic structure is evident. Other subspecies have not received taxonomic or genetic scrutiny since being described in the nineteenth and early twentieth centuries.

ECOLOGY: Demographic characteristics such as population density, sex ratios, age structure, and survival rates vary widely across eastern cottontail populations. The highest estimate of population density for eastern cottontails is 20 rabbits/ha, but densities of less than 10 rabbits/ha are more common. Annual survival of adult cottontails is generally 20% to 40%, but can be as low as 5%. Primary mortality factors include avian and mammalian predators, weather, and human harvest. Annual home range sizes of eastern cottontails vary regionally from 0.5 to 8.0 ha, but may be larger in areas of poor habitat suitability. Home ranges are generally larger for males than females due to differences in breeding behavior.

HABITAT AND DIET: Eastern cottontails are found in a wide variety of disturbed, early successional habitats containing dense ground cover and abundant forage. Most productive habitats include old fields and shrublands and areas with high interspersion of vegetation types that provide cover. Within their distributional range, site-level factors (e.g., vegetation density) tend to be more important descriptors of suitable habitat for eastern cottontails than macro-scale factors (e.g., patch size). Forms (depressions) are commonly found in the densest vegetation available; these areas provide ample thermal cover and protection from predators. Underground dens are used in northern latitudes during cold winter temperatures. Eastern cottontails co-exist with swamp rabbits and other leporids in portions of their range, but habitat use usually differs among species when in sympatry.

Eastern cottontails consume dozens of woody and herbaceous plant species, and their diets vary considerably by region. Spring and summer diets are primarily of herbaceous species such as clover, timothy, and alfalfa. Winter diets consist of woody species.

BEHAVIOR: Eastern cottontails are active foragers at dusk and dawn. Aggressive behaviors of the eastern cottontail include charges, scratches at the ground (though less exaggerated than the paw raking of the swamp rabbit), a threat posture with ears flattened, and boxing with the forepaws. Vocalizations include distress cries, squeals during copulation, and grunts made by females when the nest is approached by an intruder. A sequence of partial rushes and jumps occur during times of sexual receptivity, but these do not result in copulation and also occur in males.

Vigorous fighting resulting in injury is rare. Instead, a dominance hierarchy among males is established through aggressive behaviors, and the submissive male generally retreats or crouches. Males do not hold territories, but higher-ranking males chase or dislodge subordinates in the presence of a receptive female. A short period of intense competition among males can occur particularly just before the postpartum estrus, which is the short span of time (< 2 hours) from initial nest-building behavior (plucking of body fur), to parturition, to intense male pursuit and copulation, and return to aggression by the female. A dominance hierarchy has also been observed among females, though their movements are more limited and interactions are fewer than for males.

It has been suggested that the eastern cottontail can aggressively interfere with and outcompete congeners, but competition trials with the New England cottontail indicate outcomes are equally split among the species. The eastern cottontail can, however, detect predators in open habitats better than New England cottontails, potentially due to greater exposed surface area of the eye. This may allow the eastern cottontail to exploit higher-quality food in more open areas rather than being restricted to areas close to cover and providing a competitive advantage in fragmented and disturbed areas.

PHYSIOLOGY AND GENETICS: Diploid chromosome number = 42. Higher genetic diversity in portions of the E United States, relative to sympatric *Sylvilagus* species, has been attributed to frequent translocations of different subspecies and higher effective population size. Lineages of closely related and sympatric *Sylvilagus* show no genetic evidence of hybridization with *S. floridanus*. However, the Davis Mountains cottontail and the eastern cottontail cannot be resolved using genetic markers, likely due to more recent speciation.

The eastern cottontail responds to seasonal changes similarly to the European rabbit (*Oryctolagus cuniculus*) in which stress, as inferred by high adrenal weight, is highest in the winter. Therefore, resource shortages and severe weather may be more stressful to eastern cottontails than pressures in peak breeding season or crowding in late summer. However, relative adrenal weight is higher in males potentially due to establishment of social hierarchy and breeding pressure. Eastern cottontails increase body fat in the fall, but they lose weight in the winter despite consuming more food even in the presence of supplemental food. Thus, the eastern cottontail may be poorly adapted to survive severe winters but well adapted for the breeding season.

REPRODUCTION AND DEVELOPMENT: The onset of breeding, which is dependent on the physiological readiness of females, is brought on by milder spring weather or the presence of succulent vegetation. The breeding season can be year-round in southern latitudes. At this time, the pituitary secretes follicle-stimulating hormone, which stimulates ova to develop to a submature stage and the female is in heat. Luteinizing hormone and ovulation are induced by copulation, so eastern cottontails do not have a true estrous cycle. Despite the lack of a true estrous cycle, females exhibit breeding synchrony through consecutive pregnancies or pseudo-pregnancies, which result when ovulation occurs without fertilization. Progesterone production by the corpus luteum continues throughout the pregnancy, but decreases toward the end, allowing the maturation of follicles during pregnancy. This enables females to conceive again immediately after parturition. Litter resorption, usually early or late in the breeding season, is rare and does not appear to result from excessive lactation during pregnancy as it does for European rabbits. The mean gestation period is about 28 days. Eastern cottontails are the most fecund species of cottontail; mean litter size ranges from three to six, depending on region, with up to eight litters/year. Females create elaborate nests in slanting holes in the ground lined with grass and fur. Little is known about juvenile recruitment in wild populations.

PARASITES AND DISEASES: Several ectoparasite and endoparasite species have been reported in eastern cottontails. Ectoparasites of eastern cottontails include ticks (Ixodidae), fleas (Pulicidae and Leptopsyllidae), and warbles (Cuterebridae). Common endoparasites include nematodes (*Obeliscoides*, *Trichostrongylus*, *Longistriata*, and *Trichuris*), cestodes (*Mosgovoyia* and *Taenia*), and coccidian parasites. Cottontails are known reservoirs of tularemia, and their ectoparasites may carry rickettsial diseases of humans such as Rocky Mountain spotted fever.

CONSERVATION STATUS:

IUCN Red List Classification: Least Concern (LC)

MANAGEMENT: The eastern cottontail is hunted for sport, meat, and fur, and is overall one of the most important game species in the United States. Given declines in hunter effort, hunter harvest does not appear to significantly affect eastern cottontail populations. Populations declined considerably during the twentieth century because of habitat loss and fragmentation due to agricultural development. The primary means for increasing eastern cottontail populations is habitat management, and specific prescriptions include setting back succession in climax habitats to increase ground vegetation, establishing brush piles where ground cover is limited, and promoting the growth of dense understory vegetation.

ACCOUNT AUTHORS: Clayton K. Nielsen and Leah K. Berkman

Key References: Alroy 2013; Althoff et al. 1997; Berkman et al. 2009, 2015; Boland and Litvaitis 2008, Bond et al. 2001, 2002; Chapman and Litvaitis 2003; Chapman et at. 1977, 1980; Crawford 2014; Ecke 1955; Elbroch 2006; Ferrusquía-Villafranca et al. 2010; Fostowicz-Frelik and Meng 2013; Hershkovitz 1950; Lee et al. 2010; Litvaitis 2001; Litvaitis et al. 1997; Mankin and Warner 1999a, 1999b; Marsden and Holler 1964; Matthee et al. 2004e; Olcott and Barry 2000; Probert and Litvaitis 1996; Reid 1997; Robinson et el. 1983b; Ruedas 1998; Scharine et al. 2011; Smith and Litvaitis 2000.

Sylvilagus gabbi (J. Allen, 1877)
Gabb's Cottontail

OTHER COMMON NAMES: Central American lowland forest cottontail; Conejo de monte (Spanish; sometimes applied generically to other wild rabbit species)

DESCRIPTION: *Sylvilagus gabbi gabbi*, distributed from a narrow coastal strip on the Guatemala-Honduras border to the Isthmus of Panama, is a relatively small cottontail. The pelage is relatively dark and typical of forest *Sylvilagus*, but not as dark, e.g., as Dice's cottontail (*S. dicei*), and it is smaller than the latter. *S. g. gabbi* is predominantly brown in overall appearance; closer inspection reveals agouti pelage with bands of black and cream to buff, particularly giving a darker general appearance in the mid-dorsal region. All specimens display a bright orange nape with otherwise sparse pelage; this nuchal patch appears relatively small, generally shorter than the ears. The head differs from the body in having more orange on it, and the agouti banding narrows to give a more homogeneously colored appearance. The ears are basally furred mostly in orangeish brown and distally sparsely furred. The lower ~ 40% of the leading edge of the pinna exhibits a thin creamy-white band of hairs; the caudal aspect of the ears is sparsely furred. The eyes typically have a whitish cream supercilious band trailing and blending into the base of the ears, gaining more black hairs caudally. The ventral surface of the head is cream-colored, blending into a gular

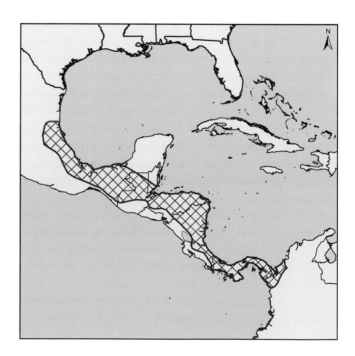

patch that is predominantly buff brown flecked with some cream. The ventral surface is cream to white. The forefeet are predominantly orangeish brown, with a more orange hue closer to the body. The hind feet are lighter in color, buffy brown with somewhat more cream dorsally, darker on the undersurface, with more gray.

Sylvilagus gabbi truei, distributed north of the coastal strip on the Guatemala-Honduras border in the south to S Veracruz in the north, primarily in lowland areas, is somewhat larger than the nominal subspecies. It is somewhat lighter in color than *S. g. gabbi*, with more gray, particularly on the flanks. The mid-dorsal region appears more black than brown, and the agouti bands have more black in them than in specimens of *S. g. gabbi*. The flanks are predominantly yellowish buff with some black bands. The nuchal patch is orange, but with a duller tinge than in *S. g. gabbi*, and generally extending the length of the ears. The ears are similar to those of *S. g. gabbi*. The forehead region lacks any white hairs. A creamish superciliary band is present in all specimens. The forefeet and hind feet are similar to those of *S. g. gabbi*. Juveniles of this subspecies are extremely dark, with very narrow agouti banding resulting in an almost homogeneously colored blackish dark brown color, lacking the white fringe in the proximal anterior edge of the pinna. Subadults are somewhat lighter relative to juveniles, with agouti banding intermediate in breadth between that of juveniles and adults, still with black but brown predominating; the rump area is more brown with no black; the ears are lighter in color than those of juveniles, but still with no white on the pinna. The nuchal patch becomes progressively larger relative to that of juveniles. External measurements are ~ 20% smaller in magnitude than of those of the eastern cottontail (*S. floridanus*).

SIZE: Head and body 302–395 mm (*S. g. gabbi*), 354–372 mm (*S. g. truei*); Tail 13–22 mm (*S. g. gabbi*), 18–22 mm (*S. g. truei*); Hind foot 67–75 mm (*S. g. gabbi*), 77–80 (*S. g. truei*); Ear 40–46 mm (*S. g. gabbi*); Greatest length of skull 71–73 mm

CURRENT DISTRIBUTION: The current distribution of Gabb's cottontail is SE Mexico (southern state of Veracruz, *S. gabbi truei*) south to NW Colombia (*S. g. messorius*). See also "Taxonomy and Geographic Variation," below.

TAXONOMY AND GEOGRAPHIC VARIATION: Gabb's cottontail is currently constituted by six subspecies. The most widely spread are *S. gabbi gabbi* (E Costa Rica, E Nicaragua, most of Honduras, and Panama east of the Isthmus) and *S. g. truei* (S Belize, most of Guatemala, and SE Mexico [extreme SE Tamaulipas, Veracruz, Tabasco,

Chiapas, and extreme SW Campeche]). There are differences in the cranial and dental morphology between the holotypes of these forms that suggest they may be distinct species; however, these should be examined at the population level. Remaining subspecies likewise deserve attention: southern and northern populations currently identified as *S. g. gabbi* are separated by *S. dicei* (see that account) and are only putatively connected along a narrow band on the Caribbean coast of Costa Rica. *S. g. consobrinus* is probably not distinct from other surrounding taxa. However, whether it is more closely allied with *S. g. gabbi* or *S. g. messorius* is an open question. Similarly, *S. g. tumacus* is likely similar to northern range *S. g. gabbi*. *S. g. incitatus* definitively bears scrutiny as it is known only from the type locality (San Miguel Island = Isla del Rey), Archipélago de las Perlas, Panama. Whether *S. g. incitatus* was brought to the island by humans (either pre- or post-Columbian) or actually constitutes an island endemic will be interesting to ascertain. *Sylvilagus g. messorius* differs ecologically from the other taxa currently subsumed within *S. gabbi* in that it is an upland form occurring as high as 550 m asl. Additionally, the cranial characters definitively identify it as taxonomically distinct from the other subspecies. There is controversy as to whether *S. g. messorius* extends into South America or is restricted to the Daríen in Panama. In any case, it is highly unlikely that *S. g. messorius* could be distributed east of the N Andes. Some regard *S. g. consobrinus*, *S. g. incitatus*, *S. g. messorius*, and *S. g. tumacus*, as synonymous with *S. g. gabbi*. Herein the subspecies have been retained because the taxonomy of *S. gabbi* is somewhat unstable and their taxonomic identity remains to be fully resolved.

ECOLOGY: Little is known of the natural history of Gabb's cottontail. It is generally restricted to forested areas in lowland tropics with high humidity. It rarely occurs in sympatry with the eastern cottontail.

PHYSIOLOGY AND GENETICS: Diploid chromosome number (from Panama, hence applies to *S. g. gabbi*) = 38

REPRODUCTION AND DEVELOPMENT: Testes lengths were variable in males of *S. g. gabbi* collected in June and July in Zelaya Province, Nicaragua. In the same study, a pregnant female was collected on 11 May carrying 4 embryos measuring 16 mm in crown-rump length; a lactating female was reported on 19 June and non-pregnant females on 20 and 30 July and 11 August. Otherwise reproduction in Gabb's cottontail is unknown.

CONSERVATION STATUS:

IUCN Red List Classification: Not Evaluated (NE)

MANAGEMENT: Although Gabb's cottontails are generally considered "forest rabbits," it also has been suggested that they prefer clearings in the forests. Notwithstanding, when forest is cleared, the areas formerly occupied by Gabb's cottontail become dominated, or even exclusively occupied, by the eastern cottontail. Because much of Central America's formerly forested area has become deforested or converted to agriculture, the conservation status of Gabb's cottontail and its subspecies should be viewed with more than a little concern. In particular, these animals are elevationally restricted to forests in the tropical lowlands, which also have seen the greatest degree of deforestation and human alteration of habitat.

ACCOUNT AUTHOR: Luis A. Ruedas

Key References: Diersing 1981; Eisenberg 1989; Hall 1981; Hershkovitz 1950; Ruedas and Salazar-Bravo 2007; Yates et al. 1979.

Sylvilagus graysoni (J. A. Allen, 1877)
Tres Marías Cottontail

OTHER COMMON NAMES: Tres Marías rabbit; Conejo de las Islas Marías, Conejo (Spanish)

DESCRIPTION: The Tres Marías cottontail has relatively short ears, and its size varies from medium to large. The dorsal pelage is rufous, and the nape and rump areas are the brightest in color. The lateral parts of the rabbit are pale reddish; the whole venter is whitish; there is a brown section on the throat.

The size of the skull is medium-large, presenting a long rostrum, a long diastema, and long incisive foramina. The maxillary tooth row is relatively short, and the basioccipital is narrow. As in the marsh rabbit (*S. palustris*), the posterior extensions of the supraorbital processes are fused to the braincase throughout most of their length.

SIZE: Head and body 437–480 mm; Tail 50–51 mm; Hind foot 91–99 mm; Ear 57–64 mm; Greatest length of skull 78–80 mm

CURRENT DISTRIBUTION: The Tres Marías cottontail occurs on the four islands of the Tres Marías Islands Archipelago, off the state of Nayarit, Mexico. The extent of its known range is less than 500 km². The subspecies *S. g. graysoni* occurs on the islands María Madre, María Magdalena, and María Cleofas, while *S. g. badistes* is found only on San Juanito Island.

TAXONOMY AND GEOGRAPHIC VARIATION: *Sylvilagus graysoni* is closely related to the adjacent mainland species the Mexican cottontail (*S. cunicularius*). It is hypothesized that the ancestral group of *S. graysoni* may have reached the islands during some early period of land connection to the island. After the population became established on all four islands, the population on San Juanito Island must have become isolated from the others, resulting in characters distinct enough to warrant subspecific recognition as *S. g. badistes*.

Sylvilagus graysoni. Photo courtesy Jorge Antonio Castrejón Pineda

ECOLOGY: The Tres Marías cottontail occurs in deciduous tropical forest up to 305 m asl where the vegetation is predominantly composed of the false tamarind (*Lysiloma divaricatum*), *Cyrtocarpa procera*, and *Bursera* spp. The Tres Marías cottontail has only a few historical predators: raccoons (*Procyon lotor*), red-tailed hawks (*Buteo jamaicensis*), and other prey birds such as caracaras (*Polyborus plancus*). A few non-native species have been introduced to the islands through human activity, threatening the existence of the cottontails, especially due to the competition for resources. White-tailed deer (*Odocoileus virginianus*), pigs (*Sus scrofa*), and domestic goats (*Capra aegagrus hircus*) were introduced deliberately to María Magdalena Island, while the introduction of the black rat (*Rattus rattus*) to all the islands was purely accidental.

HABITAT AND DIET: The habitat on Tres Marías Islands is more arid than on the mainland. Average yearly rainfall is 635 mm, most of which falls in summer, resulting in a tropical dry deciduous and moist forest habitat. Temperatures are moderate although recorded extremes are 4.6°C and 37.5°C; monthly averages range from 20.3°C in January and February to 28.1°C in July and August.

San Juanito Island is covered with dense stands of trees, bushes, and cacti 3–4 m high. The natural vegetation on coastal areas of María Madre Island is tropical deciduous forest with a canopy height of 4 m. Most of this forest was removed by humans for agricultural and governmental purposes, due to the presence of a penal colony there of 2,000–3,000 habitants. The island originally was forested with trees up to 30 m, but most of that vegetation has been logged and removed.

María Magdalena Island was designated an ecological reserve by the Mexican government in 2000. This island has a less disturbed forest and a higher vegetation density than found on María Madre Island. Since there are no permanent residents in this area, the abundance of the Tres Marías cottontail there is higher than on the other islands. On María Cleofas Island, the forest appears to be in good condition; most of the vegetation is open woodland around the coast that grows thicker than that of the center of the island. The eastern side of the island has been cleared in preparation for plantations. In addition, a small military detachment stays near their camp on that side of the island.

BEHAVIOR: The Tres Marías cottontail is an easy-to-catch rabbit since, surprisingly, it does not fear humans. Rabbits from all islands show far less fear of people than do their mainland counterparts. This lack of escape behavior is particularly apparent in the Tres Marías cottontails from San Juanito Island. They are common in old fields covered with a scattered growth of bushes. They become active around 1500 h.

PHYSIOLOGY AND GENETICS: Diploid chromosome number = 42 and FN = 78. The X chromosome is a small acrocentric or subtelocentric, whereas the Y chromosome is a small acrocentric. The Tres Marías cottontail shares the diploid chromosome number with the Mexican cottontail, the eastern cottontail (*S. floridanus*), and the desert cottontail (*S. audubonii*).

REPRODUCTION AND DEVELOPMENT: Little is known of reproduction in the Tres Marías cottontail. Eleven females were lactating and 2 had 75 mm embryos in March.

CONSERVATION STATUS:

IUCN Red List Classification: Endangered (EN)—B2ab(iii,v)

National-level Assessments: Mexico (Endangered Official Norm NOM-059-ECOL-2010 (SEMARNAT))

MANAGEMENT: The Tres Marías cottontail was reported to be abundant on Tres Marías Islands in 1897. However, almost a century later no evidence of activity of rabbits was found on María Madre, María Magdalena, and María Cleofas, although some individuals were observed on San Juanito Island.

Exotic species and farming have had strong negative ecological impacts on the rabbit and contributed to its population decline, although the main cause has been human activities on the islands. An example of this is the situation on María Cleofas, which has been stripped of about half of its native vegetation in preparation for human settlement.

Some vital actions to ensure survival of the Tres Marías cottontail are to cease all forms of hunting, turn San Juanito Island into an ecological reserve, create a biological reserve on María Madre Island, and support scientific investigation to assess the rabbit's present status and specific habitat conditions. A detailed long-term study has strongly recommended the need to identify the current state of conservation and the basic biological characteristics of the species.

Since 1905, the Tres Marías Islands have housed a federal penal colony. In 1998, the archipelago was listed as a priority area for conservation; in 2000, the archipelago was declared a protected Natural Area under the category of biosphere reserve based on the good state of conservation of its ecosystems and the degree of endemism of the species of flora and fauna that it contains.

On 2 June 2010, the International Coordinating Council of UNESCO's Man and the Biosphere Programme (MAB-ICC) added the Tres Marías Islands to the UNESCO's World Network of Biosphere Reserves, with the aim of integrating conservation of biodiversity and management of natural resources.

ACCOUNT AUTHORS: Consuelo Lorenzo, Juan Pablo Ramírez-Silva, Fernando A. Cervantes, and Ricardo Farrera-Muro

Key References: Acevedo 1995; Ceballos and Navarro 1991; Cervantes 1997; Chapman and Ceballos 1990; Chapman et al. 1990; Diersing and Wilson 1980; Dooley 1988; Hall 1981; Lorenzo et al. 1993; Nelson 1899; SEMARNAT 2010; Wilson 1991.

Sylvilagus insonus (Nelson, 1904)
Omiltemi Rabbit

OTHER COMMON NAMES: Omiltemi cottontail; Conejo de Omiltemi, Conejo (Spanish)

DESCRIPTION: The Omiltemi rabbit is a medium-sized cottontail that has a short tail, medium-sized hind feet, and medium-sized ears. The dorsal and nape pelage is rufous black. The ears are black-brown, the sides are grayish black, and the tail is reddish black on the dorsal surface and dingy buffy on the ventral surface. The sides of the nose and the orbital area have dingy buffy grayish coloration. The ventral surface of the cottontail rabbit is

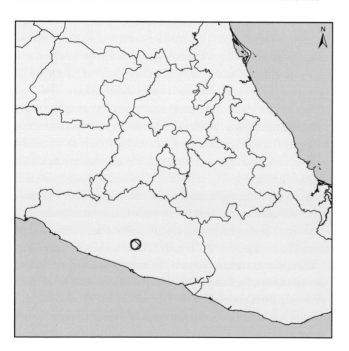

buffy brown. There is a significant amount of white on the dorsal surface of the hind feet.

The skull is medium-sized with a long palate; it has a broad braincase, a narrow breadth across the nasals, large maxillary and mandibular tooth rows, short incisive foramina, a short diastema, a narrow basioccipital, and medium-sized auditory bullae, and is broad across the carotid foramina with a narrow breadth across the infraorbital canals. The shield-bulla depth is shallow as well as the skull depth. The supraoccipital shield is squared, and the posterior section of the supraorbital process is slender and flat.

SIZE: Head and body 398–440 mm; Tail 40–45 mm; Hind foot 89–104 mm; Ear 60–76 mm; Greatest length of skull 77–78 mm

CURRENT DISTRIBUTION: The Omiltemi rabbit has a very restricted distribution, less than 500 km² within the Sierra Madre del Sur, Guerrero, Mexico. This rabbit lives in a highly fragmented area within a natural reserve, the Omiltemi State Park (Parque Ecológico Estatal Omiltemi) in the Sierra Madre del Sur. It occurs at elevations between 2,133 and 3,505 m.

TAXONOMY AND GEOGRAPHIC VARIATION: No subspecies. The phylogenetic relationship of *S. insonus* with other species of *Sylvilagus* is unclear. Morphological comparisons between the tapetí (*S. brasiliensis*), Dice's cottontail (*S. dicei*), and the Omiltemi rabbit indicate that the tapetí and Dice's cottontail are more closely related to each other than either is to the Omiltemi rabbit. The Omiltemi rabbit is one of the most poorly known of all mammals; it is represented by only five museum specimens, four complete and one with only a skin.

ECOLOGY: Very little is known about the natural history of the Omiltemi rabbit. It is sympatric with the Mexican cottontail (*S. cunicularius*). It creates runways within the dense undergrowth, and occupies burrows under rocks or other similar shelters.

HABITAT AND DIET: The preferred habitat of the Omiltemi rabbit is dense cloud forest, where pines (*Pinus*), oaks (*Quercus*), and alders (*Alnus*) are abundant.

BEHAVIOR: The Omiltemi rabbit is mainly a nocturnal species.

PHYSIOLOGY AND GENETICS: Comparative phylogenetic analysis based on partial sequences (560 bp) of the gene 16S between eight species of *Sylvilagus* showed that the Omiltemi rabbit is most closely related to the desert cottontail (*S. audubonii*) and the Mexican cottontail instead of the tapetí as had been proposed previously. The sympatric distribution between the Omiltemi rabbit and the

Mexican cottontail (in Guerrero, Mexico) seems to support the close relationship between these species. Genetic analysis using partial sequences of cytochrome *b* gene revealed an average genetic divergence of 15% between the Omiltemi rabbit and the Mexican cottontail.

CONSERVATION STATUS:

IUCN Red List Classification: Endangered (EN)—B2ab(ii,iii)

National-level Assessments: Mexico (Endangered Official Norm NOM-059-ECOL-2010 (SEMARNAT))

MANAGEMENT: Scientists consider the Omiltemi rabbit to be one of the most endangered mammals in the world. Deforestation is a serious threat to the Omiltemi rabbit, as the forests within its known range have been extensively logged, resulting in habitat fragmentation. Cattle grazing accompanies the deforestation and intensifies the threat. Hunting of the Omiltemi rabbit remains a concern even though most of the known range has been incorporated into a natural reserve area. The type locality of the Omiltemi rabbit is located within the Omiltemi State Ecological Park, which has been declared a natural reserve area (~ 3,613 ha). Intensive studies are necessary to determine whether the Omiltemi rabbit still exists and to confirm its current local distribution. Research and a risk assessment analysis should focus on habitat requirements and the survival of this threatened species. Until such information is available, a comprehensive conservation strategy for this species and its habitat cannot be developed.

ACCOUNT AUTHORS: Consuelo Lorenzo, Fernando A. Cervantes, Julieta Vargas, and Ricardo Farrera-Muro

Key References: AMCELA 2008c; Cervantes and Lorenzo 1997a; Cervantes et al. 2004; Chapman and Ceballos 1990; Diersing 1981; González-Cózatl et al. 2007; Hall 1981; Jiménez-Almaraz et al. 1993; Lorenzo et al. 2015; Nelson 1904, 1909; SEMARNAT 2010.

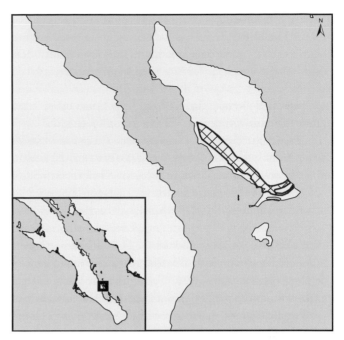

Sylvilagus mansuetus. **Photo courtesy Arturo Carrillo-Reyes**

Sylvilagus mansuetus Nelson, 1907
San José Brush Rabbit

OTHER COMMON NAMES: Conejo de la Isla San José (Spanish)

DESCRIPTION: The San José brush rabbit is a medium-sized rabbit. The dorsal part is pale brown mixed with yellowish gray at the top of the back and head, slightly obscured by the black tip of each hair. The sides of the body are paler gray than the back. The ears are brown to pale mixed with gray and have whitish hairs on the inside. The sides of the neck are paler than the sides of the body, and the throat is gray mixed with white. The front legs are ochre-colored clay, and the hind legs are yellowish white. The ventral part of the body is white mixed with gray, and the ventral part of the tail is pale white.

Proportionally, the skull of the San José brush rabbit is longer and narrower than that of the closely related brush rabbit (*S. bachmani*), and the supraorbital process is fused and the jugal longer.

SIZE: Head and body 303–340 mm; Tail 24–42 mm; Hind foot 72–82 mm; Ear 59–73 mm; Greatest length of skull 60.7–65.4 mm; Weight 620–1,300 g

CURRENT DISTRIBUTION: The San José brush rabbit is an endemic rabbit species on San José Island in the Gulf of California, Baja California Sur, Mexico. The rabbit is only known from a restricted range of approximately 20 km² along the southwestern coastal plain of San José Island.

TAXONOMY AND GEOGRAPHIC VARIATION: No subspecies. It is believed that the San José brush rabbit is closely related to the brush rabbit (*S. bachmani*) from mainland Baja California (Mexico). The San José brush rabbit may be distinguished from the brush rabbit by its larger ears and paler pelage.

ECOLOGY: In 2008, the population density of the San José brush rabbit was estimated to be 25–35/km² in an area of optimal habitat of approximately 4 km². In 2011, the average density was 11.4/km² in xeric scrub in lowlands below 200 m asl. The current estimated number of individuals in this area is 2,087 (range 786–5,547).

The San José brush rabbit shares its habitat with other species of fauna that include mammals such as the peninsula mule deer (*Odocoileus hemionus peninsulae*); several species of mice, including the San José kangaroo rat (*Dipodomys merriami insularis*), Bryant's spiny pocket mouse (*Chaetodipus spinatus bryanti*), the northern Baja deermouse (*Peromyscus fraterculus cinereus*), and the nopalera rat (*Neotoma bryanti bryanti*); carnivores, including the northern cacomistle (*Bassariscus astutus insulicola*) and cats (*Felis sylvestris*); donkeys (*Equus asinus*) and goats (*Capra aegagrus hircus*); reptiles such as whiptail lizards (*Aspidoscelis danheimae*), rattlesnakes (*Crotalus enyo enyo*, *C. mitchellii mitchellii*, *C. ruber lucasensis*), and gopher snakes (*Pituophis catenifer bimaris*, *P. vertebralis*); and birds such as ospreys (*Pandion haliaetus*), red-tailed hawks (*Buteo jamaicensis*), peregrine falcons (*Falco peregrinus*), and American kestrels (*F. sparverius*).

HABITAT AND DIET: San José Island hosts principally xeric scrub; there is also a zone of reduced size where mangrove vegetation is found. The area of highest abundance of the San José brush rabbit is characterized by a rich diversity of desert trees, cacti, and bushes: Adam's tree (*Fouquieria diguetii*), ashy limberbush (*Jatropha cinerea*), mesquite (*Prosopis* spp.), cardon cactus (*Pachycereus pringlei*), cholla (*Opuntia cholla*), copal tree (*Bursera hindsiana*), elephant tree (*B. microphylla*), goatnut (*Simmondsia chinensis*), palo verde (*Parkinsonia microphylla*), sour

Sylvilagus mansuetus habitat. Photo courtesy Tamara Rioja-Paradela

pitaya (*Stenocereus gummosus*), sweet pitaya (*S. thurberi*), wild plum (*Cyrtocarpa edulis*), a local rue (*Esenbeckia flava*), desert thorn (*Lycium* spp.), and ironwood (*Olneya tesota*). It is believed that the San José brush rabbit feeds only on green plants.

BEHAVIOR: The San José brush rabbit is solitary. The rabbit is most active between sunset and 0200 h and from about 0600 to 1000 h; therefore, it is nocturnal and crepuscular. During the day the rabbits rest under palo verde, sour pitaya, sweet pitaya, ashy limberbush, cholla, mesquite, cardon cactus, Adam's tree, and goatnut.

PHYSIOLOGY AND GENETICS: Diploid chromosome number 2n = 48 and FN = 80. Phylogenetic analysis based on partial sequences (560 bp) of the gene 16S showed that the San José brush rabbit is basal in position with respect to seven other species of *Sylvilagus*, which suggests that the San José brush rabbit followed an evolutionary path that is different from that of most species evaluated, and that it shares this quality with the brush rabbit because both species have the same diploid chromosome number (48). The basal position of the San José brush rabbit within the genus supports the proposal that the karyotype of 2n = 48 is a fundamental element in the evolution of the Leporidae, particularly with respect to the diversification of forms within the genus *Sylvilagus*.

The minor nuclear and mitochondrial differences, and color variation between *S. mansuetus* and *S. bachmani*, indicate that this form could be considered a subspecies of *S. bachmani*.

REPRODUCTION AND DEVELOPMENT: The breeding phase in the San José brush rabbit apparently takes place during the wet season, based on finding breeding sites in November, pregnant females in June, and males with scrotal testes in June. Museum specimens of animals collected in June also indicated reproductive activity in the wet season: males with large (19 × 8 mm) scrotal testes and females containing advanced embryos.

CONSERVATION STATUS:

IUCN Red List Classification: Critically Endangered (CR)—B1ab(ii,iii,v)

National-level Assessments: Mexico (Endangered Official Norm NOM-059-ECOL-2010 (SEMARNAT))

MANAGEMENT: All wildlife on San José Island is protected under Mexican law, but local people and fishers often illegally kill protected mule deer and San José brush rabbits when they are legally hunting invasive goats. This activity can be a threat to the entire rabbit population, although more information is necessary to document the extent of the hunting. Also, feral goats may negatively compete with the rabbits by eating and destroying native vegetation.

Predation by native and non-native species constitutes a threat to the entire population of the San José brush rabbit. The most serious threat is from feral cats that frequent the area of occupancy and commonly prey on rabbits. Domestic dogs in areas occupied by people may also constitute a predation threat to the rabbits. Predation by cats and maybe dogs is especially critical given the putative tameness of this insular species. It is also believed that the northern cacomistle may prey on rabbits, but this predator mainly lives in the rocky hills and not in the area of rabbit occupancy on the island; thus they may be only occasional and opportunistic predators. Several other species on San José Island could prey on rabbits, including the three species of rattlesnake, both species of gopher snake, ospreys, red-tailed hawks, peregrine falcons, and American kestrels. The gopher snake (*P. vertebralis*) has been observed eating a breeding female rabbit.

The most important threat to the San José brush rabbit is related to human development activities, in spite of its protected status on San José Island under Mexican law. Plans for a resort are being developed with an accompanying golf course, private airport, and small marina in prime habitat occupied by the rabbit. This development would cause a loss of habitat (area of occupancy) for the rabbit, including breeding and reproduction sites. Additionally, a salt mine close to the area of highest rabbit density may be activated in the near future, and this facility could be used for worker housing and equipment storage. Free-ranging dogs and cats belonging to the workers would then contribute to further predation on the rabbit.

It is urgent to assess pressures caused by the destruction of the habitat of the San José brush rabbit and illegal hunting (sport or subsistence) that have been previously reported and considered within its management programs to eradicate exotic fauna and control hunting. It is necessary to develop research projects to determine population dynamics, reproduction, ethology, and other aspects of the rabbit's biology and ecology that have been ignored.

ACCOUNT AUTHORS: Consuelo Lorenzo, Tamara Rioja-Paradela, Arturo Carrillo-Reyes, Mayra de la Paz, and Sergio Ticul Álvarez-Castañeda

Key References: Álvarez-Castañeda et al. 2006; Case and Cody 1983; Cervantes et al. 1996, 1999; Chapman 1974; Gómez-Nísino 2006; González-Cózatl et al. 2007; Hall 1981; López-Forment et al. 1996; Lorenzo et al. 2011, 2014b, 2015; Nelson 1907; SEMARNAT 2010; Thomas and Best 1994a; Worthington 1970.

Sylvilagus nuttallii (Bachman, 1837)
Mountain Cottontail

OTHER COMMON NAMES: Nuttall's cottontail

DESCRIPTION: The mountain cottontail is a small to medium-sized *Sylvilagus*. The back and rump are gray or buff, the sides are paler, and the belly is white. The nape of the neck is orange. The hind legs are long, and the hind feet are whitish or pale orange and are covered with long, dense hair. Both the front legs and the front feet are rusty-

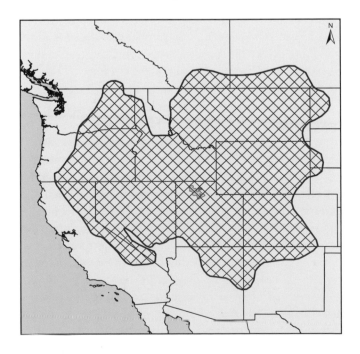

Sylvilagus nuttallii. Photo courtesy Andrew T. Smith

colored. The ears are relatively short and rounded at the tip, and there is dense hair on the inner-ear surface. The large tail is dark and grizzled on top, but white on the underside. Females are about 4% larger than males.

The frontal region, between the small supraorbital processes, is nearly flat to slightly convex. The postorbital constriction is less than 12 mm. Other cranial measurements are smaller, in general. The posterior margin of the palatal bridge is straight and has no median spine.

SIZE: Head and body 290–390 mm; Tail 30–60 mm; Hind foot 80–103 mm; Ear 54–100 mm; Greatest length of skull 56.6–72.6 mm; Weight 628.5–1,100 g

PALEONTOLOGY: *Sylvilagus nuttallii* has been reported from caves across Utah and Texas, dating back as far as ~ 35,000 ybp. The earliest appearance of *Sylvilagus* may be from the Broadwater (Nebraska) local fauna in the Early Pleistocene.

CURRENT DISTRIBUTION: Mountain cottontails occur throughout the intermountain region of North America, and have recently expanded northward into S British Columbia. They occur from 1,190 to 3,450 m in California, from 1,190 to 3,170 m in Nevada, from 1,830 to 3,500 in Colorado, and from 180 to 1,675 m in Oregon, and are restricted to elevations above 2,300 m north of the Mogollon Rim in Arizona. The species' range once extended as far southeast as Texas, but it has since been replaced there by the desert cottontail (*S. audubonii*) and the

Davis Mountains cottontail (*S. robustus*) due to climate change and habitat loss. Hybridization may occur between mountain cottontails and desert cottontails in the deserts of Arizona and New Mexico. Generally, however, the two species are found at different elevations and habitats. Where the ranges overlap, the mountain cottontail primarily inhabits woody areas, and the desert cottontail lives on the plains or more-open habitats, although both species also occasionally enter the transitional sagebrush (*Artemisia* spp.) shrublands. The mountain cottontail inhabits a transitional ecosystem above habitats occupied by lowland leporid species, but below communities inhabited by ochotonids and other leporids such as the snowshoe hare.

TAXONOMY AND GEOGRAPHIC VARIATION: Three subspecies: *S. n. grangeri* (the middle of the distributional range, extending from N Arizona to Alberta and Saskatchewan); *S. n. nuttallii* (the western range, extending from N California to British Columbia); and *S. n. pinetis* (the southeastern range, extending from NE Arizona through Colorado). Based solely on dental morphology, it has been suggested that the species complex was in fact three distinct species; follow-up research using genetic and morphometric characters is ongoing. The eastern cottontail (*S. floridanus*) and the desert cottontail are similar species.

ECOLOGY: In areas of sparse vegetation, mountain cottontails will use burrows, although they apparently do not dig their own burrows. In areas with dense vegetation, they more often live in forms (bare depressions or nests). Their densities vary from 0.06 to 2.5/ha in Oregon shrub-juniper scrublands and from 0.23 to 0.43/ha in British Columbia. Metabolic and free water determine densities, both spatially and across years. Numerous raptor and carnivore species (especially bobcats *Lynx rufus* and coyotes *Canis latrans*) prey on mountain cottontails.

HABITAT AND DIET: Habitat occupied by mountain cottontails varies over the species' geographic range. In the north, the species is associated with sagebrush, whereas in the southern part of its range (and at higher elevations throughout), it is also found in timbered areas. Rocky areas in sagebrush flats and riparian areas and gullies that are located near willows (*Salix* spp.), ponderosa pines (*Pinus ponderosa*), and spruces (*Picea* spp.) are also suitable habitats. Overall, mountain cottontails are associated with rocky, shrubby, and wooded areas. In contrast to many more-sensitive granivore species, the mountain cottontail has been observed at higher rates at horse-grazed sites compared to horse-removed areas in the W Great Basin.

The diet of mountain cottontails consists of sagebrush,

juniper (*Juniperus* spp.), and grasses. Feeding occurs under cover or within a few meters of it. They feed on sagebrush year-round, but in California during summer and spring when grasses are available, they preferentially select them. In regions with dense sagebrush, they are more likely to be found in areas where a grass/forb understory is present.

BEHAVIOR: Mountain cottontails are strongly tied to escape cover, including dense brush, rock crevices, and holes in the ground. They are able to climb trees, and are often seen climbing junipers at dawn in summer. This may be to obtain the water that is exuded from the leaves in the morning. The mountain cottontail is more solitary than its congeners and is usually seen alone or in pairs. When startled, individuals will run 5–15 m into a covered area and freeze, with their ears held erect. Active year-round, they are mainly crepuscular or nocturnal, but can be seen at any hour.

PHYSIOLOGY AND GENETICS: Diploid chromosome number = 42

REPRODUCTION AND DEVELOPMENT: The breeding season of the mountain cottontail lasts from February to July in C Oregon and from April into July in NE California. The gestation period is 28–30 days. Litter size varies according to latitude. In California, females have two litters per year with four to eight young in each. In Nevada, females may have four to five litters with litter sizes between four and eight. In Washington and Oregon, females typically produce four to five litters per year that average four to six young each. In British Columbia the mean litter size is two. Altricial young are weaned after about one month.

PARASITES AND DISEASES: Helminth species known to affect the mountain cottontail include both cestodes (e.g., *Cittotaenia pectinata*, *C. perplexa*, *C. variabilis*, *Raillietina retractilis*, and *Taenia pisiformis*) and nematodes (e.g., *Dermatoxys veligera*, *Nematodirus neomexicanus*, *Protostrongylus pulmonalis*, and *Trichostrongylus colubriformis*). Coccidia have also been reported to affect mountain cottontails.

CONSERVATION STATUS:

IUCN Red List Classification: Least Concern (LC)

MANAGEMENT: The mountain cottontail is utilized for sport and food. It is a common and important game species, with hunting seasons and bag limits controlled by individual states and provinces. It occurs in protected areas within its range, although it has been several decades since the species was last reported in several U.S. national parks (e.g., Grand Canyon, Zion, Lassen Volcanic). The mountain cottontail is abundant on the 1,400 km² Hanford Site (U.S. Department of Energy; state of Washington), where there is no farming, grazing, or predator control.

The need for conservation of this species was recognized as early as 1936. Despite this, the mountain cottontail has been little studied over much of its geographic range. If *S. nuttallii* constitutes a single species, then a species as broadly distributed as it is may have sufficient phenotypic plasticity to accommodate large changes in climate, relative to more-narrowly distributed species. In contrast, if *S. nuttallii* in fact comprises three distinct species, then the geographic distribution of each clade will be much more constrained than the whole and plasticity might be correspondingly less. The most-pervasive threats to the species' distribution include climate change, habitat fragmentation, and competition with sympatric leporids, among others.

Within states and provinces, the species is Presumed Extirpated (SX) from its former range in North Dakota (due to displacement by the eastern cottontail); considered Vulnerable (S3) in Arizona and British Columbia, at the margins of the species' range; considered Apparently Secure (S4) in New Mexico, Oregon, South Dakota, Montana, and Saskatchewan; and considered Secure (S5) in the remaining part of its range. The mountain cottontail is limited in Canada primarily because of loss of habitat to human settlement, agriculture, and cattle grazing.

ACCOUNT AUTHORS: Erik A. Beever and Johnnie French

Key References: Bailey 1936; Carter and Merkens 1994; Chapman 1975; Chapman and Ceballos 1990; Durrant 1952; Fisher 2012; Halanych and Robinson 1997; Halanych et al. 1999; Hall 1951, 1981; Harris and Hearst 2012; Hoffmann and Smith 2005; Hoffmeister and Lee 1963; MacKenzie and Kendall 2002; Matthee et al. 2004e; Newmark 1995; Orr 1940; Ruedas 1998; Verts and Carraway 1998; Verts et al. 1984; Wible 2007; Wood 1940.

Sylvilagus obscurus Chapman et al., 1992
Appalachian Cottontail

OTHER COMMON NAMES: Wood rabbit, Mountain rabbit, Mountain cottontail, Bluebelly

DESCRIPTION: The Appalachian cottontail is a medium-sized cottontail; females are slightly larger than males. The pelage is dense but silky, tan to light brown on the sides washed with gray, and darker dorsally with a

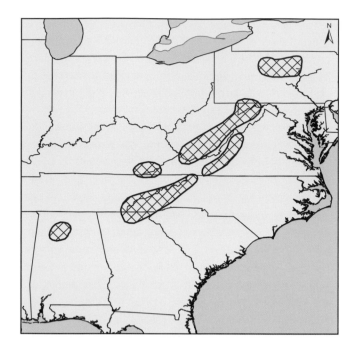

Sylvilagus obscurus. Photo courtesy Kris H. Light

supraorbital process is short or absent. The postorbital processes are slender and posteriorly generally free of, or minimally in contact with, the skull. The interparietal bone is apparent.

SIZE: Head and body 386–430 mm; Tail 22–65 mm; Hind foot 87–97 mm; Ear 54–63 mm; Greatest length of skull 66–75 mm; Weight 756–1,038 g

PALEONTOLOGY: The only fossil evidence of *S. obscurus* comes from subfossils (originally identified as the New England cottontail *S. transitionalis*) in a sink hole in New Paris, Bedford County, Pennsylvania. One hypothesis holds that an ancestral leporid moved northward along the Laurentide ice sheet (into what is now S Pennsylvania to N North Carolina) approximately 18 kya and was split into two populations by the Champlain-Hudson outflow as the ice sheet receded 12–10 kya. These two allopatric populations subsequently evolved into *S. transitionalis* to the north and northeast of the Hudson River and *S. obscurus* to the Appalachians west and south. An alternative hypothesis suggests that chromosomal rearrangements from an ancestral 2n = 48 cytotype resulted in two cytotypes, a northern one of 2n = 52 and a southern one of 2n = 46. The northern cytotype became fixed in the Late Wisconsinan (~ 12.5 kya), and rabbits with this cytotype spread northward with the oak-chestnut (*Quercus-Castanea*) forest, while rabbits with the southern cytotype remained restricted to high-elevation Appalachian habitats to the south. These two cytotypes are now recognized by most as separate species, the northern as *S. transitionalis* and the southern as *S. obscurus*.

CURRENT DISTRIBUTION: The Appalachian cottontail is a native of the E United States. This species exists as disjunct populations at high elevations (570–1,300 m) in the Appalachian Mountains from N Pennsylvania to NW Georgia and north C Alabama. The type locality is Dolly Sods Scenic Area, Grant County, West Virginia, in the Monongahela National Forest.

TAXONOMY AND GEOGRAPHIC VARIATION: Subgenus *Sylvilagus*. No subspecies. The eastern cottontail (*S. floridanus*) and the New England cottontail are similar species. Distinction of the Appalachian cottontail from the New England cottontail is based on karyotypic and craniometric differences, but is not supported by mtDNA analysis.

ECOLOGY: Daytime resting/bedding locations of the Appalachian cottontail are shallow depressions (forms) located in thick ericaceous vegetation, which provides concealment and thermal cover. Nests holding young

streaked appearance created by pencils of black hairs. The underside of the body and tail is white and the inner surface of the ears light-colored. The anterior edge of the short, rounded ear is black. A black spot lies on the top of the head between the base of the ears; very rarely is there a white spot ("star") on the forehead.

Dorsally the skull is long with a triangular rostrum. The nasals are large and widest posteriorly, the posterior margin angling anteriorly and medially and forming a jagged suture with the frontals. The braincase is rounded and large. Interorbital breadth is little constricted. The (kittens) are depressions approximately 10 cm deep and 13 cm in diameter, lined with fur and grass and covered with twigs and leaves. Individuals have been observed near openings in rocky substrate, but the use of underground dens is undocumented. Seasonal 95% adaptive kernel home ranges are 0.5–13.7 ha in area, with no difference between the sexes, although males had larger home ranges during the leaf-on (May–September) than leaf-off (October–April) season at the type locality in West Virginia. Densities reach 0.8/ha. Sex ratios vary considerably by study, from heavily male- to heavily female-biased. Mean finite daily, monthly (28-day) and yearly survival during 1997–2000 at the type locality were 0.9934, 0.8316, and 0.0904. Predation by bobcats (*Lynx rufus*), red and gray foxes (*Vulpes vulpes*, *Urocyon cinereoargenteus*), and raptors is the most frequently confirmed cause of mortality. A decline in populations in the latter part of the twentieth century is believed to have resulted from gradual climatic changes since the last glaciation, but more important, habitat maturation, destruction, and fragmentation, and reinvasion of lowland areas by introduced eastern cottontails.

HABITAT AND DIET: The Appalachian cottontail occupies mixed-oak forests (*Quercus* spp.), clearcuts with blackberry (*Rubus allegheniensis*), or dense ericaceous cover provided by mountain laurel (*Kalmia latifolia*), rhododendron (*Rhododendron* spp.), and blueberries (*Vaccinium* spp.), as well as conifer stands in boreal forests at high elevations in the C to S Appalachian Mountains.

The Appalachian cottontail feeds on ferns, grasses, forbs, shrubs, and conifer needles. Preferred winter browse includes blueberries, eastern teaberry (*Gaultheria procumbens*), black huckleberry (*Gaylussacia baccata*), serviceberries (*Amelanchier* spp.), and chokeberries (*Photinia* spp.). The most avoided browse includes mountain laurel and rhododendron.

BEHAVIOR: The Appalachian cottontail is secretive and not well known; it rarely ventures into the open. Where this rabbit is observed, its behavior differs little from that of the eastern cottontail, apart from habitat preference. Individuals move by short jumps or hops. Although presumed to be chiefly nocturnal, movement rates did not differ from dusk to dawn, from dawn to afternoon, or from afternoon to dusk at the type locality. Although the species is generally solitary, a social hierarchy may be established, with some individuals dominant and others submissive (but no allogrooming occurs). When disturbed or handled, an individual may scream or squeal and emit vocalizations that include a series of "ticks," the latter being unique among *Sylvilagus* spp. The species is presumed to be coprophagous.

PHYSIOLOGY AND GENETICS: Diploid chromosome number = 46. The condition index—W/L^3, where W is body weight in g and L is greatest length in cm—is highest during winter into spring and lowest from summer through fall (i.e., after the breeding season). The body fat index—categorical, from 1 (no trace of fat associated with stomach, small intestine, or kidneys) to 4 (much fat)—is highest in winter. The adrenal index—adrenal weight (mg)/body weight (g)—is highest during the breeding season (March–September). Collectively, these indices indicate that the Appalachian cottontail is well adapted to cold climates (i.e., its high-elevation Appalachian habitat) and that high temperatures and especially reproduction are stressful.

REPRODUCTION AND DEVELOPMENT: The breeding season of the Appalachian cottontail extends from early March through early September. Appalachian cottontails are induced ovulators, and breeding is synchronous. A small percentage of juvenile females breed (18% in West Virginia), but males typically do not breed in their first year. Gestation is 28 days, litter size is 2–8 young, and an average of 23 young per breeding female are born annually. The period of lactation is 16 days. Eyes begin to open at 7 days, and kittens are fully mobile and become independent of the nest by day 16.

PARASITES AND DISEASES: Parasites of the Appalachian cottontail include a nematode (*Dermatoxys veliger*), a botfly (*Cuterebra* spp.), and ticks (*Haemaphysalis leporispalustris* and *Ixodes dentatus*).

CONSERVATION STATUS:

IUCN Red List Classification: Near Threatened (NT); declining and nearly Vulnerable (VU)—A3c; B2ab(i,ii,v)

MANAGEMENT: Appalachian cottontails have been used for sport, meat, fur, handicrafts, jewelry, decorations, and curios. The species is threatened by destruction, fragmentation, and maturation of habitat, as well as urban and suburban development. Encroachment on habitat, and thus competition with the eastern cottontail, has led to declines in Appalachian cottontail populations. The species is also impacted by indiscriminant hunting by sportsmen who are not familiar with the existence, biology, and habitat requirements of the species or how to distinguish it from the eastern cottontail. The species would benefit from additional surveying and population monitoring, conservation of critical habitat, further development and implementation of state management plans across its range, and additional public education and awareness programs focused on the distinction between eastern cottontails and Appalachian cottontails and their habitats.

ACCOUNT AUTHOR: Ron Barry

Key References: Barry and Lazell 2008; Boyce 2001; Boyce and Barry 2007; Chapman 1975; Chapman and Litvaitis 2003; Chapman and Morgan II 1973; Chapman and Stauffer 1981; Chapman et al. 1977, 1982, 1992; Guilday and Bender 1958; Hartman and Barry 2010; Jensen 2003; Laseter 1999; Linzey 1998; Litvaitis et al. 1997; Merritt 1987; Ruedas et al. 1989; Sommer 1997; Stevens and Barry 2002; Sucke 2002; Tefft and Chapman 1983; Whitaker and Hamilton 1998; Wilson and Reeder 2005.

Sylvilagus palustris (Bachman, 1837)
Marsh Rabbit

DESCRIPTION: The marsh rabbit is a smaller rabbit characterized by short, broad ears, a small inconspicuous tail, and slender feet. Unlike the white fluffy tail of the eastern cottontail (*S. floridanus*), the underside of the tail is a dull grayish brown. The rump and flanks are darker brown or reddish brown with some black from banded guard hairs. The pelage of the nape and neck area is often a lighter rufous or cinnamon brown (yellowish brown in some individuals in the Florida Keys). The underside is gray to buff, while the mid-abdominal region can be white. Some melanistic individuals have been observed in the Cape Sable region of Everglades National Park and near Marco Island, Florida.

The braincase is relatively short, broad, and rounded, and the rostrum tapers more rapidly compared to that of the closely related swamp rabbit (*S. aquaticus*). The posterior and anterior extensions of the supraorbital processes are fused to the skull along most or all of their length. The basisphenoid is funnelform, not broadly truncated anteriorly, and not dovetailed into the pterygoids. The anterior extension of the supraorbital process and the bullar spike are reduced or even absent. The frontonasal suture is raggedly dovetailed, and the zygomatic arch is ventrally convex to flat when viewed laterally.

SIZE: Head and body 361–475 mm; Tail 20–51 mm; Hind foot 70–95 mm; Ear 45–70 mm; Greatest length of skull 62–77 mm; Weight 750–1,500 g

PALEONTOLOGY: Marsh rabbits are found exclusively in the coastal plains of the SE United States formed from erosion of the Appalachian Mountains during the Mesozoic era. The first evidence of marsh rabbits in the region came from Sangamon interglacial deposits (between 125,000 and 75,000 ybp) from Melbourne, Florida. More recently (6,500–5,000 ybp), the coastal plains region saw increased rain and the creation of wet swamps, prairies,

Sylvilagus palustris. Photo courtesy J. Rachel Smith

and pinelands that are common today, likely fostering the growth and spread of the marsh rabbit to its current range.

CURRENT DISTRIBUTION: The marsh rabbit occupies wet areas of low elevation across the SE United States from SE Virginia through the eastern portions of the Carolinas, SE Georgia, a small portion of SE Alabama, the Florida Panhandle, peninsular Florida, and the Florida Keys.

TAXONOMY AND GEOGRAPHIC VARIATION: Traditionally, three subspecies have been named: *S. p. palustris* (from the northern part of the species range: NW Florida to SE Virginia); *S. p. paludicola* (N Florida to the upper Florida Keys); and *S. p. hefneri* (the lower Florida Keys). Recent molecular evidence, however, suggests that the marsh rabbits in the lower Keys may be distinct populations or management units, but not a separate subspecies. *S. aquaticus* and *S. floridanus* are similar species, although *S. aquaticus* is considerably larger and the underside of its tail is white. *S. floridanus* is more comparable in size, but is also easily distinguished from the marsh rabbit by its white fluffy tail.

ECOLOGY: Marsh rabbits are most often found within close proximity to water. They establish home ranges and have been thought to exclude members of the same sex. However, they appear to show considerable variation (0% to 20%) in the amount of overlapping core areas from conspecifics, likely as a function of habitat quality. Estimates of average home range size from the Everglades and the Florida Keys range from 0.3 to 32.8 ha. Within their home ranges, marsh rabbits create daytime forms (depressions) in thick groundcover, either in dense herbaceous vegetation, under thickets of woody plants, or in larger tree cavities. Marsh rabbits appear to maintain several forms throughout their home ranges, and will utilize the forms of other rabbits. In addition, rabbits regularly bring items back to their forms to chew on such as crayfish molts (family Cambaridae) and Florida box turtle (*Terrapene carolina bauri*) plastrons. Recent telemetry data show that males and females of different age classes frequently make long-distance movements of up to 5 km.

A large study on marsh rabbit survival and mortality suggested they were heavily preyed on by bobcats (*Lynx rufus*) and coyotes (*Canis latrans*). Raptors, including great horned owls (*Bubo virginianus*), barred owls (*Strix varia*), barn owls (*Tyto alba*), red-tailed hawks (*Buteo jamaicensis*), and bald eagles (*Haliaeetus leucocephalus*), and reptiles, including rattlesnakes (*Crotalus* spp.), cottonmouths (*Agkistrodon piscivorus*), and alligators (*Alligator mississippiensis*), also have measurable impacts on marsh rabbit survival.

HABITAT AND DIET: Marsh rabbits occur in low-lying areas near fresh or brackish water. This includes prairies, sawgrass marshes, cypress domes, flooded forests, urban retention ponds, floating mats of vegetation, tidal zones, coastal dunes and berms, and the edges of lakes, rivers, and canals.

Marsh rabbits consume a broad variety of aquatic and terrestrial plant material, including grasses, sedges, forbs, and seeds, as well as the leaves and twigs of some woody species. Their diet varies seasonally, but they appear to consume species in proportion to their availability. To retain nutrients not absorbed on a first pass through their digestive systems, like other *Sylvilagus* species, the marsh rabbit is coprophagous.

BEHAVIOR: During the middle of the day, marsh rabbits are often found in thick cover, but they venture out on well-beaten paths, or "runs," to eat shorter vegetation between late afternoon and a couple hours after sunrise. When marsh rabbits forage, instead of hopping they often move by alternating feet (walking) or alternating their front feet and then hopping forward. When running they often scurry close to the ground. Like the swamp rabbit, the marsh rabbit often deposits fecal pellets on raised surfaces such as rocks, logs, and stumps when available. Often solitary foragers, marsh rabbits tend to remain still until approached and usually move short distances between patches or take to water. It is not uncommon to see marsh rabbits in water, and they are capable swimmers (although it is not clear that they "like" to swim).

PHYSIOLOGY AND GENETICS: Diploid chromosome number = 38

REPRODUCTION AND DEVELOPMENT: Marsh rabbits breed year-round and can have up to seven litters annually, but pregnancy rates may decrease during late fall (October–December). The gestation period is estimated to be between 30 and 40 days. Marsh rabbits average 3–5 young per litter and can be reproductively active as early as 6 months old, but do not reach full maturity until 9 months of age. Marsh rabbits raise their young in nests consisting of several smaller chambers and exit routes that are constructed of grasses or sedges and sometimes fur. Dispersal of young from their natal range is believed to occur at 8–10 months of age, with males usually dispersing farther than females.

PARASITES AND DISEASES: Marsh rabbits are often found with ticks (including *Haemaphysalis leporispalustris*), chiggers (family Trombiculidae), and fleas around their ears and nose, and can also be parasitized by botflies

(*Cuterebra* spp.). Internal parasites include a host of trematodes, tapeworms, and roundworms. Minimal work has been done on the diseases of marsh rabbits, but they have been recorded with sarcoptic mange and tularemia.

CONSERVATION STATUS:

IUCN Red List Classification: Least Concern (LC)

National-level Assessments: USA ESA for *S. p. hefneri* (Endangered)

MANAGEMENT: Marsh rabbits are common throughout most of their range. They are managed as a game species in Florida, Georgia, North and South Carolina, and Alabama and considered an agricultural pest in sugarcane fields in S Florida. There are only two areas of real conservation concern for the marsh rabbit: the Everglades and the Lower Florida Keys.

In the areas around Everglades National Park in S Florida, marsh rabbits were once the most commonly seen mammal. However, starting in the early 2000s, they appear to have mostly disappeared from the region. Experimental translocations of marsh rabbits suggested that while the habitat in the Everglades appeared to be suitable, the presence of invasive Burmese pythons (*Python molurus bivittatus*) precluded the establishment of any populations there. This conservation problem could continue to grow if these invasive pythons continue to expand northward.

Farther south on the chain of islands (Keys) south of the Everglades, *S. p. hefneri* was listed as an endangered species by the U.S. Fish and Wildlife Service in 1990. This subspecies appears to be threatened by the loss and fragmentation of wetland habitat from sea-level rise and development. Over a 47-year period from 1959 to 2006, 47% of marsh rabbit habitat was lost to sea-level rise, while 8% was directly lost to development. However, development has blocked off the inland migration of habitat triggered by sea-level rise and degraded habitats through fragmentation, fire suppression, and increased alien and invasive species, all of which have been shown to limit the recovery of this subspecies.

ACCOUNT AUTHORS: Robert McCleery, Craig Faulhaber, and Adia Sovie

Key References: Blair 1936; Caldwell 1966; Carr 1939; Chapman and Willner 1981; Crouse et al. 2009; Faulhaber et al. 2007, 2008; Forys 1995, 1999; Forys and Humphrey 1996, 1999; Holler and Conaway 1979; Lazell 1984; Markham and Webster 1993; McCleery et al. 2015; Nelson 1909; Padgett 1989; J. Schmidt et al. 2012; P. Schmidt et al. 2010; Sikes and Chamberlain 1954; Sovie 2015; Tomkins 1935; Tursi et al. 2012; Whitaker and Hamilton 1998.

Sylvilagus robustus (Bailey, 1905)
Davis Mountains Cottontail

DESCRIPTION: The Davis Mountains cottontail is a large rabbit. The dorsal pelage appears more grizzled gray than brownish, similar to that of the parapatric desert cottontail (*S. audubonii*), from which it can generally be distinguished due to its much larger size. The Davis Mountains cottontail's dorsal pelage is also distinct from that of the eastern cottontail (*S. floridanus*), which is darker, mostly brown with some black. The Davis Mountains cottontail has a gray rump patch. The white feet are large and thickly furred. The ears are also large and gray, while the legs are a rusty cinnamon in color. The hair of the Davis Mountains cottontail and the hair of the eastern cottontail are distinct; in particular, the cuticular scales of hair from the eastern cottontail are wavy and hardly visible, whereas those of the Davis Mountains cottontail form deep V-patterns, with rows that are almost vertical. In addition, the midshaft diameter of the eastern cottontail hair is ~ 187 µm, whereas that of the Davis Mountains cottontail is 60% smaller (~ 112 µm).

The skull of the Davis Mountains cottontail is grossly similar to those of the eastern cottontail and the mountain cottontail (*S. nuttallii*). These can be distinguished from the Davis Mountains cottontail only by the latter's overall larger size. The only reliable discrete morphological character to distinguish between the species appears

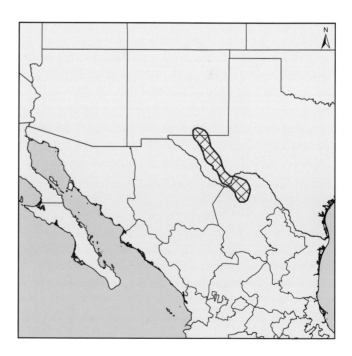

to be the enamel pattern of the third upper premolar. However, most crania of Davis Mountains cottontails examined lacked a tympanic process, which is present in the eastern cottontail, and most mandibles of Davis Mountains cottontails, in contrast to those of the eastern cottontail, had a mental foramen twice as long as high (versus half again as long as high) located on the dorsal aspect of the mandible (versus the labial aspect).

SIZE: Head and body 320–473 mm; Tail 30–71 mm; Hind foot 65–109 mm; Ear 52–86 mm; Greatest length of skull 69–79 mm; Weight 1,300–1,800 g

PALEONTOLOGY: A number of fossil specimens from Sierra Diablo Cave (Hudspeth County, Texas; with a Pleistocene horizon) have been identified as "*Sylvilagus floridanus/robustus.*" The specimens co-occur with *S. audubonii* and *Aztlanolagus agilis*, the latter being a diagnostic element of the regional Rancholabrean fauna. Two additional specimens of the species document its existence in the Middle to Late Wisconsin, and another dates to the Late Wisconsin to the Holocene.

CURRENT DISTRIBUTION: The Davis Mountains cottontail historically has occupied several mountain ranges in the SW United States and N Mexico: the Davis and Chisos Mountains (Texas), the Guadalupe Mountains (Texas and New Mexico), and possibly the Sierra de la Madera and Santa Rosa Mountains (Coahuila, Mexico) and the Sacramento Mountains (Otero County, New Mexico). The upper elevational limit of this species is unknown (although it is documented to at least 2,347 m asl at Chinati Peak, Big Bend National Park). It is not generally known to occur lower than 1,400 m, although one specimen from Big Bend National Park was collected at 853 m, on Pinnacle Mountain.

TAXONOMY AND GEOGRAPHIC VARIATION: *Sylvilagus robustus* is monotypic. The species is closely related to *S. floridanus*. Craniodental and genetic data support the taxonomic status as a distinct species.

ECOLOGY: The Davis Mountains cottontail occurs in Madrean evergreen woodland dominated by evergreen oaks (especially *Quercus grisea*) and juniper (especially *Juniperus deppeana*). Most elements of its natural history are unknown.

HABITAT AND DIET: The Davis Mountains cottontail inhabits piñon pine-oak-juniper woodlands in the Chihuahuan subprovince of the Sonoran Province of the Madrean Floristic Region.

PHYSIOLOGY AND GENETICS: Diploid chromosome number = 42

REPRODUCTION AND DEVELOPMENT: Females with reproductive data from the Davis Mountains were identified as "gravid" on 7 March, lactating on 26 and 27 May, with no embryos on 30 May, and with three embryos on 27 July. Females from Big Bend National Park were listed as lactating with three embryos (15 July) and carrying a single embryo (25 July). Males with descended (scrotal) testes were found in the Sawtooth Mountain area of the Davis Mountains on 27 July. Based on size of the testes, males were reproductively active in the Davis Mountains on 7 March, 26 and 30 May, and 10 June, and inactive on 19 and 21 October. In Big Bend National Park, presumed sexually inactive males were collected on 22 June and 1 August. A juvenile female, possibly recently born (total length 138 mm, mass 77 g), was recorded on 3 July in Big Bend National Park. Larger individuals noted on their specimen tags as juveniles were recorded on 7 March and 27 May.

CONSERVATION STATUS:

IUCN Red List Classification: Endangered (EN)— B1ac(iv); C2b

MANAGEMENT: The population of the Davis Mountains cottontail in the Guadalupe Mountains has decreased in size continually since it was initially reduced to approximately 50 individuals in the 1940s. Few individuals of the Guadalupe Mountains population have been verified since the 1960s, although one specimen was collected in May 2000. It is probable that the Guadalupe Mountains population, like others of the species, is highly dependent on precipitation levels and as a result may be at a historical low. In the Davis Mountains, recent specimens include one from 1996 and 3 road kills from 1997, collectively constituting the first record in 20 years from that area. Several specimens from the Chisos Mountains have been recovered in the past 15 years. The population status in the Sierra de la Madera is not known, but it is reported that there may be a population of *Sylvilagus* in good condition in the Sierra del Carmen, in Coahuila, Mexico, though it has not been confirmed to be *S. robustus*. The Davis Mountains cottontail probably occurred historically in low densities and population numbers, fluctuating as a function of precipitation, resulting in greater sensitivity to threats and leading to local extirpations and extinctions. The narrow elevational range suggests the species likely is sensitive to global warming.

ACCOUNT AUTHORS: Luis A. Ruedas and Robert C. Dowler

Key References: Dalquest and Stangl, Jr. 1984; Debelica and Thies 2009; Lee et al. 2010; Nalls et al. 2012; Ruedas 1998; Van Devender and Bradley 1990; Vestal 2005.

Sylvilagus transitionalis (Bangs, 1895)
New England Cottontail

OTHER COMMON NAMES: Coney or cooney, Wood rabbit
DESCRIPTION: The New England cottontail is a medium-sized cottontail with a dark brown dorsal pelage that has a blackwash giving it a penciled effect. The nape may be a lighter brown or buff. There is a distinct black spot

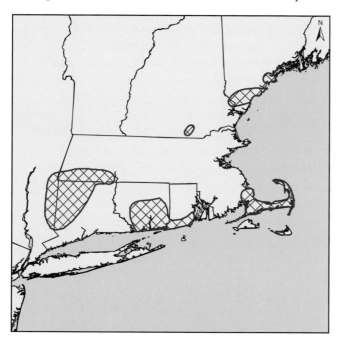

Sylvilagus transitionalis. Photo courtesy Kelly Boland/USFWS

between the ears, and the anterior edges of the ears are outlined in black fur. Females tend to be slightly larger (at least 1%) than males. In the field, the New England cottontail can be mistaken for the eastern cottontail (*S. floridanus*).

The supraorbital process of the New England cottontail is short and narrow. The suture between the nasal and the frontal bones is jagged and irregular and is often used to differentiate between the New England cottontail and the eastern cottontail. The tympanic bullae are smaller than those found in all subspecies of the eastern cottontail.

SIZE: Head and body 386–430 mm; Tail 22–65 mm; Hind foot 87–97mm; Ear 54–63 mm; Greatest length of skull 66–75 mm; Weight 756–1,038 g

CURRENT DISTRIBUTION: The current distribution of the New England cottontail is fragmented, with populations ranging from S Maine and New Hampshire, Cape Cod and W Massachusetts, coastal Rhode Island, much of Connecticut, and extreme SE New York.

TAXONOMY AND GEOGRAPHIC VARIATION: No subspecies. The eastern cottontail and the Appalachian cottontail (*S. obscurus*) are similar species. The New England cottontail and the Appalachian cottontail are considered "sister species" and were previously considered together in the same species. Although hybridization between New England cottontails and eastern cottontails has been suggested, range-wide surveys have not provided any evidence to support this.

ECOLOGY: New England cottontails will den in rock walls and stone foundations, and use burrows made by woodchucks (*Marmota monax*). Brush piles also provide escape cover. Home range size can vary substantially among individual cottontails (0.2–7.5 ha) and may be a consequence of productivity of the site. Size of the occupied habitats also has a direct influence on the vulnerability to predation, where cottontails on small patches (< 3 ha) have a winter mortality rate twice the rate of cottontails on larger patches (> 5 ha). Common predators include red and gray foxes (*Vulpes vulpes, Urocyon cinereoargenteus*), coyotes (*Canis latrans*), fishers (*Martes pennanti*), barred owls (*Strix varia*), great-horned owls (*Bubo virginianus*), and red-tailed hawks (*Buteo jamaicensis*).

HABITAT AND DIET: The New England cottontail occupies a variety of habitats and depends more on the structure of the vegetation (form, height, and density) than on specific plants. The most consistent feature is a dense woody understory (often > 20,000 stems/ha). In old fields, it is commonly associated with juniper (*Juniperus*), blackberry (*Rubus occidentalis*), spirea (*Spiraea*), dogwoods (*Cornus*), viburnums (*Viburnum*), and a variety of young deciduous trees (e.g., *Acer* and *Populus*). Former agricultural fields that are occupied can include an abundance of invasive shrubs such as honeysuckle (*Lonicera*), autumn olive (*Elaeagnus umbellata*), multiflora rose (*Rosa multiflora*), and Japanese barberry (*Berberis thunbergii*). On wetter sites, alders (*Alnus*), willows (*Salix*), and blueberries (*Vaccinium*) are frequent dominants.

Spring and summer diets include herbaceous vegetation, especially grasses such as timothy (*Phleum pratense*) and non-graminoids, including Canadian goldenrod (*Solidago canadensis*) and wild strawberry (*Waldsteinia fragarioides*). November–December is a transition period when cottontails switch to woody plants. Winter diets consist mainly of twigs from small trees, including gray birch (*Betula populifolia*), red maple (*Acer rubrum*), apple (*Malus* spp.), aspen (*Populus tremuloides*), and choke cherry (*Prunus virginiana*), and shrubs, especially blackberry and dogwoods. The incidence of woody bark consumption (considered a low-quality food) by cottontails is most common in late winter when more preferred foods have been depleted.

BEHAVIOR: New England cottontails are largely solitary outside the breeding season. Although competition with the eastern cottontail is considered to be a factor contributing to the decline of populations of the New England cottontail in much of its range, interactions between these species within enclosures did not reveal an obvious dominance hierarchy. Field and captive studies of both cottontails also revealed some differences in habitat affinities. The New England cottontail will rarely venture more than 5 m from escape cover while foraging, whereas the eastern cottontail will range much farther from cover.

PHYSIOLOGY AND GENETICS: Diploid chromosome number = 52. There has been speculation that physiological differences between New England cottontails and eastern cottontails may partially explain the ability of the New England cottontail to persist in the presence of a more adaptable eastern cottontail, where a lower metabolic rate enables the New England cottontail to occupy low-quality habitats that are unsuitable for the eastern cottontail. Environmental chamber trials revealed no difference in metabolism and conductance, but the lower critical temperature of New England cottontails was higher than that of eastern cottontails. Therefore, population trends of the two sympatric cottontails are not explained by differences in their respective metabolic capacity.

REPRODUCTION AND DEVELOPMENT: Recent initiatives to propagate New England cottontails in captivity have re-

vealed that most do not breed until the second year of life. Mating occurs from mid-March through August and yields an average of 2.5 litters/female and 4.2 young/litter.

CONSERVATION STATUS:

IUCN Red List Classification: Vulnerable (VU)—A2ace
MANAGEMENT: Although New England cottontails occupy less than 15% of their historic range, a substantial initiative is in place to create and maintain habitat within a large portion of the historic range. This effort includes federal and state natural resource agencies as well as national and regional non-governmental organizations. State agencies in Maine and New Hampshire list this species as endangered. A recent decision by the U.S. Fish and Wildlife Service indicated that the New England cottontail should not be listed as threatened or endangered. Urban expansion continues to isolate remaining populations. The role of competition between eastern cottontails and New England cottontails on the restoration of the New England cottontail should be investigated. The New England cottontail is hunted for sport and meat.

ACCOUNT AUTHORS: John A. Litvaitis, Marian Litvaitis, Adrienne Kovach, and Howard Kilpatrick

Key References: Barbour and Litvaitis 1993; Brown and Litvaitis 1995; Chapman and Litvaitis 2003; Chapman et al. 1992; Fenderson et al. 2011, 2014; Litvaitis 1993; Litvaitis and Villafuerte 1996; Litvaitis et al., 1997, 2006, 2007; Probert and Litvaitis 1996; Smith and Litvaitis 2000; Tash and Litvaitis 2007; Villafuerte et al. 1997; Warren et al. 2016.

Sylvilagus varynaensis Durant and Guevara, 2001

Venezuelan Lowland Rabbit

OTHER COMMON NAMES: Barinas wild rabbit; Conejo de Barinas, Conejo de monte (Spanish)
DESCRIPTION: The Venezuelan lowland rabbit is larger and darker than other rabbits that reside within its distribution. It is longer, is heavier, and has larger cranial measurements than the tapetí (*Sylvilagus brasiliensis sensu stricto*). Though similar in length to the eastern cottontail (*S. floridanus*) in the region, it is typically heavier, has a shorter, more uniformly colored tail, and has larger cranial measurements except for palatal length. *S. f. valenciae* has a palatal length of 40.02 ± 4.76 mm, while in *S. vary-*

naensis it measures 35.67 ± 5.43 mm. The tip of the nose is tawny, while the fur on the rostrum grades into a tawny-cinnamon that is densely mixed with black at the frontal region. The nares are white at the outer border, and the cheek and supraorbital region are buffy, although slightly bordered with black at the postero-ventral side. The orbit is ringed in white. The ear is light buffy on the outer surface. The nuchal patch is reddish and cylindrical. Dorsally the Venezuelan lowland rabbit is buff-colored, bordered with black, and the lateral coloration is light buffy. The ventral surface is white, and the gular patch is reddish cinnamon. The legs are reddish cinnamon, mixed with light cream and white hairs dorsally; the internal side of the thigh is reddish cream. The base of the tail is reddish cinnamon dorsally. Females have three pairs of mammae and are slightly heavier and longer than males.

SIZE: Head and body 425–434 mm; Tail 23–24 mm; Hind foot 86 mm; Ear 60–61 mm; Greatest length of skull 79–82 mm; Weight 1,602–1,739 g

CURRENT DISTRIBUTION: Little is known regarding the distribution of the Venezuelan lowland rabbit. Currently, it is limited to the low-altitude areas (82–240 m) of 3 Venezuelan states: Barinas, Portuguesa, and Guárico.

TAXONOMY AND GEOGRAPHIC VARIATION: No subspecies. The Venezuelan lowland rabbit is closely related to the tapetí.

ECOLOGY: The crab-eating fox (*Cerdocyon thous*) and the margay (*Leopardus wiedii*) are the main predators within the Venezuelan lowland rabbit's distribution, along with some falcons and owls. The spiny rat (*Proechimys* spp.) is considered a potential food competitor. Observation and capture rates for this rabbit were typically lowest during the rainiest months (April–October), with a noted peak in August–September that coincided with cooler temperatures. The highest observation rates occurred in February, August, and December.

HABITAT AND DIET: What little is known regarding the habitat and diet preferences of the Venezuelan lowland rabbit comes from a study conducted in 1989 in the state of Barinas, Venezuela. The study areas were characterized as being low in plant diversity, dominated by low shrubs and with segments cleared for agricultural use and timber harvest. The Venezuelan lowland rabbit was most closely associated with vegetation types "Escobillia" and "Bruscal-Urapal" than other types due to its herbaceous and shrubby nature, which provides cover and food. "Escobillia" vegetation is dominated by two shrubs in the Malvaceae (*Sida* spp. and *Malvastrum* spp.). Preference for consuming these plants has been noted by local residents. The following vegetation was also recorded in the region occupied by the Venezuelan lowland rabbit: West Indian elm (*Guazuma ulmifolia*), yellow mombin (*Spondias mombin*), crecopia (*Cecropia* spp.), fig (*Ficus insipida*), septicweed (*Senna occidentalis*), petite flamboyant bauhinia (*Bauhinia multinervia*), sensitive plant (*Mimosa* spp.), and lobster-claws (*Heliconia* spp.), along with native grasses (Poaceae) and sedges (Cyperaceae).

BEHAVIOR: The Venezuela lowland rabbit is sexually dimorphic; females are larger than males. It is assumed that this difference in size can be attributed to similar behavior observed in other small mammals in the region. Females tend to be larger because they are more sedentary, expending energy only for nest building/maintenance, reproduction, and neonate care. Males expend more energy defending their territory and preserving social hierarchy.

REPRODUCTION AND DEVELOPMENT: Reproduction peaks between September and January, tapering off during the rainy season. The highest average number of embryos (2.8) was recorded in January, with an overall average litter size of 2.6. Gestation is 35 days. The highest percentages of males with scrotal testes were captured in March, September, and December. Captured rabbits are considered adults when body length is greater than 420 mm, with subadults measuring 400–420 mm. The sex ratio among adults is essentially 1:1.

PARASITES AND DISEASES: Adult and larval *Taenia* spp. and Nematoda have been recorded as parasites of the Venezuelan lowland rabbit, but they do not seem to negatively impact infected rabbits. It is inferred that the rabbits have either adapted to these parasites or were not infected with enough of them to be deleterious.

CONSERVATION STATUS:

IUCN Red List Classification: Data Deficient (DD)

MANAGEMENT: No studies have been conducted to determine any necessary management actions. It is assumed that the correlation between population declines of other rabbit species in association with increased habitat change resulting from agricultural activities would also negatively impact this species. Illegal hunting and increasing domestic and feral animal activity are also likely to have harmful impacts.

ACCOUNT AUTHOR: Charlotte H. Johnston

Key References: Durrant and Guevara 2000a, 2000b, 2001.

Genus *Lepus* Linnaeus, 1758

Hares (genus *Lepus*) are the largest of all lagomorphs. The 32 species of hare are nearly cosmopolitan, with many species being found on most continents (Africa, Asia, Europe, and North America). They are not native to South America or Australia, although *L. europaeus* has been introduced on each of these continents. Some hare species are among the most northern in distribution of any terrestrial mammal; others inhabit deserts or open grasslands. Most features of hares represent an exaggeration of the features of the smaller lagomorphs: they have spectacularly large ears, hind feet, and eyes. They have four digits on both front and hind feet, although the first digit tends to be reduced. The dental formula is 2.0.3.3/1.0.2.3 = 28.

Lepus alleni Mearns, 1890
Antelope Jackrabbit

OTHER COMMON NAMES: Antelope hare, White-sided jackrabbit, Allen's jackrabbit; Liebre torda, Liebre blanca, Liebre appaloosa (Spanish)

DESCRIPTION: The antelope jackrabbit has been considered the handsomest and most striking of the North American hares. It was one of the last mammals in North America to be described, despite its large size. It has ex-

tremely long ears coupled with pearly white flanks, hips, and rump. The long ears and lanky legs led to the formulation of the famous ecogeographic "Allen's Rule." Dorsally it is dappled in brown, gray, and black with distinctive black nape markings and a throat patch of ochraceous buff. The white and black tail is short, and the huge ears are nearly hairless and fringed in beige with little or no black on the tips. The legs are long and slender, and the

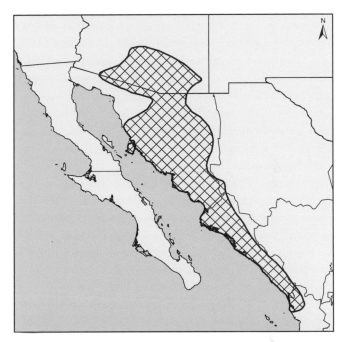

Lepus alleni. Photo courtesy David E. Brown

mostly fulvous head is large. The lower parts of the largely white feet are densely furred in cinnamon brown. The sexes are similar in appearance, and the reddish orange eyes are set high on the head, giving a motionless animal a greater than 300° view of its surroundings. The female has six mammae, two pectoral and four abdominal. Two anal glands secrete a yellowish substance that emits a musky odor when the animal is excited.

The skull of the antelope jackrabbit is proportionally large, as are the zygomatic breadth and nasal bones. Generous premaxillaries and molars contrast with the similar, but more fragile-appearing, skull of the black-tailed jackrabbit (*L. californicus*). The frontal area is broad and the rostrum is long and heavy as is the supraorbital and postorbital processes. The auditory bullae are small, and the basioccipital is long and not deeply constricted. Only one molt occurs annually, and the hairs cannot be differentiated from those of the black-tailed jackrabbit by cuticular examination.

SIZE: Head and body 430–700 mm; Tail 50–140 mm; Hind foot 113–145 mm; Ear 137–180 mm; Greatest length of skull 85–102 mm; Weight 2,500–4,400 g

PALEONTOLOGY: Determined largely on the basis of size, *L. alleni* first appeared in the Early Irvingtonian ~ 1.1 million ybp, where it might have been conspecific with the then widespread *L. giganteus*. Reported to occur as far east as Florida during the Pleistocene, animals attributed to *L. alleni* have been reported from such SW Pleistocene sites as Ventana Cave in Arizona and Burnet Cave in S New Mexico, the latter animal possibly being the white-sided jackrabbit (*L. callotis*).

CURRENT DISTRIBUTION: The antelope jackrabbit is unique to Arizona and NW Mexico. Its range in Mexico includes four states and Tiburon Island in the Gulf of California. The full extent of its geographic range is estimated at 102,000 km², with 12.5% (12,700 km²) of the animal's distribution being in Arizona, 76% (78,000 km²) in Sonora, 10.5% (9,500 km²) in Sinaloa, and the remaining 1% (1,800 km²) in W Chihuahua and N Nayarit. Within this range it is restricted to the plains, valleys, and foothills of the Sonoran and Sinaloan biotic provinces. With the exception of the similar Tehuantepec jackrabbit (*L. flavigularis*), the antelope jackrabbit has the most restricted range of any non-insular North American hare.

TAXONOMY AND GEOGRAPHIC VARIATION: Three subspecies: *L. a. alleni* (Arizona and N Sonora); *L. a. palitans* (southern extension of the species range); *L. a. tiburonensis* (isolated on Tiburon Island in the Gulf of California). Genetic comparisons are needed to verify the integrity of these taxa, as an analysis of 12 cranial characters of the

three *L. alleni* subspecies produced no clusters, indicating a weak if any differentiation between subspecies. Similar species include the white-tailed jackrabbit (*L. townsendii*), the white-sided jackrabbit, and the Tehuantepec jackrabbit, all of which are allopatric to the antelope jackrabbit.

ECOLOGY: The antelope jackrabbit is most often encountered in areas of low relief, including level ground, gently sloping plains, alluvial fans, and mesas. Fewer observations occur in broken country, and steep slopes appear to be used mainly as travel routes between mesas and arroyo bottoms. It normally occurs between 185 and 1,200 m asl, placing the animal in mostly tropic-subtropic environments receiving mean annual rainfall amounts between 200 and 450 mm, of which a minimum of 90 mm falls during April through September. Minimum temperatures appear to influence the distribution of the antelope jackrabbit; it has been documented only in areas where mean annual temperatures exceed 18°C, and where the average number of days 0°C or colder is less than 60.

Home ranges of the antelope jackrabbit in good habitat in Arizona average 29.14 ha. Males tend to have larger home ranges (35.97 ha) than females (19.57 ha). Population sizes fluctuate dramatically depending on habitat condition, predator densities, and other factors not yet identified. Most mortality appears to occur during the first six months of life. Known predators include coyotes (*Canis latrans*), bobcats (*Lynx rufus*), mountain lions (*Puma concolor*), jaguars (*Panthera onca*), the larger hawks (*Buteo* spp.), great-horned owls (*Bubo virginianus*), and golden eagles (*Aquila chrysaetos*). Probable predators of *L. a. tiburonensis* on Tiburon Island are coyotes (*C. l. jamesi*), which are present in high density, and possibly gray foxes (*Urocyon cinereoargenteus*) and rattlesnakes (*Crotalus molossus estebanensis*). The main competitor there for forage is the mule deer (*Odocoileus hemionus sheldoni*). The antelope jackrabbit is reported to be an important disperser of plant seeds via its fur and especially its feces.

The antelope jackrabbit is sympatric with the black-tailed jackrabbit in Arizona and much of N Sonora, although mixed groups of the two species are rarely encountered. The black-tailed jackrabbit ranges both lower and higher in elevation than the antelope jackrabbit. Antelope jackrabbits also occur more often in groups than do black-tailed jackrabbits. Observations on Tiburon Island included only solitary individuals and groups of two animals.

HABITAT AND DIET: Antelope jackrabbits are most commonly seen in areas of present or former Sonoran savanna grassland or a Sonoran savanna grassland / semi-desert grassland transition. Other observations occur in vegetation characteristic of either the Arizona Upland subdivision or Lower Colorado River subdivision of the Sonoran Desert. These jackrabbits also use mesquite bosques and thornscrub, and infrequently are seen in retired farmlands and buffelgrass (*Pennisetum ciliare*) pastures. Velvet mesquites (*Prosopis velutina*) are often a dominant over-story plant, and no other plant is more often associated with antelope jackrabbit observations than this species. Relatively few observations occur in temperate oak woodlands and semi-desert grasslands more than 1,220 m in elevation.

Percentages of various classes of vegetation in the diet of the antelope jackrabbit are grasses and forbs (45%), mesquite (36%), and cacti (8%). Important grasses include spidergrass (*Aristida ternipes*), purple threeawn (*A. purpurea*), cane bluestem (*Bothriochloa barbinodis*), jackrabbit grass (*Bouteloua barbata*), sideoats grama (*B. curtipendula*), Rothrock's grama (*B. rothrockii*), feather fingergrass (*Chloris virgata*), bush muhly (*Muhlenbergia porteri*), witchgrass (*Panicum capillare*), whiplash pappusgrass (*Pappophorum vaginatum*), and sand dropseed (*Sporobolus cryptandrus*), all of which are believed to provide both hiding cover and food. Interestingly, these grasses average 64 cm in height, approximately the mean length of an adult antelope jackrabbit. Antelope jackrabbits also occupy dense stands of tropical grasses found along drainages and other more mesic areas, including sideoats grama, needle grama (*Bouteloua aristidoides*), vine mesquite (*Panicum obtusum*), Bermuda grass (*Cynodon dactylon*), spear grass (*Heteropogon contortus*), sprangletops (*Leptochloa* spp.), and Arizona cottontop (*Digitaria californica*).

The principal habitat requirement for the antelope jackrabbit in Arizona appears to be a potentially herbaceous understory within a velvet mesquite or other Sonoran vegetation matrix typical of a tropic-subtropic environment. Important forbs include such winter annuals as globe mallow (*Sphaeralcea emoryi*), native and non-native mustards (*Brassica* spp., *Sisymbrium* spp., *Lesquerella* spp., *Descurainia* spp.), nievitas (*Cryptantha* spp.), plantains (*Plantago* spp.), filaree (*Erodium cicutarium*), California poppy (*Eschscholzia mexicana*), broadfruit combseed (*Pectocarya platycarpa*), and red-maids (*Calandrinia ciliata*). Summer rains result in significant areas occupied by Mexican poppy (*Kallstroemia grandiflora*), amaranths (*Amaranthus* spp.), and purselanes (*Portulaca* spp.).

Cacti are also frequently associated with the antelope jackrabbit, these succulents being present in about one-third of the observations. Based on the evidence of tracks, fecal pellets, and teeth marks at the base of the cacti, it is believed that the consumption of cacti pulp and fruits allows the hare to survive without free water and causes the animal to pass urine in white, bird-like splashes.

The diet varies with location and vegetation availability. Antelope jackrabbits feed intensely and often with green grasses and forbs being the primary food items during winter-spring and summer monsoon rains when grasses constitute more than 80% of the diet. At other times the velvet mesquite is the preferred item, the hares eating the leaves, bark, buds, and tufts of newly emerging green leaves during the appropriate seasons. So heavy is the use of this plant that browse-lines can often be seen in heavily populated areas, where the hares stand on their hind legs to reach green leaves and stem tips. Cacti, particularly barrel cacti, and prickly pear pads and fruit are fed on during dry seasons, presumably for the water content as such use appears to decline where stock tanks or catchments are present. Cactus roots, mineral dirt, and the feces of other jackrabbits are also eaten.

BEHAVIOR: The antelope jackrabbit's skipping run with head held high at first appears comical and inefficient, but an especially energetic fourth or fifth bound enables the hare's bulging eyes to survey the brushy terrain ahead while checking for any pursuers behind. When pressed, the animal pursues an all-out run that can attain a speed of more than 70 km/h and may encompass leaps of up to 1.5 m in height. Rarely does the animal run in a straight line, and if more than one animal is flushed each tends to go its own way. Otherwise, when feeding, the hare hops along in a rather ungainly manner, often holding its hindquarters higher than its forequarters.

Active year-round, the antelope jackrabbit is primarily a nocturnal and crepuscular animal that can be observed at any hour. During the day, individuals rest in shallow depressions known as forms, typically under the cover of a mesquite or other shade-providing plants. These forms, which range from 8 to 15 cm wide and 28 to 46 cm long, are typically free of rocks and sticks; these are temporary structures that may be used repeatedly or only occasionally. Here the hare may be seen resting on its haunches or laid out like a dog with feet resting fore and aft. The ears may be erect or laid back, depending on the alertness of the animal. Generally solitary, resting animals may also be seen in pairs and less often in groups of up to a half-dozen animals or more.

While usually silent, individuals may squeal or scream when in distress, and thump their hind feet when alarmed. Antelope jackrabbits are difficult to handle and maintain in captivity, the animals struggling or throwing themselves against the pen until serious injury occurs.

Males may fight by engaging in "boxing matches" on their hind legs, pummeling each other for up to a minute or more. These contests, which are really scratching matches, continue until both participants are bloodied and one of them leaves the field.

PHYSIOLOGY AND GENETICS: Diploid chromosome number = 48. Genetic analyses indicate a common ancestor of the white-sided jackrabbit clade evolved with the increasing climatic isolation attendant with the Pleistocene into the temperate white-sided jackrabbit, the tropic-subtropic antelope jackrabbit, and the tropical Tehuantepec jackrabbit.

The antelope jackrabbit possesses thermoregulatory, physiological, and behavioral adaptations for surviving in hot, arid environments and feeding on low-quality foods. Able to dissipate heat via its lanky extremities, the antelope jackrabbit also employs heat-avoiding activity patterns such as feeding at night, use of shade, and solar alignment when resting. Having a normal body temperature of ~ 37.9°C, the antelope jackrabbit is able to withstand external temperatures of up to 51°C for up to 4 hours. Its relatively high lethal body temperature of 45.4°C is avoided by an efficient mean water evaporation rate of less than 0.10% of body mass / h, which is obtained through respiration rates that may be up to 350/ minute. The tolerance of antelope jackrabbits for cold is less pronounced, animals showing symptoms of hypothermia when temperatures were below 20°C.

The antelope jackrabbit does not require preformed water and is one of the largest "desert" animals thought able to subsist entirely on metabolic water obtained from plants. This adaptation is thought to be due to the large percentage of succulent cacti in the animal's diet.

REPRODUCTION AND DEVELOPMENT: Antelope jackrabbits are thought to be promiscuous, with receptive females mating with any sexually mature male encountered. Under favorable conditions they breed as early as 6 months of age. The breeding season extends from late December to September, with one peak in spring and another in summer. Timing of breeding appears to depend on temperature and forage conditions. Mating activity consists of males fighting, chasing, and jumping, and a short copulation is sometimes accompanied by low-pitched growls or chucks. Gestation is about 6 weeks, and neonates are precocial, weighing from 133 to 184 grams at birth. The fur-lined nests are usually well concealed in a depression or vegetation. The young, which have a white spot in the center of the forehead, are more wary, more nocturnal, and more sedentary than adults. The mean litter size is 1.9 (range 1–5), with a potential of 3 to 4 litters/year. Ac-

tual recruitment rates appear much lower, however, and population size appears determined more by variation in juvenile mortality than variation in recruitment rate.

PARASITES AND DISEASES: Antelope jackrabbits are susceptible to a variety of pathogens, and animals have been observed suffering from tumors, sores, and mange. Parasite loads, especially during the warmer months, may be excessive, with up to 50% of the animals harboring botflies (*Cuterebra princeps*, *C. americana*) in the throat and rump areas. Tapeworm cysts (*Taenia multiceps*, *Cittotaenia* spp.) are also common, as are an array of nematodes (e.g., *Dermatoxys veligera*, *Nematodirus arizonensis*, *Passalurus ambiguus*). Ectoparasites include chiggers (*Hexidionis allredi*), fleas (e.g., *Hoplopsyllus glacialis affinis*), and ticks (*Haemaphysalis leporispalustris*, *Dermacentor albipictus*, *D. parumapertus*).

CONSERVATION STATUS:

IUCN Red List Classification: Least Concern (LC)

MANAGEMENT: The antelope jackrabbit remains a common animal within suitable habitat, but habitat losses within its limited range have been extensive due to agricultural and residential development. Because the species shuns open habitats, extensive croplands, clean farms maintained by chemicals, and cleared pastures can limit or eliminate populations. Additionally, vehicle collisions can constitute significant local sources of mortality, as can predator irruptions, extremely cold weather, and, probably, disease. The animal has no legal status other than as "wildlife" in either Arizona or Mexico. Antelope jackrabbits are shot for subsistence meat, sport, dog food, and bait to trap fur-bearing carnivores. In the past, live individuals have been used to train racing greyhounds. Previously the species was regarded as a pest during years of high abundance on certain rangelands and in certain agricultural fields. Such conditions are now of rare or no occurrence.

Research and monitoring are needed to address long-term status and trend of populations, the animal's long-term relationship with the black-tailed jackrabbit, and, most important, to understand the driving forces relating to its population dynamics. Understanding the relationships of individuals within groups would also be of scientific interest.

ACCOUNT AUTHORS: David E. Brown, Consuelo Lorenzo, and Maria Altemus

Key References: Alcalá-Galván and Miranda-Zarazúa 2012; Allen 1890, 1906; Altemus 2016; AMCELA 2008b; Armstrong and Jones, Jr. 1971; Arnold 1942; Best and Henry 1993; Brown et al. 2014; Caire 1978; Dawson and Schmidt-Nielsen 1966; Dixon et al. 1983;

Gray 1977; Hinds 1977; Hoffmeister 1984; Huey 1942; Kürten and Anderson 1980; Lorenzo et al. 2014a, 2014b; Madsen 1974; Mearns 1890; Palmer 1897; Roth and Cockrum 1976; Townsend 1912; Vorhies and Taylor 1933; Woolsey 1956.

Lepus americanus Erxleben, 1777
Snowshoe Hare

OTHER COMMON NAMES: Varying hare, Snowshoe rabbit

DESCRIPTION: In summer the dorsal pelage of the snowshoe hare is sandy brown and the ventral pelage is grayish white. In winter snowshoe hares molt to a white pelage of longer hair. The ears are brownish with black tips and white or creamy borders. During winter, the hare is almost entirely white, except for black eyelids and the blackened tips on the ears. Some southern or coastal populations fail to turn white in winter. The soles of the feet are densely furred, with stiff hairs that form the snowshoes on the hind feet.

Distinctive skull characteristics include an indistinct interparietal bone that is entirely fused with the parietals. Also, the supraorbital in the snowshoe hare lacks the conspicuous anterior projection that is found in most jackrabbits.

SIZE: Head and body 380–505 mm; Tail 49–52 mm; Hind foot 95–145 mm; Ear 62–70 mm; Greatest length of skull 81 mm; Weight 1,400–2,300 g

PALEONTOLOGY: Recent molecular analyses of the phylogeny of the Lagomorpha place the divergence of *L. americanus* from all the other *Lepus* species at the end of the Miocene, approximately 8–10 mya. *Lepus* species from Eurasia diversified mainly in the Pleistocene, during the past 2.5 million years. Known fossil sites for snowshoe hares are described and mapped in http://www.ucmp.berkeley.edu/neomap/use.html and indicate a broad range of fossil sites from the C United States (N California, Utah, Missouri, to Virginia) to NW Canada (near Tuktoyaktuk at the mouth of the Mackenzie River), and into C and W Alaska.

Lepus americanus summer. Photo courtesy Moose Peterson / WRP

Lepus americanus winter. Photo courtesy Alice J. Kenney

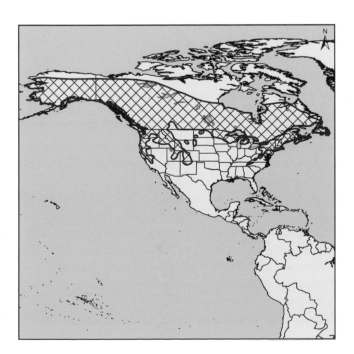

CURRENT DISTRIBUTION: The snowshoe hare may be found in boreal and mixed deciduous forests of N North America. It occurs in all provinces and territories of Canada, except Nunavut. In the United States it is present in Alaska, as well as the western mountain states of Oregon, Washington, Nevada, Idaho, Montana, Wyoming, North and South Dakota, and Colorado, and small pockets in high-elevation areas in New Mexico, Utah, and California. Its distribution also includes the Great Lakes region and the eastern states of Pennsylvania, New York, Maine, Vermont, Rhode Island, Wisconsin, Michigan, Minnesota, Massachusetts, Connecticut, and New Hampshire. Historically, snowshoe hares were found in mountain portions of West Virginia, North Carolina, Tennessee, and Virginia, but those populations currently appear to be very low.

TAXONOMY AND GEOGRAPHIC VARIATION: *Lepus americanus* is taxonomically distinct from all North American *Lepus*, and extant populations are not known to currently hybridize with any species. For the most part, *L. americanus* is isolated either geographically or in terms of its choice of habitat from congeneric species. Fifteen subspecies: *L. a. americanus*, *L. a. bairdii*, *L. a. cascadensis*, *L. a. columbiensis*, *L. a. dalli*, *L. a. klamathensis*, *L. a. oregonus*, *L. a. pallidus*, *L. a. phaeonotus*, *L. a. pineus*, *L. a. seclusus*, *L. a. struthopus*, *L. a. tahoensis*, *L. a. virginianus*, and *L. a. washingtonii*.

ECOLOGY: Snowshoe hare populations typically fluctuate in density over an 8- to 11-year cycle, although the amplitude of the cycle varies greatly and may decrease in southern populations. In northern populations the amplitude of the cycle averages about 15 to 1, but on occasion may be as high as 40 to 1. As the southern edge of the snowshoe hare distribution is fragmented by agriculture and human habitations, cycles tend to disappear, presumably because of generalist predators that invade patchy forest. The hare cyclic peak tends to be synchronous over large spatial areas of approximately 300 × 300 km, but is not in phase across all of North America. In Alaska, spatial variation in hare cycles may be greater because of forest fire history and mountain barriers.

HABITAT AND DIET: Snowshoe hares live in boreal forest and do not colonize tundra areas. Within the boreal forest their distribution and abundance are patchy, largely dictated by cover and winter food supplies. In winter they eat the terminal twigs of a variety of shrubs and small trees, with local food choice dictated in part by secondary chemical defenses in the available plants. When peak densities are high, bark stripping of both shrubs and trees is highly visible, and extensive vegetation damage and mortality of both shrubs and small trees may occur. In summer snowshoe hares eat a variety of leafy plants, and summer food is not limiting. Captive snowshoe hares must be fed a mixed diet; a single-species diet, even if composed of the most preferred plant species, results in weight loss and mortality.

BEHAVIOR: Snowshoe hares are not territorial but occupy home ranges of 3–5 ha, with males having larger home ranges than females. Aggression occurs only during the mating period when several males may chase a female in estrus. No records of infanticide have been recorded in snowshoe hares. Snowshoe hares are nocturnal and are readily seen in the evening in summer when day lengths are maximal.

PHYSIOLOGY AND GENETICS: Diploid chromosome number = 48. Little genetic work has been done on snowshoe hares. One analysis of 8 microsatellite loci and mitochondrial DNA sequences (cytochrome *b* and control region) in almost 1,000 snowshoe hares showed a hierarchical structure suggesting initial subdivision in two groups: boreal and southwestern. The southwestern group further split into Greater Pacific Northwest and U.S. Rockies. These three groups correspond with evolutionarily significant units that might have evolved in separate refugia south and east of the Pleistocene ice sheets. Genetic diversity appears highest at mid-latitudes of the species' range, and genetic uniqueness was greatest in southern populations. Snowshoe hares in the Greater Pacific Northwest mtDNA lineage were more closely related to black-tailed jackrabbits (*L. californicus*) than to other snowshoe hares, which may result from secondary introgression or shared ancestral polymorphisms. Unlike their northern counterparts, some snowshoe hares in the Greater Pacific Northwest remain brown year-round. Given the genetic distinctiveness of southern populations and minimal gene flow with northern snowshoe hares, fragmentation and loss of southern boreal habitats could mean loss of many unique alleles and reduced evolutionary potential.

Physiological research on snowshoe hares has previously concentrated on their cold tolerance and more recently focused on stress responses to predator attacks. Stress has now been shown to reduce the reproductive rate of females as well as to be transmissible as an epigenetic trait from mothers to daughters. For snowshoe hares, stress originates from predator attacks that are unsuccessful, and thus varies cyclically with hare and predator numbers.

REPRODUCTION AND DEVELOPMENT: In northern parts of their range, snowshoe hares begin breeding in April and have the first litter when the snow is melting in mid-May.

Subsequent litters follow with a gestation period of 35–37 days and postpartum estrus. Females tend to synchronize their breeding periods. Up to four litters can be produced in a summer, and the reproductive output varies with the cyclic phase. During the increase females may produce 16–19 leverets per female per year, while during the decline phase only 6–8 leverets per female will be produced, largely because the third and fourth litters are absent. Snowshoe hares do not reach maturity until they approach age 1, and they do not breed in the season of their birth. Leverets are 55–65 g at birth and are precocious in development. By 7–10 days of age they begin feeding on green plants. They are fed once per day by the female early in the night, and are weaned by 24–28 days of age. Birth weight varies from 54 to 79 g, depending on the season of birth.

PARASITES AND DISEASES: Snowshoe hares are parasitized by many nematode and cestode species. *Trichuris leporis*, *Obeliscoides cuniculi*, *Dirofilaria scapiceps*, *Trichostrongylus* spp., *Nematodirus triangularis*, *Passalurus nonanulatus*, *Taenia pisiformis*, and *Eimeria* spp. can have relatively high infestation rates and show cyclic fluctuations. Juvenile hares often have heavier parasite infestations than adults. There is limited evidence that parasite loads affect survival or reproduction in snowshoe hares, although experimental studies are few.

CONSERVATION STATUS:

IUCN Red List Classification: Least Concern (LC)

MANAGEMENT: The snowshoe hare is the key species in the food web for several furbearers that have been and still are part of the fur trade in Canada. They form a large fraction of the diet of Canadian lynx (*Lynx canadensis*) and coyotes (*Canis latrans*), and part of the diet of wolverines (*Gulo gulo*) and wolves (*C. lupus*). They were used extensively for food by First Nations people before the twentieth century. At the present time they are hunted only a little in most of Canada. The exception is Newfoundland, where they are an important food source for many people and harvesting is sufficiently strong to affect the cyclic period because of harvesting in the low phase. Hunting in the New England states is permitted, but with low bag limits. Because the abundance of snowshoe hares varies in a cyclic manner, few populations are currently overharvested and management is minimal. Overall abundance in the southern part of their geographic range is primarily affected by the amount of forest harvesting, which produces dense successional vegetation beneficial to hares.

ACCOUNT AUTHORS: Charles J. Krebs and Dennis L. Murray

Key References: Bloomer et al. 1995; Boutin 1984; Boutin et al. 1985; Bryant et al. 1994; Burton and Krebs 2003; Burton et al. 2002; Cary and Keith 1979; Cheng et al. 2014; Hodges 2000; Hodges et al. 2001; Joyce 2002; Keith 1963, 1990; Keith et al. 1985; Krebs et al. 1986, 1995, 2001, 2013, 2014; Murray 1999, 2000, 2003; Murray et al. 1998; O'Donoghue 1994; O'Donoghue and Bergman 1992; O'Donoghue and Boutin 1995; O'Donoghue and Krebs 1992; Sheriff et al. 2009, 2015; Sinclair et al. 1988, 2003; Smith et al. 1988; Stefan and Krebs 2001.

Lepus arcticus Ross, 1819
Arctic Hare

DESCRIPTION: The fur of the Arctic hare in summer is gray in southern populations and white in northern populations. This species is completely white in winter apart from the black ear tips. The winter fur is long and

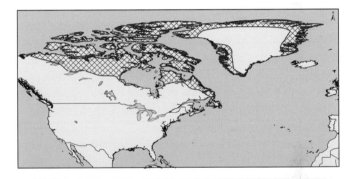

Lepus arcticus. Photo courtesy Moose Peterson/WRP

smooth. The hares strip off loose hanks of hair by rolling in the snow during their molt. The eyes have a yellowish brown color. The ears are anterior blackish and posterior white, with a whitish stripe delineating the black tips. The large feet are cushioned with a dense yellowish brush. The front and hind feet have long claws that are well adapted for digging through snow. Northern Arctic hares are larger than those in the south.

SIZE: Head and body 560–660 mm; Tail 45–100 mm; Hind foot 146–164 mm; Ear 80–90 mm; Greatest length of skull 103 mm; Weight 2,500–6,800 g, depending on geographical location

PALEONTOLOGY: Historical demographic estimations indicate that *Lepus arcticus* and the Alaskan hare (*L. othus*) persevered in two separate North American arctic refugia (Beringia and High Canadian Arctic) during glacial advances of the Pleistocene. The Arctic hare distribution was changed radically subsequent to the post-Pleistocene recession of the continental ice sheets. At least two colonizations across the Bering Strait region are probable for this hare species. The oldest evidence of the Arctic hare living in Greenland was found at Zackenberg in the form of 7,960-year-old droppings. Remains of the Arctic hare have been found in 34,000-year-old sediments on Banks Island, Northwest Territories, Canada; in a 1,200-year-old Eskimo site in N Greenland; and in a Norse waste heap in Greenland.

CURRENT DISTRIBUTION: The Arctic hare inhabits the ice-free coastal area of Greenland and the Canadian Arctic islands southward to the western-central shoe of the Hudson Bay, onward northwest to the west of Fort Anderson on the coast of the Arctic Ocean. Isolated populations live in N Quebec, Labrador, and Newfoundland (Canada). Many islands inhabited by this species are accessible across the ice in winter. This species occurs at elevations from sea level up to 900 m. The geographic distributions of the Arctic hare and the snowshoe hare (*L. americanus*) marginally overlap, but the two species vary in habitat. The Arctic hare inhabits treeless barrens or tundra, whereas the snowshoe hare occurs in woodlands. The replacement of the Arctic hare by the snowshoe hare in Newfoundland may be an instance of competitive exclusion. Other hypotheses list insufficient appropriate food or predation by the Canadian lynx (*Lynx canadensis*) to be the cause of the restricted range of the Arctic hare in Newfoundland.

TAXONOMY AND GEOGRAPHIC VARIATION: Previously the three Arctic hare species, the mountain hare (*L. timidus*), the Arctic hare, and the Alaskan hare, were placed in *L. timidus* based on morphological characteristics on the grounds that they form a circumpolar "ring species." This is also supported by genetic analysis of mtDNA, although evidence based only on mtDNA should be treated thoughtfully. Another hypothesis states that two species exist: the mountain hare in the Old World and the Arctic hare in Greenland, N Canada, Alaska, and the Chukchi Peninsula, Russia. Other lagomorph taxonomists think that the Arctic hare is conspecific with the mountain hare and different from the Alaskan hare. Awaiting conclusive evidence, the three species are currently assumed to be independent species, with the mountain hare in the Old World, the Alaskan hare in Alaska, and the Arctic hare in N Canada and Greenland. As taxonomists are still trying to clarify the species differentiation in *Lepus*, the subspecific taxonomy has not yet been elaborated. The original descriptions of the subspecies are often not very helpful as they are mostly based on a few exterior characteristics and on a small number of individuals. It has been shown that the variability among subspecies can be clinal; hence, the distinction in subspecies might be arbitrary and unreasonable. There are currently nine recognized subspecies: *Lepus arcticus arcticus, L. a. andersoni, L. a. bangsii, L. a. banksicola, L. a. groenlandicus, L. a. hubbardi, L. a. labradorius, L. a. monstrabilis,* and *L. a. porsildi.*

ECOLOGY: Peak densities of the Arctic hare seem high since these animals are large and frequently very noticeable with white coats against a snowless background, but vast regions may have no hares at all and numbers fluctuate from year to year. Therefore, densities documented differ greatly such as 0.2/100 ha on Kugong Island, 1/100 ha in Newfoundland, 8–15/100 ha on Banks Island, and 100/100 ha on Brunette Island. The Arctic hare may be migratory as most hares vanish during summer, seemingly by migrating northward. The species might migrate southward in November. Nonetheless, others disbelieve that mass migrations are undertaken by the Arctic hare. Home range size is variable and ranges from 9 to 290 ha depending on habitat quality. The sizes of male home ranges are double those of female home ranges in Newfoundland in summer. Movements increase in March and April with the beginning of the breeding season. The hares might preserve themselves from extreme cold in winter by burrowing into the snow. These dens are made up of a channel approximately 10 cm in diameter and 30 cm in depth with an enlarged terminal cavity.

HABITAT AND DIET: The Arctic hare inhabits tundra principally north of the tree line in the Arctic life zone, arctic-alpine, or exposed coastal barren regions. Hillsides or rock-strewn plateaus are favored rather than plain bog habitat. The hare's habitat is subject to unbroken darkness for numerous weeks in winter. The hares pass the summer north of the tree line throughout most of their distribution, but in winter they may move more than 160 km into the northern border of trees. Regions with slight snow cover and wind exposure are preferred, and groups of several hundred animals might gather in these areas during winter, moving from one place to another in search of fodder. The Arctic hare's diet consists primarily of woody plants throughout the year. Arctic willow (*Salix arctica*) is the main species consumed in all seasons and makes up to 95% of the winter diet. The hares consume the foliage, buds, bark, and roots of the willow. It feeds on mosses, lichens, buds and berries of crowberry (*Empetrum nigrum*), young blooms of saxifrage (*Saxifraga oppositifolia*), mountain sorrel (*Oxyria digyna*), and numerous species of grasses. This hare species' diet is highly variable in summer, but includes Arctic willow, mountain avens (*Dryas integrifolia*), and grasses. Up to 70% of the hare's summer diet is composed of the legumes alpine milkvetch (*Astragalus alpinus*) and Maydell's oxytrope (*Oxytropis maydelliana*) on Banks Island. Dwarf willows (*S. arctica* and *S. herbacea*) and crowberry form the main food on Baffin Island. The hare's diet consists of Arctic willow and mountain sorrel in Greenland. The Arctic hare frequently feeds on stomach contents of eviscerated caribou (*Rangifer tarandus*). Water is acquired by eating snow. It makes tracks from one browsing place to the next. Often high places are chosen to feed. This hare digs through the snow to find food. It stamps on the crust with its forelegs to make a hole if the snow has a hard crust. Feeding typically takes place in the morning and the evening.

BEHAVIOR: The Arctic hare rests during daytime when the sun is shining, but during the dark winter it has no fixed time for resting. When resting, it regularly sits close to a big stone, dozing or sleeping, protected from wind, concealed from aerial predators, and warmed by the sun. Typically, it selects a place a little way up a hill. Arctic hares habitually are immobile during resting time, sitting crouched with the ears half upright and the eyes almost closed. Frequently two to four hares rest together. The Arctic hare habitually leaves the resting place and begins feeding in the afternoon. Hopping on the hind feet without touching the ground with the forefeet has been described for disturbed and scared Arctic hares in northern populations. It is uncertain how common this behavior is. This species can run up to 64 km/h, and can swim freely across small watercourses that cross the Arctic barrens. The Arctic hare normally is silent, but lactating females might emit a short series of low growls as they come close to their nursing place. The hares are generally solitary, but flocking behavior is very characteristic and animals might gather in groups of 100–300 individuals. Most individuals might be asleep in large groups, but one typically is awake and attentive, and keeps a lookout to all sides. This species forms groups of about 15–20 individuals from early winter until early spring. Animals often move among groups. Adults dominate the juveniles, and dominance is unrelated to sex or breeding condition. Groups dissolve when the breeding season begins, pairs are formed, and each pair establishes a small territory. The male typically leaves the female after the birth of the leverets, but occasionally stays close.

PHYSIOLOGY AND GENETICS: Diploid chromosome number = 48. Recent genetic work focused on taxonomic and phylogenetic clarification analyzing mtDNA only.

The animal's white coloration is explained as camouflage against predators. However, the white coloration may actually be an adaptation to prevent overheating as the basic thermoregulatory adaptation in cold environment is insulation thickness. This insulation must protect the animals from excessive cold when they are at rest. Hence, the heavy insulation might be a disadvantage at other times of the day as problems with excessive heat might occur during activity. The Arctic hare can maintain a normal body temperature (38.9°C) with a low basal metabolic rate (0.36 cm³ O_2/g·h, only 62% to 83% of the values predicted from its body weight) by having a low surface area to volume ratio and effective insulation. The reduction of metabolism is energetically adaptive for a species inhabiting exclusively a cold and typically barren habitat. Kidney mass, overall body mass, and heart mass are significantly less in winter than in summer. This may be a physiological response to a decline in food quality and quantity and to a reduction in basal metabolic rate.

REPRODUCTION AND DEVELOPMENT: The reproductive season of the Arctic hare takes place in April and May. The male follows the female incessantly. During copulation the male bites the female ferociously in the neck and back so that she often is stained with blood. Mating corresponds with molting; therefore, fur is flying during copulation. Gestation lasts 53 days. The leverets are born in May, June, or July in nests lined with dry grass, moss, and the fur of their mother. Nests are either a depression among mosses and grasses, or under or between rocks. The females have litter sizes of 2 to 8 with an average of

5.4. Females in the far north have one litter per year with an average litter size of three recorded in Newfoundland. The newborns have a blackish brown color, whereas the belly, chin, and throat are white. The young flatten out on the soil with their eyes closed and ears pressed firmly to their backs when they perceive a threat. The mother stays with the leverets and shields them from danger during the first two to three days. Nursing happens at an interval of 18–20 hours, and the nursing bout takes 1–4 minutes. By the third days the leverets are able to conceal themselves among stones when danger approaches. The young disperse and hide behind rocks, appearing only when nibbling on vegetation or nursing during the first two weeks of life. The mother remains near the nursing site. Progressively, the leverets enlarge their range to a maximum of approximately 1 km. The juvenile fur is darker than the summer fur of adults. The young leave the mother by the third week of life and form nursery groups comprising up to 20 animals. The leverets are weaned at an age of 8–9 weeks. Subadults are almost as large as their parents and have turned white by late July. The young breed for the first time as yearlings. The adult sex ratio is biased toward males.

PARASITES AND DISEASES: No information on diseases affecting the Arctic hare is available. The Arctic hare is host to several parasites, including protozoans (*Eimeria exigua, E. magna, E. perforans,* and *E. sculpta*), nematodes (*Oxyuris ambigua* and Filarioidea type), other unidentified nematodes, and lice (*Haemodipsus lyriocephalus* and *H. setoni*). Further parasites are an extraordinary number of fleas, including *Hoplopsyllus glacialis, Euhoplopsyllus glacialis glacialis,* and *Megabothris groenlandicus*.

CONSERVATION STATUS:

IUCN Red List Classification: Least Concern (LC)

MANAGEMENT: The Arctic hare is a widespread species, and its population status seems to be constant even though little monitoring takes place. The densities are so low over much of its range that it seems to be at risk, yet there is no indication that this condition is atypical or that it has become altered in historical times. Its high densities on small islands free of competitors and predators probably provides a good conservation method for the Arctic hare. This species is modestly used for food and fur by natives. Southern populations might be subject to habitat loss and perhaps climate change as well, although this is highly hypothetical. The disturbing influence of airplanes and snowmobiles ought to be scrutinized as tourism intensifies.

ACCOUNT AUTHORS: Stéphanie Schai-Braun and Klaus Hackländer

Key References: Angermann 1967; Aniśkowicz et al. 1990; Audubon 2005; Baker et al. 1983; Barta et al. 1989; Bennike et al. 1989; Bergerud 1967; Best and Henry 1994; Bittner and Rongstad 1982; Cameron 1958; Christiansen et al. 2002; Corbet 1983; Dixon et al. 1983; Feilden 1877; Fitzgerald and Keith 1990; Flux and Angermann 1990; Gray 1993; Hall 1951; Hall 1916; Hearn et al. 1987; Hoffmann and Smith 2005; Larter 1999; Loukashkin 1943; Mercer et al. 1981; Murray 2003; Murray and Smith 2008; Parker 1977; Waltari and Cook 2005; Waltari et al. 2004; Wang et al. 1973; Wu et al. 2005.

Lepus brachyurus Temminck, 1844
Japanese Hare

OTHER COMMON NAMES: Nihon nousagi (Japanese)

DESCRIPTION: The Japanese hare is a relatively small species within *Lepus*. Several color forms are recognized, varying from dark brownish gray to reddish brown, with variable amounts of white on the head and legs. In populations in the snow zones of Honshu (on the Sea of Japan side), the fur color turns white in winter. The body color begins to change to white from the middle of September to the end of November, except for the black color on the

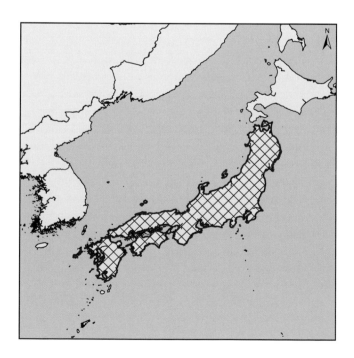

Lepus brachyurus. Photo courtesy Yasuko Segawa

tips of the ears. From the end of January to May, it begins to change back to brown, depending on the photoperiod.

The skull is long and has a triangular rostrum. Laterally, the skull shape is arched with the occipital sinking downward. The braincase is rounded and relatively large. The interorbital region is broad. The supraorbital processes are long and have projections that are short at the anterior. The nasals are long, and the tips of the nasals extend anteriorly a little to the premaxilla. The posteriors of the nasals are wide. The interparietal bone after approximately two months after birth becomes difficult to recognize owing to sutural ossification between the bone and surrounding bones. The bony plate is short. The incisive foramina are long, and their posteriors are wide; the ends of the incisive foramina extend backward to the anterior edge of the premolar. The zygoma is moderately heavy, and its posterior free extremity is moderately long. The auditory bullae are large and round. The mandible has a very large, rounded angular process, which is separated from the condyle by a small, shallow notch.

SIZE: Head and body 450–540 mm; Tail 20–50 mm; Hind foot 120–150 mm; Ear 76–83 mm; Greatest length of skull 79–96 mm; Weight 2,100–2,600 g

PALEONTOLOGY: Fossils of *L. brachyurus* were found from the Middle Pleistocene at Yamaguchi Prefecture in W Honshu and from the Late Pleistocene at Tochigi and Gifu Prefectures in middle Honshu and Hiroshima Prefecture in W Honshu.

CURRENT DISTRIBUTION: Endemic to Japan, the Japanese hare occupies Honshu, Shikoku, Kyushu, and

their islands, including Sado, Oki, Goto, Amakusa, and Shimokoshiki, with the exception of Hokkaido.

TAXONOMY AND GEOGRAPHIC VARIATION: Four subspecies: *L. b. angustidens* (northwestern half of Honshu); *L. b. brachyurus* (southeastern half of Honshu, Shikoku, and Kyushu); *L. b. lyoni* (Sado Island); and *L. b. okiensis* (Oki Islands). The published year of the species authority of *L. brachyurus* is 1844, but some authors have used the wrong year (1845).

ECOLOGY: The home range size of the Japanese hare is 10–30 ha, and the distance of movement at night is 1,176 m (range 841–1,729 m). The red fox (*Vulpes vulpes*), the Japanese marten (*Martes melampus*), weasels (*Mustela* spp.), and raptors are the main predators. Bones and hair are found in 36% of red fox fecal pellets and 13.8% of Japanese marten fecal pellets.

HABITAT AND DIET: The habitat of the Japanese hare is variable—from seaside forests to mountainous lands, including agricultural land, forests, and meadows. They prefer open fields, edges of forests, and young forests that supply undergrowth as food and cover in which to hide. However, the area of preferable habitat in mountainous areas has become smaller recently due to 40- to 50-year-old artificial conifer forests. Here food and cover are insufficient; these forests occupy most of the Japanese hare's distribution range within these zones.

The Japanese hare consumes a wide variety of plants, with more than 150 species recorded. It eats more than 64% to 75% of all plant species (53–59 species) found in forest habitats. Almost all available grasses and forbs are eaten by this hare. Approximately 30% of twigs and shoots clipped by it are not eaten because it chooses twigs and shoots that have stem diameters of 3 mm or less. Hares begin to debark stems once they are more than 7 mm in diameter. They eat 200–500 g of plants per day, which is equivalent to 10% to 20% of their body weight. The Japanese hare is coprophagic and even eats some hard feces if food is scarce while in captivity.

BEHAVIOR: The Japanese hare is solitary and nocturnal. It becomes active at about 1900 h and returns to a form (depression) at 0700–0800 h, after a period of restless running. The mating system is promiscuous. Males chase females and box to repel rivals. The time of copulation, including mating behavior, is very short (1–2 minutes).

PHYSIOLOGY AND GENETICS: Diploid chromosome number = 48

REPRODUCTION AND DEVELOPMENT: Breeding season and litter size vary between northern and southern populations of the Japanese hare. In the north (Yamagata Prefecture), the breeding season extends from February to July with young born between April and August, mostly in May and June, while in the south (Kagoshima Prefecture) the breeding season is year-round. The average litter size is 1.86 (range 1–4) in the north and 1.16 (range 1–3) in the south. Gestation period is 42–43 days in the north and 45–48 days in the south. Weight at birth is 77–165 g in the north and 125–150 g in the south. Females become sexually mature at 8 to 10 months. Growth rate is faster and adult body weight is 20% heavier in the north than in the south in order to survive winters. Females are induced ovulators, and ovulation occurs 12–15 hours after stimulation of copulation. Parturition takes place in a shallow den (e.g., 5 cm in depth and 20 cm in diameter) dug by the female, and the duration of parturition is short (2 minutes). Newborns can run approximately 1 hour after delivery, owing to precocity. Young begin to feed on plants on day 8, but are nursed until 1 month after birth, suckling occurring only at midnight for approximately 2 minutes. Average life span is 1 year, and the maximum is around 4 years. Thirty percent of newborns survive to year 1.

PARASITES AND DISEASES: The Japanese hare is known to be infected by several cestodes, including *Mosgovoyia pectinata* and *Taenia pisiformis*. Ectoparasites include ticks (*Haemaphysalis flava*, *Ixodes nipponensis*, and *I. persulcatus*). The Japanese hare can be infected by *Francisella tularensis* causing tularemia, and it is an important zoonosis from the hare to humans.

CONSERVATION STATUS:

IUCN Red List Classification: Least Concern (LC)

National-level Assessments: Japan Ministry of the Environment for *L. b. lyoni* (Near Threatened (NT)—criteria a and b)

MANAGEMENT: The Japanese hare is a game species. The current status of the Japanese hare is declining due to the recent reduction of preferable habitat. One subspecies, *L. b. lyoni*, endemic to Sado Island (855 km²) in Niigata Prefecture, is experiencing a severe decline due to the reduction of preferable habitat and the predation impact of introduced Japanese martens. During the 1950s and 1960s, 53 Japanese martens and seven red foxes were introduced from Honshu to the island for biological control to reduce feeding damage of tree seedlings by the hares. The martens covered all of the island, but the foxes did not successfully colonize.

ACCOUNT AUTHOR: Fumio Yamada

Key References: Abe et al. 2005; Fujimoto et al.1986; Fujita 2004; Hirakawa 2001, Hirakawa et al. 1992; Imaizumi 1960; Kawamura et al.1989; Nunome et al. 2010, 2014; Otsu 1974; Shimizu and Shimano 2010; Takada and Yamaguchi 1974; Taniguchi 1986; Temminck 1844; Torii 1989, 1990; Tsuchiya 1979; Wu et al. 2005; Yamada 1987, 1990, 2014, 2015a; Yamada et al. 1988, 1989, 1990, 2002; Yamaguchi et al. 2008; Yamaguti 1935; Yatake et al. 2003.

Lepus californicus Gray, 1837
Black-tailed Jackrabbit

OTHER COMMON NAMES: California jackrabbit, Gray-sided jackrabbit, Narrow-gauge mule, Jackass-hare, Texas jack, Texan hare, Great Plains jackrabbit; Liebre de cola negra, Liebre (Spanish)

DESCRIPTION: The black-tailed jackrabbit is the most widely distributed North American jackrabbit and is medium-sized for all measurements, relative to other *Lepus* species. The pelage is buffy gray or sandy-colored above, peppered with black, but white below. The tail is whitish or pale gray underneath and black above; the black coloration extends onto the rump. The ears are very long, conspicuously tipped with black, and white on the back. Eye-shine at night is reddish.

The interparietal bone is fused to and indistinguishable from the parietal bones. The rostrum is narrow and long, and tapers anteriorly. Occipito-nasal length is greater than

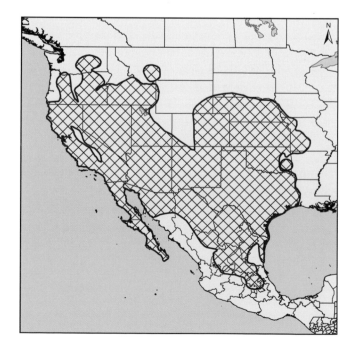

Lepus californicus. Photo courtesy Randall D. Babb

75 mm. The pronounced supraorbital process has an anterior projection, and its posterior projection is usually long, tapering, and fused to the cranium.

SIZE: Head and body 430–700 mm; Tail 50–140 mm; Hind foot 113–145 mm; Ear 137–180 mm; Greatest length of skull 85.4–101.9 mm; Weight 1,300–3,600 g

PALEONTOLOGY: Fossil remains and DNA analysis show that the black-tailed jackrabbit is genetically distinct from the sympatric white-tailed jackrabbit (*L. townsendii*) and more closely related to the white-sided jackrabbit (*L. callotis*) clade, this taxon evolving from a common ancestor some 1.1 million ybp. Centered in the Intermountain Region of W North America, the species greatly expanded its range during the Holocene.

CURRENT DISTRIBUTION: The black-tailed jackrabbit occurs across the southern two-thirds of the W and C United States, as well as in 15 states in Mexico (down Baja California and the central northern portion of Mexico). Disjunct populations occur in SW Montana, SE Oklahoma, and several islands off Baja California. The black-tailed jackrabbit has been successfully introduced in New Jersey, Massachusetts, Maryland, Virginia, and Florida. The species ranges in elevation from −84 m to 3,750 m. Another form recently introduced (1960–1991) to Cerralvo Island, Baja California Sur, Mexico, is considered *L. c. xanti*.

TAXONOMY AND GEOGRAPHIC VARIATION: Seventeen subspecies: *L. c. altamirae*, *L. c. asellus*, *L. c. bennettii*, *L. c. californicus*, *L. c. curti*, *L. c. deserticola*, *L. c. eremicus*, *L. c. festinus*, *L. c. magdalenae*, *L. c. martirensis*, *L. c. melano-*

tis, *L. c. merriami*, *L. c. richardsonii*, *L. c. sheldoni*, *L. c. texianus*, *L. c. wallawalla*, and *L. c. xanti*. The integrity of some of these "ecotypes" is questionable, and the phylogenetic relationships of these subspecies need to be evaluated. Genetic analyses highlight the distribution of subspecies of *L. californicus* on the Baja California Peninsula, Mexico, into three groups: *L. c. xanti* from La Paz, Santa Anita, and San Jose del Cabo; *L. c. magdalenae* from the La Paz Isthmus to S Vizcaino Desert; and *L. c. martirensis* from the Vizcaino Desert and northward.

Similar species include the white-tailed jackrabbit, the antelope jackrabbit (*L. alleni*), and the white-sided jackrabbit. The black-tailed jackrabbit does not interbreed with the closely related and sympatric antelope and white-sided jackrabbits. The specific status of the black jackrabbit (*L. insularis*) has been questioned by researchers who consider that it may be only a melanistic form of *L. californicus*. The nuclear and mitochondrial differences, color variation, and small differences in the structure of a single bone in the skull (jugal) between *L. insularis* and *L. californicus* suggest that the two lineages may have diverged recently. Ongoing research may further clarify the relationship between these two forms.

ECOLOGY: The black-tailed jackrabbit is a shrub-associated hare that has extraordinary thermoregulatory, physiological, and behavioral adaptations for surviving in hot, dry regions on low-quality foods. Population densities can vary from 0.1 to 1.2 animals/ha in natural habitats and from 1.5 to 208/ha adjacent to and on agricultural lands. Home range sizes typically vary between 10 and 20 ha, but can range from 4 to 140 ha. Population sizes can fluctuate dramatically depending on weather, land-use practices, and predator-control activities. Although population fluctuations may be influenced by precipitation amounts and variations in recruitment rate, most fluctuations are the result of changes in mortality rate. The black-tailed jackrabbit is a keystone prey species for coyotes (*Canis latrans*), foxes (*Vulpes* spp., *Urocyon cinereoargenteus*), badgers (*Taxidea taxus*), bobcats (*Lynx rufus*), large buteos (*Buteo* spp.), great-horned owls (*Bubo virginianus*), and especially golden eagles (*Aquila chrysaetos*).

The black-tailed jackrabbit molts only once annually. It is also an important disperser of plant seeds, on its fur and especially via its feces.

HABITAT AND DIET: The black-tailed jackrabbit inhabits more-arid areas as diverse as dry forests, savannas, grasslands, deserts, croplands, shrublands, and dune areas (in island populations). It typically inhabits open country with scattered thickets or patches of shrubs; consequently, its numbers are often high in intensively grazed areas adjacent to cultivated fields. This animal is one of the most widespread and numerically common inhabitants of Great Basin, Mohave, Sonoran, and Chihuahuan desert-scrub communities, as well as plains, intermountain, and semidesert grasslands.

The diet varies with vegetation availability and location. This jackrabbit forages on grasses, forbs, crops, and hay in summer, and buds, bark, and leaves of woody plants in winter. Individuals obtain water both from vegetation and by re-ingesting soft fecal pellets; drinking free water has also been reported.

BEHAVIOR: Active year-round, the black-tailed jackrabbit is often crepuscular. Most feeding occurs at night, yet the species can be observed at any hour. During the day, individuals rest in shallow depressions (forms), typically under the shade of a shrub or small tree. In the hottest regions, these are extended to shallow burrows, which are used for 3–5 hours per day. Exceptional individuals may travel up to 16 km round-trip to feed in fields. Generally solitary yet feeding in loose and surprisingly large groups, black-tailed jackrabbit individuals will typically "freeze" when approached. However, when one knows it has been seen, it will jump up suddenly and flee at up to 64 km/h, occasionally stopping to see if it is pursued. When chased by predators at moderate speeds, black-tailed jackrabbit individuals will periodically bounce in an exceptionally high "observation leap," which may reach 6 m in height. Although usually silent, individuals may squeal or scream when in distress and thump their hind feet when alarmed.

PHYSIOLOGY AND GENETICS: Diploid chromosome number = 48 and FN = 82

REPRODUCTION AND DEVELOPMENT: Breeding is promiscuous, with females accepting the first interested male. The breeding season varies with latitude and environmental factors. In Idaho, breeding is restricted to February to May; in the SW United States, it can last from December to September. Gestation ranges from 40 to 47 days. Litter size varies from 3.8 to 4.9 in the north to 1.8 to 2.2 in the south, giving a total output per female of about 10–14 precocial offspring per year. Actual recruitment rates, however, are much lower; the ratio of young to adult animals does not usually exceed 1:1. Early-born females can breed in their year of birth. Average life span averages 1.4–1.8 years, but one individual in captivity lived 6.75 years.

PARASITES AND DISEASES: Diseases and parasites have long been suspected as being agents capable of reducing black-tailed jackrabbit numbers, and rabbit fever, or tularemia (*Francisella tularensis*), and other bacterial pathogens have been found in a number of western U.S. populations. Jackrabbits are prone to harbor the tick *Dermacentor parumapertus* and other vectors for tularemia (Q-fever, *Coxiella burnetii*). Rocky Mountain spotted fever (*Rickettsia rickettsii*) has been reported as occurring in western U.S. populations, although actual investigations have failed to find any serious disease outbreaks in western U.S. jackrabbits. An investigation of a high-density population of black-tailed jackrabbits near Battle Mountain, Nevada, in 1951 failed to incriminate either tularemia or plague in the "die-off." The only pathogens noted were antibodies for Colorado tick fever and western equine encephalomyelitis, neither of which could be implicated as a serious cause of mortality.

Parasite infestations have also been suggested as contributing to reduced jackrabbit numbers, especially after a population experiences a rapid decline. But although high nematode infestations and the presence of botflies (*Cuterebra* spp.), tapeworms (*Taenia* spp.), ticks, and lice often accompany high jackrabbit densities, most parasite infestations wax and then wane with the cessation of warm weather.

CONSERVATION STATUS:

IUCN Red List Classification: Least Concern (LC)

MANAGEMENT: The black-tailed jackrabbit is hunted for subsistence, sport, and dog food. Skins have been used for hats, and live individuals have been used to train racing dogs. Because during years of high abundance the species is regarded as a pest in certain areas, due to damage to rangeland, hayfields, and cultivated crops, local control efforts may reduce numbers temporarily. From the late 1800s to the mid-1950s, tens of thousands of black-tailed jackrabbits were poisoned or killed in "jackrabbit drives" to protect crops and provide more forage for livestock. However, these efforts generally ceased or were much reduced after 1980.

Because the black-tailed jackrabbit is not designated a game animal in most states, trends in annual harvests and hunt success trends are only available regionally. Declines in hunter and harvest numbers have been reported in California (strongly so), Colorado, Nebraska, Nevada, and Oklahoma. Abundance surveys in Nebraska suggest a declining trend in the black-tailed jackrabbit there, and Washington closed the season on all jackrabbits in 2001.

Vehicle collisions can constitute significant local sources of mortality, as can forest fires, drought, hail, tularemia outbreaks, and extremely cold weather. Because the species favors open habitats such as croplands, increasingly intensive chemical farming, pest control, and removal of all native cover may limit populations. There is concern about low encounter rates of the species during thousands of kilometers of roadside surveys in western Oregon. Rates of sightings along dirt roads across Nevada during 1995–2015 crepuscular and nocturnal surveys were two-thirds to one-sixth of those reported in field notes from the 1930s to 1950s.

Research and monitoring are needed to address the long-term status and trend of the black-tailed jackrabbit, including its relationship with management objectives and broad-scale drivers of population dynamics, genetics (e.g., more-resolved subspecies definitions), and gaps in distribution and synecological relationships between the black-tailed jackrabbit and other sympatric jackrabbits.

ACCOUNT AUTHORS: Erik A. Beever, David E. Brown, and Consuelo Lorenzo

Key References: AMCELA 2008a; Best 1996; Bowen et al. 1960; Cervantes et al. 1999–2000; Clemons et al. 2000; Desmond 2004; Fagerstone et al. 1980; Flinders and Chapman 2003; Flux and Angermann 1990; French et al. 1965; Grayson 1977; Hall 1981; Hinds 1977; Johnson and Anderson 1984; Knick and Dyer 1997; Lechleitner 1958a, 1958b, 1959; Lorenzo et al. 2014b; Marín et al. 2003; Nelson 1909; Palmer 1897; Rice and Westoby 1978; Schmidt-Nielsen et al. 1965; Simes et al. 2015; Smith 1990; Smith and Nydegger 1985; Smith et al. 2002; Stoddart 1985; Vorhies and Taylor 1933; Wywialowski and Stoddart 1988.

Lepus callotis Wagler, 1830
White-sided Jackrabbit

OTHER COMMON NAMES: Antelope rabbit, Beautiful-eared jackrabbit, Gaillard's jackrabbit, Snowsides; Liebre torda, Liebre pinta (Spanish)

DESCRIPTION: The white-sided jackrabbit is a medium to large hare with a buffy cinnamon brown dorsum merging to iron gray on the rump and hips, with distinctive white sides and underparts below the median line. The ears, while large, are not exaggerated and are tipped with a white fringe and sport dusky patches along the posterior edges. A nape patch ranges from brown to black, and the limbs are white with buff on the upper surfaces. The gular area is cinnamon to ochraceous with much white on the head. The tail is moderately short with a black upper surface and white underparts. The eye-shine at night is reddish. The white-sided jackrabbit undergoes one molt a year during early summer from front to rear. Except for the shorter pelage in summer, there is little difference between the summer and winter coats. Females are larger than males.

The skull is similar to that of the more common black-tailed jackrabbit (*L. californicus*), but with a higher nasal aperture, a smaller and more inclined supraorbital surface, and a lesser breadth across the auditory bullae.

SIZE: Head and body 432–598 mm; Tail 47–90 mm; Hind foot 113–145 mm; Ear 137–180; Greatest length of skull 85–93 mm; Weight 1,500–3,200 g

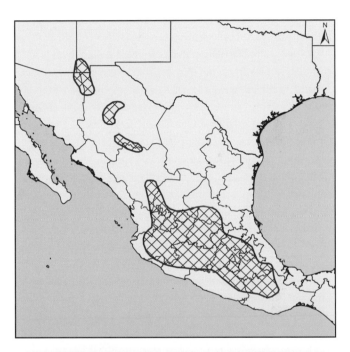

Lepus callotis. Photo courtesy Martha Marina Gomez Sapiens

PALEONTOLOGY: No fossil remains of this hare have been reported, a situation probably influenced by the difficulty in differentiating the bones of the white-sided jackrabbit from those of the black-tailed jackrabbit. One hare from outside the present range of the white-sided jackrabbit in Burnet Cave in Eddy County, New Mexico (age, 7,432 ± 300 years), and attributed to the antelope jackrabbit (*L. alleni*), may be this species.

CURRENT DISTRIBUTION: Two subspecies: *L. c. callotis* and *L. c. gaillardi*. These forms are separated along parallel 25° N, the Nazas River being a significant geographical barrier. *L. c. callotis* occurs in grassland and savanna habitats south of the Nazas River from C Durango southward through the northwestern half of Oaxaca and northern half of Guerrero at elevations ranging from ~ 750 to 2,550 m asl. North of the Nazas River, *L. c. gaillardi* is found in N Durango and W Chihuahua northward to less than 120 km² of habitat in the "bootheel" region of extreme SW New Mexico in the United States at elevations from 1,350 to 2,100 m asl.

TAXONOMY AND GEOGRAPHIC VARIATION: *Lepus callotis gaillardi* differs from *L. c. callotis* in having plainer and buffier pelage including a paler rump and less ochraceous throat patch. The white flanks of *L. c. gaillardi* also show less contrast with the upper body fur, while the skulls are typically larger and have a more elevated supraorbital process. *L. c. gaillardi* also has a brown rather than black nape and has larger body, foot, and ear measurements.

Similar species include the smaller sympatric black-tailed jackrabbit and the larger and disjunct antelope jackrabbit and the Tehuantepec jackrabbit (*L. flavigularis*).

ECOLOGY: Despite having been described as early as 1830 and being a popular game animal in much of Mexico, this large hare's natural history and ecological requirements remain largely undocumented. *L. c. gaillardi* is affiliated with warm temperate grasslands having from 75 to 100 mean days with temperatures below 0°C per annum, while *L. c. callotis* inhabits tropic-subtropic grasslands with less than 35 mean nights a year with temperatures below 0°C. With mean annual temperatures ranging from 3.6°C to 33.8°C, the grassland habitats of *L. c. gaillardi* are subject to encroachment by honey mesquites (*Prosopis glandulosa*), woody plants, and stiff-stemmed shrubs, and those of *L. c. callotis* to rapid and pronounced successional changes to thorny vegetation and cacti due to grazing pressures, fire suppression, and climate change.

Most accounts describe white-sided jackrabbits as preferring level or gentle terrain clothed in herbaceous vegetation such as blue grama (*Bouteloua gracilis*) and to-bosa (*Pleuraphis mutica*) grasslands. Their highest densities occur in a dense cover of grasses and herbs, and most collections of this species have been from open grasslands. Hills and brushy terrain are avoided, as are dense woodlands and forests. Precipitation values near white-sided jackrabbit collection sites range from 340 to 530 mm for *L. c. gaillardi* and from 430 to 1,050 mm for *L. c. callotis*. From 71% to 90% of this precipitation falls between April and September.

Significant predators include bobcats (*Lynx rufus*), coyotes (*Canis latrans*), kit foxes (*Vulpes macrotis*), red-tailed hawks (*Buteo jamaicensis*), Swainson's hawks (*B. swainsoni*), and possibly golden eagles (*Aquila chrysaetos*).

HABITAT AND DIET: One of the few habitat studies conducted in Mexico describes *L. c. gaillardi* occurring in open plains grasslands dominated by blue grama grass. In New Mexico, *L. c. gaillardi* depends on level well-developed grasslands with a low density of shrubs, and occurrences are positively correlated with buffalo grass (*B. dactyloides*) and negatively correlated with shrub cover. Open patches of semi-desert grassland are important to the animal's persistence, and from Jalisco to Puebla, white-sided jackrabbits occur in valleys and basins containing grasslands while avoiding pine-oak forests, thorn-scrub, and desert scrub.

Unlike *L. c. gaillardi*, *L. c. callotis* in Guerrero and possibly other areas where black-tailed jackrabbits are missing appears to also use cultivated lands and heavily grazed grasslands populated by thorny shrubs.

Few diet studies have been conducted, but white-sided jackrabbits are reported to feed primarily on green grass and forbs and to subsist without free water. They usually forage by cutting and pulling up grass blades, feeding on the leaves, roots, and nodes. Grass and forb particles are chewed while sitting upright and watching for predators. Discarded grass stems are not retrieved. The forepaws are used to dig and excavate nutgrass (*Cyperus rotundus*) tubers and grass rhizomes, resulting in oval feeding pits or depressions 7–19 cm long, 5–15 cm wide, and 1–3 cm deep. Fecal pellets are commonly found in and around these pits.

BEHAVIOR: The white-sided jackrabbit is primarily nocturnal in its activities and typically occurs in pairs. In New Mexico, most activity takes place from 2200 to 0500 h, with less activity during the early morning and in the late evening. Less movement occurs on nights having cloud cover, precipitation, or wind. Temperature appears to have little effect on movements.

Animals flushed from dense stands of tobosa or other

grass flee to other grassland sites that are usually out of sight of the pursuer. Hares hunted with dogs are reported to go to ground in dense grass when pursued. When flushed, white-sided jackrabbits alternately flash their white sides while running and employ an escape behavior of leaping straight upward, extending the hind legs, and flashing its white flanks. Pairs flushed in front of observers at distances from 5 to 25 m will run together for distances up to 0.5 km.

Male-female pairs of white-sided jackrabbits are commonly observed, and these pair bonds are strongest in the April through October breeding season. During this time the male guards the female from intruding males, the pairmates usually remaining within 5 m of each other.

The white-sided jackrabbit employs more elaborate forms (depressions) than most hares. Form shelters average 37 cm in length, 18.3 cm in width, and 6.3 cm in depth. Dense stands of grass usually surround the form, which can extend beneath the soil surface. These sites are frequently characterized by concentrations of jackrabbit fecal pellets.

These hares are generally silent except for the high-pitched screams of a stressed individual or the huffs and grunts of interacting animals during the breeding season.

PHYSIOLOGY AND GENETICS: Diploid chromosome number = 48. A member of the white-sided clade, *L. callotis* is closely related to *L. flavigularis* and *L. alleni*. The genetic relationship between *L. callotis* and *L. californicus* is in need of further investigation. The form *altamirae*, formerly identified as being in the white-sided clade, has been assigned subspecific status under *L. californicus* based on morphological criteria, although this classification is open to interpretation.

REPRODUCTION AND DEVELOPMENT: Little is known about the breeding habits of this seemingly monogamous hare. The presence of embryos and lactating females in collections of white-sided jackrabbits indicates a breeding season from mid-March to mid-August. Mean litter size is no more than 2.2, and probably less. All indications are that recruitment rates are low.

PARASITES AND DISEASES: Unlike other western jackrabbits, this low-density animal does not appear prone to heavy parasite loads of botfly larva (*Cuterebra* spp.) and tapeworm cysts. Ectoparasites and possible disease vectors include a flea (*Pulex simulans*) and a tick (*Dermacentor parumapertus*). Microorganisms from a limited sample of *L. c. gaillardi* in New Mexico include a coccidian, *Staphylococcus aureus*, *Pneumococcus* spp., *Streptococcus* spp., *Bacillus* spp., *Pseudomonas pseudomallei*, *Alcaligenes denitri-*

ficans, *Pantoea agglomerans*, *Klebsiella ozaenae*, *Escherichia coli*, and *Yersinia pseudotuberculosis*.

CONSERVATION STATUS:

IUCN Red List Classification: Near Threatened (NT); decreasing

MANAGEMENT: Most of the information available on white-sided jackrabbits comes from anecdotal observations made by museum collectors and scientists conducting general zoological inventories. Although a few life history studies of *L. c. gaillardi* have been conducted, only one study has investigated the status of *L. c. callotis*. Nonetheless, available information suggests that *L. c. gaillardi* is in serious decline due to environmental changes resulting from overgrazing, shrub invasion, and other habitat changes. The declining remnant population in New Mexico is also subject to road kills from significant U.S. Border Patrol activities.

The status of *L. c. callotis* is less clear, but the limited information available indicates that it may also be in trouble due to habitat alteration, localized hunting activity, disturbance by dogs, and vehicle collisions. Additionally, a study that modeled the effects of climate change on grassland mammals in Mexico predicted an 80% reduction in range and habitat of *L. callotis* by 2050.

Several investigators have reported the white-sided jackrabbit to be uncommon in both New Mexico and Mexico (Chihuahua, Guanajuato, Guerrero, Michoacán, SE Morelos, San Luis Potosi, and Zacatecas). Populations of white-sided jackrabbits in many areas have been diminishing for years, and in some areas this species is now said to be rare where it was formerly common. In other areas white-sided jackrabbits are reported to have been replaced by the highly adaptable black-tailed jackrabbit.

In the United States *L. callotis* has been classified as "Threatened" by the state of New Mexico since 1975; however, it is not afforded any protection by the U.S. Endangered Species Act (ESA). In 2009, the U.S. Fish and Wildlife Service was petitioned to list the species under the ESA, but it was rejected after a 12-month review due to a lack of information on the status of the species in Mexico, where neither *L. c. gaillardi* nor *L. c. callotis* is considered a mammal "at risk." This decision runs counter to the overwhelming number of publications and proceedings that have recommended the species be considered as endangered throughout its entire range and in need of research and protection.

The white-sided jackrabbit is hunted for subsistence and sport in Mexico. The preferred method is with dogs such as beagles, and the species is reported to be susceptible to this kind of take. Nonetheless, no systematic surveys or harvest estimates have been conducted on Mexican jackrabbits.

Research and monitoring are needed to address the long-term status and trends of the white-sided jackrabbit, including its relationship to broad-scale drivers of population dynamics, genetics (e.g., more-resolved subspecies definitions), and gaps in distribution and synecological relationships between subspecies of the black-tailed jackrabbit.

Although vehicle collisions can constitute significant local sources of mortality, as can forest fires, drought, disease outbreaks, and possibly extremely cold weather, the principal factors of concern with this species are habitat alteration and condition. Because the species appears to favor open grasslands devoid of woody vegetation, it is increasingly prone to intensive grazing, chemical farming, pest control, and removal of native cover. Most important, the white-sided jackrabbit is in much need of study so that its limiting factors can be identified and addressed.

ACCOUNT AUTHORS: David E. Brown, Consuelo Lorenzo, and Myles B. Traphagen

Key References: Anderson and Gaunt 1962; Bednarz and Cook 1984; Bello-Sánchez 2010; Bogan and Jones 1975; Cook 1986; Dalquest 1953; Davis and Lukens 1958; Davis and Russell 1954; Delgadillo-Quezada 2011; Desmond 2004; Hall and Villa 1949; Leopold 1972; Matson and Baker 1986; Mearns 1895; Nelson 1909; SEMARNAT 2010; Traphagen 2011.

Lepus capensis Linnaeus, 1758
Cape Hare

OTHER COMMON NAMES: Arabian hare, Brown hare, Desert hare

DESCRIPTION: The Cape hare is a middle-sized hare with smooth and straight fur. The dorsal pelage and head are silvery-gray grizzled with black. The hairs are white at the base with a wide black sub-terminal band, a whitish terminal band, and a black or white tip. The underfur is white or grayish white. The ventral fur is pure white and long. The lower flanks are pale buff. The lateral profile of the head is markedly angular. The Cape hare has white eye rings and frequently rufous marks above and below the eye-rings. The upper lips are pale rufous, and the chin and throat are white. Typically, the hares have a buffy-

white collar. The ears are comparatively long with the inner fringe lined with long white hairs. The ear tips are rounded and lined with short black hairs, particularly on the external surface. The nuchal patch is brownish pink. The forelimbs are pale rufous above and white below, whereas the hind limbs are pale rufous. The soles of all feet have a buffy-brown fur. The tail is relatively long, downy, black above, and white on the sides and below. Both morphological characteristics and pelage color vary widely throughout the species' geographical range. The specimens from arid and semi-arid areas are paler colored than those from more temperate areas. The ear tips in des-

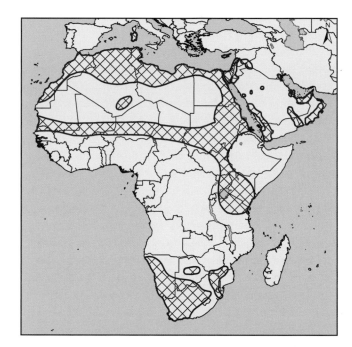

Lepus capensis. **Photo courtesy David E. Brown**

ert forms may be pale, whereas others have a thin black margin. The length of the ear and hind foot increase with advancing aridity of the habitat. The most diminutive form of the Cape hare inhabits the south of the Arabian Peninsula. The fur color of these animals is distinctly adapted to the environment. Specimens inhabiting the central Arabic sand deserts are rufous, whereas those living in the stony savannah of W and E Arabia are gray colored.

SIZE: Head and body 450–550 mm; Tail 100–145 mm; Hind foot 110–138 mm; Ear 110–140 mm; Greatest length of skull 88 mm; Weight 1,700–2,500 g

PALEONTOLOGY: A supermatrix analysis including published nucleotide sequence data, unique insertion/deletion events, morphological characters, and presumed geographical centers of origin of each genus within the Leporidae proposed that the shared ancestor of the European, African, and Asian hares spread approximately five million ybp from North America via the Bering Strait to Eurasia, founding an ancient *Lepus* lineage of which the Cape scrub hare (*L. saxatilis*) and the Cape hare are still extant today.

CURRENT DISTRIBUTION: The Cape hare has a remarkably large geographical distribution. It is divided in two separate expanses in Africa: first, in the savannah and semi-desert regions of South Africa, Lesotho, Swaziland, Namibia, Botswana, S Zimbabwe, and S Mozambique; second, in Tanzania, Kenya, Uganda, and Ethiopia, and throughout most of the dry savannah regions of C, W, and N Africa, including parts of the Sahara Desert. There is a break in its range in E Africa, and the species does not exist in wooded regions. The Cape hare is found in isolated populations dispersed throughout most of the Arabian Peninsula. It also occurs on the islands Sardinia and Cypress. Until recently the species was believed to extend across the Middle East and into China, Mongolia, and Russia. Nevertheless, the consideration of the Cape hare inhabiting this region is arbitrary as long as the desert hare (*L. tibetanus*) and the tolai hare (*L. tolai*) are recognized as valid species (see those accounts).

TAXONOMY AND GEOGRAPHIC VARIATION: Previously, the Cape hare included the European hare (*Lepus europaeus*), the Corsican hare (*L. corsicanus*), the Iberian hare (*L. granatensis*), the Abyssinian hare (*L. habessinicus*), and the tolai hare (including the desert hare) as subspecies. Currently these forms are considered to be true species. Nevertheless, no morphological characteristics have been found to separate the *tolai-tibetanus* group from the Cape hare, and some taxonomists still consider

that the Abyssinian hare might be conspecific with the Cape hare. The taxonomy of the Cape hare throughout its distribution is under debate. As construed in previous times, one single species (*capensis sensu lato*) lives in Africa in two distinct zones: the southern distribution and the northern distribution. There is no indication of gene flow between the two distributions, and the intervening zones are occupied by other *Lepus* species. Therefore, an informal subdivision of *capensis sensu lato* into four groups constructed around their geographic localities (South Africa, East Africa, Arabia and Near East, NW Africa) is broadly acknowledged and may embody separate species. In this analysis, *L. capensis* would become limited to the South African hare. Unfortunately, a formal revision of these taxa is not currently possible owing to a lack of data. The 13 recognized subspecies can be categorized into the 4 groups: South Africa (*capensis, aquilo, carpi, granti*), E Africa (*aegyptius, hawkeri, isabellinus, sinaiticus*), Arabia and Near East (*arabicus*), and NW Africa (*atlanticus, mediterraneus, schlumbergeri, whitakeri*). The taxonomic position of the Sardinian hare (*L. c. mediterraneus*) is uncertain. Examination of mtDNA shows that Sardinian hares configure a monophyletic clade with North African hares, whereas a phylogenetic analysis of mtDNA from Tunisian and Egyptian hares regarded them as monophyletic and distinct from the Cape hare. Nevertheless, an investigation of the nuclear gene pool of the Cape hare, the European hare, and the North African hare showed that the North African hare, as well as the European hare, belong to the Cape hare, affirming a theory of the incorporation of the European hare in the Cape hare. The genetic discrimination between the Cape hare and the European hare may be ascribed to geographic remoteness rather than divergence. It has been assumed that gene flow may be occurring in the Near East where ranges meet, eventuating in intergraded populations that have locally different gene pools. Nonetheless, a joint phylogenetic, phylogeographic, and population genetic approach based on several nuclear and mitochondrial markers and comprising other biological characters such as phenotypic and morphometric data is required for decisive confirmation of a single species complex. Considering this ongoing incertitude concerning the taxonomic status of Sardinian and North African hares, both will remain included in the Cape hare, and the European hare preserves its taxonomic status as a separate species. Furthermore, the affiliation of the Cape hare with the African savanna hare (*L. victoriae*) is uncertain, as in Somalia individuals seem to display characters intermediate between the two species, and in

Kenya specimens of the Cape hare and the African savanna hare are almost identical. Forms inhabiting Arabia may be different species. Because of its extensive range and inter-population dissimilarity, in total 38 forms of the Cape hare have been designated. These forms differ in overall size, fur color, and ear length. The forms were initially contemplated as separate species, but are currently assumed to be synonyms, although some of them may yet prove to be true species. As taxonomists are still trying to clarify the species differentiation in *Lepus*, the subspecific taxonomy has not yet been elaborated. First, the original descriptions of the subspecies are often not very helpful as they are mostly based on a few exterior characteristics and on a small number of individuals. Second, the extent of variation has not been studied yet. As a result, it is difficult to understand if any feature represents an important diagnostic mark, or individual or age-specific variation. Moreover, it has been shown that the variability is clinal in more careful investigations. Hence, the distinction in subspecies might be arbitrary and unreasonable.

ECOLOGY: Densities diverge depending on the habitat and are higher in agricultural areas and valleys than on mountain pastures and steppes. Therefore, assessed densities vary between 4 and 25 hares / 100 ha. No consistent seasonal or annual cycle in Cape hare numbers has been documented for this hare species. The home range differs depending on the habitat type in which it occurs. The Cape hare is mostly solitary, and only sporadically is seen in small groups. Out of 800 individuals studied in Kenya, only 4 groups of 3 hares were encountered.

HABITAT AND DIET: The Cape hare occurs in grasslands and open habitats, including Acacia (*Acacia*) and miombo (*Brachystegia*) savannah, Sahel and Sudan savannah, steppe, semi-desert, desert, and mountain valleys up to 2,400 m on alpine meadows. This species is particularly adaptable and inhabits numerous environments. However, it favors open habitats and avoids scrub habitats. Pastures that have been overgrazed by domestic stock are a preferred habitat. The hares move into burned areas as soon as grasses begin to grow. Therefore, the distribution of the Cape hare has increased with scrub clearance and widespread savannah fires. On the Arabian Peninsula, the Cape hare selects bushes rather than grasses to shelter under in summer. In the Namib Desert, hares dig short tunnels to shelter from the sun. Cape hares are herbivorous and browse at night on grasses. There are few data on the diet of this species, but it apparently differs according to its habitat. Fecal analysis in Kenya discovered that 19% of their diet was dicotyledons, 40% grasses, 1% sedge, and

32% stem fibers. The main grasses fed on by the hares were sacaton (*Sporobolus* spp.), three-awns (*Aristida* spp.), windmill grass (*Chloris* spp.), Bermuda grass (*Cynodon dactylon*), pappas grass (*Enneapogon* spp.), and canegrass (*Eragrostis* spp.). The proportion of each grass species eaten differs at distinct locations, but seasonal variability in the diet, including wet and dry seasons, may be minor. It appears as if the Cape hare is rather opportunistic, feeding on different grass species according to availability. This species is increasing its distribution at the cost of less adaptable hares as it prospers on overgrazed pastures.

BEHAVIOR: The Cape hare is nocturnal. Cape hares rest in forms (depressions) during the day, but might feed during the day when the weather conditions are cloudy. The hares run into the open when disturbed.

PHYSIOLOGY AND GENETICS: Diploid chromosome number = 48. Recent genetic work focused on taxonomic clarification by analyzing nuclear and mtDNA. Furthermore, the complete mtDNA sequence of the hare subspecies *pamirensis* inhabiting the Pamir Mountains has been determined recently. However, whether this subspecies belongs to the desert hare or to the Cape hare is still under debate.

Cape hares have relatively large hearts (14.0 g), which agrees with their typical behavior of running in the open when disturbed. In comparison, African savanna hares (*L. victoriae*) have smaller hearts (10.3 g) and tend to run for cover in scrub.

REPRODUCTION AND DEVELOPMENT: Reproduction in the Cape hare varies according to the locality. Near the equator, males are fertile and females are gravid throughout the year. Gravidity rate is 80% to 100% for most of the year excluding April (wettest month of the year) and June and July (end of the wet season). Females have 6 to 8 litters/year, with an average litter size of 1.5; thus females produce an average of 11.6 young per year. The mean litter size differs seasonally, possibly in response to fluctuations in rainfall and food (1.0 in September to 1.9 in January). Additionally, litter size differs by elevation; specimens at higher elevations have smaller litters (mean 1.24 at 1,800 m) than those at lower elevations (mean 1.75 at 600 m). The weight of the young at birth is between 80 and 130 g. The growth rate of the leverets is approximately 10 g/day, and the young reach adult size in 4 to 5 months.

PARASITES AND DISEASES: No information on parasites and diseases affecting the Cape hare is available.

CONSERVATION STATUS:

IUCN Red List Classification: Least Concern (LC), decreasing

MANAGEMENT: The Cape hare is a widespread species with a large population. In the southern extent of the African range, a population decline of less than 10% since the year 1904 has been stated. The total number of this species is presently predicted to be greater than 10,000 individuals. No population information is available for the N African region of this species. The population trend for the Arabian distribution is considered to be in decline. The present population trend of the Cape hare on islands in the Persian Gulf, specifically Masirah Island and Bahrain, is under investigation as it is probable that the tiny Cape hare form inhabiting these islands is a distinct species. Some indications exist that on Masirah Island and Bahrain the Cape hare no longer occurs. The population on Sardinia has been declining as well, although the Cape hare is considered as locally common. The Cape hare has experienced habitat loss on the Arabian Peninsula since the 1950s, mostly caused by urbanization, habitat fragmentation, overgrazing, livestock competition, agricultural encroachment, recreational activities, harvest/hunting, and infrastructure associated with tourism. These threats are likely to lead to a continuing population decline in the Arabian Peninsula. In Africa a loss of habitat due to agricultural practices and hunting poses a threat to the Cape hare.

ACCOUNT AUTHORS: Stéphanie Schai-Braun and Klaus Hackländer

Key References: Allen 1939; Angermann 1972; Azzaroli-Puccetti 1987a; Ben Slimen et al. 2005, 2006, 2008a, 2008b; Boitani et al. 1999; Dixon 1975; Drew et al. 2008; Ellerman and Morrison-Scott 1951; Ellerman et al. 1953; Flux 1969, 1981a, 1981b; Flux and Angermann 1990; Flux and Flux 1983; Flux and Jarvis 1970; Happold 2013; Hoffmann and Smith 2005; Kryger et al. 2004; Matthee et al. 2004e; Mitchell-Jones et al. 1999; Petter 1959, 1961; Scandura et al. 2007; Shan and Liu 2015; Stewart 1971.

Lepus castroviejoi Palacios, 1977

Broom Hare

OTHER COMMON NAMES: Liebre de piornal (Spanish)
DESCRIPTION: The broom hare is intermediate in size between the Iberian hare (*L. granatensis*) and the European hare (*L. europaeus*), the other two species occurring on the Iberian Peninsula. The dorsal and lateral pelage is grayish brown, the ventral pelage white, and the tail black

and white. The line between the dorsal and ventral colors is crisp. The ears are brown with black tips. The white ventral part is more extended than in the European hare, in some cases reaching the upper part of the forelegs, but not as extended as in the Iberian hare. The broom hare has a distinctive facial design, with a gray fringe between the ears and the cheeks. There are no sexual or seasonal variations in pelage coloration.

Skull characteristics in the broom hare are the V-shaped nasofrontal suture, a small orbital process, small dimples in the upper bones of the skull, and tiny palatine foramina.

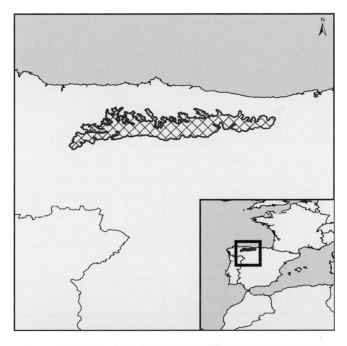

Lepus castroviejoi. Photo courtesy Héctor Ruiz Villar

SIZE: Head and body 410–590 mm; Hind foot 130–147 mm; Ear 85–96 mm; Greatest length of skull 93–97 mm; Weight 2,000–3,500 g

PALEONTOLOGY: This species is very closely related to the Corsican hare (*L. corsicanus*) from the Italian Peninsula. Both seem to have shared a common ancestor during the Late Pleistocene, occupying a large range in Europe, and then breaking away in two different refugia on the Iberian and Italian Peninsulas as a consequence of climatic change. Competition with other *Lepus* species could have maintained *L. castroviejoi* and *L. corsicanus* in their respective small and allopatric ranges until the present. Two different events of introgression of the mountain hare (*L. timidus*) mtDNA into *L. castroviejoi* have been described, replacing its aboriginal mtDNA. This situation reflects contact and hybridization between these species in two different times, the recent one during the last deglaciation, also affecting the Iberian hare and European hare populations of N Spain. No fossil information exists about the broom hare or its ancestors.

CURRENT DISTRIBUTION: The broom hare is an endemic species restricted to the Cantabrian Mountains in NW Spain, where it occupies a distribution range of about 5,000 km² (230 km long, 25–40 km wide), from Sierra de los Ancares (between Lugo and León) to Sierra de Pena Labra (between Palencia and Cantabria). It occurs at elevations ranging from 1,000 to 1,900 m, with a highly fragmented distribution, as it occupies specialized patches of habitat within pastures and shrubland, environments that are sparse in a diverse mountain landscape.

TAXONOMY AND GEOGRAPHIC VARIATION: No subspecies or geographic variation has been described. Previously it has been considered as a form of *L. granatensis*. It was described as a new species in 1976. Subsequent genetic evidence has supported recognition of *L. granatensis* and *L. castroviejoi* as distinct species. Recent mtDNA and nuclear DNA analyses have shown that *L. castroviejoi* is closely related and a sister taxa to *L. corsicanus*, sharing not only genetic similarities but also phenotypic features and ecological niche properties. The range of *L. castroviejoi* partially overlaps with those of *L. granatensis* and *L. europaeus*, but no hybridization has been detected so far.

ECOLOGY: There is not much information about the ecology of the broom hare. Nocturnal censuses with spotlights have determined average densities of 4.83/100 ha in Liébana (Cantabria), 6.89/100 ha in Picos de Europa National Park (León), and between 8.83 and 23.32/100 ha in Somiedo (Asturias). Given the dispersed distribution of its suitable habitat, it seems that broom hares function as a metapopulation with different small local populations in good habitats, connected by dispersal through unfavorable areas (such as forests, rocks, and human settlements).

Adult and juvenile broom hares could be preyed on by carnivores such as red foxes (*Vulpes vulpes*), Iberian wolves (*Canis lupus*), European wildcats (*Felis silvestris*), or pine martens (*Martes martes*), or raptors such as golden eagles (*Aquila chrysaetos*) or goshawks (*Accipiter gentilis*), all of which are well established in the broom hare's range.

No big fluctuations or cycles in number or density have been documented, but some data from abundance monitoring in a protected area (Picos de Europa National Park) suggest abundance variations between years, probably as a consequence of variation in reproductive success.

HABITAT AND DIET: The broom hare occupies a characteristic habitat in the higher parts (1,300–1,900 m) of the Cantabrian Mountains, consisting of open pastures or small herbaceous clearings surrounded by broom formations, including Scotch broom (*Cytisus scoparius*), white broom (*C. multiflorus*), and gorse (*Genista florida*); heathland including Irish heath (*Daboecia cantabrica*), winter heather (*Erica* spp.), Scotch heather (*Calluna vulgaris*), and blueberries (*Vaccinium* spp.); other shrubs; and deciduous forest of European beech (*Fagus sylvatica*), sessile oak (*Quercus petraea*), Pyrenean oak (*Q. pyrenaica*), silver birch (*Betula pendula*), and others. In the past, the use of small grain cultures and orchards near mountain villages has been described, although these areas are practically nonexistent currently. Habitat suitability models have determined that broom hares select areas characterized by a high percentage of broom and heather scrublands, high altitude and slope, and limited human accessibility. Other studies have also shown preference for the smallest clearings that exist in forest and scrublands as well as recently burned scrublands.

Although the broom hare is herbivorous and feeds on grasses and other herbaceous plants, there is no detailed scientific information about its diet. Feeding areas overlap extensively with that of domestic animals, especially cows, but also sheep and horses. There is a strong niche overlap between that of the broom hare and that of the endangered population of the gray partridge (*Perdix perdix*) in the Cantabrian Mountains.

BEHAVIOR: Broom hares are mainly nocturnal. At dusk and during the night they come to feed in open pastures or clearings in shrublands, where they tend to stay relatively close to the broom or heath formations as an antipredatory strategy (distance to protecting shrubs is about 30 m). Contacts during nocturnal census surveys show a clearly contagious distribution, for both specific habitat selection and social interactions.

Home range size of 2 radio-tracked female broom hares have been reported as 53.2 and 60.3 ha, in 4 months of monitoring, although both hares spent most of the time in a preferred area of ~ 7–8 ha. According to other hare populations, it seems that broom hares living in a high-quality and very diverse mountainous habitat could use smaller home ranges than that of other hares living in more extensive agricultural lands.

PHYSIOLOGY AND GENETICS: Diploid chromosome number = 48. Overall genetic variability in the broom hare (mean He = 0.284; n = 11; six microsatellites) is lower than that found in the other Iberian hare species, probably reflecting the restricted geographic range and population numbers. The broom hare has a close genetic relationship to the Corsican hare. No native mtDNA of the broom hare has been detected so far, only two existing mitochondrial lineages, both belonging to the mountain hare.

REPRODUCTION AND DEVELOPMENT: There are few data about reproduction and development in the broom hare. Courtship behavior has been observed in April and May. The sex ratio in some samples seems to be biased in favor of males (0.55–0.86 females/male). Litter size ranges from one to three young.

PARASITES AND DISEASES: No detailed information is available about parasites and diseases of *L. castroviejoi*. Some helminths (*Dicrocoelium dendriticum*, *Mosgovoyia pectinata*, and *Trichuris leporis*) have been described. A recent study has detected a widespread presence of *Leishmania infantum* in both broom and European hares in Spain, although a low number of samples prevented the confirmation of this parasite in broom hares.

CONSERVATION STATUS:

IUCN Red List Classification: Vulnerable (VU)—B1ab(iii)+2ab(iii)

National Red List: Spain (Vulnerable (VU)—B1ab(iii)+2ab(iii))

MANAGEMENT: The broom hare is an endemic species in Spain with a very restricted distribution range and specific habitat requirements. Although no decreasing trend can be assumed for the whole population, some peripheral areas have experienced significant regression or even the local disappearance of the species (Sierra de Peña Labra in Cantabria, Sierra de Pando, Peña Manteca, and Sierra del Aramo in Asturias). Fragmentation and metapopulation dynamics resulting from the patchy distribution of suitable habitat is another risk factor for this species. A study based on ranger surveys indicated that broom hare popu-

lations have decreased in 16.67% of their area of occupancy, increased in 4.17%, and remained stable in 70.82%.

Much of the range of the broom hare is included in natural parks and game reserves under regional government management as well as European Natura 2000 sites. Although the broom hare is listed as Vulnerable in Spain, it remains a hunted species and is not included in the Spanish Catalogue of Endangered Species or the Spanish List of Wildlife Species of Special Protection Regime. There have been some local bans, but currently the broom hare is hunted across all of its range, albeit with a low extraction level. In most areas the species is hunted in regional game reserves with a very low quota, although excessive hunting and poaching could be a relevant risk.

In addition to overhunting and poaching vulnerability, the main risk for broom hare conservation is probably the habitat change process as a consequence of climate change and rural abandonment. Traditional grazing by autochthonous livestock breeds has been helpful for the maintenance of favorable habitats for the hares.

Recommended conservation measures must include an increase in research and monitoring, especially to determine the effects of habitat change and hunting, as well as improvement in the protection against poaching, the establishment of a global hunting plan for all regional reserves and private hunting areas, and the establishment of a habitat management strategy taking into account grazing management, fire prevention, habitat restoration, preservation of most relevant habitats, and improvement of corridors.

ACCOUNT AUTHOR: Fernando Ballesteros

Key References: Acevedo et al. 2007, 2014; Alves and Melo-Ferreira 2007; Alves and Niethammer 2003; Alves et al. 2008b, 2008c; Ballesteros 2003, 2007; Ballesteros and Palacios 2009; Estonba et al. 2006; Koutsogiannouli et al. 2012; Melo-Ferreira et al. 2005, 2007, 2012; 2014; Palacios 1976; Palacios and Meijide 1979; Ruíz-Fons et al. 2013; Vila et al. 1999.

Lepus comus Allen, 1927
Yunnan Hare

OTHER COMMON NAMES: Yunnan tu (Chinese)
DESCRIPTION: The Yunnan hare is distinguishable from the woolly hare (*L. oiostolus*) by its small size. The dorsal pelage is grayish brown. The flanks, forelegs, and outer sides of its hind legs are an ochraceous buff color. The

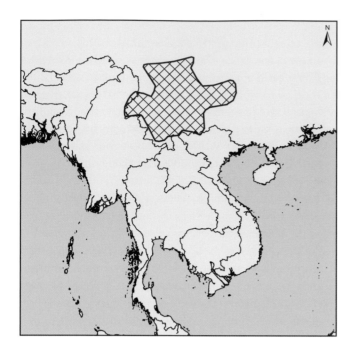

ventral pelage is white. The muzzle has a whitish band that includes the eye-ring and extends to the base of the ear. The nape is gray-brown, and the ears are long, tipped in black, while the interior portion is a pale gray. The rump is gray, and the tail is dorsally dark lacking a stripe, gray and tinged with yellow on the sides and below.

The skull has a slender profile, and the supraorbital process is low, flat, and not flared upward. The nasals are relatively short, and the posterior portion of the nasals is broad.

SIZE: Head and body 322–480 mm; Tail 95–110 mm; Hind foot 98–130 mm; Ear 97–135 mm; Greatest length of skull 84–95 mm; Weight 1,800–2,500 g

CURRENT DISTRIBUTION: The Yunnan hare occurs in the Chinese provinces of Yunnan, W Guizhou, and S Sichuan, and perhaps extending into NE Myanmar. It can be found at elevations between 1,300 and 3,200 m asl.

TAXONOMY AND GEOGRAPHIC VARIATION: Three subspecies: *L. c. comus* (W Yunnan); *L. c. peni* (E and N Yunnan into Guizhou); and *L. c. pygmaeus* (S Sichuan to N Yunnan). Apparently, all three subspecies share a recent maternal common ancestor. The species shows very little intraspecific genetic variation. *L. comus* is most closely related to the woolly hare, and it has been included with that species in some earlier treatments.

ECOLOGY: Very little is known about the ecology of the Yunnan hare. It has been reported that it uses burrows. Male hares typically have smaller, straighter, and shallower burrows, while female hares occupy burrows that are bigger and more oval in shape.

HABITAT AND DIET: The Yunnan hare occurs in high montane pastures like the woolly hare on the Tibetan plateau, as well as in shrublands. It has also been recorded in forest edges and sparse forests, but it prefers bunchgrass, hilly areas, and ravines.

This species consumes forbs and tender shrub leaves, but will also forage for young wheat, corn, and bean crops.

BEHAVIOR: Foraging typically occurs at night, but Yunnan hares may also be active diurnally.

PHYSIOLOGY AND GENETICS: Diploid chromosome number = 48

REPRODUCTION AND DEVELOPMENT: Litter size for this hare is one to four young, with two to three litters produced each year. The breeding season begins in April and extends through October, with the first litter appearing in May.

CONSERVATION STATUS:

IUCN Red List Classification: Least Concern (LC)

National-level Assessments: China Red List (Near Threatened—NT)

MANAGEMENT: The Yunnan hare is hunted by local inhabitants for subsistence. Current threats to this species are unknown, but human encroachment into valleys that surround the mountainous habitat of this species may result in populations becoming increasingly fragmented.

ACCOUNT AUTHOR: Andrew T. Smith

Key References: Allen 1938; Jiang et al. 2016; Liu et al. 2011b; Smith and Xie 2008; Wu et al. 2000, 2005; Yu 2004.

Lepus coreanus Thomas, 1892
Korean Hare

OTHER COMMON NAMES: Mettokki, Santokki (Korean); Gaoli tu (Chinese)

DESCRIPTION: The Korean hare is a medium to large hare. The dorsal pelage is grayish yellow with grizzled black tips. Most animals sport a white spot on the forehead. The ventral pelage is a pinkish gray. The tail is light brown above and at the tip and white underneath.

The skull is stouter and heavier than that of the Chinese hare (*L. sinensis*), and the greatest skull length does not reach 90 mm. The nasals are of equal breadth throughout, thus not markedly compressed and pointed anteriorly, nor significantly bowed in profile. The frontal

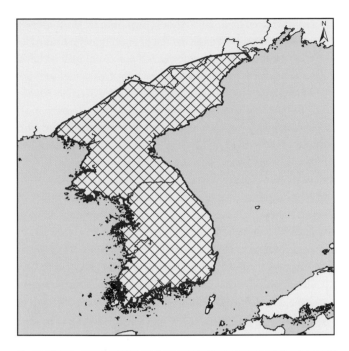

Lepus coreanus. Photo courtesy Ki-Young Hwang

region is broad and posteriorly bulging. The postorbital processes are small and project out from the skull. The incisive foramen is large, and the tympanic bullae are small.
SIZE: Head and body 425–565 mm; Tail 60–75 mm; Hind foot 108–122 mm; Ear 73–95 mm; Greatest length of skull 82–86 mm; Weight 2,100–2,600 g
CURRENT DISTRIBUTION: The Korean hare occurs throughout the Korean Peninsula, except the extreme northeast and extremely high mountainous regions in the south. There is a minor distribution in S Jilin, China.

TAXONOMY AND GEOGRAPHIC VARIATION: No subspecies. The Korean hare has generated significant confusion taxonomically, being at times included in the Chinese hare (*L. sinensis*), the Japanese hare (*L. brachyurus*), and the mountain hare (*L. timidus*), while some consider that it is most closely related to the Manchurian hare (*L. mandshuricus*). Recent molecular studies, however, have clearly supported the independent species status of *L. coreanus*, and shown that its closest leporid relative is *L. mandshuricus*.
ECOLOGY: The population density of the Korean hare varies according to habitat, 4.1/km^2 in hilly regions to 5.1/km^2 in mountainous regions. Recent studies have found that Korean hares are most common in Japanese larch plantations compared with natural deciduous forests. Abundance in winter appears correlated with the amount of shrub cover and presence of fallen logs. They do not dig or occupy burrows.
HABITAT AND DIET: The Korean hare occurs in a variety of habitats, including hillsides, grassy forest edges, woodlands, and forests. Little is known of the diet of this species, but it does consume tree bark, barley, and likely other cultivated crops. They forage in cultivated crops, and the species has been viewed as a pest due to the damage it causes.
BEHAVIOR: Korean hares are generally nocturnal, although they may be seen active early in the morning and before sunset. This species is solitary except during the mating season. Individuals occupy home ranges of 10–20 ha, and they habitually use the same game trails.
PHYSIOLOGY AND GENETICS: Diploid chromosome number = 48
REPRODUCTION AND DEVELOPMENT: The mating season extends from February to July. Gestation is 42–47 days in length, and 2–3 litters of 1–4 young are produced. Young are precocial at birth, being fully furred and with their eyes open.
PARASITES AND DISEASES: Korean hares are heavily parasitized, and the infection rate by parasites may reach 97.5%. Nematodes (*Strongyloides* spp., *Trichostrongylus* spp., and *Trichuris leporis*) are major parasites, along with cestodes (*Cittotaenia* spp.) and protozoa (*Eimeria* spp.).
CONSERVATION STATUS:
    IUCN Red List Classification: Least Concern (LC)
    National-level Assessments: China (Critically Endangered (CR)—B1ab(i,ii,iii)+2ab(i,ii,iii))
MANAGEMENT: This species has a history of being hunted for meat and fur. However, loss and degradation of habitat and overharvesting have caused recent declines in

abundance. As a result, the Korean hare has been delisted as a game species. Otherwise, the wide distribution of the species has not led to significant management initiatives.

ACCOUNT AUTHORS: Andrew T. Smith and Yeong-Seok Jo

Key References: Allen and Andrews 1913; Hwang et al. 2014; Jiang et al. 2016; Jones and Johnson 1965; Kim and Kim 1974; Koh and Jang 2010; Koh et al. 2001; NIER 2006; Thomas 1892; Won 1968; Won and Smith 1999.

Lepus corsicanus de Winton, 1898
Corsican Hare

OTHER COMMON NAMES: Apennine hare, Italian hare
DESCRIPTION: The general appearance of the Corsican hare and the European hare (*L. europaeus*) seems to be fairly similar. Nevertheless, the two species vary in numerous pelage characteristics of taxonomic significance. Among the discriminant features is the color of the basal fringe of the dorsal underfur of adult specimens, which is always white in the European hare and gray in the Corsican hare. The Corsican hare is slightly smaller than the European hare. This can be assessed by measuring the variables head and body, tail, ear, and hind foot length.

The two species are also distinguishable by skull measurements. With regard to dental characters, the difference between the two species is ascertainable in the shape of the posterior contour of the cross section of the first upper incisor, which is concave in the European hare and smooth or convex in the Corsican hare.
SIZE: Head and body 550–610 mm; Tail 70–120 mm; Hind foot 124–141 mm; Ear 90–100 mm; Greatest length of skull 90 mm; Weight 3,500–5,000 g
PALEONTOLOGY: Evolutionary lineages of the Corsican hare and the European hare are assessed to have split around 3 mya. During the Pleistocene glaciations, the Mediterranean peninsulas (Iberia, Italy, and Balkans) were characterized by a warmer climate than that of N Europe, and they offered a retreat for many temperate species. Furthermore, geographical obstacles, such as the Alps and Pyrenees, could have prohibited range expansions of terrestrial species. These isolation periods had a significant importance in shaping differentiation in mammals and causing the formation of numerous species endemic to the Mediterranean peninsulas, including the Corsican hare in Italy. Secondary interactions with cold-adapted species such as the European hare and the mountain hare (*L. timidus*) led to widespread introgression waves. Reconstruction of ancient distribution proposes that natural populations of Corsican and European hares were living in allopatry since in the Late Pleistocene, when the European hare invaded the Italian Peninsula. Potential contact areas extended from C Tus-

Lepus corsicanus. Photo © Alessandro Calabrese

cany to the Gargano Promontory. Colonization of Sicily by the Corsican hare might have happened between 45,000 and 121,000 years ago because during this period the level of the Mediterranean Sea was around 110 m lower than at the present and Sicily was joined to the mainland.

CURRENT DISTRIBUTION: The Corsican hare is endemic to Italy. It inhabits C and S Italy, as well as Sicily. The species was introduced into Corsica around the sixteenth century. The exact range is still not entirely known. The range seems to be continuous in Sicily, whereas on the Italian Peninsula populations are known only in Tuscany, Latium, Abruzzo, Molise, Apulia, Campania, Basilicata, and Calabria. The distribution of the Corsican hare in Corsica covers three distinct parts on the island: 97% of the total range in Corsica is situated in Haute-Corse, while the other two parts of the island where the Corsican hare sparsely occurs are Cap Corse and Sagone. The Corsican hare lives at elevations from sea level to 2,400 m asl on Mount Etna.

TAXONOMY AND GEOGRAPHIC VARIATION: No subspecies. Formerly, the Corsican hare was included in the Cape hare (*L. capensis*) or the European hare, but has received species status due to distinctness in genetics and morphological characters. Phylogenetical analyses suggest that the Corsican hare and the European hare are not closely related, but belong to separate evolutionary lineages that spread in W Europe in different periods during the Early Pleistocene. The Corsican hare is morphologically and genetically similar to the broom hare (*L. castroviejoi*) from the Cantabrian Mountains. The two species are sister taxa and may have had a shared predecessor inhabiting a large distribution area in SE Europe between Italy and Spain before the expansion of the European hare. The Corsican hare possibly differentiated in isolated areas during the last glaciation. Nonetheless, another recently conducted genetic study based on nuclear DNA renews the debate about the species status of these two species, as the results propose that the Corsican hare and the broom hare might be conspecific. Consequently, both forms might presently be in the speciation process due to their range fragmentation. An investigation examining nuclear DNA and mtDNA indicates that present populations of the Corsican hare are not hybridizing with other hare species and, most probably, they did not hybridize in the recent past, suggesting reproductive isolation with the other species currently distributed in Italy. However, two interspecific hybrids in the Corsican hare have now been found on the Italian Peninsula, where it occurs in sympatry with the European hare in some areas. In each case mtDNA of the Corsican hare was found in a European hare individual. In Corsica where frequent introductions of thousands of individuals from other hare species, explicitly the European hare and the Iberian hare (*Lepus granatensis*), have occurred, a recently conducted genetic study found two cases of interspecific hybrids including the Corsican hare. In the first case, European hare mtDNA introgression was found in a Corsican hare individual; in the other case, the analysis of the nuclear marker suggested that one part of the ancestry of this individual was the Iberian hare and the other part was the Corsican hare. These few cases of hybridization propose that the Corsican hare is not genetically isolated. Therefore, a tendency for the deterioration of the genetic integrity of the Corsican hare may exist.

ECOLOGY: Hare densities are assessed to be lower in hunting areas with 0.5 hares / 100 ha and higher in protected areas with 11 hares / 100 ha. However, in Sicily, where the European hare is absent, the Corsican hare has density estimates of 10 hares / 100 ha in protected areas and 2 hares / 100 ha in hunting areas.

HABITAT AND DIET: Due to competition with the European hare, the remaining populations of the Corsican hare are currently limited to the mountainous regions of the peninsula, and in plains and hills where the density of introduced European hares is low. In Sicily where the European hare is absent, the species lives also in the plains. The Corsican hare lives in the Mediterranean maquis, the assortment of clearances, bushy zones, and broad-leaved forests, and coastal dune habitat. In sympatry with the European hare, the Corsican hare dwells practically only in pastures and grasslands, whereas the European hare tends to be more of a habitat generalist. The Corsican hare occupies a variety of natural and artificial habitats such as open grassland, bushy pastures, garigue, and cultivated areas in Sicily.

The diet of the Corsican hare in Sicily differs seasonally according to the variation of the obtainable vegetation. The species eats 70 different plant species during the year, showing the aptitude of using an extensive variety of vegetation. In the Corsican hare, the consumption of Poaceae (20.5%) does not appear to reach the importance detected in other *Lepus* species (36% to 80%). The species feeds on Monocotyledons, Cyperaceae, and Juncaceae throughout the year, whereas Poaceae and Lamiaceae are fed on during spring and summer, respectively. In winter, leaves, buds and bark of arboreal species such as Fagaceae and Pinaceae are frequently consumed at high elevations where a thick snow coat is normally present, whereas at lower elevations where weather conditions are more favorable, the Corsican hare does not feed on twigs and bark from trees. In summer, the diet is also oriented toward taxa adapted to xeric climate such as leaves of French sorrel (*Rumex scutatus*), Mount Etna broom (*Genista aetnensis*), Sicilian milkvetch (*Astragalus siculus*), and assorted Asteraceae. Dicotyledons included in the diet year-round are Leguminosae and Asteraceae. The species complements its diet seasonally with some high-value nutritive foods such as the fruits of blackthorn (*Prunus spinosa*), European wild pear (*Pyrus pyraster*), and European crab apple (*Malus sylvestris*).

BEHAVIOR: No behavioral studies have been conducted on the Corsican hare. However, the behavior of the Corsican hare is probably similar to that of the European hare, as this species was formerly included in the European hare.

PHYSIOLOGY AND GENETICS: Diploid chromosome number = 48. Recent genetic work has focused on taxonomic and phylogenetic clarification analyzing nuclear DNA and mtDNA.

REPRODUCTION AND DEVELOPMENT: The Corsican hare's reproductive activity takes place throughout the year, peaking in spring and at a minimum during winter and summer. The reproductive activity is modest in autumn. Most adult females (63%) become reproductively active. Although the data are preliminary, it appears that the dry and cold seasons have a negative effect on reproduction in this species. Females have on average two litters per year with a maximum of four. Litter size is on average 1.5. Therefore, females have on average three and a maximum of six leverets per year.

PARASITES AND DISEASES: Corsican hares are hosts for several tick species (*Ixodes ricinus*, *Rhipicephalus turanicus*, and *Hyalomma marginatum*). A study on gastrointestinal helminths in the Corsican hare showed that 86% were positive for at least one parasite. Among the helminths were two cestode species (*Cittotaenia pectinata*, prevalence 3%, and *Paranoplocephala* spp., 3%) and four nematode species (*Trichostrongylus retortaeformis*, 86%; *Graphidium strigosum*, 14%; *Trichuris* spp., 10%; and *Teladorsagia circumcincta*, 7%). The introduction of the European hare is a threat to the Corsican hare regarding disease transmission, as the Corsican hare is fully susceptible to the European brown hare syndrome. Rabbit hemorrhagic disease virus (RHDV), a member of the genus *Lagovirus*, causes rabbit hemorrhagic disease (RHD), a fatal hepatitis of rabbits that has previously not been reported in hares. Recently, a new RHDV-related virus emerged,

called RHDV2. This lagovirus has been described to cause a RHD-like syndrome in the Corsican hare and the Cape hare.

CONSERVATION STATUS:
   IUCN Red List Classification: Vulnerable (VU)—A2bcde+3bcde), decreasing
   National-level Assessments: Europe (Vulnerable (VU)—A2bcde+3bcde)

MANAGEMENT: The Corsican hare has been monitored uninterruptedly since 1997 in Italy, including Sicily. The species is fragmented and rare on the peninsula, but numerous protected areas have been created that will help the populations to improve. The species is widespread and locally abundant in Sicily. The local government there officially recognized the Corsican hare as a true species in 1998 with the result that hunting is prohibited on the island. The status of the species in Corsica is uncertain, but in a recently conducted genetic study 70% of the collected specimens were Corsican hares. Population and range declines have been assessed to be 50% for the continental populations and 38% for the species as a whole. The population and range reduction possibly have occurred due to overhunting, as well as the introduction of the European hare. For the past 50 years, the frequent use of restocking with European hares from Eastern European countries, from N Europe, and even from South American countries for hunting purposes has been common in Italy, even though these restocking procedures are generally futile. The restocking programs have been exceedingly intense with thousands of captive individuals released every year into the wild. The presence of foreign specimens caused dilution of the Corsican hare probably due to the occurrence of interspecific competition. Another problem concerning Corsican hares relates to the high pressure inferred from hunting of free-ranging populations. The Corsican hare is legally protected in continental Italy and Sicily. However, the problem of distinguishing between the European hare and the Corsican hare in the field makes it challenging to protect this species from hunting. Moreover, in Sicily the ban to hunt the Corsican hare was lifted during the hunting seasons of 2004 and 2005. The Corsican hare is still a game species in Corsica because in the French hunting act the species is considered conspecific with the European hare.

   Major threats to the species are habitat degradation, fragmentation that causes low or absent gene flow between populations, low population densities, competition with the introduced European hare, and overhunting. Further threats are infectious diseases, poaching, and predation by foxes and feral dogs. Captive breeding programs are being carried out in Italy and Sicily. Conservation actions to preserve viable populations of the Corsican hare might include the ban of European hare introductions inside the historical range of the Corsican hare, the establishment of a network of protected areas for the peninsula, and the introduction of a harvest system modeled on the sustained yield theory in Sicily. The Corsican hare is listed in the Bern Convention, Appendix III, as part of *L. capensis sensu lato*.

ACCOUNT AUTHORS: Stéphanie Schai-Braun and Klaus Hackländer

Key References: Alves et al. 2003, 2008b; Angelici and Luiselli 2001, 2007; Angelici and Spagnesi. 2008a; Angelici et al. 2008, 2010; Camarda et al. 2014; Dantas-Torres et al. 2011; De Battisti et al. 2004; De Marinis et al. 2007a, 2007b; Ellerman and Morrison-Scott 1951; Freschi et al. 2014, 2015, 2016; Hoffmann and Smith 2005; Koutsogiannouli et al. 2012; Lorusso et al. 2011; Melo-Ferreira et al. 2007; Mengoni et al. 2015; Mitchell-Jones et al. 1999; Palacios 1996; Petter 1961; Pierpaoli et al. 1999, 2003; Pietri 2015; Pietri et al. 2011; Riga et al. 2001; Usai et al. 2012; Vigne 1988.

Lepus europaeus Pallas, 1778
European Hare

OTHER COMMON NAMES: Brown hare, European brown hare, Common hare

DESCRIPTION: The European hare is the largest *Lepus* species in W Europe with larger animals in continental climates and smaller ones in oceanic climates. It stays brownish year-round, with variable colorations (ochraceous gray to yellowish gray) across the distribution range. The belly, the inner parts of the legs, and the ventral side of the tail are whitish. The underfur hairs (15 mm) are dark at the tip (black to black-brown) and silvery or white at the base. The tail is relatively long with a black stripe on the dorsal side. The ears are comparatively long with a light (white or grayish) posterior side and a distinct black tip. When flapped to the front, the ears extend past the muzzle. The eyes are large with yellow to orange irises. The winter pelage is dense and fluffy, but varies in coloration by latitude. In the north, only the ears, head, and anterior portion of the back are dark; the remainder turns grayish or even whitish. Specimens of southern latitudes show very little change in coloration, but may

appear more reddish. Geographic variation is low among the mainland populations due to the species' high mobility and occupation of continuous habitats. Moreover, the geographic variation is obscured by translocations by humans.

SIZE: Head and body 550–675 mm; Tail 75–140 mm; Hind foot 124–167 mm; Ear 100–140 mm; Greatest length of skull 102 mm; Weight 3,500–5,000 g

PALEONTOLOGY: European hares derive from *L. capensis sensu lato* in the Near East and were spreading into Europe from the Middle Pleistocene on. The oldest fossil records of *L. europaeus* were found in the Caucasus region. The species was quite abundant in S Europe and Asia in the Late Pleistocene, with records ranging from Spain to Kazakhstan.

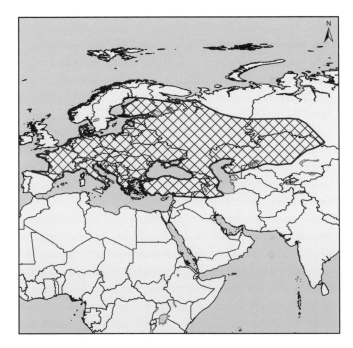

Lepus europaeus. Photo courtesy Rolf Giger

CURRENT DISTRIBUTION: The European hare's origin is the steppe habitat in SE and C Europe, the Middle East, Caucasus, the Russian lowlands, and the W Caucasus. The range expanded to the north when humans introduced farming in the Neolithic period. The current distribution is heavily influenced by human translocations. Today, the European hare is spread all over Europe except for the southwest (most of the Iberian Peninsula) and north (N Scandinavia and N Finland). European hares have been introduced (mainly for hunting purposes) to N Ireland, S Sweden, E Siberia, Far East Russia, E Australia, New Zealand, South America up to 28°S, Canada, and the United States (around the Great Lakes area), as well as numerous islands in the Indian and Atlantic Oceans. Due to global warming and land-use changes, European hares are extending their European range farther north and in Russia/Kazakhstan farther east.

TAXONOMY AND GEOGRAPHIC VARIATION: As with other hare species, the taxonomy of *L. europaeus* is still under debate. Clear distinction from other taxa is difficult as the species is comparatively young and still expanding its range, leading to sympatric occurrences of *Lepus* species. The European hare was formerly included in the Cape hare (*L. capensis*) based on a cline in morphological characters (mainly size) from NE Africa eastward across the N Arabian Peninsula and the Middle East and northward through Israel to Turkey. In Kazakhstan, where the large "*europaeus*" and the small "*capensis*" (= *tolai*) live in sympatry without hybridization, the interpretation was of the overlapping ends of a Rassenkreis (circle of races). A reanalysis showed a discontinuity between the smaller "*capensis*" and the larger "*europaeus*" running from the E Mediterranean coast through Iran. This is the basis on which the European hare is separated from the Cape hare and the tolai hare (*L. tolai*). In Iran, east of the border of the range of the European hare, the tolai hare occurs apparently in allo- or parapatry with the European hare. However, recent evidence suggests that the hypothesis of conspecificity of the European hare and the Cape hare may be correct after all. A study of the nuclear gene pool of the Cape hare, the European hare, and hares in North Africa indicated that North African hares, as well as the European hare, belong to the Cape hare. However, a study of the mtDNA of these three groups indicated a significant degree of divergence supporting species-specific designation. These studies show that genetic differentiation between the Cape hare and the European hare might be attributed to geographic distance rather than divergence. It has been speculated that gene flow may be occurring in the Near East where distributions meet, resulting in the potential for intergraded populations. Until conclusive evidence of a single species complex is available in order to support a change in the taxonomic status of the European hare, it remains a true species. Molecular phylogenetic studies in Spain have shown that the Cantabric population has unique mtDNA in relation to other European populations. As taxonomists are still trying to clarify the species differentiation in *Lepus*, the subspecific taxonomy is not elaborated yet. The original descriptions of the subspecies are often not very helpful, as they are mostly based on a few exterior characteristics and on a small numbers of individuals. It has been shown that the variability is clinal in more careful investigations. Hence, the distinction in subspecies might be arbitrary and unreasonable.

Sixteen subspecies: *L. e. europaeus*, *L. e. caspicus*, *L. e. connori*, *L. e. creticus*, *L. e. cyprius*, *L. e. cyrensis*, *L. e. hybridus*, *L. e. judeae*, *L. e. karpathorum*, *L. e. medius*, *L. e. occidentalis*, *L. e. parnassius*, *L. e. ponticus*, *L. e. rhodius*, *L. e. syriacus*, and *L. e. transsylvanicus*.

ECOLOGY: In its original habitats, the European hare has average densities of about 2/100 ha. In more suitable habitats with milder climates and more fertile soils, densities can be very high, up to 275/100 ha. Population dynamics are mainly affected by juvenile mortality. Usually, less than 10% of juveniles survive. The main reasons for mortality are mechanical activities in agricultural land, diseases, and predation. Home range sizes of European hares depend on habitat suitability. In arable land, home range size increases with increasing field size. Small field size with high crop variation leads to smaller home ranges. In intensively used arable land, set-asides help to reduce home range size. Well-structured habitats lead to a yearly home range of less than 20 ha, whereas European hares in intensively used agricultural land might have home ranges of up to 300 ha.

HABITAT AND DIET: Originating from steppe habitats, European hares are now typically found in agricultural areas, ranging from sea level to alpine regions. They occupy arable land, grasslands, moorlands, alpine grasslands, steppe, saltmarsh, and near desert, and in some cases they will venture deeply into forests, although confining activities to clearings and forest roads. They tend to avoid regions that have heavy snow cover, where they are replaced by the mountain hare (*L. timidus*). In Chile, Argentina, Australia, and New Zealand, where no other *Lepus* species are present, European hares also inhabit pampas, sand dunes, marshes, and alpine fellfields. Many surveys have investigated the habitat preferences of European hares. As the study areas differ greatly, the results also diverge considerably. In all studies, set-asides or fallow land are preferred consistently, whereas settlement areas are avoided. European hares feed predominantly on cultivated crops, weeds, and grasses. A study investigating the diet preferences of the European hare showed that this species selected weed and grasses. Furthermore, chemical analysis revealed that the European hare selects its food for high-energy content (crude fat and crude protein), and avoids crude fiber. During food shortage and patchy food availability, a dominance hierarchy among individual hares has been demonstrated. Coprophagy plays an important role in protein supply and provides sustenance during periods of food shortages.

BEHAVIOR: The species is active mainly during the night. In winter, European hare activity starts with regularity shortly after sunset and ends shortly before sunrise. In summer, hare activity is less consistently tied to the dark period so that active hares can regularly be observed in full daylight. When sunset is early and sunrise late, evening and morning activity peaks occur during the dark phase. This is the case in late spring and early fall, when nights are longer. When sunset is late and sunrise early, European hares show activity peaks in full daylight. When inactive, the European hare rests in forms (depressions), holes, or even burrows dug by foxes and marmots. Exceptions from the crepuscular and nocturnal activity pattern can be observed in spring, when mating leads to agglomerations of usually solitary hares (so-called "March madness"). European hares are not territorial.

PHYSIOLOGY AND GENETICS: Diploid chromosome number = 48. Recent genetic work has focused on taxonomic clarification, including allozymes, Y chromosomes, and mtDNA. The high variation in mtDNA across the range of the European hare is due to interspecific ancient and recent hybridization. It is difficult to find distinct genetic differences across the species' range as the European hare is still on a course of expansion. In addition, translocations over huge distances (for hunting purposes) obscure taxonomic and phylogenetic relationships.

European hares are adapted to arid environments and thus do not need open water to sustain their metabolism. With the help of their relatively large ears, they can dissipate heat. Hares have relatively large hearts, typical of fast-running species (the European hare runs up to 70 km/h). Running speed is enabled by selective transportation of polyunsaturated fatty acids from the intestines to the muscles.

REPRODUCTION AND DEVELOPMENT: In Europe the main period of reproduction lasts from February until September. In spring, the European hare also aggregates for mating during the day. Males chase females, and the latter box with males prior to copulations. The reproduc-

tive period might be shorter in colder climates (northern range or higher altitude), whereas it lasts year-round in the Mediterranean area or in oceanic climates. The gestation period is about 42 days, with shorter inter-birth intervals (36 days) in case of superfetation. A female gives birth to about 11 young per year. The number of litters per year varies with climatic conditions (the longer the season, the smaller the litters). Young are precocial and weigh about 100 g at birth. Leverets are suckled once each day for a few minutes. As mothers produce milk rich in fat, young are weaned after 4–5 weeks (when they have reached about 1 kg). They are fertile at an age of 4–5 months of age and reach their adult body weight within 8 months. Reproduction in the year of birth is more likely in Mediterranean or oceanic climates. In continental climates, hares invest in body size to survive winter and start to reproduce after their first winter. Physiological age expectancy is 8–12 years. The oldest hare found in the wild reached an age of 12.5 years.

PARASITES AND DISEASES: Some diseases related to heavy parasite burden, bacteria, or viruses might lead to epidemic outbreaks with massive losses in some years or regions. These diseases include brucellosis, tularemia, pseudotuberculosis, pasteurellosis, European brown hare syndrome, and lung worm infestation.

CONSERVATION STATUS:

IUCN Red List Classification: Least Concern (LC)

MANAGEMENT: The European hare is widespread and abundant across its geographic range. European hare populations have been on the decrease throughout Europe since about 1910 due to agricultural intensification. This species is listed under Appendix III of the Bern Convention in Europe. Several countries have classified the European hare on national levels as "near threatened" or "threatened." A meta-study, reviewing literature from 12 European countries comparing population densities in relation with habitat characteristics, concluded that the primary cause of the European hare decline was agricultural intensification. Negative associations in respect to abundance were predation and precipitation. Field size, temperature, precipitation, and hunting had no effect on hare density. European hares are an important game species in Europe with hundreds of thousands harvested each year. The human-hare relationship has been intensive since antiquity, and many cultural aspects (such as Easter, symbol of fertility) in Europe are related to hares. Hence, people are interested in hare conservation and programs for improving habitat of European hares are widespread within the native range. Where hares have been introduced, eradication programs are ongoing to avoid negative impacts, e.g., hybridization with the Irish hare (*L. timidus hibernicus*). In Greece, Spain, France, and Denmark, the restocking of hares from other regions or countries to supplement hare densities for hunting has been identified as a threat to the regional gene pools.

ACCOUNT AUTHORS: Klaus Hackländer and Stéphanie Schai-Braun

Key References: Alves and Hackländer 2008; Angermann 1983; Averianov et al. 2003; Ben Slimen et al. 2006; Broekhuizen and Maaskamp 1982; Flux 1967; Flux and Angermann 1990; Hackländer et al. 2001, 2002, 2008, 2011; Hoffmann and Smith 2005; Holley 2001; Jennings et al. 2006; Lincoln 1974; Mamuris et al. 2001; Marboutin et al. 2003; Panek and Kamieniarz 1999; Petter 1961; Pielowski 1972; Schai-Braun and Hackländer 2014; Schai-Braun et al. 2012, 2013, 2015; Slamečka et al. 1997; Smith et al. 2005a, 2005b; Sokolov et al. 2009; Tapper and Barnes 1986; Zörner 1996.

Lepus fagani Thomas, 1903
Ethiopian Hare

OTHER COMMON NAMES: Lièvre éthiopien (French); Äthiopischer Hase (German)

DESCRIPTION: The Ethiopian hare is of medium size with long, dense, harsh fur. The color of the upper parts is buff ochraceous brown, while the sides, nape, and chest

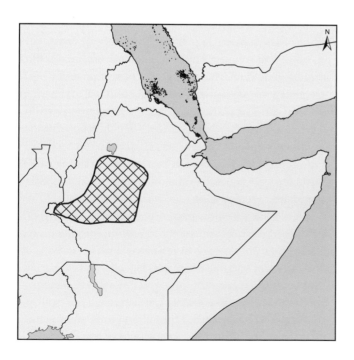

are tawny. A cinnamon-ginger nuchal patch spreads to the sides of the neck. The ears are generally shorter than those of the Abyssinian hare (*L. habessinicus*) and the Ethiopian highland hare (*L. starcki*) with a fawn rim and narrow black edge at the interior tip. The chin is grayish underneath and the abdomen white. The legs are brown, with the hind limbs more buff and the forelimbs more cinnamon. The tail is white or buff below and yellow on the sides, with a large blackish dorsal stripe above.

SIZE: Head and body 420–500 mm; Tail 70–107 mm; Hind foot 90–110 mm; Ear 60–95 mm; Weight: 1,704–2,278 g

CURRENT DISTRIBUTION: The Ethiopian hare is distributed along the plateau in N and W Ethiopia, west of the Rift Valley. It is sympatric with the Ethiopian highland hare in the Shewa Highlands and possibly sympatric with the Abyssinian hare at the eastern edge of its distribution. It occurs at elevations from 500 to 2,500 m.

TAXONOMY AND GEOGRAPHIC VARIATION: The Ethiopian hare was initially considered to be closely related to the African savanna hare (*L. victoriae*) / Cape scrub hare (*L. saxatilis*) complex. The species has been separated as a distinct species based on morphological characteristics within its range. Additional molecular and morphological data have now revealed that *L. fagani* is distinct from the *L. victoriae* / *L. saxatilis* complex. These molecular data now support a closer relation between *L. fagani* and *L. habessinicus*.

HABITAT AND DIET: The Ethiopian hare occupies scrub and forest edge habitat, similar to that of the African savanna hare.

PHYSIOLOGY AND GENETICS: The Ethiopian hare likely has the same diploid number shared by all other *Lepus* species, 48.

CONSERVATION STATUS:
   IUCN Red List Classification: Data Deficient (DD)

MANAGEMENT: The Ethiopian hare has been recorded in Abijatta-Shalla and Gambela National Parks. Studies should be undertaken to address the lack of data available on the population size/density, reproductive biology, diseases, and ecology of the Ethiopian hare.

ACCOUNT AUTHORS: Zelalem Gebremariam Tolesa and Charlotte H. Johnston

Key References: Angermann 1983; Azzaroli-Puccetti 1987a, 1987b; Azzaroli-Puccetti et al. 1996; Flux and Angermann 1990; Happold 2013; Hoffmann and Smith 2008; Mekonnen et al. 2011; Petter 1963; Pierpaoli et al. 1999; Tolesa 2014; Yalden et al. 1986, 1996.

Lepus flavigularis Wagner, 1844
Tehuantepec Jackrabbit

OTHER COMMON NAMES: Liebre tropical, Liebre de Tehuantepec (Spanish)

DESCRIPTION: The Tehuantepec jackrabbit is a large, slender hare. It has long, pointed ears of yellowish brown color. There are two longitudinal and dorsal black stripes that extend from the base of each ear to the base of the neck. The dorsal pelage is a gray pale yellow with dark brown; the hindquarters and the limbs are pale gray. The ventral pelage and flanks are white, with the exception of the throat, which is pale yellow. The black color on the back of the tail extends toward the hindquarters; the ventral part of the tail is white.

The skull is elongated, and the tympanic bullae are the smallest among *Lepus* in Mexico.

SIZE: Head and body 499–640 mm; Tail 50–118 mm; Hind foot 109–130 mm; Ear 103–145 mm; Greatest length of skull 97 mm; Weight 1,700–2,900 g

CURRENT DISTRIBUTION: A Mexican endemic, the Tehuantepec jackrabbit comprises only four geographically isolated populations (Montecillo Santa Cruz, San Francisco del Mar Pueblo Viejo, Aguachil, and Santa María del Mar), around the Laguna Superior and Laguna Inferior in the Tehuantepec Isthmus, Oaxaca, Mexico.

TAXONOMY AND GEOGRAPHIC VARIATION: No subspecies. Phylogenetic relationships based on molecular analyses indicate that white-sided jackrabbits are grouped in a subclade, where the white-sided jackrabbit (*L. callotis*) and the Tehuantepec jackrabbit are closely related, and the black-tailed jackrabbit (*L. californicus*) is separated from this group in another subclade. The most differentiated species is the white-tailed jackrabbit (*L. townsendii*).

ECOLOGY: Average population density of the Tehuantepec jackrabbit varies from 4/km² at Montecillo Santa Cruz to 26/km² at Santa María del Mar. The total population of the species is small, with estimates ranging from 44 in a sampling area of 2.68 km² at Aguachil, 55/4.79 km² at San Francisco del Mar Pueblo Viejo, 132/33.16 km² at Montecillo Santa Cruz, and 377/14.33 km² at Santa María del Mar. A viability population analysis of the Tehuantepec jackrabbit at Santa Maria del Mar showed that it is at high risk of extinction.

Average home range size of the Tehuantepec jackrabbit at Montecillo Santa Cruz is 55 ha, and at Santa María del Mar shows wide intraspecific variation, with a range from

Lepus flavigularis. Photo courtesy Arturo Carrillo-Reyes

0.20 to 152 ha. There is a high level of overlap between the sexes (> 70%); therefore, home ranges are not exclusive.

The Tehuantepec jackrabbit shares its habitat with several other mammals: common opossums (*Didelphis marsupialis*), nine-banded armadillos (*Dasypus novemcinctus*), eastern cottontails (*Sylvilagus floridanus*), hooded skunks (*Mephitis macroura*), American hog-nosed skunks (*Conepatus leuconotus*), spotted skunks (*Spilogale gracilis*),

Lepus flavigularis habitat. Photo courtesy Eugenia C. Sántiz-López

raccoons (*Procyon lotor*), gray foxes (*Urocyon cinereoargenteus*), and coyotes (*Canis latrans*).

HABITAT AND DIET: The Tehuantepec jackrabbit lives in thorny scrub and grassland-nanchal (*Byrsonima crassifolia*) and grassland-morro (*Crescentia* spp.) associations at Montecillo Santa Cruz, and grassland (with predominance of the grass *Jouvea pilosa*), coastal dunes, and xeric shrubs at Santa María del Mar, San Francisco del Mar Pueblo Viejo, and Aguachil.

The diet of the Tehuantepec jackrabbit is highly diverse; 18 plant species are consumed in the dry season and 16 plant species in the rainy season (primarily from the families Poaceae, Cyperaceae, and herbs). Favored plants during both wet and dry seasons are buffalograss (*Bouteloua dactyloides*), zacale (*Cathestecum brevifolium*), and southern crabgrass (*Digitaria ciliaris*).

BEHAVIOR: The Tehuantepec jackrabbit is not territorial. Groups of up to 12 jackrabbits have been observed resting on beds among tillers of grasses and shrubs and under nopal (*Opuntia* spp.) located within a radius of no more than 50 m, without any evidence of aggressive territorial behavior. These jackrabbits are nocturnal, becoming most active at 18:40 ± 2 h. The greater part of this period (12 ± 3 hours), is occupied in feeding, interspersed with periods of grooming (average duration 12 ± 9 minutes) and socialization. The jackrabbits feed in a solitary manner, as well as within large groups.

PHYSIOLOGY AND GENETICS: Diploid chromosome number = 48 and FN = 88. The Tehuantepec jackrabbit displays low genetic diversity. Additionally, there is a phylogeographic pattern in that the species can be clearly differentiated in two clades (one that corresponds to the populations of Montecillo Santa Cruz, San Francisco del Mar Pueblo Viejo, and Aguachil, and another that corresponds to the population of Santa María del Mar). The gene flow between these groups is low due to the presence

of a channel of water and anthropogenic features that constitute a barrier between these sites. Microsatellite variation indicates that the Montecillo Santa Cruz population has low levels of polymorphism (43%) and heterozygosity (Ho = 0.18), in contrast to the San Francisco del Mar Pueblo Viejo population (polymorphism = 71%; Ho = 0.51).

REPRODUCTION AND DEVELOPMENT: Tehuantepec jackrabbits are sexually active (estrous phase) in the dry season (November–April). The lactation phase takes place during the wet season (May–November). The breeding season of the Tehuantepec jackrabbit extends 250 days a year, with an increase in reproductive activity during the wet season (May–October). Reproduction sites are located almost exclusively in open grassland in pasture areas. Females produce four offspring per breeding season. They exhibit a polygynous reproductive breeding system.

PARASITES AND DISEASES: The Tehuantepec jackrabbit is parasitized by a nematode of the family Onchocercidae (*Pelecitus meridionaleporinus*), which occurs in the subcutaneous tissue at the base of ears. This is the second species of *Pelecitus* that is known among New World lagomorphs, and the third found infecting mammals.

CONSERVATION STATUS:

IUCN Red List Classification: Endangered (EN): A2b+3c; B1ab(i,ii,iii,v)+2ab(i,ii,iii,v); C1; D; decreasing

National-level Assessments: Mexico (Endangered Official Norm NOM-059-ECOL-2010 (SEMARNAT))

MANAGEMENT: Habitat fragmentation caused by changes in local land use, an increase in settlements, and activities such as ranching and burning pastures for seasonal agriculture has almost completely isolated the subpopulations of the Tehuantepec jackrabbit. This isolation has in turn led to a decrease in genetic viability in these populations. Historically, excessive hunting was another anthropogenic factor that caused a decrease in the population of this species.

Possible species competing for food with the Tehuantepec jackrabbit are eastern cottontails, cattle, and horses. These species may also transmit diseases to the jackrabbits. There are also several predators of the jackrabbits that may further be affecting their populations: domestic dogs and cats (associated with the camps of fishermen), coyotes, gray foxes, Neotropical whip snakes (*Masticophis mentovarius*), and western lyre snakes (*Trimorphodon biscutatus*).

We recommend continuous monitoring of the populations of the Tehuantepec jackrabbit. In addition, it is important to carry out activities within the social sector to recommend actions for conservation of the jackrabbit: proper use and management of paddock grasslands, proper burning strategies within grasslands, cessation of hunting of jackrabbits, and institution of environmental education programs. Long-term studies of the effect of cattle grazing on the composition of plant species in the pastures are required to detect the possible occurrence of overgrazing that might affect the carrying capacity and therefore the Tehuantepec jackrabbit population. An improved understanding of this aspect would provide the necessary tools to establish an appropriate program of rotational grazing that is currently absent.

ACCOUNT AUTHORS: Consuelo Lorenzo, Tamara Rioja-Paradela, Arturo Carrillo-Reyes, Eugenia C. Sántiz-López, Jorge Bolaños

Key References: Álvarez del Toro 1991; Carillo-Reyes 2009; Carillo-Reyes et al. 2010; Cervantes 1993; Cervantes and Lorenzo 1997b; Farías 2004; Farías et al. 2006; Flux and Angermann 1990; Hall 1981; Jiménez-Ruiz et al. 2004; López et al. 2009; Lorenzo et al. 2006, 2008, 2011, 2014a, 2014b, 2015; Reid 1997; Rico et al. 2008; Rioja et al. 2008, 2011, 2012; Sántiz et al. 2012; SEMARNAT 2010; Uribe-Alcocer et al. 1989; Vargas 2000.

Lepus granatensis Rosenhauer, 1856

Iberian Hare

OTHER COMMON NAMES: Granada hare; Liebre Ibérica (Spanish); Lebre Ibérica (Portuguese)

DESCRIPTION: The Iberian hare is smaller than other European hare species. In contrast to the uniformity of the pelage pattern of other European hares, in the Iberian hare the white ventral area is extensive, and there is a clear contrast between the ochraceous brown (gray-brown) color of the back and the white belly pelage that extends in a white strip to the forefeet and hind feet. The external faces of the hips are bright reddish in contrast with the gray-brown of the dorsal part of the back. The Iberian hare has a typical white coloration on the inside of the legs, which continues with the upper side of the legs and reaches to the tip of the toes. There are no white stripes on the face. The Iberian hare is smaller than the other sympatric hare species, the European hare (*L. europaeus*) and the broom hare (*L. castroviejoi*), with a mean body weight ranging from 2,000 to 2,600 g and with

hind feet smaller than 130 mm. Female Iberian hares are larger than males.

The Iberian hare possesses a developed posterior arm of the supraorbital process, which reaches the temporal tubercule.

SIZE: Head and body 445–473 mm; Tail 93–112 mm; Hind foot 116.6–127.4 mm; Ear 92.5–102.7 mm; Greatest length of skull 86.2–91.2 mm; Weight: 2,129–2,832 g

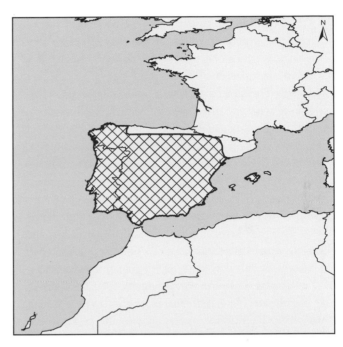

Lepus granatensis. Photo courtesy Rolf Giger

PALEONTOLOGY: The Iberian hare was first documented in the Middle Pleistocene at Cúllar de Baza (Granada, Spain). It appears to have been consistently present in the Iberian Peninsula in the Upper Pleistocene, with remains at Cova de les Cendres and Cueva Murciélagos. Fossil records were also found in the transition between the Pleistocene and the Holocene, and in different places during the Holocene. Yet, several fossil records identified in the Pleistocene deposits have been attributed to the African Cape hare (*L. capensis*). Nevertheless, the presence of the Cape hare in Iberia warrants some caution because of taxonomical confusion, and this evidence may actually represent the Iberian hare or the European hare. The lack of consistent paleontological data, associated with the presence of multiple hare species in the Iberian Peninsula (the mountain hare *L. timidus* has been also recorded), complicates the establishment of a robust evolutionary history of the Iberian hare. Further studies are needed, revisiting the existing paleontological findings.

CURRENT DISTRIBUTION: The Iberian hare is endemic on the Iberian Peninsula (Portugal and Spain). It mainly occurs in the Mediterranean continental bioregion in Iberia, where the European hare is not present, and on Majorca Island (Balearic Archipelago, Spain), where it was introduced. The distribution of the Iberian hare extends from the Mediterranean coast south of Ebro River to the Atlantic coast, and from the northern mountains of Burgos-Palencia and the southern slopes of Navarra-Huesca to the Andalusian coast. In the northwest it extends to the north of the Cantabrian Mountains as far as C Asturias. The Iberian hare has recently been introduced in S France (Perpignan) for game purposes, where there is now a stable and expanding population.

TAXONOMY AND GEOGRAPHIC VARIATION: The Iberian hare is often confounded with the Cape hare, mainly due to its smaller size and longer ears in comparison to the other hare species in Europe. However, morphological and genetic data clearly demonstrate that the Iberian hare is a valid species. Three subspecies: *L. g. granatensis*, described for all of the Iberian Peninsula, except the northwest; *L. g. gallaecius*, for NW Spain, Galiza, and C Asturias, with measurements similar to *L. g. granatensis*, but with clearly darker coat coloration; and *L. g. solisi* for Majorca Island, somewhat smaller than *L. g. granatensis*, with short hind feet and condylobasal length.

ECOLOGY: The Iberian hare typically occupies open agroecosystems with high levels of habitat heterogeneity, although it may occur in extremely variable habitats across the entire Iberian range. A seasonal habitat selection study in NW Spain showed a rotation between the different habitats of the landscape, which were linked to the annual farming cycle. In NW Spain the relative density, based on night count transects, varied from 0 to 13 hares / 100 ha. Higher densities are usually reached in the south, for instance, 24.2 hares / 100 ha in Doñana National Park, but it can reach even higher densities in S Iberia (79.8 hares / 100 ha). In the past decade this species showed a general positive trend in both N and S Spain. However, currently some local populations are declining, probably due to the generalized use of herbicides.

HABITAT AND DIET: The Iberian hare is present at low densities in N Iberia (Asturias, Galicia, and NW Portugal), where it occupies humid mountain areas with Atlantic-type climate (annual rainfall 1,500–2,000 mm), natural pastures, meadows, agricultural fields (rye, wheat, cork, etc.), and woods. In C Iberia (annual rainfall 300–800 mm), it occupies mainly open areas of agricultural landscapes such as wheat, barley, oats, vineyards, and alfalfa. In S Iberia it inhabits dry areas (annual rainfall of 300–600 mm, some places < 200 mm) with olive groves, vineyards, wheat, barley, and natural scrubland vegetation. It can also occur in marshy areas or dunes with sparse vegetation.

The Iberian hare is herbivorous, like other lagomorphs. A wide range of plant species has been observed in the Iberian hare's diet. However, grasses represent the basis of the diet, always presenting frequencies higher than 50%. Most existing grasses (91.43%) have been identified in fecal pellets, though only six species were found in proportions greater than 5%: sweet vernal grass (*Anthoxanthum odoratum*), rye grass (*Secale cereale*), and bent grasses (*Agrostis* spp.). In summer, grass consumption decreases to values of about 55%, concomitant with a rise in the ingestion of other plant groups, like herbs and some shrubs.

BEHAVIOR: The Iberian hare is mainly nocturnal and solitary, spending most of the daytime in resting places. The home range estimated in an open area of NW Spain (Zamora), using introduced animals, was 123 ha ± 66.3 for males and 237 ha ± 260 for females. In Doñana National Park the home ranges were substantially smaller, 28 ha ± 1.6 for males and 24 ha ± 10 for females. In Zamora, 2,621 m (maximum 7,700 m) was the mean distance covered at night, and 334 m (maximum 950 m) the mean distance between resting places. Iberian hares make shallow depressions in the ground (forms) for resting and raising leverets.

PHYSIOLOGY AND GENETICS: Diploid chromosome number = 48. There is no evidence of genetic structure. However, striking evidence of historical hybridization with the mountain hare, resulting in subsequent mitochondrial introgression, has been reported and is well documented. The presence of mountain hare mitochondrial lineage in the Iberian hare has a clear gradient from C to N Iberia, where it can be predominant. Traces of nuclear DNA introgression are generally sporadic and found across its distribution (but introgression at one X-linked marker is abundant).

Evidence of contemporaneous hybridization has been found in the northern contact zone with the European hare. The complete mitogenome (17,765 bp), as well as transcriptome and whole genome data, are available.

Negative relationships between escape duration and internal parasite burdens (intestinal coccidian *Taenia pisiformis*), as well as with parasite diversity, have been reported. Seasonal variation in the kidney fat index (KFI) followed a similar pattern in males and females, with an increase in mid-fall and a decline at the end of winter.

REPRODUCTION AND DEVELOPMENT: Breeding in the Iberian hare occurs throughout the year, with a peak in March–April. Mean litter size (based on embryo counts) is 1.58 (range 1 to 4). The largest litter size reported in the wild was seven. A mean ovary weight of 1.14 g (range 0.21–3.1 g) and a mean testis and epididymis weight of 7.85 g (range 3.6–12.2 g) and 1.49 g (range 0.53–2.28 g), respectively, have been recorded. Mean daily sperm production is $362 \times 10^6 \pm 115 \times 10^6$. Gestation period is ~ 42 days. Like in other hare species, the leverets are born with full pelage and open eyes. The mean weight of newborn leverets in captivity is 128.6 g (range 123–140 g). No data are available on superfetation. The prenatal mortality rate is 37.8%, based on the difference of the number of corpora luteum and healthy embryos.

PARASITES AND DISEASES: The following parasites have been reported: *Ixodes ricinus*, *Spilopsyllus cuniculi*, *Eimeria europaea*, *E. hungarica*, *E. leporis*, *T. pisiformis*, *Mosgovoyia pectinata*, *M. ctenoides*, *Trichostrongylus retortaeformis*, *Graphidium strigosum*, *Nematodiroides zembrae*, *Passalurus nonanulatus*, *P. ambiguus*, *Micipsella numidica*, *Hyalomma marginatum*, *Haemaphysalis hispanica*, *Rhipicephalus sanguineus*, *Haemodipsus lyriocephalus*, *Linguatula serrata*, and *Dicrocoelium dendriticum*.

Bacteria causing diseases include *Brucella suis*, *Treponema paraluis-cuniculi*, *Yersinia pseudotuberculosis*, *Listeria monocytogenes*, *Pasteurella multocida*, *P. haemolytica*, and *Bordetella bronchiseptica*. Some cases of tularemia (*Francisella tularensis*) have been detected in Spain.

European brown hare syndrome (EBHS) has never been detected in the Iberian hare. Rabbit hemorrhagic disease virus (RHDV) strains were reported in dead Iberian hares collected in the 1990s in Portugal, revealing the earliest evidence of rabbit lagovirus cross-species infection.

CONSERVATION STATUS:

IUCN Red List Classification: Least Concern (LC)

MANAGEMENT: The Iberian hare is an important game species, and has significant economic relevance in some regions of S Spain and Portugal. Populations in some localities have suffered from overhunting and poaching. Habitat fragmentation and change, as a consequence of climate change and rural abandonment (at a global scale), and the use of herbicides (at a local scale), may also have had an important effect in declining populations. In general, it is a vulnerable species to any habitat change.

ACCOUNT AUTHORS: Paulo C. Alves and Pelayo Acevedo

Key References: Acevedo et al. 2012a, 2012b; Alves and Niethammer 2003; Alves and Rocha 2003; Alves et al. 2002, 2008; Alzaga et al. 2008; Carro and Soriguer 2010; Duarte 2000; Farfán et al. 2004, 2012; Lopes et al. 2014; López-Martínez 1989; Melo-Ferreira et al. 2005, 2009, 2011, 2012, 2014; Palacios 1983, 1989; Palacios and Fernandez 1992; Palacios and Lopez 1980; Palacios and Meijide 1979; Paupério and Alves 2008; Pinheiro et al. 2015; Rodriguez et al. 1997; Romero-Rodriguez 1976; Seixas et al. 2014; Tapia et al. 2010.

Lepus habessinicus Hemprich and Ehrenberg, 1832
Abyssinian Hare

DESCRIPTION: The Abyssinian hare is a small hare with a body weight of about 2 kg, with a dorsal pelage of cinereous color marked with black or grizzled black and buff. The sides are lighter than the dorsal fur and are separated from the more sparsely furred, white venter by a rufous or cinnamon stripe. Coloration of the head is similar to that of the dorsal fur, with silvery white cheeks and muzzle. There may be a whitish eye-ring. Its wide ears are longer than those of both the Ethiopian hare (*L. fagani*) and the Ethiopian highland hare (*L. starcki*). Posteriorly the ears are silvery brown to brown; the anterior fur is whitish buff with black-edged tips. There is a brownish cinnamon nuchal patch. The hind limbs are grizzled silvery gray; the forelimbs are brownish cinnamon. The tail is black above and white below.

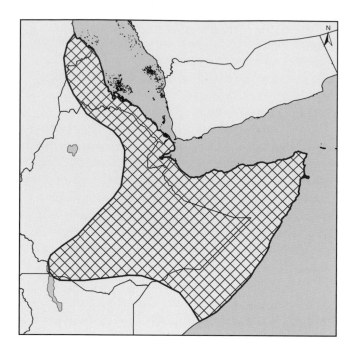

SIZE: Head and body 435–530 mm; Tail 50–111 mm; Hind foot 95–114 mm; Ear 105–121 mm; Greatest length of skull 82.2–91.8 mm; Weight 1,415–2,440 g

CURRENT DISTRIBUTION: The Abyssinian hare has been recorded in C and N Somalia, E Ethiopia, Eritrea, and Djibouti, with a minor range in E Sudan. Abyssinian hares have recently been recorded from the Mount Kaka area in SE Ethiopia. The type locality is from NE Eritrea. It occurs from sea level up to 2,000 m, but may occur as high as 2,500 m on the northeast plateau of Ethiopia and Eritrea.

TAXONOMY AND GEOGRAPHIC VARIATION: Previously *L. habessinicus* has been considered a subspecies of the Cape hare (*L. capensis*), and there remains unresolved consensus about the species status of this taxon. Some authors accept the species status of *L. habessincus*, while acknowledging the distinctive absence of the interparietal bone in some populations (the Berbera population of N Somalia). Others have suggested that it is a subspecies of Cape hare, widespread and abundant in its range. A molecular analysis of hare species from Ethiopia indicated that *L. habessinicus* and *L. starcki* are closer to each other than other taxa included in that analysis. Nevertheless, *L. habessinicus* carries mtDNA distinct that of from south and north African *L. capensis*, with no sign of introgression, contrary to earlier suggestions to include it in *L. capensis*.

ECOLOGY: The Abyssinian hare shows preference for open landscapes, but requires some scattered scrub for shade by day and cover from predators. Areas with thicker scrub are taken over by other hares in the Ethiopian hare complex.

HABITAT AND DIET: Preferred habitats of the Abyssinian hare are open grassland, steppe, savanna, and desert areas. In these areas it replaces the Cape hare. Like the Cape hare, it is spreading due to overgrazing of domestic stock.

BEHAVIOR: Little or no data are available on the behavior of the Abyssinian hare. However, it is possibly nocturnal like the Cape hare.

PHYSIOLOGY AND GENETICS: Diploid chromosome number = 48. Chromosome morphology as well as G-banding are similar to that of *L. starcki* and the European hare (*L. europaeus*). In addition, the autosomal complements in this taxon are composed of 21 pairs of bi-armed chromosomes (i.e., 8 pairs of metacentrics and submentacentrics and 13 pairs of telocentric and subtelocentrics) and 2 pairs of acrocentric chromosomes. Furthermore, the X-chromosome is a large submetacentric, one of the largest of the group.

PARASITES AND DISEASES: There are no data on the parasites and diseases of the Abyssinian hare, but field studies have reported the presence of ticks in some specimens.

CONSERVATION STATUS:

IUCN Red List Classification: Least Concern (LC)

MANAGEMENT: There are no known management practices in place for the Abyssinian hare. It has been recorded from Awash National Park, Mago National Park, and possibly Yangudi-Rassa National Park in Ethiopia. The lack of data available on population size/density, reproductive biology, diseases, and ecology of the Abyssinian hare should prompt research in these areas.

ACCOUNT AUTHOR: Zelalem Gebremariam Tolesa

Key References: Angermann 1983; Azzaroli-Puccetti 1987a, 1987b; Azzaroli-Puccetti et al. 1996; Flux and Angermann 1990; Girma et al. 2012; Happold 2013; Hoffmann and Smith 2005; Pierpaoli et al. 1999; Tolesa 2014; Yalden et al. 1986, 1996.

Lepus hainanus Swinhoe, 1870
Hainan Hare

OTHER COMMON NAMES: Hainan tu, Ye tu (Chinese)
DESCRIPTION: The Hainan hare is a small hare with soft pelage that is brighter in winter. Dorsally, it is pale brown

sea level to ~ 300 m. This habitat is distinctly different from that of the closely related Burmese hare, which is largely a mountain form and often found above the tree line. Currently the Hainan hare can be found only on established Eld's deer (*Rucervus eldii*) ranches that provide a measure of incidental protection for the species. These ranches were once heavily forested areas, but today constitute novel hare habitat. In optimal habitat at these localities (totaling no more than 2 km^2), its density ranges from approximately 60 to 160 hares per km^2. Thus the total population is estimated to be no more than 250–500 animals.

HABITAT AND DIET: The Hainan hare occupies lowland shrub forest and grasslands on the island; essentially it is a species of dry, open country in the coastal lowlands.

BEHAVIOR: Though occasionally seen active during the day, the Hainan hare is primarily nocturnal with the greatest activity occurring before midnight, tapering off thereafter. It has not been observed burrowing. Although it is a shy animal, it is characterized as relatively tame likely due to a lack of native carnivores.

CONSERVATION STATUS:

IUCN Red List Classification: Vulnerable (VU)—A2ac+3cd; B1ab(iii)

National-level Assessments: China Red List (Near Threatened—NT)

MANAGEMENT: The Hainan hare is under threat from overharvesting for consumption and skins. Being an insular form, it is particularly tame and easily approached for killing. No medicinal properties have been attributed to the species.

The main cause of decline of the Hainan hare has been habitat loss and degradation due to conversion. Thorough surveys at known localities have failed to detect any hares, in both the far north and south of Hainan Island. The two deer ranches, Bang Xi and Da Tien, remain the only known localities of the species at the present time.

Although the Hainan hare was listed on the China Key List—II in 1991, granting it protection, no enforcement leading to its protection has been undertaken.

ACCOUNT AUTHOR: Andrew T. Smith

Key References: Allen 1938; Jiang et al. 2016; Kong et al. 2016; Lazell et al. 1995; Liu et al. 2011b; Smith and Xie 2008; Wu et al. 2005; Yu 2004.

and tinged with chestnut brown and black. The chin and ventral pelage are white, while the sides are a mix of pale brown and brownish white. The forearms are bright ochraceous, and anteriorly this coloring deepens slightly in a broad band across the lower throat. The upper part of the tail is black, bordered with white. The eye-ring is whitish and extends toward the base of the ear and forward to the snout. The feet are pale brown and have white spots.

The Hainan hare has a rounded cranium, and the supraorbital process is warped upward. The auditory bullae are small, and it has a short and broad rostrum. The upper incisors possess a Y-shaped groove in comparison to the typical V-shaped groove in most leporids.

SIZE: Head and body 350–394 mm; Tail 45–70 mm; Hind foot 76–96 mm; Ear 76–98 mm; Greatest length of skull 73–84 mm; Weight 1,250–1,750 g

CURRENT DISTRIBUTION: Current distribution is on Hainan Island. The Hainan hare is endemic to China.

TAXONOMY AND GEOGRAPHIC VARIATION: No subspecies. The taxonomic position of *L. hainanus* is uncertain. Earlier treatments considered it to be a subspecies of the Burmese hare (*L. peguensis*), and recent molecular analyses document it as a sister species to *peguensis*. Biogeographically, it appears to be most closely related to other SE Asian leporids, rather than any of the *Lepus* species to the north in China.

ECOLOGY: At one time the Hainan hare was widespread and abundant throughout Hainan Island at elevations from

Lepus insularis W. Bryant, 1891
Black Jackrabbit

OTHER COMMON NAMES: Espiritu Santo jackrabbit; Liebre negra, Liebre de Espíritu Santo, Liebre prieta (Spanish)

DESCRIPTION: The black jackrabbit is a large and slender jackrabbit; the ears are long, are black at the edge of the superior tip, and have grayish hair on the front. The dorsal part of the body is black mixed with dark cinnamon or brown fur; the shoulders, sides, and forelimbs are yellow to dark. The black back is distinctly different from that of the black-tailed jackrabbit (*L. californicus*). The dorsal part of the tail is black and brown, fading to pale on the ventral side. The top of the head is black, and white rings surround the eyes. The ventral part of the body is cinnamon color mixed with light yellow. The dorsal part of the hind legs is yellowish white.

The skull is long; the cheeks are thicker than those of any subspecies of the black-tailed jackrabbit.

SIZE: Head and body 542–608 mm; Tail 64–111 mm; Hind foot 107–122 mm; Ear 110–122 mm; Greatest length of skull 89–97 mm; Weight 2,200–3,400 g

CURRENT DISTRIBUTION: The black jackrabbit is endemic to the Espiritu Santo Archipelago in the Gulf of California, Mexico, including Espiritu Santo Island and La Partida Island. It occurs from sea level to 300 m asl, and its available potential habitat is not greater than 112 km².

TAXONOMY AND GEOGRAPHIC VARIATION: No subspecies. *L. insularis* is closely related to *L. californicus*, and it has been suggested that *L. insularis* derived from a common ancestor of *L. californicus* as a result of a vicariant event of a population of *L. c. xanti*. The isolation of *L. insularis* from peninsular forms occurred approximately 11,000 years ago when Espiritu Santo Island separated from the peninsula. The specific status of *L. insularis* has been questioned by researchers who consider that it may be only a melanistic form of *L. californicus*. A comparative study of allozyme variation involving 26 loci between the two species showed that *L. insularis* has lower intraspecific genetic variation, and that little genetic differentiation separates *L. insularis* from *L. californicus* forms. However, a karyotypic and G-banding analysis reported that *L. insularis* is chromosomally differentiated from *L. californicus*, thus supporting the hypothesis that they are genetically distinctive species. On the other hand, it has been reported that *L. californicus* and *L. insularis*

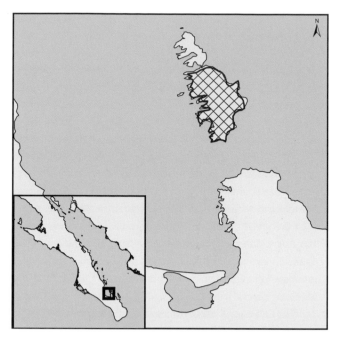

Lepus insularis. Photo courtesy Arturo Carrillo-Reyes

are sister taxa although *L. californicus* is paraphyletic in relation to *L. insularis*. The nuclear and mitochondrial differences, color variation, and small differences in the structure of a single bone in the skull (jugal) between *L. insularis* and *L. californicus* do not support the specific level of *L. insularis*; thus this form may be considered a subspecies of *L. californicus*. Definite clustering using morphological characteristics separates *L. insularis* from the subspecies of *L. californicus* from Baja California (*L. c. martirensis, L. c. xanti*, and the insular subspecies *L. c. magdalenae*).

ECOLOGY: The population density of the black jackrabbit averages 11.4/km², and the total number of animals has been estimated at 923 (range 537–1,586) in a sampling area of 81 km² in xeric shrub. The island has many cliffs and canyons that are not optimal habitat for the species.

The black jackrabbit shares its habitat with other endemic mammals on Espiritu Santo Island: Lamb's spiny pocket mouse (*Chaetodipus spinatus lambi*), the northern Baja deermouse (*Peromyscus fraterculus insulicola*), Bryant's woodrat (*Neotoma bryanti bryanti*), the Espiritu Santo antelope squirrel (*Ammospermophilus leucurus insularis*), and the ringtail (*Bassariscus astutus saxicola*). Introduced goats (*Capra aegagrus hircus*) also occur on the island and may compete with the black jackrabbit for forage. Other vertebrates recorded are the American kestrel (*Falco sparverius*), the red-tailed hawk (*Buteo jamaicensis*), the crested caracara (*Caracara cheriway*), the speckled rattlesnake (*Crotalus mitchellii*), the variable sandsnake (*Chilomeniscus stramineus*), the Baja California striped whip snake (*Masticophis barbouri*), and the Espiritu Santo orange-throated whiptail (*Aspidoscelis hyperythra espiritensis*).

HABITAT AND DIET: The black jackrabbit lives in xeric shrub with several dominant plant species: palo adán (*Fouquieria diguetii*), matacora (*Jatropha cuneata*), lomboy (*Jatropha cinerea*), jojoba (*Simmondsia chinensis*), acacia (*Acacia cymbispina*), palo blanco (*Lysiloma candida*), chain-link cholla (*Cylindropuntia cholla*), and cardón (*Pachycereus pringlei*). Black jackrabbits avoid areas covered by mangrove.

The diet of the black jackrabbit is composed mainly of grasses (more than 55% of the diet), including three-awns (*Aristida* spp.), grama grasses (*Bouteloua* spp.), panic-grasses (*Panicum* spp.), and bentgrasses (*Agrostis* spp.). As generalized herbivores, however, they are known to consume 52 different plant species; the majority of non-grass plants include Engelmann prickly pear (*Opuntia engelmannii*), cardons (*Pachycereus* spp.), pincushion cacti (*Mammillaria* spp.), pitaya (*Stenocereus* spp.), and tender branches of mesquite (*Prosopis* spp.). The availability of each of these sources varies seasonally.

BEHAVIOR: The black jackrabbit is a solitary species; occasionally it may be found in groups of two individuals. Black jackrabbits are active from the early twilight into the night and then again at dawn; thus they are considered nocturnal and crepuscular. This jackrabbit uses beds

Lepus insularis habitat. Photo courtesy Arturo Carrillo-Reyes

of sand and dry grass (15 × 24 × 3 cm) surrounded by slipper plants (*Pedilanthus macrocarpus*) and bushes with thorns for resting and protection from the sun and hot temperatures. They can stand up on their hind legs to reach and eat leaves of acacias (*Acacia* spp.).

PHYSIOLOGY AND GENETICS: Diploid chromosome number = 48 and FN = 80

REPRODUCTION AND DEVELOPMENT: The reproductive season of the black jackrabbit can be pieced together by scattered observations of males with scrotal testes and pregnant or lactating females. These observations indicate that the breeding season occurs primarily during the wet season (May–October). Some additional data show that breeding may also extend into the dry season, both before (as early as March) and after (as late as November) the wet season. The only six adult females collected that were pregnant contained two embryos each, one in each uterine horn.

CONSERVATION STATUS:

IUCN Red List Classification: Near Threatened (NT)—qualifies for B1b(iii), but does not qualify for second sub-criteria for listing as Vulnerable (VU)

National-level Assessments: Mexico (Subject to special protection in the Mexican Official Norm NOM-059-2010 [SEMARNAT])

MANAGEMENT: Espiritu Santo Island is uninhabited and has been set aside as a protected area. It represents an important tourist destination, and many trails have been constructed on the island. Many of the beaches are used for recreation or by fishers. In spite of these impacts, the habitat of the jackrabbit has not been fragmented or heavily modified. One negative impact on the jackrabbit population on the island is the presence of introduced goats, which compete with the jackrabbits for food and habitat. Similarly, fishers hunt goats as well as jackrabbits, therefore they also constitute a risk to the jackrabbits. We recommend continuous monitoring of the populations of the black jackrabbit, as well as assessment of the effects of goat competitors and natural predators (birds of prey and snakes) on such a small landscape.

ACCOUNT AUTHORS: Consuelo Lorenzo, Tamara Rioja-Paradela, Arturo Carrillo-Reyes, Mayra de la Paz, Fernando A. Cervantes, and Sergio Ticul Álvarez-Castañeda

Key References: Álvarez-Castañeda and Patton 1999, 2000; Álvarez-Castañeda et al. 2006; Casas-Andreu 1992; Cervantes and Castañeda 2012; Cervantes et al. 1996, 1999–2000; Dixon et al. 1983; Gastil et al. 1983; Hall 1981; Hoffmann and Smith 2005; Lorenzo et al. 2012, 2014b; Nelson 1909; Orr 1960; Ramírez-Silva et al. 2010; SEMARNAT 2010; Thomas and Best 1994b; Townsend 1912; Whitaker and Morales-Malacara 2005.

Lepus mandshuricus Radde, 1861
Manchurian Hare

OTHER COMMON NAMES: Dongbei tu (Chinese); Manzhurskiy zayats (Russian); Manjutokki (Korean)

DESCRIPTION: The dorsal pelage of the Manchurian hare, including the top of the head, is ochraceous brown or

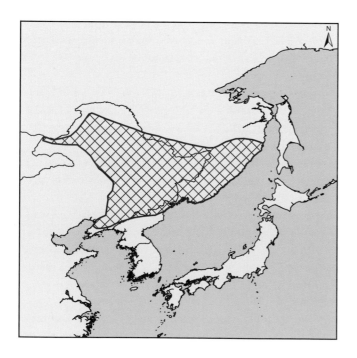

ochraceous gray. The sides are light yellow, and the ventral pelage is dirty white. A dark band is visible below the eyes. The sides of the head are lighter than the top and have white spots along the anterior and lower portions. The tail is black-brown on top and dingy white below. There is a melanistic form, which is shiny black and tinged with brown on the back and flanks; the belly is white. The Manchurian hare's throat and chest are cinnamon-buff, and there is a white spot on the head.

The skull is relatively narrow and the cranium is weakly bulged. The supraorbital process is short, narrow, and not warped upward. The zygomatic arches are massive and wide. The palatal bridge is broad, and the auditory bullae are laterally compressed and not inflated.

SIZE: Head and body 410–540 mm; Tail 50–80 mm; Hind foot 110–145 mm; Ear 75–118 mm; Greatest length of skull 79–89 mm; Weight 1,400–2,600 g

CURRENT DISTRIBUTION: The Manchurian hare occurs in SE Russia (Amur Oblast, Primorsky Krai, and Khabarovsk Krai), NE China (Jilin, Nei Mongol, Liaoning, and Heilongjiang Provinces), and the extreme northern region of the Korean Peninsula where it is potentially parapatric with the Korean hare (*L. coreanus*). It occupies elevations of 300–900 m asl.

TAXONOMY AND GEOGRAPHIC VARIATION: *Lepus mandshuricus* is a monotypic species, although melanistic forms have been previously designated as *L. melainus*. *Lepus mandshuricus* appears to be generally similar to the Japanese hare (*L. brachyurus*).

ECOLOGY: Little is known about the ecology of the Manchurian hare, but it is thought to be the ecological equivalent of the North American snowshoe hare (*L. americanus*). In Russia individual territories surrounding permanent shelters do not exceed a few hundred square meters.

HABITAT AND DIET: Manchurian hares occur in broadleaf and coniferous forest habitats, particularly areas with Manchurian hazelnut (*Corylus sieboldiana mandshurica*) undergrowth in Mongolian oak (*Quercus mongolica*) stands. They do not enter the zone of fir-spruce forests in high mountains, and neither do they like open valleys or grasslands.

The diet consists of herbaceous plants, shrubs, fallen fruit, and twigs from a variety of trees: willow (*Salix* spp.), linden (*Tilia* spp.), wild apple (*Malus* spp.), birch (*Betula* spp.), and elm (*Ulmus* spp.).

BEHAVIOR: The Manchurian hare is a nocturnal species that is known to display some activity at dawn. It is shy and likely solitary. It typically beds down in closed shelters rather than lairs in the open. Holes in fallen trees, old badger nests, and covered stony outcroppings tend to be permanent shelters. Open resting areas are used temporarily and are found in shrub thickets. Manchurian hares will cover their tracks before leaving a shelter. Unlike most hares, they are reticent to leave their shelters in the event of danger. When scared they will utter audible sounds similar to sneezing.

PHYSIOLOGY AND GENETICS: Diploid chromosome number = 48

REPRODUCTION AND DEVELOPMENT: The number of litters per year tends to be two to three. The female typically has small litters of one to two leverets, occasionally up to six. Depending on location, breeding can begin sometime between March and April. Reproduction has been recorded in late August, indicating that the breeding season extends approximately over a six-month period.

PARASITES AND DISEASES: Seven species of helminth have been recorded in Manchurian hares, and the rate of infection may approach 95%. They may also serve as host to many species of mite, and up to 1,127 individual mites have been recorded on a single hare. In winter Manchurian hares are the primary host for the tick *Hyaemaphysalis japonica*. Lung (*Protostrongylus terminalis*) and stomach (*Obeliscoides lepuris*) nematodes may cause mortality in these hares. Other cestodes and nematodes may also frequent Manchurian hares.

CONSERVATION STATUS:

IUCN Red List Classification: Least Concern (LC)

MANAGEMENT: Subsistence hunting occurs, but the Manchurian hare is typically not targeted for its pelt, as the skin is thin and readily ripped. Some commercial harvesting has taken place on the Korean Peninsula. Forest clearing results in loss of habitat that promotes replacement by the tolai hare (*L. tolai*). In some localities the high density of Manchurian hares and their selective foraging may demonstrably influence the local composition of vegetation.

ACCOUNT AUTHOR: Andrew T. Smith

Key References: Averianov 1994b; Flux and Angermann 1990; Ge et al. 2012; Liu et al. 2011a, 2011b; Loukashkin 1943; Smith and Xie 2008; Sokolov et al. 2009; Wu et al. 2005.

Lepus nigricollis F. Cuvier, 1823
Indian Hare

OTHER COMMON NAMES: Black-napped hare; Kāṭṭmuyal (Malayalam); Kharghosh (Hindi/Bengali); Choura pilli (Telugu); Muyal (Tamil); Sassa (Marathi); Mola (Kannada); Saslo (Gujarati); Soha pohu (Assamese); Thekua (Odia)

DESCRIPTION: The Indian hare is a medium-sized hare, with size increasing in the southern extent of its range.

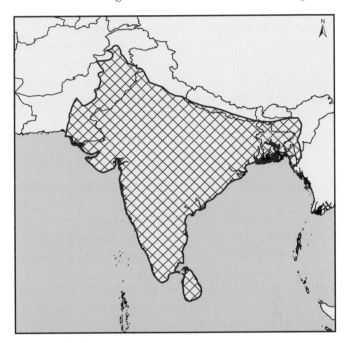

Lepus nigricollis. Photo courtesy Clement Francis

The dorsal and facial pelage is rufous black mixed with black, the breast and legs are rufous, and the chin and ventral pelage are white. The tail is dorsally brown and white on the sides and underside. There are variations seen in the subspecies: *L. n. nigricollis* has a black patch on the back of the neck and the tail is dorsally black; *L. n. dayanus* tends to be paler and have more yellow-sandy fur; *L. n. ruficaudatus* has a dark brown patch on the back of the neck, and the tail is dorsally rufous.

The average occipitonasal length of the skull of the Indian hare is 95 mm, the diastema is about one-third of the skull length, and the palate is almost the same length as the diastema.

SIZE: Head and body 330–530 mm; Tail 64 mm; Hind foot 85–115 mm; Ear 80–120 mm; Greatest length of skull 80–100 mm; Weight 1,800–3,600 g

CURRENT DISTRIBUTION: The Indian hare is distributed throughout the Indian subcontinent except in the upper ranges of the Himalayas and mangroves of the Sundarbans. This range includes India, E Pakistan, S Nepal, Bangladesh, and Sri Lanka, and it is thought to occur in Bhutan although exact locations are not known. It has been introduced to many islands of the Indian Ocean. It occupies elevations ranging from 50 to 4,500 m asl.

TAXONOMY AND GEOGRAPHIC VARIATION: Thirteen subspecies: *L. n. aryabertensis, L. n. cutchensis, L. n. dayanus, L. n. joongshaiensis, L. n. macrotus, L. n. mahadeva, L. n. nigricollis, L. n. rajput, L. n. ruficaudatus, L. n. sadiya, L. n. simcoxi, L. n. singhala,* and *L. n. tytleri.* The Burmese hare (*L. peguensis*) is a similar species, and some consider it to be synonymous with *L. nigricollis.*

ECOLOGY: Recorded home ranges for the Indian hare in Nepal and on Cousin Island were reported as 1–10 ha and 0.7–1.6 ha, respectively. Hares that occupy more open habitat are likely to have greater ranges. When located on isolated islands without predation pressure, the population density can become quite high. A density of 4.7/km^2 was recorded in the lowland dry zone of Sri Lanka.

HABITAT AND DIET: The Indian hare occupies many habitat types, including deserts, grasslands, shrublands, cultivated areas, and forests. It predominantly feeds on grass and forbs; in wet regions the diet may contain up to 73% grasses.

BEHAVIOR: The Indian hare is usually a crepuscular to nocturnal species; during the day it seeks shelter using a series of forms (depressions). These can be found in tall grasses, low shrubs, and young plants. When threatened Indian hares may retreat to ditches or animal burrows.

PHYSIOLOGY AND GENETICS: Diploid chromosome number = 48

REPRODUCTION AND DEVELOPMENT: The breeding season is year-round, but a peak occurs during the monsoon season. Litter size ranges from 1 to 4 (average 1.8), with fewer offspring in winter and more during summer.

CONSERVATION STATUS:

IUCN Red List Classification: Least Concern (LC)

MANAGEMENT: The Indian hare is used for human and animal consumption, research, and sport hunting and as pets. Hunting for subsistence, habitat destruction, conversion of prime forests to cultivated land, predation by native and domesticated animals, livestock competition, and human-induced fires all pose a threat to this species.

ACCOUNT AUTHORS: Sanjay Molur and P. O. Nameer

Key References: Eisenberg and Lockhart 1972; Eisenberg et al. 2015; Flux and Angermann 1990; Ghose 1971; Gurung and Singh 1996; Kirk and Bathe 1994; Manakadan and Rahmani 1999; Menon 2009, 2014; Prakash and Taneja 1969; Suchentrunk 2004; Suchentrunk and Davidovic 2004.

Lepus oiostolus Hodgson, 1840
Woolly Hare

OTHER COMMON NAMES: Highland hare, Gray-tailed hare; Gaoyuan tu (Chinese)

DESCRIPTION: The woolly hare has thick, soft hair that is long and curly, giving it a "woolly" appearance. The dorsal pelage coloration is variable, from sandy yellow to light brown. The hip area is lighter than the back and rump, varying from silvery gray to brownish gray. Its long ears are the largest of the Chinese hares and are dark and tipped with black. The eye-ring is whitish. The bushy tail is entirely white except for a narrow brown-gray stripe on top.

The muzzle is elongated and narrow, and the anterior and posterior branches of the supraorbital process are well developed and triangular in shape. The supraorbital process is clearly warped upward. The auditory bullae are small.

SIZE: Head and body 400–580 mm; Tail 65–125 mm; Hind foot 102–140 mm; Ear 105–155 mm; Greatest length of skull 84–100 mm; Weight 1,500–4,250 g

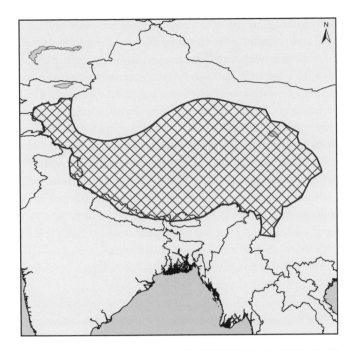

Lepus oiostolus. Photo courtesy Andrew T. Smith

CURRENT DISTRIBUTION: The range of the woolly hare encompasses the bulk of the Tibetan Plateau. This includes a wide distribution in China (Qinghai, Sichuan, Xizang, Xinjiang, and Gansu Provinces) as well as the northern border regions of Nepal and India (Sikkim and Ladak) along the Himalayan massif. It occurs at elevations between 2,500 and 5,400 m.

TAXONOMY AND GEOGRAPHIC VARIATION: *Lepus oiostolus* is most similar in morphology and systematically to the Yunnan hare (*L. comus*). Several subspecies of *L. oiostolus* have been identified (*grahami, kozlovi, przewalskii,*

qinghaiensis, qusongensis, sechuenensis), but these appear to intergrade and to be continuously distributed across the species' range; thus they are not herein recognized. Some authors have assigned the form *przewalskii* in the tolai hare (*L. tolai*—as *capensis*); again, this designation does not appear warranted.

ECOLOGY: Populations of woolly hares are composed equally of males and females, and young animals may make up as much as half of populations. Reported densities of woolly hares range from 13 to 27 hares/km². Major predators of woolly hares are golden eagles (*Aquila chrysaetos*), eagle owls (*Bubo bubo*), and Siberian weasels (*Mustela sibirica*).

HABITAT AND DIET: Woolly hares inhabit rocky areas, montane meadows, shrub meadows, and cold alpine desert regions. They tend to avoid dense shrubby cover. They forage predominantly on grasses and herbaceous plants, but also consume the leaves and stems of small shrubs.

BEHAVIOR: Woolly hares are active during all hours of the day, but tend to be most active in the evenings and in the mornings. They are typically solitary animals, but small groups may form during the mating season when foraging. They are known to seek shelter among rocks and in marmot burrows.

PHYSIOLOGY AND GENETICS: Diploid chromosome number = 48

REPRODUCTION AND DEVELOPMENT: Two litters are produced annually, with one to two leverets produced per litter. The reproductive season begins in April and extends until July.

PARASITES AND DISEASES: Woolly hares may serve as an intermediate host for the tapeworm *Echinococcus multilocularis.*

CONSERVATION STATUS:

IUCN Red List Classification: Least Concern (LC)

National-level Assessments: India (Endangered (EN)—B1ab(ii,iii)+2ab(ii,iii), the marginal range within the country)

MANAGEMENT: Habitat loss, deterioration, and fragmentation have resulted in population declines. In some areas local subsistence hunting negatively impacts populations. A significant portion of the range of woolly hares overlaps with protected areas.

ACCOUNT AUTHOR: Andrew T. Smith

Key References: Cai and Feng 1982; Chakraborty et al. 2005; Gao and Feng 1964; Lu 2010, 2011; Smith and Xie 2008; Wu et al. 2005; Xiao et al. 2004.

Lepus othus Merriam, 1900
Alaskan Hare

OTHER COMMON NAMES: Tundra hare, Alaska tundra hare, Tundra polar hare, St. Michael's hare, Swift hare, Alaska Arctic hare, Peninsula Arctic hare, Alaska Peninsula hare, Northern hare, Jackrabbit; Ukalisukruk, Ugalishugruk, Ushkánuk, Zaisch, Okhotsk, Gichiga, Marcova, Oo-skon (Alaskan Native languages)

DESCRIPTION: The Alaskan hare is one of the largest species of hare. Its winter pelage is white except for the extreme tips of the ears, which are black. Its summer pelage is brownish orange on the nose, sides of the face, and top of the head, with the latter being the darkest. The edges of the ears and rings around the eyes have white pelage. Claws are stout for digging. Both sexes are similar in size.

The skull is massive relative to those of most other species of *Lepus*. Nonetheless, the Alaskan hare, the mountain hare (*L. timidus*), and the Arctic hare (*L. arcticus*) have apparently indistinguishable skull morphology.

SIZE: Head and body 500–600 mm; Tail 65–104 mm; Hind foot 170–189 mm; Ear 70–91 mm; Greatest length of skull 97–109 mm; Weight 3,500–7,200 g

CURRENT DISTRIBUTION: The Alaskan hare has an isolated and somewhat restricted distribution. The species

occupies the Arctic tundra region along the coast for the entire western half of Alaska, as well as to the western tip of the Alaska Peninsula. On the southern coast, distribution does not exceed the far western tip of the Cook Inlet. Although the northern edge of Alaska still appears in most range maps, there have been no verifiable records from there since at least 1951. Records of the Alaskan hare exist from locations at sea level to 660 m asl.

TAXONOMY AND GEOGRAPHIC VARIATION: Two subspecies: *L. o. othus* (northern form) and *L. o. poadromus* (Alaskan Peninsula). The Arctic hare, the mountain hare, and the snowshoe hare (*L. americanus*) are similar species. The taxonomic status of *L. othus* remains unclear, with some authorities suggesting, based on cranial or other morphometric measurements, that they are conspecific with *L. arcticus*, *L. timidus*, or both. *Lepus othus* and *L. arcticus* also share similar behavioral and ecological characteristics, yet *L. othus* is geographically isolated from *L. arcticus* and thus may warrant distinct taxonomic status. Individuals from E Siberia may pertain to *L. timidus* or *L. othus*, but a recent phylogenetic study using mtDNA inconclusively determined their greater affiliation to *timidus*.

ECOLOGY: Predators include several raptor species (bald eagles *Haliaeetus leucocephalus*, golden eagles *Aquila chrysaetos*), owls (snowy owls *Bubo scandiacus*, short-eared owls *Asio flammeus*), ermines (*Mustela erminea*), wolverines (*Gulo gulo*), gray wolves (*Canis lupis*), lynx (*Lynx canadensis*), and Arctic foxes (*Vulpes lagopus*). Red foxes (*Vulpes vulpes*) are likely the most important predator of the Alaskan hare.

HABITAT AND DIET: The Alaskan hare is found primarily in open tundra habitats along the southwest and western coast of Alaska.

Foods consumed include dwarf willow (*Salix herbacea*), grasses, sedges, and other tundra plants. In April and May, the diet primarily includes shrubs, most commonly Alaska willow (*S. alaxensis*) and leaves and berries of black crowberry (*Empetrum nigrum*). In winter and early spring, the diet consists of bark, twigs, and shoots of woody vegetation. Individuals seldom drink water.

BEHAVIOR: Individuals are solitary except during the April–May mating season, when aggregations of more than 20 individuals may occur. Individuals associated with dense alder thickets venture out from the thickets in the evenings to feed. When attacked by a snowy owl, Alaskan hares may attempt to strike the owl with their forefeet when the bird is at its low point, and run to cover between owl swoops. Vocalizations include puffing and hissing sounds, as well as a "hoo-hoo" call uttered exclusively during the mating season.

PHYSIOLOGY AND GENETICS: Diploid chromosome number = 48

REPRODUCTION AND DEVELOPMENT: The Alaskan hare has a single litter per year. After a gestation period of about 46 days, females give birth to 5–7 (average 6.3) precocial young, in mid-April to early June. Birth timing appears to coincide with loss of snow cover. Young are born in shallow natural depressions, lined nests, or the thick shelter of brush. Weighing ~ 105 g at birth, juveniles grow an average of 37.2 g/day over a 102-day growth period.

PARASITES AND DISEASE: No ectoparasites of the Alaskan hare have been observed on captured individuals, but 1 adult female contained 12 nematodes in its small intestine.

CONSERVATION STATUS:

IUCN Red List Classification: Least Concern (LC)

MANAGEMENT: A small percentage of Alaskan hare populations may be used by Alaskan natives for subsistence food and occasionally for fur. To the degree that mining expansion, oil-development projects, and anthropogenic infrastructure alter tundra habitats, these may contribute to local declines in the Alaskan hare. Southern populations may be affected by habitat loss or climate change–induced alterations, but this speculation is not yet supported by data. Although evidence in support is only circumstantial, interference competition with snowshoe hares may affect Alaskan hares. This competition may be direct (e.g., via parasites, diseases, or limited food) or indirect (e.g., increased numbers of predators during snowshoe hare population highs).

Research to determine the taxonomic status of the Alaskan hare relative to the Arctic hare and the mountain hare is needed, as are research and monitoring regarding geographic distribution, relationship to proximate and broad-scale drivers, harvest levels, and status and trends of populations and habitat.

ACCOUNT AUTHOR: Erik A. Beever

Key References: Anderson and Lent 1977; Baker et al. 1983; Bee and Hall 1956; Best 1999; Best and Henry 1994; Flux and Angermann 1990; Hall 1981; Hoffmann and Smith 2005; Klein 1995; Murray and Smith 2008; Reid 2006; Waltari and Cook 2005; Waltari et al. 2004.

Lepus peguensis Blyth, 1855
Burmese Hare

OTHER COMMON NAMES: Siamese hare

DESCRIPTION: The Burmese hare is a middle-sized hare closely resembling the Indian hare (*L. nigricollis*). The dorsal fur is reddish gray mixed with black. Its rump has a grayer color than the back, and the posterior neck is rufous. The ventral fur is white with a light rufous upper breast. The ears are rather large and brown with pronounced black tips. The tail is black above and white below. The front legs are rufous, while the hind legs are fulvous on the outside and white on the inside. The feet of Burmese individuals are white, and those of Thailand individuals fulvous.

SIZE: Head and body 400–590 mm; Tail 55–84 mm; Hind foot 96–110 mm; Ear 80–90 mm; Greatest length of skull 88 mm; Weight 2,000–2,500 g

CURRENT DISTRIBUTION: The Burmese hare inhabits C and S Myanmar from the Chindwin River valley, east through Thailand, Cambodia, S Laos, and S Vietnam, and south into the upper Malay Peninsula (Myanmar, Thailand). The distribution possibly includes the northern and central regions of Laos, as indicators of the hare's presence in the wild and fresh kills in markets have been found, but the identification has only been conducted to the genus level. The distribution in Myanmar is mostly

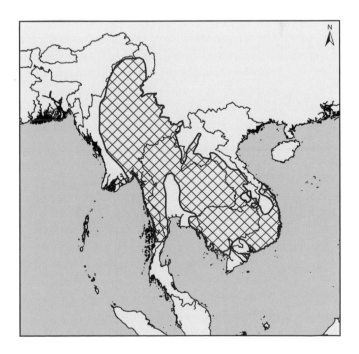

hypothetical, and the species might be more restricted within this country. A collection expedition in Thailand in 1940 documented the Burmese hare at an elevation of 1,300 m on Mount Inthanon. Nevertheless, plentiful field surveys for the species have not found it above 700 m; therefore, the upper boundary of 1,300 m is dubious.

TAXONOMY AND GEOGRAPHIC VARIATION: Two subspecies: *L. peguensis peguensis* and *L. P. vassali*. A third subspecies (*siamensis*) is under consideration as a true species.

The Burmese hare was initially placed in *Caprolagus* (*Indolagus*). Previously, the Burmese hare incorporated the Hainan hare (*L. hainanus*), which subsequently has received species status. A recently conducted study using rigorous molecular species delimitation methods suggested that based on genetic distance and the phylogenetic network, the Hainan hare and the Burmese hare are conspecific. However, a combined mtDNA analysis, the phylogenetic species concept, and Bayesian species delimitation analysis supported the hypothesis that these two forms are different species. Thus, the relationship of the Burmese hare and the Hainan hare requires taxonomic clarification.

It has also been proposed that the Burmese hare is conspecific with the Indian hare (*L. nigricollis*) because of its similar appearance to *L. n. ruficaudatus*. Nonetheless, *L. n. ruficaudatus* seems to live in allopatry with the Burmese hare from W Burma to E India. A study examining dental characters between these forms failed to resolve the separate species status of the Burmese hare. Thus, the Indian hare and the Burmese hare are either conspecific with a somewhat increased gene pool separation of the Burmese hare, or the Indian hare and the Burmese hare represent a tightly connected pair of very young species. However, the epigenetic results are more in favor of the hypothesis of the two species' conspecificity. The lack of data indicates that the Burmese hare has never been revised properly regarding its taxonomy.

ECOLOGY: The Burmese hare is typically solitary and territorial.

HABITAT AND DIET: The Burmese hare lives in the plains and frequently dwells in land cleared for cultivated plants and dry fallow land. It inhabits sandy areas along the shore in Malaysia and S Vietnam, whereas it subsists in forest clearances, in lalang grass (*Imperata cylindrica*), or around hill tribe villages in Thailand. Hare numbers are particularly high in the grass and bush vegetation of seasonally exposed large river canals. The Burmese hare occurs in rain/flood-fed low-intensity rice fields, but avoids watered multi-crop rice fields that cover much of

Thailand. The hares spend the day in forms (depressions) located under scrubs or in high sward. The species' diet comprises grass, bark, and branches.

BEHAVIOR: The Burmese hare is crepuscular and nocturnal. The animals become very aggressive during the breeding season, boxing with their forefeet or kicking with their hind feet. The females might be quite seriously assaulted by the males, who bite and kick while mating.

PHYSIOLOGY AND GENETICS: Diploid chromosome number = 48. Recent genetic work focused on taxonomic and phylogenetic clarification between the Burmese hare and the Hainan hare by analyzing mtDNA.

REPRODUCTION AND DEVELOPMENT: The Burmese hare may have multiple litters per year. Litter size varies between one and seven, with an average of three to four leverets. Gestation lasts 35–40 days. The young are born in open grassy areas. They are entirely furred and have their eyes open.

CONSERVATION STATUS:

IUCN Red List Classification: Least Concern (LC)

MANAGEMENT: The Burmese hare is widespread and common. The habitat is possibly increasing due to forest clearance. Nevertheless, this species lives in populations isolated from each other by forests. The species is severely hunted, but this does not seem to cause any major danger. The expansion of watered rice fields destroys habitat in certain regions. Habitat in Laos and Vietnam is frequently burned during the dry season (February–May), which might pose a threat to existing young. Recommendations are to resolve the taxonomy, range, and behavior of this species.

ACCOUNT AUTHORS: Stéphanie Schai-Braun and Klaus Hackländer

Key References: Allen and Coolidge, Jr. 1940; Duckworth 1996; Duckworth et al. 1994, 2008; Ellerman and Morrison-Scott 1951; Evans et al. 2000; Flux and Angermann 1990; Gyldenstolpe 1917; Hoffmann and Smith 2005; Kloss 1919; Kong et al. 2016; Lekagul and McNeely 1977; Matthee et al. 2004e; Petter 1961; Pfeffer 1969; Suchentrunk 2004; Van Peenen 1969; Wroughton 1915.

Lepus saxatilis F. Cuvier, 1823
Cape Scrub Hare

OTHER COMMON NAMES: Ribbokhaas (Afrikaans)
DESCRIPTION: The Cape scrub hare is similar to the African savannah hare (*L. victoriae*; see below) except for

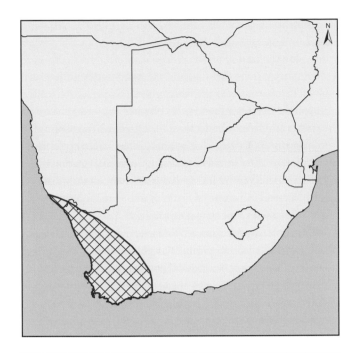

Lepus saxatilis. Photo courtesy Guy Palmer

its large size, large ears, and the simple enamel groove of the principal incisors. The fur is coarse; the forehead, back, and sides are a grizzled buffy grayish (grizzling as in the Cape hare *L. capensis*) and the underfur is gray. The chin, chest, anal region, and inside of the limbs are white. The soles of the feet are well furred. The ears are sparsely haired, their tips narrowly black, and the eyes are encircled in off-white. The sides of the face and muzzle are grayer than the rest of the upper parts, and the nape patch is a pink rufous. The tail is black above and white below. Females are heavier than males.

The ratio of the mesopterygoid space expressed as a percentage of hard palate length or as a percentage of total length of skull may be diagnostic for the Cape scrub hare as defined herein.

SIZE: Head and body 545–595 mm; Tail 110–160 mm; Hind foot 138–155 mm; Ear 130–150 mm; Greatest length of skull 99.3–107.7 mm; Weight 3,200–4,000 g

PALEONTOLOGY: Molecular clock estimates place the divergence of *Lepus* at ~ 11.8 mya with the development of grasslands at ~ 7–5 mya, the trigger for the global dispersal and diversification of the genus. The fossil record suggests the appearance of *Lepus* in Africa in the Late Pliocene or Early Pleistocene ~ 4–1.5 mya.

CURRENT DISTRIBUTION: As defined here, the Cape scrub hare is a South African, and more specifically, a W and N Cape Province endemic. In large part, its range is thought to encompass the Cape Floristic region in the southwest and the Succulent Karoo areas of Namaqualand, an area where diversification of biota has been considered to reflect the physical complexity of the Cape environment. The Cape scrub hare occurs in reasonable numbers in these regions (i.e., the Cape Floristic region and Namaqualand), but densities fall off markedly in the extreme northwestern parts of the W Cape Province and beyond the western fringes of the N Cape Province where the Cape hare is dominant. In the more central, northern, and eastern parts of the country it is replaced by a smaller scrub hare, probably the African savannah hare, which extends into Namibia, Botswana, Zimbabwe, and possibly Mozambique and beyond.

TAXONOMY AND REGIONAL VARIATION: The taxonomy of African *Lepus* is unsettled, and particularly so in the S African subregion (south of the Cunene and Zambezi Rivers). Initially *Lepus* was proposed to be represented by two species in South Africa, *L. saxatilis* and *L. capensis*, a treatment followed by many subsequent authors. Petter, on the other hand, recognized four species: *L. capensis*, *L. saxatilis*, *L. crawshayi*, and what appeared to have been a Malawian variant, *L. whytei* (the status of the latter has been questioned because the diagnostic character, the width of the mesopterygoid space, is variable). In turn, *L. crawshayi* was subsumed in *L. victoriae*. As a consequence, persuasive arguments have been made for three southern African hare species: *L. capensis*, *L. saxatilis*, and *L. victoriae*. However, the geographic delimitation of all three hares is vague.

It has been argued, based principally on evidence from cross-sections through the upper incisors and multivariate morphometrics of skull, teeth, and standard body measurements, that there exists a clinal variation within what appeared to be a single species, *L. saxatilis*, which extended from the extreme southwestern region of the Cape Province of South Africa northeastward that included *L. victoriae*—in other words, *L. saxatilis sensu lato*. Although there was no evidence of a step-cline in the characters scrutinized, sampling was not dense by current standards (particularly so in the SW Cape Province). More recently a molecular phylogeographic study of *L. saxatilis sensu lato* from South Africa and neighboring areas (Namibia, Botswana, and Zimbabwe) based on the hypervariable mtDNA control region (mtDNA CR-1) showed 3 major mtDNA lineages in a sample of 159 specimens. The study material was drawn principally from collection localities in the central and northern parts of South Africa but, importantly, included three sample sites in the SW Cape Province (called the SW group). These findings have been interpreted as suggesting subspecific differences among the three phylogeographic assemblages. Importantly, there was a marked discontinuity separating the SW cluster from the other two groups. Based on these data it seems reasonable to argue that *L. saxatilis* should be considered a distinct species with a relatively limited distribution. The two remaining mtDNA haplotypic groups are not as well delimited; the northern matrigroup is broadly distributed in the northern part of South Africa and adjacent regions (Namibia, Botswana, and Zimbabwe), with the third lineage predominating in C South Africa. If one accepts *L. saxatilis* as a monotypic entity, the taxonomy of northern and central matrigroups will need to be revisited. These forms may be referable to a single species, *L. victoriae* (with two or more subspecies), but acceptance of this alignment would require a detailed analysis of specimens previously referred to as *L. crawshayi*, *L. victoriae*, and *L. whytei* so as to establish the distributional limits, as well as to determine the prior name for the African savannah hare complex. The two groups (defined herein as *L. saxatilis* and *L. victoriae*) have diverged recently, well within the period in which incomplete lineage sorting could have an effect, and consequently reciprocal monophyly for characters may not always be evident.

ECOLOGY: There is no literature specifically limited to the Cape scrub hare as defined herein. However, species in the genus *Lepus* are fairly catholic in habitat preference. *L. saxatilis* (*sensu lato*) occurs in open scrubland with scattered rocks and perturbed habitats associated with agriculture. They tend to avoid forest and steeply mountainous situations but occur at the ecotone of these areas.

HABITAT AND DIET: The Cape scrub hare selects leaves, stems, and rhizomes of dry and green grass, but has a preference for the latter.

BEHAVIOR: Very little is known of the behavior of the Cape scrub hare, but it is considered to be generally solitary, although sometimes found in pairs.

PHYSIOLOGY AND GENETICS: Diploid chromosome number = 48. The X chromosome is submetacentric, and the Y a small acrocentric. Small amounts of pericentromeic heterochromatin are found on almost all chromosomes.

REPRODUCTION AND DEVELOPMENT: The Cape scrub hare (as with other S African species of the genus) is a seasonal breeder, with young being born in early and late summer. Gestation (based on other species) is approximately 40 days.

PARASITES AND DISEASES: Ixodid ticks have been detected on the Cape scrub hare.

CONSERVATION STATUS:

IUCN Red List Classification: Least Concern (LC); decreasing

MANAGEMENT: The Cape scrub hare occurs in national parks and reserves in both the W and N Cape Provinces of South Africa and is abundant outside protected areas. There are no management implications at this stage. The species is generally quite common throughout the core region (SW Cape Province and Namaqualand), but there is no detailed information available. The International Union for Conservation of Nature (IUCN) and National Level information is based on a single species, *L. saxatilis*, which includes specimens referable to *L. victoriae*.

ACCOUNT AUTHOR: Terry J. Robinson

Key References: Allen 1939; Angermann et al. 1990; Carroll 1988; Flux and Angermann 1990; Horak and Fourie 1991; Kingdon 1974; Kryger 2002; Kryger et al. 2004; Lavocat 1978; Matthee et al. 2004e; Petter 1972; Roberts 1951; Robinson 1981b, 1986; Robinson and Dippenaar 1983, 1987; Robinson et al. 1983, 2002; Skinner and Chimimba 2005; Smithers 1971.

Lepus sinensis Gray, 1832
Chinese Hare

OTHER COMMON NAMES: Huanan tu (Chinese)

DESCRIPTION: The Chinese hare is relatively small compared with most other hares in China. The dorsal pelage coloration is variable, from sandy brown to grayish yellow,

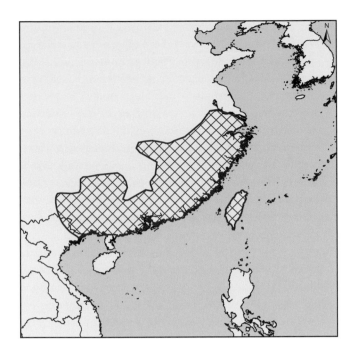

BEHAVIOR: The Chinese hare is primarily nocturnal, although it is sometimes active during the day. A burrow-dweller, it will occupy burrows dug by other animals.

PHYSIOLOGY AND GENETICS: Diploid chromosome number = 48

REPRODUCTION AND DEVELOPMENT: The breeding season of the Chinese hare extends from April to August, with females averaging three young per litter.

CONSERVATION STATUS:
IUCN Red List Classification: Least Concern (LC)

MANAGEMENT: The Chinese hare is threatened by loss of habitat throughout its range, and overutilization (hunting or trapping) may pose a threat to this species. Given its geographic range, it is remarkable how little attention has been given to this species.

ACCOUNT AUTHOR: Andrew T. Smith

Key References: Liu et al. 2011b; Smith and Xie 2008; Wu et al. 2005; Yu 2004.

but often has a rufous or chestnut tone. The belly fur is generally paler than that of the dorsum. The ears are short and have black triangular marks at the tips. The head is similar in color to the back and an eye-ring is present. The nape is rufous in color. The short tail is dorsally brown, the sides are gray-brown, and the underside is off-white. The winter pelage is typically a yellow color, with a mix of black-tipped hairs.

The supraorbital process of the Chinese hare is diminutive, and it lacks a deep anterior notch; the tips do not extend back to the braincase. The postorbital constriction is narrower as well, so that when viewed from above much of the orbit is visible. The bullae are smaller than those of most *Lepus* species. The Chinese hare has elongated nasals. The groove on the anterior face of the upper incisors is indistinct and not filled with cement.

SIZE: Head and body 350–450 mm; Tail 40–57 mm; Hind foot 81–111 mm; Ear 60–82 mm; Greatest length of skull 67–93 mm; Weight 1,025–1,938 g

CURRENT DISTRIBUTION: The Chinese hare extends across most of SE China as well as on Taiwan and possibly into N Vietnam.

TAXONOMY AND GEOGRAPHIC VARIATION: Two subspecies: *L. s. sinensis* (forms extending across SE China) and *L. s. formosus* (the form found on Taiwan).

ECOLOGY: The Chinese hare is found in a variety of habitats from lowland up to 4,000–5000 m asl.

HABITAT AND DIET: The Chinese hare inhabits hilly regions associated with edge grassland and scrubby vegetation. It consumes leafy plants, green shoots, and twigs.

Lepus starcki Petter, 1963
Ethiopian Highland Hare

OTHER COMMON NAMES: Lièvre des haut plateaux d'Èthiopiens (French); Äthiopischer Hochland Hase (German)

DESCRIPTION: The Ethiopian highland hare is a medium-sized hare. The pelage is soft and thick; the upper parts are tawny mixed with black, becoming grayer toward the rump. The nape, breast, and flanks are tawny. The ventral pelage is white. The head has similar coloration to the dorsal pelage. The nuchal patch is rufous or cinnamon. The Ethiopian highland hare has a whitish chin that becomes cinnamon toward the lips. The ears are intermediate in length compared to those of the Abyssinian hare (*L. habessinicus*) and the Ethiopian hare (*L. fagani*). The posterior of the ears is primarily light gray with whitish edges except at the tip, where there is a noticeable black patch on both sides. The limbs are generally cinnamon or tawny in color, but the hind limbs are anteriorly whitish gray. There is geographic variation in tail coloration; the population at Shewa has entirely white tails, while in the Bale Mountains the tails have a black mid-dorsal stripe.

SIZE: Head and body 540 mm; Tail 95–130 mm; Hind foot 105–123 mm; Ear 100–119 mm; Weight 1,692–2,963 g

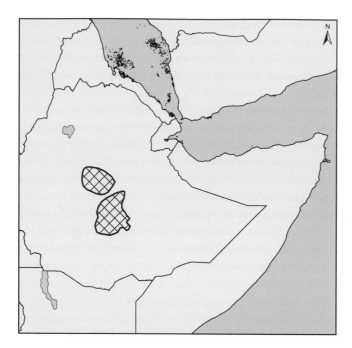

Lepus starcki. Photo courtesy Jeff Kerby

CURRENT DISTRIBUTION: The Ethiopian highland hare is endemic to Ethiopia and is distributed in the central highland of Shewa and the mountains of the Bale region, from 2,500 to 4,000 m asl. The type locality is Geldu Liben, C Ethiopia. Part of its range is sympatric with that of the Abyssinian hare.

TAXONOMY AND GEOGRAPHIC VARIATION: Although the Ethiopian highland hare has a well-established species status, its phylogenetic position relative to other taxa in the genus is questionable. Using morphological, dental, and cytogenetic characters, *L. starcki* has previously been categorized as a subspecies of the Cape hare (*L. capensis*)

or the European hare (*L. europaeus*). However, a molecular analysis indicates a closer relationship of *L. starcki* with *L. habessinicus*. In addition, a recent study using both molecular and morphological data concluded that *L. starcki* is distinct from *L. europaeus*.

ECOLOGY: There are few data regarding the ecology of the Ethiopian highland hare. Densities in the Bale Mountains have been estimated at 0.17/ha (grasslands of Web Valley), 0.2/ha (*Helichrysum* scrub of Mount Tullu Deemtu), and 0.3/ha (afroalpine grasslands of the Sanetti Plateau). The Ethiopian highland hare has been identified as part of the prey base for the tawny eagle (*Aquila rapax*) and the endangered Ethiopian wolf (*Canis simensis*).

HABITAT AND DIET: The Ethiopian highland hare is mostly restricted to rocky grassland and montaine moorlands, but will feed in wetlands during the dry season. It has recently been recorded from Mount Kaka and Mount Hunkulo.

Analysis of fecal pellets and direct observations indicate that the diet consists of 27 plant species (21 herbs, 4 grasses, and 2 shrub species), with a preference for grasses. Monocotyledons are the most frequently used plants by the Ethiopian highland hare. During the wet season only monocots are consumed, but during the dry season consumption of some dicots does occur.

PHYSIOLOGY AND GENETICS: Diploid chromosome number = 48. Chromosome morphology as well as G-banding are similar to that of *L. habessinicus* and *L. europaeus*. The Ethiopian highland hare has 21 pairs of bi-armed autosomal chromosomes (i.e., 8 pairs of metacentrics and submentacentrics and 13 pairs of telocentric and subtelocentrics) and 2 pairs of acrocentric chromosomes. Furthermore, the X-chromosome is a large submetacentric, one of the largest of the group, while the Y-chromosome is acrocentric and smaller in size.

REPRODUCTION AND DEVELOPMENT: Little is known of the reproduction and development of the Ethiopian highland hare. The reproductive period is suspected to be during the dry season. Two females, one lactating and one pregnant (one embryo), were recorded in December.

PARASITES AND DISEASES: There are no data on the parasites and diseases of this species, but the species is known to be susceptible to tick infestations.

CONSERVATION STATUS:

IUCN Red List Classification: Least Concern (LC)

MANAGEMENT: There are no known management measures in place for the Ethiopian highland hare, although it does occur in protected areas within its range. Populations may be found in restricted and fragmented habitats in SE Ethiopia and the highlands of C Ethiopia, and consequently need greater conservation attention. The scarcity or absence of available data on the population size/density, reproductive biology, diseases, and ecology of the Ethiopian highland hare should be addressed.

ACCOUNT AUTHORS: Zelalem Gebremariam Tolesa and Charlotte H. Johnston

Key References: Angermann 1983; Azzaroli-Puccetti 1987a, 1987b; Azzaroli-Puccetti et al. 1996; Flux and Angermann 1990; Happold 2013; Hoffmann and Smith 2005; Mekonnen et al. 2011; Petter 1963; Pierpaoli et al. 1999; Tolesa 2014; Yalden and Largen 1992; Yalden et al. 1986, 1996.

Lepus tibetanus Waterhouse, 1841
Desert Hare

OTHER COMMON NAMES: Zang tu (Chinese); Zayats peschannik (Russian)

DESCRIPTION: The desert hare is slender and has a relatively small head. The pelage varies according to location, and generally is sandy brown or drab and has traces of black dorsally. The ears are tipped with black-brown, and the anterior hairs are tufted. There is a light eye-ring. The exterior side of the hind legs and forefeet are white, and the hip is gray. The ventral pelage is pale yellow to white. The tail is white with a black to black-brown dorsal stripe. The winter pelage tends to be thicker and grayer than the summer coat, but does not turn white.

The premaxillary is relatively long in the desert hare, coupled with short nasal bones. The supraorbital process is warped upward, and the zygomatic arch is broad. The auditory bullae are inflated.

SIZE: Head and body 401–480 mm; Tail 87–109 mm; Hind foot 109–135 mm; Ear 81–110 mm; Greatest length of skull 84–92 mm; Weight 1,625–2,500 g

CURRENT DISTRIBUTION: The range of the desert hare extends from W and N Pakistan through much of Afghanistan, north through E Tajikistan and SE Kyrgyzstan, into NW China with a minor distribution in Mongolia. It occurs up to 4,000 m asl, but the lower limit of its distribution is not known. The range of the desert hare is typically allo- to parapatric with that of the tolai hare (*L. tolai*), although the two species may be sympatric in some reaches of the Tian Shan (mountains). The actual extent of its range is unknown due to the remote landscape it occupies and the lack of studies within its range.

TAXONOMY AND GEOGRAPHIC VARIATION: The desert hare is monotypic. Considerable controversy has surrounded the taxonomy of this species. It was considered independent from the time of its description until the 1930s, after which it was aligned with the *L. tolai*, the European hare (*L. europaeus*), and the Cape hare (*L. capensis;* in various combinations). Most recently it was thought to be composed of the forms *centrasiaticus* and *pamirensis;* however, molecular studies have clarified that *centrasiaticus* is a subspecies of *L. tolai*. Morphologically it shares several characteristics with the woolly hare (*L. oiostolus*), but not *L. tolai*.

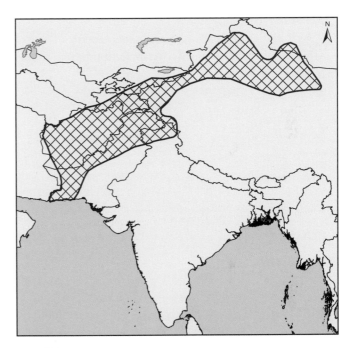

Lepus tibetanus. Photo courtesy Weidong Li

ECOLOGY, HABITAT, AND DIET: The desert hare occurs in grassland and shrubby areas along rivers within desert, semi-desert, and steppe habitats. While it does not occupy the alpine, its range may extend into the subalpine belt between 3,500 and 4,000 m. The species is known to eat plants, seeds, berries, roots, and twigs.

BEHAVIOR: Although it is sometimes seen during the day, the desert hare is primarily active at dusk. It does not construct its own burrow.

PHYSIOLOGY AND GENETICS: Diploid chromosome number = 48

REPRODUCTION AND DEVELOPMENT: Desert hares produce 1 to 3 litters per year, with 3 to 10 leverets per litter.

CONSERVATION STATUS:

IUCN Red List Classification: Least Concern (LC)

MANAGEMENT: The desert hare may be hunted for food throughout its range.

ACCOUNT AUTHOR: Andrew T. Smith

Key References: Cheng et al. 2012; Smith and Xie 2008; Wu et al. 2005.

Lepus timidus Linnaeus, 1758

Mountain Hare

OTHER COMMON NAMES: Mountain hare, Varying hare, Arctic hare (also refers to *L. arcticus*), Snow hare, Irish hare, Blue hare

DESCRIPTION: Mountain hares are smaller than European hares (*L. europaeus*), with shorter front legs and ears but longer hind feet. The relatively large head has prominent eyes with a yellowish iris in the adult and a dark brown one in the young. The tail is shorter and white below, and the back is paler. Compared with European hares, mountain hares have shorter faces. The ears are

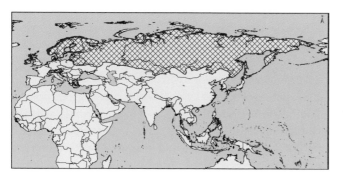

short and narrow, with a black 30 mm long tip. The hind feet are long, with hairy soles and widely spreading toes. The head and body are dusky brown in summer, and often richer brown on the head, with gray-blue underfur showing through especially on the flanks. The legs, lower throat, and upper breast are lighter than the back. The chin, upper throat, and insides of the legs are dirty white. The stomach and tail are white. The coat tends to be reddish brown at low latitudes, but is grayish brown at high altitudes and toward the arctic. There are two morphs of winter coat, white and blue. In white morphs the coat is finer, woollier, and almost pure white except for a small

Lepus timidus summer. Photo courtesy Mike Brown

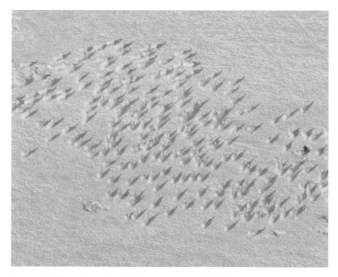

Lepus timidus winter herd. Photo courtesy Joel Berger

brown spot on the nose and above the eyes and black ear tips. In blue forms, the fur is 15 mm long with an underfur 25 mm at mid-back, black underfur hair 25 mm long with white or pale buff bands, white or white-tipped guard hairs, and white or black and white vibrissae. The nose and forehead are rust brown with a brown spot over each eye. The hind feet are white, but rust brown between the toes; the back, sides, and flanks are pale blue gray. In European Russia the underfur is snow white. The winter fur of mountain hares in Ireland is partially white and patchy, with the upper back and head remaining brown. A sprinkling of reddish hair is always present. The molt to the white winter coat begins on the back and is relatively rapid; the spring molt, in contrast, begins ventrally. Females turn brown about two weeks before males, most strongly influenced by photoperiod. Northern populations of mountain hares are consistently larger than more southerly forms.

SIZE: Head and body 509 mm (*L. t. varronis*; Alps), 546 mm (*L. t. timidus*; Siberia, Russia); Tail 53–70 mm; Hind foot 159–164 mm; Ears 86–93 mm; Greatest length of skull 86–98 mm; Weight 2,397 g (Alps), 2,750 g (Scotland), 3,000 g (Hokkiado, Japan), 3,190 g (Ireland), 3,395 g (Norway), 3,438 g (Siberia)

PALEONTOLOGY: *Lepus* arose in the early Middle Pleistocene, and mountain hares appear in the F-Eemian (e.g., Lambrecht Cave, Hungary) and are common in 4-Wiirm, with numerous Magdalenian sites from the Pyrenees to SE Europe. *L. timidus* was in general common in the Pleistocene. In C Europe, they appeared at the end of the Riss glaciation in the Bockstein-höhle (cave) in Württemberg. Mountain hares were eaten by humans 28,000 years ago in Belgium, and remains are common in Pleistocene middens. The general pattern of distribution suggests an origin in Asia or even North America, spreading eastward to Greenland and westward to Britain; there are many remains in S England, Italy, Spain, and glacial deposits in the Crimea.

CURRENT DISTRIBUTION: The geographic distribution of the mountain hare is circumpolar in tundra and taiga habitats from Britain and Norway to Japan. They occur over the whole of Norway and Finland, and in Sweden south to 56°N, including Öland and Gotland, with both ancient and recent introductions to many small islands. The mountain hare occurs in Russia from the Kola Peninsula south to 53°N with isolated pockets to 49°N; in Mongolia, China, and eastward to Sakhalin and Kamchatka (Russia) and Hokkaido (Japan); and in Estonia, Latvia, Lithuania, and parts of Poland. Isolated alpine populations occur above 1,300 m in the mountains of S Germany, France, Italy, the Slovenian Alps, and all of Switzerland except the Canton of Jura. It has never been present on Iceland. *Lepus a. groenlandicus* and *L. t. timidus* were introduced into Spitzbergen (Norway) in 1930 and 1931, but there are none there now. *Lepus t. timidus* X *L. t. sylvaticus* was introduced into the Faroe Islands (Denmark) in 1854. It is indigenous to Scotland, Ireland, and the Isle of Man. It has been introduced (mainly in the nineteenth century) to the Shetland, Orkney, Outer Hebrides, Skye, Raasay, Scalpay, Mull, Arran, Bute, Jura, and Eigg islands of Scotland. Reports of mountain hares from the Pyrenees are not substantiated.

TAXONOMY AND GEOGRAPHIC VARIATION: Sixteen subspecies: *L. timidus ainu, L. t. begitschevi, L. t. gichiganus, L. t. hibernicus, L. t. kamtschaticus, L. t. kolymensis, L. t. kozhevnikovi, L. t. lugubris, L. t. mordeni, L. t. orii, L. t. scoticus, L. t. sibiricorum, L. t. sylvaticus, L. t. timidus, L. t. transbaicalicus,* and *L. t. varronis*. However, this classification is based especially on fur characteristics and has not been reviewed with modern methods. The subspecies *L. t. timidus* range includes all of Norway, Sweden south to 59°N, all Finland, and Russia south to 57°N and east of the Ural Mountains. *Lepus t. sylvaticus* is the hare of S Sweden, extending in hybrid zones with *L. t. timidus* to W Latvia, the south coast of Norway, and the Faroes (introduced). *Lepus t. kozhevnikovi,* the C Russian hare, ranges from 57°N to 58°N, where it apparently intergrades with *L. t. timidus,* south to 53°N. The western boundary is unclear; most authors refer to E Baltic hares as *L. t. timidus*. The distribution of the other Asian subspecies is unclear. To the east, *L. t. sibiricorum* is found on the plains of W Siberia, and in the Altai Mountains, *L. t. lugubris*. Farther to the east, in Transbiakalia, is *L. t. transbaicalicus,* and along the coast south to Manchuria (China), *L. t. mordeni*. In N Siberia *L. t. begitschevi* is found in Taimyr (Russia), and east of that *L. t. kolymensis* at the Kolyma River. *Lepus t. gichiganus* occurs in Kamchatka, along the coast of the Sea of Okhotsk, and in C Sakha (Russia), southward to the range *of L. t. mordeni*. *L. t. ainu* is restricted to Hokkaido, and *L. t. orii* to Sakhalin Island, but might also be present on the mainland along the coast. The original distribution of *L. t. scoticus* in the Scottish Highlands has been extended by introduction to SW Scotland in 1834–1862, the English Peak District, N Wales, and many islands. *Lepus t. varronis* consists of relict populations in the Alps above ~ 1,300 m, in Germany, France, Switzerland, Italy, Austria, and the former Yugoslavia. *Lepus t. hibernicus* occurs over the whole of Ireland on moorland and pasture down to sea level. *Lepus t. hibernicus* was introduced to SW Scotland, Mull, and the English Lake District about 1890. Some consider *hibernicus* sufficiently distinct as to be an independent species—the Irish hare.

ECOLOGY: The most important mammalian predator of the mountain hare is the red fox (*Vulpes vulpes*; up to 90% of its diet). In forested areas pine martens (*Martes martes*; up to 35%), wolves (*Canis lupus*; 70%), lynx (*Lynx lynx*; 70%), and ermines (*Mustela erminea*) frequently kill mountain hares, and above the tree line also Arctic foxes (*Vulpes lagopus*) and wolverines (*Gulo gulo*). Golden eagles (*Aquila chrysaetos*; up to 35%) and goshawks (*Accipiter gentilis*) are important bird of prey predators, but common buzzards (*Buteo buteo*), rough legged buzzards (*B. lagopus*), hen harriers (*Circus cyaneus*), and eagle owls (*Bubo bubo*) sometimes kill hares.

The average winter survival for adult mountain hares is 42% at low predation pressure and 19% under high predation. Juvenile survivorship is 75% from weaning to fall, 36% during a winter with low predation pressure, and 16% during a winter with high predation. Overall juvenile survivorship from birth to the following spring averages 20%. A few mountain hares in Scotland reach an age of 9 years in the wild, and one was shot after 18 years. Densities are variable: 1 hare/km^2 over large parts of Sweden and European Russia, 2–6/km^2 in C Finland, 25/km^2 on Baltic Islands, and up to 200–400/km^2 in Scotland and on islands. In Europe numbers fluctuate less than in Russia where peaks normally are 8 to 12 years apart, as in Scotland, but 3 to 4 years in Fennoscandia. The periodic crashes are related to parasitism, predation, or starvation.

Populations are stable across much of the geographic distribution of the mountain hare, with fluctuations occurring in N Europe and possible declines in the Alps. Population declines have been observed in Russia and in the extreme, southern portions of Sweden. In N Ireland, historical game bag records indicate a substantial decline in hare abundance.

HABITAT AND DIET: The mountain hare lives mostly in mixed forests, reaching highest densities in transition zones with open clearings, and in swamps, river valleys, and patches of regrowth among fallen trees. Bogs with willows (*Salix*) and blueberries (*Vaccinium*) are favored, followed by stands of spruce (*Picea*) and birch (*Betula*). Fewer hares live in pines (*Pinus*) and mature forests, but less favorable habitats, such as birch forests on windy slopes, may be occupied when populations are high. A second major habitat is tundra, where hares can survive to the limits of vegetation if there is cover such as cliffs or

rocks. Hares in Scotland reach high densities (~ 50/km^2) on well-managed heather moorland. Hares in Ireland occupy moorland and agricultural land at all altitudes. In the Alps they live on alpine pastures up to 4,000 m asl in summer and at 3,200 m in January, but most move to shelter in woodland escarpments. On Hokkaido, mountain hares are common in grassy fields, scrublands, and open forests from sea level up to the mountains.

The food eaten by hares in forest habitats includes leaves and twigs of most deciduous trees or shrubs: willows, birch, mountain ash (*Sorbus*), cottonwoods (*Populus*), and junipers (*Juniperus*). In tundra areas especially dwarf willow (*Salix herbacea*) is preferred. On Scottish moorlands and Swedish islands, common heather (*Calluna vulgaris*) constitutes the most important plant. Palatable grasses and clovers are eaten when available, and hares prefer plants growing on fertilized soil with a high nutrient content. Snow cover restricts the diet to twigs, bark, moss, and lichens. The diets of both sexes are similar (September–November), although reproductively active females eat more grasses and less heather than do males, and young eat more bentgrass (*Agrostis*) and less matgrass (*Nardus*) than do adults. They seldom drink water but may eat snow. At least 500 g/day of birch and willow is eaten.

Studies on captive *L. t. timidus* demonstrate that dry matter digestibility is the basic factor in meeting energy requirements. Secondary plant compounds in the twigs of *Calluna* or *Betula* may reduce digestibility by causing hares to excrete sodium, but they maintained a sodium balance on a mixed diet. Plant phenolic compounds have been suggested to be involved in this reduced digestibility, but they are not reliable as a predictor of food consumption.

The winter food of mountain hares in Scotland and Ireland is dominated by heather, 30% to 90% frequency from stomach analyses, whereas grasses, especially bentgrass and tussock grass (*Deschampsia*), sedges (*Eriophorum*), and dicotyledons like bedstraw (*Galium*) are more frequent in the summer diet. In Fennoscandia, the winter diet (from browsed plants) is dominated by *Betula* (15% to 62%), *Sorbus* (5% to 72%), and *Salix* (10% to 34%), whereas in Russia *Salix* (20% to 36%), *Populus* (6% to 20%), and *Vaccinium* (5% to 25%) are more important. Little pine or spruce is eaten (0% to 3% in all areas), but more junipers (3% to 10% in all areas). In summer, captive hares in Finland avoided many of the winter foods, instead preferring star sedge (*Carex echinata*), couch grass (*Elymus repens*), common hazel (*Corylus avellana*), meadow pea (*Lathyrus pratensis*), bird vetch (*Vicia cracca*), red clover (*Trifolium pratense*), and horsetail (*Equisetum* spp.). Winter trials indicate the preference order of *Salix, Betula, Populus,* and *Sorbus.* In Russia, seasonal food consumption is largely determined by availability rather than preference. Other species eaten in low frequencies include those in the following genera: *Molina, Nardus, Scirpus, Carex, Juncus, Alnus, Rubus, Ribes, Prunus, Rosa, Hippophae, Erica, Arctostaphylos, Empetrum, Myrica, Ledum, Potentilla, Filipendula, Chamaenerion, Saxifraga, Gentiana, Ranunculus, Achillea, Rumex, Fragaria, Taraxacum, Quercus, Acer, Fagus, Larix,* and *Rhododendron.*

BEHAVIOR: The mountain hare is primarily nocturnal, but shows increased daylight activity in summer when nights are short. During the mating period from April to June more hares are day active. The hare rests by day in a form (depression) with ears back and eyes half closed, but sleeps for only a few minutes. Mountain hares groom themselves carefully, especially in early morning and evening, and can be seen rolling in the dust in dry weather.

The main feeding period begins around sunset and ends about sunrise, but this varies over the season. Hares feed with their back to the wind and can clear snow from vegetation with their forepaws. In open country, many hares may gather on good feeding areas. On continental Europe they confuse tracks and jump to the side before resting for the day. From January to July male hares often follow 2–20 m behind a female for hours. Males obviously use scent to track females; the activity of the inguinal glands increases during reproduction. Females dominate males and may strike with their forepaws with ears laid back if the male approaches too closely. Copulation can involve five or more males and a single female, and males can be seen in long fights. Mountain hares appear to dominate European rabbits, but are subdominant to European hares with competitive exclusion.

The size of the home range in Scotland is around 100 ha, and in Finland up to 300 ha, but it is probably much larger in northern areas. The daytime range for adult males could be 15 ha, with travels of 1–3 km to feed. Ranges may overlap between both males and females, but to a lesser extent between male-male and female-female dyads.

Hares make several forms, by trimming the vegetation, to sit in by day. If undisturbed, the same form may be used for weeks, or may be changed depending on the weather. On Scottish moors forms may last 25 years. Hares use deep vegetation for concealment in the summer and exposed positions or rocks in winter. In snow, hares burrow to reach or make forms and enter rock crevices for shelter or escape. They sometimes dig permanent burrows 1–2 m long in the ground, often used by young, but seldom by adults.

PHYSIOLOGY AND GENETICS: Diploid number of chromosomes = 48. *Lepus* is highly conservative karyotypically, with G-banding patterns identical in most species. Hybrids with *L. europaeus* occur in Sweden where predominately mountain hare females mate with European hare males. There is genetic evidence of similar hybridization on the Iberian Peninsula, where mountain hare DNA can be traced in both the Iberian hare and the broom hare. First-generation hybrids are intermediate in form and fertile.

The energy requirement for maintenance in winter is about 105 kcal/kg/day, with a maximum intake of 150 kcal. Re-ingestion of soft fresh fecal pellets during the day increases digestibility up to 25%. Between 200 and 450 hard fecal pellets are produced daily.

REPRODUCTION AND DEVELOPMENT: The gestation period of mountain hares is 50 days (range 47–55), with copulation occurring a few hours after parturition. The number of litters per year varies from one in northern areas to three or even four in southern areas. Litter size increases with age, body weight, and season from one to two, but up to seven in areas where only a single litter is produced. At birth the young are fully furred, their eyes are open, and they start suckling at once. Birth weight varies between 70 and 150 g. The growth rate is 14–30 g/day and leverets depend on milk for 2–6 weeks. Young hares can be aged by size up to 3 months of age. They reach adult size in about 4 months, but can be distinguished up to 8–10 months by the notch between the tibia and humerus. Fat deposits in both sexes are very low from June to October, increasing to a peak in January to March, and declining when breeding starts.

PARASITES AND DISEASES: Mountain hares carry several ectoparasites, including fleas, lice, and ticks. Of the endoparasites, trematodes, nematodes, and lungworms can be problematic. Mountain hares can be infected with several species of coccidia and bacteria such as *Toxoplasma gondii, Pasteurella,* and *Listeria.* There have been isolated cases of myxomatosis. Heavily parasitized hares are more frequent in high-density populations, and have enlarged spleens, less coronary fat, and less body mass; these populations might be controlled by parasitism. Tularemia has been reported in mountain hares in Asia, Sweden, and Finland, and toxoplasmosis in Japan. The calicivirus in European brown hare syndrome (EHBS) can also infect mountain hares. There is also a strain in the related RHD2 (rabbit hemorrhagic disease) that might infect mountain hares.

CONSERVATION STATUS:

IUCN Red List Classification: Least Concern (LC)

MANAGEMENT: The mountain hare is a popular game species in most parts of the distribution range. Release and restocking take place in many areas for hunting opportunities, but often with little effect on wild populations. However, these practices are associated with the risks of spreading diseases and parasites as well as genetic swamping. Supplemental feeding during winters can have positive effects on survival and reproduction. Hares may damage cereals, brassica crops, fruit trees, and tree plantations, especially in winter, and ring-bark pines.

Although populations are stable across much of the geographic distribution of the mountain hare, with strong fluctuations occurring in N Europe, there are declines in several areas, e.g., Russia, southern portions of Sweden, the Alps, and Ireland. Potential threats are competitive exclusion and hybridization with European hares, European brown hare syndrome (EHBS), and tularemia. The discovery of an introduced population of European hare to Ireland could pose a threat to *L. t. hibernicus*. The mountain hare is listed under the Bern Convention, Appendix III, as well as the European Union Habitats and Species Directive, Annex V.

ACCOUNT AUTHOR: Anders Angerbjörn

Key References: Ahlgren et al. 2016; Angerbjörn 1986, 1989; Angerbjörn and Flux 1995; Baker et al. 1983; Bergengren 1969; Caravaggi et al. 2015; Dingerkus and Montgomery 2002; Flux 1970, 1987; Gustavsson 1971; Hewson 1962, 1985; Hewson and Hinge 1990; Höglund 1957; Iason 1989, 1990; Iason and Boag 1988; Lind 1963; Lindström et al. 1986; Marcström et al. 1989; Melo-Ferreira et al. 2005; Naumov 1947; Newey and Thirgood 2009; Pehrson 1983; Pehrson and Lindlöf 1984; Reid 2011; Reid et al. 2007; Thulin 2003; Thulin et al. 1997; Walhovd 1965; Watson and Hewson 1973.

Lepus tolai Pallas, 1778
Tolai Hare

OTHER COMMON NAMES: Tuoshi tu (Chinese); Bor tuulai (Mongolian); Zayats tolai (Russian)

DESCRIPTION: The pelage of the tolai hare is soft, and the coloration varies across the wide range of the species. In general, the dorsal pelage is sandy yellow, brownish gray, or gray with a dark ripple. The color of the upper part of the head is usually darker than the back. The eyes are set off by a whitish area that extends forward onto the muzzle, and there is a thin light collar stripe. The hip may be grayish to ochraceous, and the tail has a broad black or blackish-brown stripe above and white beneath. The tips of the ears are black. The ventral pelage is pure white.

The skull of the tolai hare is of light construction and not large, and possesses a bulged cranium and a narrow rostrum. The supraorbital process is well developed and

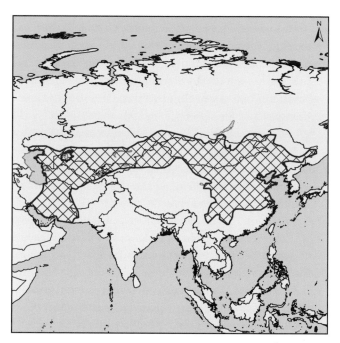

Lepus tolai. Photo courtesy Paul B. Jones

almost triangular in shape. The auditory bullae are round and large.

SIZE: Head and body 387–590 mm; Tail 72–116 mm; Hind foot 106–135 mm; Ear 80–119 mm; Greatest length of skull 80–88 mm; Weight 1,650–2,650 g

CURRENT DISTRIBUTION: The geographic distribution of the tolai hare extends from the northern and eastern shores of the Caspian Sea, south through C and E Iran with a marginal range in Afghanistan, east through Kazakhstan and S Siberia, as well as to Mongolia and W, C, and NE China. It is known to range from low to high elevations (4,900 m). The actual extent of its range is unknown due to the remote landscape it occupies and the lack of studies within its range.

TAXONOMY AND GEOGRAPHIC VARIATION: *Lepus tolai* has generated considerable taxonomic confusion; it has been included in the Cape hare (*L. capensis*), the European hare (*L. europaeus*), and the desert hare (*L. tibetanus*). Particularly problematic has been the assignment of subspecies among these various taxa. It has included the form *przewalskii*, which we have included in the woolly hare (*L. oiostolus*). The form *centrasiaticus* has been assigned to *L. tibetanus*, but appears to represent *L. tolai* based on molecular analyses. Some separate the form *swinhoei* as an independent species. But, parsimoniously, it appears that *L. tolai* can be represented by seven subspecies in China: *L. t. aurigineus* (Jiangxi, Anhui, Hubei, Shaanxi, Sichuan, Guizhou), *L. t. centrasiaticus* (Xinjiang, Gansu, Nei Mongol), *L. t. cinnamomeus* (Sichuan, Yunnan), *L. t. huangshuiensis* (Qinghai), *L. t. lehmanni* (Xinjiang), *L. t. swinhoei* (Heilongjiang, Jilin, Liaoning, Nei Mongol, Hebei, Beijing, Henan, Shaanxi, Shanxi, Shandong), and *L. t. tolai* (Nei Mongol, Gansu). Which of these forms extend north into C Asia and Siberia is unknown, as well as the range of additionally named subspecies (*L. t. buchariensis*, *L. t. cheybani*, *L. t. filchneri*) in these regions or extending into the Iranian portion of the species range.

ECOLOGY: The tolai hare experiences fluctuations in population density. Several studies have recorded that, at its peak, density can be between 10 and 30/ha. These hares do not undertake long-distance migrations, but are known to engage in seasonal altitudinal migrations to facilitate foraging.

HABITAT AND DIET: The tolai hare occurs in a variety of habitats, including desert, semi-desert, mountain-steppe, forest steppe, rocky habitats, and grasslands. It does not occupy true forests. It is an herbivore that consumes roots, grass, sedges, shrubs, bark, wood, seeds, shoots, inflorescences, and bulbs.

BEHAVIOR: Adults are predominantly nocturnal, but under favorable conditions will forage during the daytime. Young hares are more active during daylight hours. Daytime shelters vary according to situation. Some dig oval pits or holes under shrubs; others shelter under shrubs, under rock fragments, or in meadows and fields.

PHYSIOLOGY AND GENETICS: Diploid chromosome number = 48

REPRODUCTION AND DEVELOPMENT: The tolai hare produces two to three litters per year and has one to nine young per litter. The first litter is smaller than the second and the third (if there is one) is smaller than the second. In the desert part of the tolai hare's range, reproduction may begin by the end of January; in mountainous environments the reproductive cycle usually starts at the end of February to the beginning of March. The reproductive season may extend for as long as five months throughout much of the species' range.

PARASITES AND DISEASES: Twenty-two species of ixodid mites have been recorded on tolai hares, most commonly *Dermacentor daghestanicus*, *D. marginatus*, and *Rhipicephalus pumilio*. Gamasid and trombuculid mites may also be found on the hares. Infestation by mites depends on weather conditions and habitat; tolai hares are most likely to be parasitized in spring and through the summer months; they are generally free of mites from November to March. Other ectoparasites include fleas, which at times can be numerous on tolai hares. Tolai hares can also be intensively infected with endoparasites, in particular, trematodes (most commonly *Dicrocoelium lanceatum*), cestodes, and nematodes.

Tolai hares are a source of the tularemia pathogen, and their rate of infectivity fluctuates from 0.9% to 14%. Percent infection is highest in spring and lowest in fall. They are also known to be casual carriers of the plague microbe, as well as the pathogen of brucellosis.

CONSERVATION STATUS:

IUCN Red List Classification: Least Concern (LC)

MANAGEMENT: Tolai hares are hunted for meat and pelts. Overharvesting may become a potential threat in some regions of its geographic range, although a consistent commercial harvest of tolai hares in E Mongolia has remained constant at about 1,000/year for several decades and appears to be sustainable.

ACCOUNT AUTHOR: Andrew T. Smith

Key References: Allen 1938; Cheng et al. 2012; Reading et al. 1998; Smith and Xie 2008; Sokolov et al. 2009.

Lepus townsendii Bachman, 1839
White-tailed Jackrabbit

DESCRIPTION: The winter pelage of the white-tailed jackrabbit is monochromatic white in northern regions except for the black tips on the ears. In warmer months (and year-round in southern regions), the pelage is grayish-brown dorsally and white or gray ventrally. The tail is relatively large and white in all seasons, on both top and bottom, but has a small dorsal black stripe in some areas. In Nevada, the species has bright, fiery-red eye-shine, compared to that of the black-tailed jackrabbit (*L. californicus*). The white-tailed jackrabbit is well adapted to cold, and has thicker fur, a higher metabolic rate, and a higher body temperature than the black-tailed jackrabbit. White-tailed jackrabbit females are on average 1% to 7% larger than males.

The skull of the white-tailed jackrabbit is relatively short, broad, and arched, possessing molariform teeth and prominent flanges that project laterally from the dorsal margins of relatively broad jugals. Compared to those of the black-tailed jackrabbit, the bullae are proportionately smaller. The supraorbital process has anterior and posterior projections; the latter is usually short and obtusely angled, and rarely fused to the cranium.

SIZE: Head and body 440–700 mm; Tail 66–105 mm; Hind foot 130–170 mm; Ear 100–113 mm; Greatest length of skull 77–103 mm; Weight 2,179–4,400 g

PALEONTOLOGY: The dominant hare in the Intermountain Region prior to the Holocene, *L. townsendii* began giving way to the newer *L. californicus* with warming temperatures and changing vegetation about 7,000–8,000 years ago. This replacement, although subject to fluctuations, continues today.

CURRENT DISTRIBUTION: Centered in the north C United States, the distribution of the white-tailed jackrabbit extends west to the Cascade and Sierra Nevada Mountains, south through Utah and N New Mexico, east to the western edges of Lakes Superior and Michigan, and north into the southern edge of five Canadian provinces (particularly near the Canadian Rocky Mountains). This wide distribution represents an expansion to the north and the east from the historical distribution, associated with the creation of suitable habitat. Following this expansion, however, the species' range has contracted in the Central Plains region (Illinois, Missouri, and Kansas, as well as in parts of Minnesota), E Washington, and a

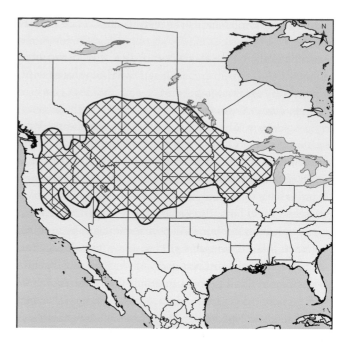

Lepus townsendii. Photo courtesy Justine A. Smith

very restricted area in S British Columbia (Canada). The white-tailed jackrabbit has been recorded at elevations from 30 to 4,319 m asl, the greatest span of any jackrabbit. It is fairly common in some parts of the west, but is more common in northern areas.

TAXONOMY AND GEOGRAPHIC VARIATION: Two subspecies: *L. t. townsendii*, and *L. t. campanius*. *L. townsendii* was historically placed (as *L. t. campestris*) in the subgenus *Proeulagus* by Gureev or in *Eulagos* by Averianov; *L. campestris* is a recent synonym. A comparison of the variation in life history traits (e.g., sex ratios, age distributions, growth rates) from individuals across Wyoming (*L. t. townsendii*) with individuals from numerous eastern populations (*L. t. campanius*) indicated that observable differences in phenotype may be reduced due to plasticity, or that considerable genetic similarity exists among populations across the range of the species.

ECOLOGY: The density of white-tailed jackrabbit populations varies markedly across space and time, and has been reported as 2.2/km² in the shortgrass prairie of Colorado, 3.5–43/km² in Minnesota, 3.9–71/km² in Iowa, and 6.2–9.0/km² in Wyoming. The white-tailed jackrabbit is preyed on by all North American canid and felid species, as well as American martens (*Martes americana*), weasels (*Mustela* spp.), golden eagles (*Aquila chrysaetos*), hawks (*Accipiter* spp., *Buteo* spp.), and great-horned owls (*Bubo virginianus*). When sympatric with black-tailed jackrabbits, white-tailed jackrabbits are more selective in the plant species they eat, placing them at a competitive disadvantage, are predominant on the slopes and ridges, while black-tailed jackrabbits are more common on valley floors, and use sagebrush (*Artemisia* spp.) and rabbitbrush (*Ericameria* spp.) habitats less frequently than do black-tailed jackrabbits. White-tailed jackrabbits have been detected in five of 16 mountain-range complexes in the Great Basin, the second-fewest among the 15 montane mammal species studied.

HABITAT AND DIET: The white-tailed jackrabbit inhabits open prairies and plains, agricultural lands, grasslands, meadows, shrublands, and alpine habitats, as well as forest edges during severe weather. Habitat associations can vary across seasons, especially as these jackrabbits move to lower elevations during winter. They are most common at higher (> 1,500 m) elevations, especially farther south in the species' range.

The diet varies seasonally according to food availability. Individuals consume mostly succulent grasses and forbs during snow-free seasons, and browse on twigs, buds, and bark in winter. In late winter in Utah, individuals even eat prickly pear cactus (*Opuntia* spp.) and juniper (*Juniperus* spp.) berries. Individuals consume some agricultural crops, when these are the most nutritious food available.

BEHAVIOR: White-tailed jackrabbits are typically solitary, but up to 100 or more individuals may occasionally aggregate, especially at dense food resources such as in agricultural fields. Individuals are nocturnal, and hide in forms (depressions) during the day unless flushed. When fleeing, the species can reach speeds of 55 km/h, jumping in bounds of 3.7–6.0 m. White-tailed jackrabbits tend to be more sedentary and have larger home ranges than black-tailed jackrabbits.

PHYSIOLOGY AND GENETICS: Diploid chromosome number = 48

REPRODUCTION AND DEVELOPMENT: The breeding season of the white-tailed jackrabbit lasts from late February until mid-July in North Dakota and Wyoming, yet only from May through early July in Canada. Juveniles do not breed in the year of their birth. Gestation period may range from 30 to 47 days. White-tailed jackrabbits produce 1 to 4 litters per year (usually 2) and litter size ranges from 1 to 11, but 4 or 5 is most common. The species is polygamous, and *L. t. campanius* exhibits breeding synchrony. Males fight furiously over females during the breeding season, by biting and especially by kicking with their hind feet. The maximum recorded lifespan is 8 years.

PARASITES AND DISEASES: Endoparasites include protozoans (*Eimeria* spp.) and tapeworms (*Cittotaenia* spp., *Raillietina loeweni*, *Taenia pisiformis*, and *Multiceps* spp.). Roundworm filariid species have been found in the circulatory system of some white-tailed jackrabbits. Ectoparasites include fleas (*Cediopsylla inaequalis*, *Hoplopsyllus* spp., and *Pulex irritans*), ticks (*Dermacentor andersoni*), and lice (*Haemodipsus setoni*). Diseases carried by white-tailed jackrabbits include Colorado tick fever, equine encephalitis, tularemia, papillomas, and fever caused by heavy infestation of botfly (*Cuterebra*) larvae. In a study of livers from 314 white-tailed jackrabbits collected from 44 counties in South Dakota, there was no evidence of disease or parasitism being the cause for the species' decline within the state.

CONSERVATION STATUS:

IUCN Red List Classification: Least Concern (LC)

MANAGEMENT: The white-tailed jackrabbit is hunted for subsistence and sport, and its fur is sold commercially; hunting limits exist in some states and provinces. Otherwise, nothing other than local conservation measures exists. White-tailed jackrabbits occur in protected areas within their range. They do appear to be missing from many areas, especially at the edge of their range, apparently due to range expansions of the black-tailed jackrabbit, conversion of habitat to agriculture, overgrazing by livestock, their persecution as a pest, drought, warming climate, or a combination of these factors. In the Greater Yellowstone Ecosystem, severe winters, diseases, predation, human persecution, habitat change, competition with high ungulate densities, and climate change have been hypothesized to contribute to declines of the white-tailed jackrabbit.

Overall, the status of the white-tailed jackrabbit is most secure in the center of the range and more marginal at the fringes. The white-tailed jackrabbit is ranked S5 (Secure) in Alberta, Idaho, and Nevada, and S4 (Apparently Secure) in Manitoba, Saskatchewan, Montana, Wyoming, Oregon, South Dakota, Colorado, and Nebraska. It is considered S3 (Vulnerable) in Iowa and California and S3S4 in Utah. The white-tailed jackrabbit is ranked S2 (Imperiled) in New Mexico, S2S3 in Washington, and S1 (Critically Imperiled) in Wisconsin and Ontario. Additionally, it is considered SH (Possibly Extirpated) in British Columbia and SX (Presumed Extirpated) in Illinois, Missouri, and Kansas. Greater understanding of the synecology of this species and effects of the threats listed above, particularly the factors that affect the species' distribution and local abundance, would better inform conservation.

White-tailed jackrabbits appear to require tall grass. In research in Colorado, white-tailed jackrabbits experienced high mortality rates in grazed areas; few animals survived more than one year. Past fluctuations in density and range occupied have been attributed to changes in land use and predator control, with most-pervasive declines occurring during the past 25 years. Reports of past cycles and "die-offs" due to disease are poorly documented and need verification.

ACCOUNT AUTHORS: Erik A. Beever, David E. Brown, and Joel Berger

Key References: Armstrong 1972; Berger 2008; Berger et al. 2005; Braun and Streeter 1968; Brown 1940, 1947; Burnett 1926; Carter 1939; Clanton and Johnson 1954; Dalquest 1948; DeVos 1964; Donoho 1971; Grayson 1977; Gunther et al. 2009; Kline 1963; Lim 1987; Mohr and Mohr 1936; Rogowitz 1990, 1992, 1997; Rogowitz and Wolfe 1991; Simes et al. 2015; Watkins and Nowak 1973.

Lepus victoriae Thomas, 1893
African Savanna Hare

OTHER COMMON NAMES: Africanischer savannen Hase (German); Lièvres des savanes (French)

DESCRIPTION: The African savanna hare covers vast areas with quite dissimilar habitats, which leads to variability in size and pelage color. The African savanna hare is of medium size, about 2 kg. The dorsal pelage is brown, grizzled (agouti) with black, and the ventral pelage is white or buff. The African savanna hare is characterized by the russet area on the nape of the neck in addition to the russet pelage on the chest sides and legs. The ears are of medium length (~ 114% of GLS) with black color toward the tip, and the tail is black above and white below. The lips are grayish without black-tipped hairs. The soles of the feet are covered with dense brown hairs, and the hind feet are cinnamon-brown above and brown below. The fur texture is coarser than in the Cape hare (*L. capensis*), but where the two species co-exist in E Africa they may be almost identical in color. Several minor characters together with dental traits aid in separating the two species in E Africa.

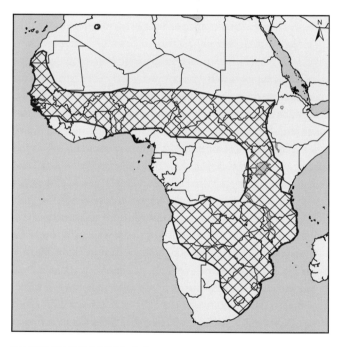

Lepus victoriae. Photo courtesy Thierry Aebischer / Chinko Project

SIZE: Head and body 415–575 mm; Tail 68–121 mm; Hind foot 103–127 mm; Ear 93–119 mm; Greatest length of skull 84.9–93.6 mm; Weight 1,500–3,000 g

PALEONTOLOGY: The African continent has experienced various climatic changes. However, the most relevant for the genus *Lepus* are those changes that occurred during the Pleistocene (1.8–0.11 mya) and the Holocene (0.11 mya to the present). Various terrestrial records indicate broad arid phases in the Last Glacial Maximum (LGM). Fossil sand dunes indicate that the Sahara expanded southward during this time. The Early Holocene was anomalously wet in most of Africa as indicated by almost uniformly higher lakes across Africa; at this time there was a 150–400 mm annual increase in precipitation in the Sahara. The Late Holocene was comparatively arid in much of Africa. Thus Quaternary climatic changes on the African continent consisted of cycles of cold/arid and warm/moist phases that, in turn, led to cycles of contraction and expansion of major vegetation zones. The African fossil record suggests that the paleoenvironmental shifts toward more arid, open conditions near 2.8 mya, 1.7 mya, and 1.0 mya were coincident with major evolutionary steps in African hominids and other vertebrates. Leporids are first reported from Africa (usually as *Lepus* spp.) in the Early Pliocene of E Africa. The fossil record of the genus of *Lepus* in Africa is the most speciose, numerically abundant, and geographically widespread of the extant African leporids. *Lepus* was reported in Ethiopia (Late Miocene), Kenya (Early Pliocene), and Tanzania (Late Pliocene). For the Late Pleistocene, the Cape scrub hare (*L. saxatilis*) has been recorded from Blombos Cave and *L. victoriae* (as *L. crawshayi*) from Border Cave. However, none of these materials has been adequately described, and these assignments should be considered extremely tentative.

CURRENT DISTRIBUTION: The African savanna hare is endemic to Africa where its distribution covers most of Africa south of the Sahara and north of South Africa. It occurs in montane regions up to about 3,600 m on Mount Kenya. There is an isolated population around Beni Abbes in the Sahara Desert, in W Algeria. Recent investigation suggests occurrence of the African savanna hare in S Morocco (specimen collected near Tantan, North Moroccan Atlantic Sahara).

TAXONOMY AND GEOGRAPHIC VARIATION: The status of *L. victoriae* and its various synonyms (including *microtis*, *crawshayi*, *canopus*, *zechi*, and *whytei*) is still unclear and under debate. This confusion is due to high variation in external phenotypic characters (such as coat color,

body size, ear length) across the vast geographic range of the species. Indeed, montane forms are more rufous and darker than lowland forms. The African savanna hare has been considered conspecific with *L. saxatilis*. For the present we consider these species separately, and we consider the status of *L. microtis*, given to the African savanna hare, as *nomina dubium*. No molecular data are available.

ECOLOGY: African savanna hares are mostly solitary, but can be seen in groups of two to three in favorable areas.

The African savanna hare has an important role in plant species richness and disseminule propagation. In Uganda, 436 disseminules belonging to 22 different species were extracted from the fur of 96 hares. Most of them were grass and herbaceous disseminules, and tambookie grass (*Hyparrhenia filipendula*) and bindii (*Tribulus terrestris*), respectively, were the grass and herb species most frequently carried. African savanna hares also serve as the primary dietary item of many carnivores in the wild such as birds and snakes.

HABITAT AND DIET: The African savanna hare occupies different habitats across its large distribution area from semi-desert regions to rainforest-savanna mosaics. It prefers shrubby grasslands and montane areas.

The African savanna hare is herbivorous. The food eaten varies widely with the habitat and the season. The few studies of hares from Kenya and Uganda showed that main plant items in the diet were grasses, herbs, and shrubs. African savanna hares are also known to gnaw on exposed roots, bark, shoots, the pulp of fallen fruit, and berries, and to occasionally pluck leaves or eat fungi.

BEHAVIOR: The African savanna hare is mostly a solitary, nocturnal species. In E Africa, groups of two to three African savanna hare were seen while feeding. Incidence of hares in Uganda increased following burning or overgrazing. The home range is between 5 and 10 km².

African savanna hares have very good sight, hearing, and sense of smell. They rely more on their hearing and sense of smell than on sight. In addition, they use their ears in signaling, with different positions for different moods. They use secretions from glands, urine, and feces as transmitters of olfactory messages. They communicate vocally by producing high-pitched contact calls as well as ultrasonic noises made by rubbing their lower incisors against their peg teeth.

PHYSIOLOGY AND GENETICS: Diploid chromosome number = 48. Cytogenetic analysis with G- and C-banding of *L. victoriae* (as *L. crawshayi*) have shown that most chromosomes are characterized by small amounts of heterochromatin. The genus *Lepus* has shown to be highly conservative karyotypically, which suggests an ancestral karyotype from which all leporids may have been derived.

Like all hares of the genus *Lepus,* African savanna hares circulate their food twice; this means they produce soft fecal pellets during the night that they consume again, to obtain the remaining nutrients. They then produce dry fecal pellets during the day, which have few remaining nutrients.

REPRODUCTION AND DEVELOPMENT: During breeding, multiple males will pursue one female. Fights and chasing are common between males during breeding times. Males and females also fight as a way to stimulate sexual behavior. Young African savanna hares are born in the open, not in a nest. They can open their eyes and are mobile within 48 hours. After one month, leverets are fully weaned and independent. After eight months, they are sexually mature. The reproductive rate is high. Breeding extends throughout the year, as evidenced from two-thirds of adult females being pregnant in all months of the year. These results were observed in females examined from Botswana, Uganda, and Kenya. Six to eight litters of usually a small litter size, averaging to 1.6 young, are normally produced. The annual production of young ranges between 8–14/female.

PARASITES AND DISEASES: Ectoparasites (759 ticks belonging to 13 species) have been collected from African savanna hares from Kenya and Uganda. These tick species are agents of human or animal infections in nature. Selected examples of epidemiological importance include Nairobi sheep disease virus, Kadam virus, East Coast fever, and other *Theileria* infections. In several countries lagomorphs and ticks are implicated in the spread of disease.

CONSERVATION STATUS:

IUCN Red List Classification: Least Concern (LC)

MANAGEMENT: Little is known about conservation and management actions taken in the different countries across the distribution range of the African savanna hare. However, available data on the ecology of the species have been included in many treatments, but have been restricted to selected areas of its distribution. Moreover, much of the range of the African savanna hare overlaps with large natural nature reserves (~ 10% of Africa), which contributes to the protection of this species. More ecological research should be conducted through the distribution areas of this species, including the isolated population at Beni Abbes in W Algeria.

ACCOUNT AUTHORS: Asma Awadi and Hichem Ben Slimen

Key References: Angerman 1983; Angerman et al. 1990; Clifford et al. 1976; Flux 1983; Flux and Angerman 1990; Flux and Flux 1983; Happold 2013; Ogen-Odoi and Dilworth 1984, 1985; Petter 1959, 1972; Richard and Daniel 2013; Robinson et al. 1981; Rondinini et al. 2006; Schneider 1990; Suchentrunk and Flux 1996; Suchentrunk et al. 2007; Werdelin and Sanders 2010.

Lepus yarkandensis Günther, 1875
Yarkand Hare

OTHER COMMON NAMES: Talimu tu (Chinese)

DESCRIPTION: The Yarkand hare is a smallish hare, with short and straight pelage. The dorsal pelage is sandy brown mixed with grayish black stripes, becoming lighter during winter. The ventral pelage is entirely white. The relatively long ears do not have a black tip and are white or creamy white along the sides.

The small skull has narrow nasals, the posterior portion being flat and straight. The palatal bridge is similarly narrow. The auditory bullae are well developed, being round and tall.

SIZE: Head and body 285–430 mm; Tail 55–86 mm; Hind foot 90–110 mm; Ear 90–110 mm; Greatest length of skull 76–88 mm; Weight 1,100–1,900 g

PALEONTOLOGY: The Tarim Basin has been vulnerable throughout the Quaternary climatic oscillations, affecting its ecosystems and thus the distribution of the endemic

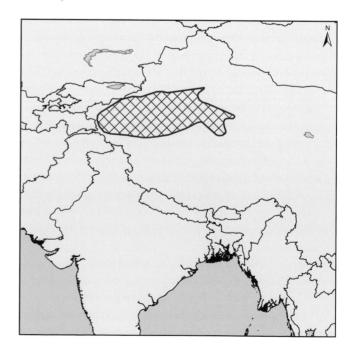

Lepus yarkandensis. Photo courtesy Weidong Li

Yarkand hare. Northern and eastern populations have shared a similar history, as have those from the western and southern regions. These latter regions apparently served as glacial refugia during the extreme oscillations in climate, whereas those in the north and east reflect three separate postglacial colonization events. These patterns are reflected in the genetic subdivisions present among populations of Yarkand hares today.

CURRENT DISTRIBUTION: The Yarkand hare occupies the interior desert Tarim Basin of NW China (Xinjiang Province). It occurs up to 4,000 m asl, but the lower limit of its distribution is not known. The range of the Yarkand hare is typically allo- to parapatric with the tolai hare (*L. tolai*), although the two species may be sympatric in some reaches bordering the Tian Shan (mountains); the Yarkand hare is sympatric over much of its range with the desert hare (*L. tibetanus*).

TAXONOMY AND GEOGRAPHIC VARIATION: No subspecies. *L. yarkandensis* exhibits a high level of genetic diversity, with significant genetic subdivision and a low level of gene flow among fragmented populations, as commonly occurs throughout its range.

HABITAT AND DIET: Yarkand hares occupy the internal drainages of the Tarim Basin, primarily in tamarisk (*Tamarix* spp.) and poplar (*Populus* spp.) forests along the edges of rivers. They are general herbivores, concentrating on graminoids and crops.

BEHAVIOR: Yarkand hare activity primarily occurs from dawn to mid-morning and late afternoon; foraging may also occur at night. When not active during the day, animals hide in shallow depressions under vegetation. Hares typically forage along routes that can extend up to 2 km.

PHYSIOLOGY AND GENETICS: Diploid chromosome number = 48. The riverine systems occupied by the Yarkand hare are widely separated, and as a result there is significant genetic subdivision among the resulting isolated populations. Continuous populations show high levels of genetic similarity, whereas isolated populations express low rates of genetic migration. In addition, *L. yarkandensis* is known to hybridize with the sympatric *L. tibetanus*, and the movement of genes between these species is apparently bidirectional.

REPRODUCTION AND DEVELOPMENT: Litters of between two and five young are born between February and July; two to three litters may be reproduced seasonally.

CONSERVATION STATUS:

IUCN Red List Classification: Near Threatened (NT)

National-level Assessments: China Red List (Near Threatened—NT)

MANAGEMENT: The Yarkand hare faces several threats: overharvesting (human consumption, pelt, and traditional Chinese medicine), infrastructure development, petroleum exploration, and issues associated with isolated populations such as genetic drift and inbreeding depression.

ACCOUNT AUTHOR: Andrew T. Smith

Key References: Flux and Angermann 1990; Jiang et al. 2016; Li et al. 2005, 2006; Shan et al. 2011; Smith and Xie 2008; Wu et al. 2010, 2011; Xu 1986.

REFERENCES

Abe, H. 1971. Small mammals of central Nepal. Journal of the Faculty of Agriculture, Hokkaido University, Sapporo 56:367–423.

Abe, H., N. Ishii, T. Itoo, Y. Kaneko, K. Maeda, S. Miura, et al. 2005. A Guide to the Mammals of Japan. Tokai University Press: Hadano, Japan. [In Japanese].

Abramov, A., R. J. Timmins, D. Touk, J. W. Duckworth, and R. Steinmetz. 2008. *Nesolagus timminsi*. The IUCN Red List of Threatened Species 2008: e.T41209A10412274. http://dx.doi.org/10.2305/IUCN.UK.2008.RLTS.T41209A10412274.en.

Abrantes, J., C. R. Carmo, C. A. Matthee, F. Yamada, W. van der Loo, and P. J. Esteves. 2011. A shared unusual genetic change at the chemokine receptor type 5 between *Oryctolagus*, *Bunolagus* and *Pentalagus*. Conservation Genetics 12:325–330.

Abrantes, J., W. van der Loo, J. Le Pendu, and P. J. Esteves. 2012. Rabbit haemorrhagic disease (RHD) and rabbit haemorrhagic disease virus (RHDV): a review. Veterinary Research 43:12.

Acevedo, F. J. 1995. Islas Marías II. México Desconocido 226:36–46.

Acevedo, P., V. Alzaga, J. Cassinello, and C. Gortazar. 2007. Habitat suitability modelling reveals a strong niche overlap between two poorly known species, the broom hare and the Pyrenean grey partridge, in the north of Spain. Acta Oecologica 31:174–184.

Acevedo, P., J. Melo-Ferreira, R. Real, and P. C. Alves. 2012a. Past, present and future distributions of an Iberian endemic, *Lepus granatensis*: ecological and evolutionary clues from species distribution models. PLoS ONE 7:e51529.

———. 2014. Evidence for niche similarities in the allopatric sister species *Lepus castroviejoi* and *Lepus corsicanus*. Journal of Biogeography 41:977–986.

Acevedo, P., A. Jiménez-Valverde, J. Melo-Ferreira, R. Real, and P. C. Alves. 2012b. Parapatric species and the implications for climate change studies: a case study on hares in Europe. Global Change Biology 18:1509–1519.

Adams, J. R., C. S. Goldberg, B. Bosworth, J. L. Rachlow, and L. P. Waits. 2011. Rapid species identification of pygmy rabbits (*Brachylagus idahoensis*) from pellet DNA. Molecular Ecology Resources 11: 808–812.

Adams, R. 1975. Watership Down. Avon Books: New York.

Aguilar, F., H. G. Rödel, J. Vázquez, L. Nicolas, L. Rodríguez-Martínez, A. Bautista, et al. 2014. Seasonal changes in testosterone levels in wild Mexican cottontails *Sylvilagus cunicularius*. Mammalian Biology 79:225–229.

Ahlgren, H., K. Norén, A. Angerbjörn, and K. Lidén. 2016. Multiple prehistoric introductions of the mountain hare (*Lepus timidus*) on a remote island, as revealed by ancient DNA. Journal of Biogeography 43:1786–1796.

Ahlmann, V., K. Collins, and U. S. Seal (editors). 2000. Riverine Rabbit (*Bunolagus monticularis*): A Population and Habitat Viability Assessment Workshop. Conservation Breeding Specialist Group: Apple Valley, Minnesota.

Alcalá-Galván, C. H., and H. Miranda-Zarazuá. 2012. Incidence of antelope jackrabbits (*Lepus alleni*) on buffelgrass pastures of central Sonora, Mexico. Report of Centro de Investigaciones Pecuarias del Estado de Sonora, A. C. Livestock Research Center for the State of Sonora, Mexico. [In Spanish].

Allen, A. W. 1985. Habitat suitability index models: swamp rabbit. U.S. Fish and Wildlife Service, Biological Report 82(10.107): Washington, D.C.

Allen, G. A., and H. J. Coolidge, Jr. 1940. Mammal and bird collections of the Asiatic primate expedition: mammals. Bulletin of the Museum of Comparative Zoology at Harvard College 87:131–166.

Allen, G. M. 1938. The Mammals of China and Mongolia. Part 1. American Museum of Natural History: New York.

———. 1939. A checklist of African mammals. Bulletin of the Museum of Comparative Zoology 83:1–763.

Allen, J. A. 1890. *Lepus alleni* Mearns: antelope jack rabbit. Bulletin of the American Museum of Natural History 2:294–297.

———. 1906. Mammals from the states of Sinaloa and Jalisco, Mexico, collected by J. H. Batty during 1904 and 1905. Bulletin of the American Museum of Natural History 22:191–262.

Allen, J. A., and R. C. Andrews. 1913. Mammals collected in Korea. Bulletin of the American Museum of Natural History 32:427–436.

Almeida, T., A. M. Lopes, M. J. Magalhães, F. Neves, A. Pinheiro, D. Gonçalves, et al. 2015. Tracking the evolution of the G1/RHDVb recombinant strains introduced from the Iberian Peninsula to the Azores islands, Portugal. Infection, Genetics and Evolution 34: 307–313.

Alroy, J. 2013. North American Fossil Mammal Systematics. Paleobiology Database. http://paleobiodb.org/cgi-bin/bridge.pl?page=OSA_3 _North_American_mammals.

Altemus, M. M. 2016. Antelope Jackrabbit (Lepus alleni) Spatial Ecology, Habitat Characteristics, and Overlap with the Endangered Pima Pineapple Cactus (Coryphantha scheeri var. robustispina). Unpublished M.S. thesis, University of Arizona, Tucson, Arizona.

Althoff, D. P., G. L. Storm, and D. R. DeWalle. 1997. Daytime habitat selection by cottontails in central Pennsylvania. Journal of Wildlife Management 61:450–459.

Álvarez, T. 1969. Restos fósiles de mamíferos de Tlapacoya, Estado de México (Pleistoceno-Reciente). Pp. 93–112. In: J. K. Jones, Jr. (editor). Contributions in Mammalogy. Miscellaneous Publications, Museum of Natural History, University of Kansas: Lawrence, Kansas.

Álvarez, T., J. Arroyo-Cabrales, and M. González-Escamilla. 1987. Mamíferos (excepto Chiroptera) de la Costa de Michoacán, México. Anales de la Escuela Nacional de Ciencias Biológicas, México 31: 13–62.

Álvarez-Castañeda, S. T., and J. L. Patton. 1999. Mamíferos del Noroeste de México I. Centro de Investigaciones Biológicas del Noroeste: La Paz, Baja California Sur, México.

———. 2000. Mamíferos del Noroeste de México II. Centro de Investigaciones Biológicas del Noroeste: La Paz, Baja California Sur, México.

Álvarez-Castañeda, S. T., P. Cortés-Calva, L. Méndez, and A. Ortega-Rubio. 2006. Development in the Sea of Cortés calls for mitigation. BioSciences 56:825–829.

Álvarez del Toro, M. 1991. Los Mamíferos de Chiapas. 2nd edition. Gobierno del Estado de Chiapas: México City, México.

Alves, P. C., and K. Hackländer. 2008. Lagomorph species: geographical distribution and conservation status. Pp. 395–406. In: P. C. Alves, N. Ferrand, and K. Hackländer (editors). Lagomorph Biology: Evolution, Ecology, and Conservation. Springer: Berlin.

Alves, P. C., and J. Melo-Ferreira. 2007. Are Lepus corsicanus and L. castroviejoi conspecific? Evidence from the analysis of nuclear markers. Pp. 45–51. In: G. De Filippo, L. De Riso, and F. Riga (editors). Conservazione di Lepus corsicanus de Winton (1898) e stato delle conoscenze. IGF Publishing: Naples, Italy.

Alves, P. C., and J. Niethammer. 2003. Die iberischen Hasen. Pp. 137–154. In: J. Niethammer and F. Krapp (editors). Handbuch der Säugetiere Europas, Band 3/II: Lagomorpha Hasentiere. Aula Verlag: Wiebelsheim, Germany.

Alves, P. C., and A. Rocha. 2003. Environmental factors have little influence on the reproductive activity of the Iberian hare (Lepus granatensis). Wildlife Research 30:639–647.

Alves, P. C., N. Ferrand, and K. Hackländer. 2008a. Lagomorph Biology: Evolution, Ecology, and Conservation. Springer: Berlin.

Alves, P. C., N. Ferrand, F. Suchentrunk, and D. J. Harris. 2003. Ancient introgression of Lepus timidus mtDNA into L. granatensis and L. europaeus in the Iberian Peninsula. Molecular Phylogenetics and Evolution 27:70–80.

Alves, P. C., H. Gonçalves, M. Santos, and A. Rocha. 2002. Reproductive biology of the Iberian hare, Lepus granatensis, in Portugal. Mammalian Biology 67:358–371.

Alves, P. C., D. J. Harris, J. Melo-Ferreira, M. Branco, F. Suchentrunk, and N. Ferrand. 2006. Hares on thin ice: introgression of mitochondrial DNA in hares and its implications for recent phylogenetic analyses. Molecular Phylogenetics and Evolution 40:640–641.

Alves, P. C., D. J. Harris, and F. Suchentrunk. 2002. Molecular data pertinent to the phylogenetic relationships between Lepus corsicanus and L. castroviejoi (Leporidae, Lagomorpha). Mammalian Biology Supplement 67:5–6.

Alves, P. C., J. Melo-Ferreira, M. Branco, F. Suchentrunk, N. Ferrand, and D. J. Harris. 2008b. Evidence for genetic similarity of two allopatric European hares (Lepus corsicanus and L. castroviejoi) inferred from nuclear DNA sequences. Molecular Phylogenetics and Evolution 46:1191–1197.

Alves, P. C., J. Melo-Ferreira, H. Freitas, and P. Boursot 2008c. The ubiquitous mountain hare mitochondria: multiple introgressive hybridization in hares, genus Lepus. Philosophical Transactions of the Royal Society B 363:2831–2839.

Alzaga, V., J. Vicente, D. Villanua, P. Acevedo, F. Casas, and C. Gortazar. 2008. Body condition and parasite intensity correlates with escape capacity in Iberian hares (Lepus granatensis). Behavioral Ecology and Sociobiology 62:769–775.

AMCELA (Mexican Association for the Conservation and Study of Lagomorphs), F. J. Romero Malpica, and H. Rangel Cordero. 2008a. Lepus californicus. The IUCN Red List of Threatened Species 2008: e.T41276A10412537. http://dx.doi.org/10.2305/IUCN.UK.2008 .RLTS.T41276A10412537.en.

———. 2008b. Lepus alleni. The IUCN Red List of Threatened Species 2008: e.T41272A10410552. http://dx.doi.org/10.2305/IUCN.UK .2008.RLTS.T41272A10410552.en.

———. 2008c. Sylvilagus insonus. The IUCN Red List of Threatened Species 2008: e.T21207A9256528. http://dx.doi.org/10.2305/IUCN.UK .2008.RLTS.T21207A9256528.en.

Anderson, H. L., and P. C. Lent. 1977. Reproduction and growth of the tundra hare (Lepus othus). Journal of Mammalogy 58:53–57.

Anderson, S., and A. S. Gaunt. 1962. A classification of the white-sided jackrabbits of Mexico. American Museum Novitiates 2088:1–16.

Angelici, F. M., and L. Luiselli. 2001. Distribution and status of the Apennine hare Lepus corsicanus in continental Italy and Sicily. Oryx 35:245–249.

———. 2007. Body size and altitude partitioning of the hares Lepus europaeus and L. corsicanus living in sympatry and allopatry in Italy. Wildlife Biology 13:251–257.

Angelici, F. M., and M. Spagnesi. 2008a. Lepus corsicanus de Winton, 1898. Pp. 254–266. In: G. Amori, L. Contoli, and A. Nappi (editors). Fauna d'Italia, Mammalia, Erinaceomorpha, Soricomorpha, Lagomorpha, Rodentia (II edizione). Calderini: Bologna, Italy.

———. 2008b. Lepus capensis Linnaeus, 1758. Pp. 248–253. In: G. Amori, L. Contoli, and A. Nappi (editors). Fauna d'Italia, Mammalia, Erinaceomorpha, Soricomorpha, Lagomorpha, Rodentia (II edizione). Calderini: Bologna, Italy.

Angelici, F. M., F. Petrozzi, and A. Galli. 2010. The Apennine hare *Lepus corsicanus* in Latium, central Italy: a habitat suitability model and comparison with its current range. Hystrix, the Italian Journal of Mammalogy 21:177–182.

Angelici, F. M., E. Randi, F. Riga, and V. Trocchi. 2008. *Lepus corsicanus*: The IUCN Red List of Threatened Species 2008: e.T41305A10436746. http://dx.doi.org/10.2305/IUCN.UK.2008 .RLTS.T41305A10436746.en.

Angerbjörn, A. 1986. Reproduction of mountain hares (*Lepus timidus*) in relation to density and physical condition. Journal of Zoology (London) 208:559–568.

———. 1989. Mountain hare populations on islands: effects of predation by red fox. Oecologia 81:335–340.

Angerbjörn, A., and J. E. C. Flux. 1995. *Lepus timidus*. Mammalian Species 495:1–11.

Angermann, R. 1967. Beiträge zur Kenntnis der Gattung *Lepus* (Lagomorpha, Leporidae): III. Zur Variabilität palaearktischer Schneehasen. Mitteilungen aus dem Zoologischen Museum in Berlin 43: 161–178.

———. 1972. Die Hasentiere. Pp. 419–465. In: B. Grzimek (editor). Grzimeks Tierleben: Säugetiere 3. Volume 12. Kindler Verlag AG: Zürich, Switzerland.

———. 1983. The taxonomy of Old World *Lepus*. Acta Zoologica Fennica 174:17–21.

Angermann, R., J. E. C. Flux, J. A. Chapman, and A. T. Smith. 1990. Lagomorph classification. Pp. 7–13. In: J. A. Chapman and J. E. C. Flux (editors). Rabbits, Hares and Pikas: Status Survey and Conservation Action Plan. International Union for Conservation of Nature: Gland, Switzerland.

Angulo, E., and J. Bárcena. 2007. Towards a unique and transmissible vaccine against myxomatosis and rabbit haemorrhagic disease for rabbit populations. Wildlife Research 34:567–577.

Aniśkowicz, B. T., H. Hamilton, D. R. Gray, and C. Downes. 1990. Nursing behaviour of Arctic hares (*Lepus arcticus*). Pp. 643–664. In: C. R. Harington (editor). Canada's Missing Dimension: Science and History in the Canadian Arctic islands. Canadian Museum of Nature: Ottawa, Ontario, Canada.

Anthony, H. E. 1941. Mammals collected by the Vernay-Cutting Burma expedition. Field Museum of Natural History (Zoology Series) 27: 37–123.

Arias-Del Razo, I., L. Hernández, J. W. Laundré, and O. Myers. 2011. Do predator and prey foraging activity patterns match? A study of coyotes (*Canis latrans*) and lagomorphs (*Lepus californicus* and *Sylvilagus audubonii*). Journal of Arid Environments 75:112–118.

Armstrong, D. M. 1972. Distribution of mammals in Colorado. University of Kansas Museum Natural History Monograph 3:1–415.

Armstrong, D. M., and J. K. Jones, Jr. 1971. Mammals from the Mexican state of Sinaloa. 1. Marsupialia, Insectivora, Edentata, Lagomorpha. Journal of Mammalogy 52:747–757.

Arnold, J. F. 1942. Forage consumption and preferences of experimentally fed Arizona and antelope jack rabbits. University of Arizona College of Agriculture, Agricultural Experimental Station Technical Bulletin 9:50–86.

Arthur, A. D., R. P. Pech, C. Davey, Jiebu, Y. Zhang, and L. Hui. 2008. Livestock grazing, plateau pikas and the conservation of avian biodiversity on the Tibetan plateau. Biological Conservation 141: 1972–1981.

Aryal, A., D. Brunton, W. Ji, H. K. Yadav, B. Adhikari, and D. Raubenheimer. 2012. Diet and habitat use of hispid hare *Caprolagus hispidus* in Shuklaphanta Wildlife Reserve, Nepal. Mammal Study 37: 147–154.

Asher, R. J., N. Bennett, and T. Lehmann. 2009. The new framework for understanding placental mammal evolution. BioEssays 31: 853–864.

Asher, R. J., J. Meng, J. R. Wible, M. C. McKenna, G. W. Rougier, D. Dashzeveg, and M. J. Novacek. 2005. Stem Lagomorpha and the antiquity of Glires. Science 307:1091–1094.

Audubon, J. J. 2005. Audubon's Mammals: The Quadrupeds of North America. Wellfleet Press: Secaucus, New Jersey.

Averianov, A. O. 1994a. Early Eocene mimotonids of Kyrgyzstan and the problem of Mixodontia. Acta Paleontologica Polonica 39:393–411.

Averianov, A. O. 1994b. Morphologic variation of the skull and teeth in East-Asian species of the genus *Lepus* (Lagomorpha, Leporidae). Zoologicheskii Zhurnal 73:132–139.

Averianov, A. O. 1999. Phylogeny and classification of Leporidae (Mammalia, Lagomorpha). Vestnik Zoologii 33:41–48.

Averianov, A. O., and A. V. Lopatin. 2005. Eocene lagomorphs (Mammalia) of Asia: 1. Aktashmys (Strenulagidae fam. nov.). Paleontological Journal 39:308–317.

Averianov, A. O., A. V. Abramov, and A. Tikhonov. 2000. A new species of *Nesolagus* (Lagomorpha, Leporidae) from Vietnam with osteological description. Contributions from the Zoological Institute, St. Petersburg 3:1–22.

Averianov, A. O., J. Niethammer, and M. Pegel. 2003. *Lepus europaeus* Pallas, 1778—Feldhase. Pp. 35–104. In: J. Niethammer and F. Krapp (editors). Handbuch der Säugetiere Europas, Band 3/II: Lagomorpha Hasentiere. Aula Verlag: Wiebelsheim, Germany.

Avise, J. C. 2004. Molecular Markers, Natural History, and Evolution. 2nd edition. Sinauer Associates: Sunderland, Massachusetts.

Azzaroli-Puccetti, M. L. 1987a. The systematic relationships of hares (genus *Lepus*) of the horn of Africa. Cimbebasia (A) 9:1–22.

———. 1987b. On the hares of Ethiopia and Somalia and the systematic position of *Lepus whytei* Thomas, 1894 (Mammalia, Lagomorpha). Accademia Nazionale dei Lincei: Rome, Italy.

Azzaroli-Puccetti, M. L., M. Corti, A. Scanzani, M. V. Civitelli, and E. Capanna. 1996. Karyotypes of two endemic species of hares from Ethiopia, *Lepus habessinicus* and *L. starcki* (Lagomorpha, Leporidae). A comparison with *L. europaeus*. Mammalia 60:223–230.

Badingqiuying, A. T. Smith, J. Senko, and M. U. Siladan. 2016. Plateau pika *Ochotona curzoniae* poisoning campaign reduces carnivore abundance in southern Qinghai, China. Mammal Research 41:1–8.

Bagchi, S., T. Nagail, and M. E. Ritchie. 2006. Small mammalian herbivores as mediators of plant community dynamics in the high-altitude arid rangelands of Trans-Himalaya. Biological Conservation 127:438–442.

Bailey, V. 1936. The mammals and life zones of Oregon. No. 55. U.S. Fish and Wildlife Service: Portland, Oregon.

Baker, A. J., J. L. Eger, R. L. Peterson, and T. H. Manning. 1983. Geographic variation and taxonomy of Arctic hares. Acta Zoologica Fennica 174:45–48.

Ballesteros, F. 2003. Liebre de piornal, *Lepus castroviejoi* (Palacios, 1976). Galemys 15:3–13.

Ballesteros, F. 2007. *Lepus castroviejoi* (Palacios 1977). Pp. 479–483. In: L. J. Palomo, J. Gisbert, and J. C. Blanco (editors). Atlas y Libro Rojo de los Mamíferos Terrestres de España. Ministerio de Medio Ambiente: Madrid, Spain.

Ballesteros, F., and B. Palacios. 2009. Situación y Conservación de la Liebre de Piornal *Lepus castroviejoi* en la Cordillera Cantábrica. Serie técnica Naturaleza y Parques Nacionales. Ministerio de Medio Ambiente y Medio Rural y Marino: Madrid, Spain.

Barbour, M. S., and J. A. Litvaitis. 1993. Niche dimensions of New England cottontails in relation to habitat patch size. Oecologia 95: 321–327.

Barrio, I. C., and D. S. Hik. 2013. Good neighbors? Determinants of aggregation and segregation among alpine herbivores. Ecoscience 20:276–282.

Barry, R., and J. Lazell. 2008. *Sylvilagus obscurus*. The IUCN Red List of Threatened Species 2008: e.T41301A10434606. http://dx.doi.org/10.2305/IUCN.UK.2008.RLTS.T41301A10434606.en.

Barta, R. M., L. B. Keith, and S. M. Fitzgerald. 1989. Demography of sympatric arctic and snowshoe hare populations: an experimental assessment of interspecific competition. Canadian Journal of Zoology 67:2762–2775.

Batsaikhan, N., R. Samiya, S. Shar, and S. R. B. King. 2010. A Field Guide to the Mammals of Mongolia. Zoological Society of London: London.

Becker, P. A., and S. M. DeMay. 2016. Re-introduction of the Columbia Basin pygmy rabbit in central Washington, USA. Pp. 195–199. In: P. S. Soorae (editor). Global Re-introduction Perspectives: 2016. Case Studies from around the Globe. IUCN/SSC Re-Introduction Specialist Group: Gland, Switzerland.

Bednarz, J., and J. Cook. 1984. Distribution and numbers of the white-sided jackrabbit (*Lepus callotis gaillardi*) in New Mexico. The Southwestern Naturalist 29:358–360.

Bee, J. W., and E. R. Hall. 1956. Mammals of northern Alaska on the Arctic Slope. University of Kansas Museum of Natural History Miscellaneous Publications 8:309.

Beever, E. A., P. F. Brussard, and J. Berger. 2003. Patterns of apparent extirpation among isolated populations of pikas (*Ochotona princeps*) in the Great Basin. Journal of Mammalogy 84:37–54.

Beever, E. A., C. Ray, J. L. Wilkening, P. F. Brussard, and P. W. Mote. 2011. Contemporary climate change alters the pace and drivers of extinction. Global Change Biology 17:2054–2070.

Bell, D. J. 1986. A study of the hispid hare *Caprolagus hispidus* in Royal Suklaphanta Wildlife reserve, Western Nepal: a summary report. Dodo 23:24–31.

Bell, D. J., J. Hoth, A. Velázquez, F. J. Romero, L. León, and M. Aranda. 1985. A survey of the distribution of the volcano rabbit (*Romerolagus diazi*): an endangered Mexican endemic. Dodo (Jersey Wildlife Preservation Trust) 22:42–48.

Bell, D. J., W. L. R. Oliver, and R. K. Ghose. 1990. The hispid hare *Caprolagus hispidus*. Pp. 128–136. In: J. A. Chapman and J. E. C. Flux (editors). Rabbits, Hares and Pikas: Status Survey and Conservation Action Plan. International Union for Conservation of Nature: Gland, Switzerland.

Bello-Sánchez, R. A. 2010. Distribución y Abundancia de la Liebre Torda *Lepus callotis* (Wagler, 1830) en el Valle de Perote, Veracruz. Unpublished Ph.D. dissertation, Universidad Veracruzana, Veracruz, Mexico.

Ben Slimen, H., F. Suchentrunk, and A. Ben Ammar Elgaaied. 2008a. On shortcomings of using mtDNA sequence divergence for the systematics of hares (genus *Lepus*): an example from cape hares. Mammalian Biology 73:25–32.

Ben Slimen, H., F. Suchentrunk, A. Memmi, and A. Ben Ammar Elgaaied. 2005. Biochemical genetic relationships among Tunisian hares (*Lepus* sp.), South African cape hares (*L. capensis*), and European brown hares (*L. europaeus*). Biochemical Genetics 43:577–596.

Ben Slimen, H., F. Suchentrunk, A. Memmi, H. Sert, U. Kryger, P. C. Alves, et al. 2006. Evolutionary relationships among hares from North Africa (*Lepus* sp. or *Lepus* spp.), cape hares (*L. capensis*) from South Africa, and brown hares (*L. europaeus*), as inferred from mtDNA PCR-RFLP and allozyme data. Journal of Zoological Systematics and Evolutionary Research 44:88–99.

Ben Slimen, H., F. Suchentrunk, C. Stamatis, Z. Mamuris, H. Sert, P. C. Alves, et al. 2008b. Population genetics of cape and brown hares (*Lepus capensis* and *L. europaeus*): a test of Petter's hypothesis of conspecificity. Biochemical Systematics and Ecology 36:22–39.

Bennike, O., A. K. Higgins, and M. Kelly. 1989. Mammals of central North Greenland. Polar Record 25:43–49.

Benton, M. J., and P. C. J. Donoghue. 2007. Paleontological evidence to date the Tree of Life. Molecular Biology and Evolution 24:26–53.

Bergengren, A. 1969. On genetics, evolution and history of distribution of the heath-hare, a distinct population of the Arctic hare, *Lepus timidus* Lin. Viltrevy 6:381–460.

Berger, J. 2008. Undetected species losses, food webs, and ecological baselines: a cautionary tale from the Greater Yellowstone Ecosystem, USA. Oryx 42:139–142.

Berger, J., K. M. Berger, P. F. Brussard, R. Gibson, J. Rachlow, and A. T. Smith. 2005. Where Have All the Rabbits Gone? Wildlife Conservation Society: New York.

Bergerud, A. T. 1967. The distribution and abundance of arctic hares in Newfoundland. The Canadian Field-Naturalist 81:242–248.

Bergin, I. L., A. G. Wise, S. R. Bolin, T. P. Mullaney, M. Kiupel, and R. K. Maes. 2009. Novel Calicivirus identified in rabbits, Michigan, USA. Emerging Infectious Diseases 15:1955–1962.

Berkman, L. K., C. K. Nielsen, C. L. Roy, and E. J. Heist. 2015. Comparative genetic structure of sympatric leporids in southern Illinois. Journal of Mammalogy 96:552–563.

Berkman, L. K., M. J. Saltzgiver, E. J. Heist, C. K. Nielsen, C. L. Roy, and P. D. Scharine. 2009. Hybridization and polymorphic microsatellite markers for two lagomorph species (genus *Sylvilagus*): implications for conservation. Conservation Genetics Resources 1:419–424.

Bernstein, A. D. 1963. Data on the ecology of red pika (*Ochotona rutila* Sev.) in Zailisk Alatau (Tian Shan). Way of life and feeding. Byulleten' Moskovskovo Obshchestva Ispytatelei Prirodym, Otdel Biologicheski 68:24–36. [In Russian].

———. 1970. The ecology of the big-eared pika in the Tien Shan. Fauna Ecology Rodents (Moscow) 9:62–109. [In Russian].

Bertolino, S., A. Perrone, L. Gola, and R. Viterbi. 2011. Population density and habitat use of the introduced eastern cottontail (*Sylvilagus floridanus*) compared to the native European hare (*Lepus europaeus*). Zoological Studies 50:315–326.

Best, T. L. 1996. *Lepus californicus*. Mammalian Species 530:1–10.

———. 1999. Alaskan hare / *Lepus othus*. Pp. 702–704. In: D. E. Wilson and S. Ruff (editors). The Smithsonian Book of North American Mammals. Smithsonian Institution Press: Washington, D.C.

Best, T. L., and T. H. Henry. 1993. *Lepus alleni*. Mammalian Species 424:1–8.

———. 1994. *Lepus othus*. Mammalian Species 458:1–5.

Bhattacharyya, S. 2013. Habitat Ecology of Royle's Pika (*Ochotona roylei*) along an Altitudinal Gradient with Special Reference to Foraging Behaviour in Western Himalaya. Unpublished Ph.D. dissertation, Department of Wildlife Sciences, Saurashtra University, India.

Bhattacharyya, S., B. S. Adhikari, and G. S. Rawat. 2009. Abundance of Royle's pika (*Ochotona roylei*) along an altitudinal gradient in Uttarakhand, Western Himalaya. Hystrix, the Italian Journal of Mammalogy 20:111–119.

———. 2013. Forage selection by Royle's pika (*Ochotona roylei*) in the western Himalaya, India. Zoology 116:300–306.

———. 2014a. Influence of snow, food, and rock cover on Royle's pika abundance in western Himalaya. Arctic, Antarctic, and Alpine Research 46:558–567.

———. 2014b. Influence of microclimate on the activity of Royle's pika in the western Himalaya, India. Zoological Studies 53:73.

Bhattacharyya, S., S. Dutta, B. S. Adhikari, and G. S. Rawat. 2015. Presence of a small mammalian prey species in open habitat is dependent on refuge availability. Mammal Research 60:293–300.

Biju-Duval, C., H. Ennafaa, N. Dennebouy, M. Monnerot, F. Mignotte, R. C. Soriguer, et al. 1991. Mitochondrial DNA evolution in lagomorphs: origin of systematic heteroplasmy and organization of diversity in European rabbits. Journal of Molecular Evolution 33:92–102.

Bittner, S. L., and O. J. Rongstad. 1982. Snowshoe hare and allies: *Lepus americanus* and allies. Pp. 146–163. In: J. A. Chapman and G. A. Feldhamer (editors). Wild Mammals of North America: Biology, Management, and Economics. Johns Hopkins University Press: Baltimore, Maryland.

Black, D. M., K. V. K. Gilardi, L. P. Hamilton, E. Williams, D. F. Williams, P. A. Kelly, et al. 2009. Hematologic and biochemistry reference values for the endangered riparian brush rabbit (*Sylvilagus bachmani riparius*). Journal of Wildlife Diseases 45:491–496.

Blair, W. F. 1936. The Florida marsh rabbit. Journal of Mammalogy 17:197–207.

Blanford, W. T. 1879. Scientific Results of the Second Yarkand Mission; Based upon the Collections and Notes of the Late Ferdinand Stolickza, Ph.D. Mammalia. Office of the Superintendent of Government Printing: Calcutta, India.

———. 1888. The Fauna of British India, Including Ceylon and Burma: Mammalia. Taylor and Francis: London.

Bloomer, S. E. M., T. Willebrand, I. M. Keith, and L. B. Keith. 1995. Impact of helminth parasitism on a snowshoe hare population in central Wisconsin: a field experiment. Canadian Journal of Zoology 73:1891–1898.

Bock, C. E., Z. F. Jones, and J. H. Bock. 2006. Abundance of cottontails (*Sylvilagus*) in an exurbanizing southwestern savanna. The Southwestern Naturalist 51:352–357.

Bogan, M. A, and C. Jones. 1975. Observations on *Lepus callotis* in New Mexico. Proceedings of the Biological Society of Washington 88:45–50.

Boitani, L., F. Corsi, A. De Biase, I. D'Inzillo Caranza, M. Ravagli, G. Reggiani, et al. 1999. A Databank for the Conservation and Management of the African Mammals. Istituto di Ecologia Applicata: Rome, Italy.

Boland, K. M., and J. A. Litvaitis. 2008. Role of predation and hunting on eastern cottontail mortality at Cape Cod National Seashore. Canadian Journal of Zoology 86:918–927.

Bomford, M. 2003. Risk Assessment for the Import and Keeping of Exotic Vertebrates in Australia. Bureau of Rural Sciences: Canberra, Australia.

Bond, B. T., L. W. Burger, B. D. Leopold, and K. D. Godwin. 2001. Survival of cottontail rabbits (*Sylvilagus floridanus*) in Mississippi and the examination of latitudinal variation. The American Midland Naturalist 145:127–136.

Bond, B. T., L. W. Burger, D. B. Leopold, J. C. Jones, and K. D. Godwin. 2002. Habitat use by cottontail rabbits across multiple spatial scales in Mississippi. Journal of Wildlife Management 66:1171–1178.

Bonhote, J. L. 1904. On the mouse-hares of the genus *Ochotona*. Proceedings of the Zoological Society of London 2:205–220.

Bonino, N., and R. Gader. 1989. Expansión del conejo silvestre Europeo (*Oryctolagus cuniculus*) en la Argentina y perspectivas futuras. Annales de Museo de Historia Natural de Valparaíso 18:157–162.

Bonino, N., and R. C. Soriguer. 2004. Distribución actual y dispersión del conejo europeo (*Oryctolagus cuniculus*) en Mendoza (Argentina). Mastozoologia Neotropical 11:237–241.

———. 2009. The invasion of Argentina by the European wild rabbit. Mammal Review 39:159–166.

Bonino, N., D. Cossios, and J. Menegheti. 2010. Dispersal of the European hare, *Lepus europaeus*, in South America. Folia Zoologica 59:9–15.

Bonvicino, C. R., A. N. Menezes, A. Lazar, V. Penna Firme, C. Bueno, M. C. Viana, P. S. D'Andrea, and A. Langguth. 2015. Chromosomes and phylogeography of *Sylvilagus* (Mammalia, Leporidae) from eastern Brazil. Oecologia Australis 19:158–172.

Borisova, N. G., L. V. Rudneva, and A. I. Starkov. 2008. Interpopulation variability of vocalizations in the Daurian pika (*Ochotona daurica*). Zoologichesky Zhurnal 87:850–861.

Boutin, S. A. 1984. The effect of conspecifics on juvenile survival and recruitment of snowshoe hares. Journal of Animal Ecology 53:623–637.

Boutin, S., B. S. Gilbert, C. J. Krebs, A. R. E. Sinclair, and J. N. M. Smith. 1985. The role of dispersal in the population dynamics of snowshoe hares. Canadian Journal of Zoology 63:106–115.

Bowen, R. E., K. J. McMahon, and R. W. Mitchell. 1960. Infectious diseases in a black-tailed jackrabbit (*Lepus californicus melanotis*, Mearns) population in southwestern Kansas. Transactions of the Kansas Academy of Sciences 63:276–284.

Boyce, K. A. 2001. Distribution, Seasonal Home Range, Movements and Habitat of the Appalachian Cottontail, *Sylvilagus obscurus*, at Dolly Sods, West Virginia. Unpublished M.S. thesis, Frostburg State University, Frostburg, Maryland.

Boyce, K. A., and R. E. Barry. 2007. Seasonal home range and diurnal movements of *Sylvilagus obscurus* (Appalachian cottontail) at Dolly Sods, West Virginia. Northeastern Naturalist 14:99–110.

Branco, M., M. Monnerot, N. Ferrand, and A. R Templeton. 2002. Postglacial dispersal of the European rabbit (*Oryctolagus cuniculus*) on the Iberian Peninsula reconstructed from nested clade and mismatch analyses of mitochondrial DNA genetic variation. Evolution 56:792–803.

Brandt, J. F. 1855. Beiträge zur nahern Kenntniss der Säugethiere Russlands. Mémoires Académie Impériale des Sciences de Saint Pétersbourg, Series 6, Volume 9, Part 5: Untersuchungen über die craniologischen Entwickelungsstufen und die davon herzuleitenden Verwandtschaften und Classificationen der Nager der Jetztwelt, mit besonderer Beziehung auf die Gattung Castor. Pp. 125–365 + 11 pl. Buchdruckerei der Kaiserlichen Akademie der Wissenschaften.

Braun, C. E., and R. G. Streeter. 1968. Observations on the occurrence of white-tailed jackrabbits in the alpine zone. Journal of Mammalogy 49:160–161.

Bray, D. B., and A. Velázquez. 2009. From displacement-based conservation to place-based conservation. Conservation and Society 7:11–14.

Broadbrooks, H. E. 1965. Ecology and distribution of the pikas of Washington and Alaska. The American Midland Naturalist 73:299–335.

Broekhuizen, S., and F. Maaskamp. 1982. Movement, home range and clustering in the European hare (*Lepus europaeus* Pallas) in The Netherlands. Zeitschrift für Säugetierkunde 47:22–32.

Bronner, G. N., M. Hoffmann, P. J. Taylor, C. T. Chimimba, P. B. Best, C. A. Matthee, and T. J. Robinson. 2003. A revised systematic checklist of the extant mammals of the southern African subregion. Durban Museum Novitates 28:56–95.

Brown, A. L., and J. A. Litvaitis. 1995. Habitat features associated with predation of New England cottontails: what scale is appropriate? Canadian Journal of Zoology 73:1005–1011.

Brown, C. F., and P. R. Krausman. 2003. Habitat characteristics of 3 leporid species in southeastern Arizona. Journal of Wildlife Management 67:83–89.

Brown, D. E., R. D. Babb, C. Lorenzo, and M. M. Altemus. 2014. Ecology of the antelope jackrabbit (*Lepus alleni*). The Southwestern Naturalist 59:575–587.

Brown, L. 1940. The distribution of the white-tailed jack rabbit (*Lepus townsendii campanius* Hollister) in Kansas. Transactions of the Kansas Academy of Science 43:385–389.

———. 1947. Why has the white-tailed jackrabbit (*Lepus townsendii campanius*) become scarce in Kansas? Transactions of the Kansas Academy of Science 49:455–456.

Bryant, J. P., R. K. Swihart, P. B. Reichardt, and L. Newton. 1994. Biogeography of woody plant chemical defense against snowshoe hare browsing: comparison of Alaska and eastern North America. Oikos 70:385–395.

Bunch, M., S. G. Platt, and S. Miller. 2012. Swamp rabbit. Supplemental volume: species of conservation concern, South Carolina Department of Natural Resources. Retrieved June 2014 from http://www.dnr.sc.gov/swap/supplemental/mammals/swamprabbit2015.pdf.

Burgess, E. C., and L. A. Windberg. 1989. *Borrelia* sp. infection in coyotes, black-tailed jack rabbits and desert cottontails in southern Texas. Journal of Wildlife Diseases 25:47–51.

Burke, J. J. 1941. New fossil Leporidae from Mongolia. American Museum Novitates 1117:1–23.

Burnett, W. L. 1926. Jack rabbits of eastern Colorado. Technical Report of Office of State Entomologist, Circular 52:1–18.

Burton, C., and C. J. Krebs. 2003. Influence of relatedness on snowshoe hare spacing behavior. Journal of Mammalogy 84:1100–1111.

Burton, C., C. J. Krebs, and E. B. Taylor. 2002. Population genetic structure of the cyclic snowshoe hare (*Lepus americanus*) in southwestern Yukon, Canada. Molecular Ecology 11:1689–1701.

Cabrera-García, L., J. A. Velázquez, and M. E. E. Weinmann. 2006. Identification of priority habitats for conservation of the Sierra Madre sparrow *Xenospiza baileyi* in Mexico. Oryx 40:211–217.

Cai, G. Q., and Z. J. Feng. 1982. A systematic revision of the subspecies of highland hare (*Lepus oiostolus*), including two new subspecies. Acta Theriologica Sinica 2:167–182. [In Chinese].

Caire, W. 1978. A Distribution and Zoogeography of the Mammals of Sonora, Mexico. Unpublished Ph.D. dissertation, University of New Mexico, Albuquerque, New Mexico.

Caldwell, L. D. 1966. Marsh rabbit development and ectoparasites. Journal of Mammalogy 47:527–528.

Callou, C. 2003. De la Garenne au Clapier: Etude Archéozoologique du Lapin en Europe Occidentale. Mémoires du Muséum National d'Histoire Naturelle 189:1–358. [In French].

Calvete, C. 2006. Modeling the effect of population dynamics on the impact of rabbit hemorrhagic disease. Conservation Biology 20:1232–1241.

Camarda, A., N. Pugliese, P. Cavadini, E. Circella, L. Capucci, A. Caroli, et al. 2014. Detection of the new emerging rabbit haemorrhagic disease type 2 virus (RHDV2) in Sicily from rabbit (*Oryctolagus cuniculus*) and Italian hare (*Lepus corsicanus*). Research in Veterinary Science 97:642–645.

Cameron, A. W. 1958. Mammals of the islands in the Gulf of St. Lawrence. National Museum of Canada Bulletin 154, Biological Series 53:1–165.

Camp, M. J., J. L. Rachlow, L. A. Shipley, T. R. Johnson, and K. D. Bockting. 2014. Grazing in sagebrush rangelands in western North America: implications for habitat quality for a sagebrush specialist, the pygmy rabbit. The Rangeland Journal 36:151–159.

Can, D. N., A. V. Abramov, A. N. Tikhonov, and A. O. Averianov. 2001. Annamite striped rabbit *Nesolagus timminsi* in Vietnam. Acta Theriologica 46:437–440.

Capanna, E., M. Bonomo, M. V. Civitelli, A. Simonetta, and E. Capanna. 1991. The chromosomes of Royle's pika, *Ochotona roylei*, (Mammalia, Lagomorpha). Rendiconti Lincei 2:59–67.

Capucci, L., F. Fallacara, S. Grazioli, A. Lavazza, M. L. Pacciarini, and E. Brocchi. 1998. A further step in the evolution of rabbit hemorrhagic disease virus: the appearance of the first consistent antigenic variant. Virus Research 58:115–126.

Capucci, L., P. Fusi, A. Lavazza, M. L. Pacciarini, and C. Rossi. 1996. Detection and preliminary characterization of a new rabbit calicivirus related to rabbit haemorrhagic disease virus but non pathogenic. Journal of Virology 70:8614–8623.

Caravaggi, A., W. I. Montgomery, and N. Reid. 2016. Range expansion and comparative habitat use of insular, congeneric lagomorphs: invasive European hares *Lepus europaeus* and endemic Irish hares *Lepus timidus hibernicus*. Biological Invasions 18:1217–1218.

Caravaggi, A., W. I. Montgomery, and N. Reid. 2015. Range expansion and comparative habitat use of insular, congeneric lagomorphs: invasive European hares *Lepus europaeus* and endemic Irish hares *Lepus timidus hibernicus*. Biological Invasions 17:687–698.

Carneiro, M., F. W. Albert, S. Afonso, R. J. Pereira, H. Burbano, R. Campos, et al. 2014a. The genomic architecture of population divergence between subspecies of the European rabbit. PLoS Genetics 10:e1003519.

Carneiro, M., R. Carl-Johan, F. Di Palma, F. W. Albert, J. Alföldi, A. Martinez Barrio, et al. 2014b. Rabbit genome analysis reveals a polygenic basis for phenotypic change during domestication. Science 345:1074–1079.

Carnell, S. 2010. Hare. Reaktion Books: London.

Carr, A. F., Jr. 1939. Notes on escape behavior in the Florida marsh rabbit. Journal of Mammalogy 20:322–325.

Carrillo-Reyes, A. 2009. Selección de Hábitat, Ámbito Hogareño, Sitios de Alimentación y Sitios de Descanso de la Liebre de Tehuantepec (*Lepus flavigularis*) en Santa María del Mar, Oaxaca. Unpublished

Ph.D. dissertation, El Colegio de la Frontera Sur San Cristóbal de Las Casas, Chiapas, México.

Carrillo-Reyes, A., C. Lorenzo, E. J. Naranjo, M. Pando, and T. Rioja. 2010. Home range dynamics of the Tehuantepec Jackrabbit in Oaxaca, Mexico. Revista Mexicana de Biodiversidad 81:143–151.

Carro, F., and R. C. Soriguer. 2010. La Liebre Ibérica. Organismo Autónomo de Parques Nacionales: Madrid, Spain.

Carroll, R. C. 1988. Vertebrate Paleontology and Evolution. Freeman and Company: New York.

Carter, D., and M. Merkens. 1994. Updated status report on the Nuttalls' cottontail, Sylvilagus nuttallii and Sylvilagus nuttallii pinetis, in Canada. Committee on the Status of Endangered Wildlife in Canada: Ottawa, Ontario, Canada.

Carter, F. L. 1939. A study in jackrabbit shifts in range in western Kansas. Transactions of the Kansas Academy of Science 42:431–435.

Cary, J. R., and L. B. Keith. 1979. Reproductive change in the 10-year cycle of snowshoe hares. Canadian Journal of Zoology 57:375–390.

Casas-Andreu, G. 1992. Anfibios y reptiles de las islas Marías y otras islas adyacentes de la costa de Nayarit, México: aspectos sobre su biogeografía y conservación. Anales Instituto de Biología de la Universidad Nacional Autónoma de México Serie Zoología 63: 95–112.

Casc, T. J., and M. L. Cody. 1983. Island Biogeography in the Sea of Cortéz. University of California Press: Berkeley, California.

Ceballos, G., and D. Navarro. 1991. Diversity and conservation of Mexican mammals. Pp. 167–198. In: M. A. Mares and D. J. Schmidly (editors). Latin American Mammalogy: History, Biodiversity and Conservation. University of Oklahoma Press: Norman, Oklahoma.

Čermák, S., J. Obuch, and P. Benda. 2006. Notes on the genus Ochotona in the Middle East (Lagomorpha: Ochotonidae). Lynx 37:51–66.

Černý, V., and H. Hoogstraal. 1977. Haemaphysalis (Allophysalis) danieli, sp. n. (Ixodoidea: Ixodidae), female and tentatively associated immature stages from high mountains of Northern Pakistan and Afghanistan. Journal of Parasitology 63:567–574.

Cervantes, F. A. 1993. Lepus flavigularis. Mammalian Species 423:1–3.

———. 1997. Sylvilagus graysoni. Mammalian Species 559:1–3.

Cervantes, F. A., and M. Castañeda. 2012. Efecto genético del aislamiento geográfico de la liebre negra (Lepus insularis), endémica de Isla Espíritu Santo, Baja California Sur, México. Therya 3:151–170.

Cervantes, F. A., and C. Lorenzo. 1997a. Sylvilagus insonus. Mammalian Species 568:1–4.

———. 1997b. Morphometric differentiation of rabbits (Romerolagus and Sylvilagus) and jackrabbits (Lepus) of Mexico. Game and Wildlife / Gibier Faune Sauvage France 14:405–425.

Cervantes, F. A., and J. Vázquez. 2008. Conejos silvestres (Sylvilagus: Leporidae) del Municipio de Ixtacuixtla, Tlaxcala, México. Pp. 89–102. In: C. Lorenzo, E. Espinoza, and J. Ortega (editors). Avances en el Estudio de los Mamíferos de México II. Asociación Mexicana de Mastozoología, A. C.: México.

Cervantes, F. A., S. T. Álvarez-Castañeda, B. Villa, C. Lorenzo, and J. Vargas. 1996. Natural history of the black jackrabbit (Lepus insularis) from Espiritu Santo Island, Baja California Sur, Mexico. The Southwestern Naturalist 41:186–189.

Cervantes, F. A., C. Lorenzo, S. T. Álvarez-Castañeda, A. Rojas-V., and J. Vargas. 1996. Chromosomal study of the insular San José brush

rabbit (Sylvilagus mansuetus) from México. The Southwestern Naturalist 41:455–457.

Cervantes, F. A., C. Lorenzo, and F. X. González-Cózatl. 2004. The Omiltemi rabbit (Sylvilagus insonus) is not extinct. Mammalian Biology 69:61–64.

Cervantes, F. A., C. Lorenzo, and R. S. Hoffmann. 1990. Romerolagus diazi. Mammalian Species 360:1–7.

Cervantes, F. A., C. Lorenzo, and J. Vargas. Familia Leporidae. 1999. Pp. 199–23. In: S. T. Álvarez-Castañeda and J. Patton (editors). Mamíferos del Noroeste de México I. Centro de Investigaciones Biológicas del Noroeste: La Paz, Baja California Sur, México.

Cervantes, F. A., C. Lorenzo, J. Vargas, and T. Holmes. 1992. Sylvilagus cunicularius. Mammalian Species 412:1–4.

Cervantes, F. A., A. Rojas-Viloria, C. Lorenzo, and S. T. Álvarez-Castañeda. 1999–2000. Chromosomal differentiation between the jackrabbits Lepus insularis and Lepus californicus from Baja California Sur, Mexico. Revista Mexicana de Mastozoología 4:41–53.

Chakraborty, S., T. P. Bhattacharyya, and C. Srinivasulu. 2005. Lepus oiostolus Hodgson, 1840. P. 618. In: S. Molur, C. Srinivasulu, B. Srinivasulu, S. Walker, P. O. Nameer and L. Ravikumar (editors). Status of South Asian Non-volant Small Mammals: Conservation Assessment and Management Plan Workshop Report. CAMP: Coimbatore, India.

Chapman, J. A. 1974. Sylvilagus bachmani. Mammalian Species 34:1–4.

———. 1975. Sylvilagus transitionalis. Mammalian Species 55:1–4.

Chapman, J. A., and G. Ceballos. 1990. The cottontails. Pp. 95–110. In: J. A. Chapman and J. E. C. Flux (editors). Rabbits, Hares and Pikas: Status Survey and Conservation Action Plan. International Union for Conservation of Nature: Gland, Switzerland.

Chapman, J. A., and G. A. Feldhamer. 1981. Sylvilagus aquaticus. Mammalian Species 151:1–4.

Chapman, J. A., and J. E. C. Flux. 2008. Introduction to the Lagomorpha. Pp. 1–9. In: P. C. Alves, N. Ferrand, and K. Hackländer (editors). Lagomorph Biology: Evolution, Ecology, and Conservation. Springer: Berlin.

Chapman, J. A., and J. A. Litvaitis. 2003. Eastern cottontail: Sylvilagus floridanus and allies. Pp. 101–125. In: G. A. Feldhamer, B. C. Thompson, and J. A. Chapman (editors). Wild Mammals of North America: Biology, Management, and Conservation. 2nd edition. Johns Hopkins University Press: Baltimore, Maryland.

Chapman, J. A., and R. P. Morgan, II. 1973. Systematic status of the cottontail complex in western Maryland and nearby West Virginia. Wildlife Monographs 36:1–54.

Chapman, J. A., and J. R. Stauffer. 1981. The status and distribution of the New England cottontail. Pp. 973–983. In: K. Meyers and C. D. MacInnes (editors). Proceedings of the World Lagomorph Conference. University of Guelph: Guelph, Ontario, Canada.

Chapman, J. A., and G. A. Willner. 1978. Sylvilagus audubonii. Mammalian Species 106:1–4.

———. 1981. Sylvilagus palustris. Mammalian Species 153:1–3.

Chapman, J. A., K. L. Cramer, N. J. Dippenaar, and T. J. Robinson. 1992. Systematics and biogeography of the New England cottontail, Sylvilagus transitionalis (Bangs, 1895), with the description of a new species from the Appalachian Mountains. Proceedings of the Biological Society of Washington 105:841–866.

Chapman, J. A., J. E. C. Flux, A. T. Smith, D. J. Bell, G. G. Ceballos, K. R. Dixon, et al. 1990. Conservation action needed for rabbits, hares and pikas. Pp.154–168. In: J. A. Chapman and J. E. C. Flux (editors). Rabbits, Hares and Pikas, Status Survey and Conservation Action Plan. International Union for Conservation of Nature: Gland, Switzerland.

Chapman, J. A., A. L. Harman, and D. E. Samuel. 1977. Reproductive and physiological cycles in the cottontail complex in western Maryland and nearby West Virginia. Wildlife Monographs 56:1–73.

Chapman, J. A., J. G. Hockman, and W. R. Edwards. 1982. Cottontails. Pp. 83–123. In: J. A. Chapman and G. A. Feldhamer (editors). Wild Mammals of North America. Johns Hopkins University Press: Baltimore, Maryland.

Chapman, J. A., J. G. Hockman, and M. M. Ojeda. 1980. *Sylvilagus floridanus*. Mammalian Species 136:1–8.

Chen, J. C., Z. Y. Yao, and C. H. Liao. 1982. Report on the damage caused by rodent pest in the grasslands of south Gansu, China. Chinese Journal of Zoology 3:25. [In Chinese].

Cheng, C., D. Y. Ge, L. Xia, C. Q. Zhou, and Q. S. Yang. 2012. Morphometrics study on the so-called "Cape hare" (Lagomorpha: Leporidae: *Lepus*) in China. Acta Theriologica Sinica 32:275–286.

Cheng, E., K. E. Hodges, J. Melo-Ferreira, P. C. Alves, and L. S. Mills. 2014. Conservation implications of the evolutionary history and genetic diversity hotspots of the snowshoe hare. Molecular Ecology 23:2929–2942.

Chiari, M., N. Ferrari, D. Giardiello, D. Avisani, M. Zanoni, L. G. Alborali, et al. 2014. Temporal dynamics of European brown hare syndrome infection in Northern Italian brown hares (*Lepus europaeus*). European Journal of Wildlife Research 60:891–896.

Christiansen, H. H., O. Bennike, J. Böcher, B. Eberling, O. Humlum, and B. H. Jakobsen. 2002. Holocene environmental reconstruction from deltaic deposits in northeast Greenland. Journal of Quaternary Science 17:145–160.

Ci, H. X., G. H. Lin, Z. Y. Cai, L. Z. Tang, J. P. Su, and J. Q. Liu. 2009. Population history of the plateau pika endemic to the Qinghai-Tibetan plateau based on mtDNA sequence data. Journal of Zoology 279:396–403.

Clanton, C. W., and M. L. Johnson. 1954. White tailed jack rabbit in Washington. The Murrelet 35:15.

Clemons, C., L. G. Rickard, J. E. Kierans, and R. G. Botzler. 2000. Evaluation of host preferences by helminths and ectoparasites among black-tailed jackrabbits in northern California. Journal of Wildlife Diseases 36:555–558.

Clifford, C., J. E. C. Flux, and H. Hoogstraal. 1976. Seasonal and regional abundance of ticks (Ixodidae) on hares (Leporidae) in Kenya. Journal of Medical Entomology 13:40–47.

Coetzee, K. 1994. The Riverine Rabbit (*Bunolagus monticularis*) and its Habitat: Conservation Implications of an Unnaturally Fragmented Distribution. M. Tech. Diploma, Port Elizabeth Technicon, Port Elizabeth, South Africa.

Collins, K. 2001. Habitat Availability for the Riverine Rabbit, *Bunolagus monticularis*. Unpublished M.S. thesis, University of Pretoria, Pretoria, South Africa.

Collins, K. 2005. *Bunolagus monticularis*: riverine rabbit. Pp. 75–76. In: J. D. Skinner and C. T. Chimimba (editors). The Mammals of the Southern African Subregion. Cambridge University Press: Cambridge.

Collins, K., and J. T. du Toit. 2016. Population status and distribution modelling of the critically endangered riverine rabbit (*Bunolagus monticularis*). African Journal of Ecology 54:195–206.

Collins, K., V. Ahlman, C. Matthee, P. J. Taylor, M. Keith, G. Palmer, et al. 2004. *Bunolagus monticularis*. Pp. 412–413. In: Y. Friedman and B. Daly (editors). Red Data Book of the Mammals of South Africa: A Conservation Assessment. CBSG Southern Africa, Conservation Breeding Specialist Group (SCC/IUCN). Endangered Wildlife Trust: South Africa.

Conaway, C. H., T. S. Baskett, and J. E. Toll. 1960. Embryo resorption in the swamp rabbit. Journal of Wildlife Management 24:197–202.

Conner, D. A. 1985. Analysis of the vocal repertoire of adult pikas: ecological and evolutionary perspectives. Animal Behavior 33: 124–134.

Cook, J. A. 1986. The mammals of the Animas Mountains and adjacent areas, Hidalgo County, New Mexico. Occasional Papers Museum Southwest Biology 4:1–5.

Cooke, B. D. 2012a. Modelling climate constraints on the biology and distribution of the European rabbit. 4th World Lagomorph Conference, Vienna, Austria, July 23–27, 2012.

——. 2012b. Rabbits: manageable environmental pests or participants in new Australian ecosystems? Wildlife Research 39:279–289.

——. 2014. Daily food intake of free-ranging wild rabbits in inland South Australia. Wildlife Research 41:141–148.

Corbet, G. B. 1983. A review of classification in the family Leporidae. Acta Zoologica Fennica 174:11–15.

Cordell, L. S. 1977. Late Anasazi farming and hunting strategies: one example of a problem in congruence. American Antiquity 42: 449–461.

COSEWIC (Committee on the Status of Endangered Wildlife in Canada). 2011. COSEWIC Assessment and Status Report on the Collared Pika *Ochotona collaris* in Canada. Ottawa, Ontario, Canada.

Cossios, D. 2004. La liebre Europea, *Lepus europaeus*, especie invasora en el sur del Perú. Revista Peruana de Biología 11:209–212.

Coyne J. A., and H. A. Orr. 2004. Speciation. Sinauer Associates: Sunderland, Massachusetts.

Crawford, J. A., R. G. Anthony, J. T. Forbes, and G. A. Lorton. 2010. Survival and causes of mortality for pygmy rabbits (*Brachylagus idahoensis*) in Oregon and Nevada. Journal of Mammalogy 91:838–847.

Crawford, J. C. 2014. Ecology of the Swamp Rabbit and Eastern Cottontail in Bottomland Hardwood Forests in Southern Illinois. Unpublished Ph.D. dissertation, Southern Illinois University, Carbondale, Illinois.

Crouse, A. L., R. L. Honeycutt, R. A. McCleery, C. A. Faulhaber, N. D. Perry, and R. R. Lopez. 2009. Population structure of the Lower Keys marsh rabbit as determined by mitochondrial DNA analysis. Journal of Wildlife Management 73:362–367.

Dalquest, W. W. 1948. Mammals of Washington. Washington Agricultural Experiment Station Technical Bulletin 62:1–131.

——. 1953. Mammals of the Mexican State of San Luis Potosí. Louisiana State University Studies Biological Science Series 1, Louisiana State University Press: Baton Rouge, Louisiana.

——. 1961. *Sylvilagus cunicularius* in the Pleistocene of Mexico. Journal of Mammalogy 25:370–403.

Dalquest, W. W., and F. B. Stangl, Jr. 1984. Late Pleistocene and early Recent mammals from Fowlkes Cave, southern Culberson County, Texas. Pp. 432–455. In: H. H. Genoways and M. R. Dawson (editors). Contributions in Quaternary Vertebrate Paleontology: A Volume in Memorial to John E. Guilday. Carnegie Museum of Natural History, Special Publication 8:1–538.

Dalton, K. P., I. Nicieza, A. Balseiro, M. A. Muguerza, J. M. Rosell,

R. Casais, et al. 2012. Variant rabbit hemorrhagic disease virus in young rabbits, Spain. Emerging Infectious Diseases 18:2009–2012.

Dantas-Torres, F., G. Testini, P. M. DiGeronimo, V. Lorusso, E. Mallia, and D. Otranto. 2011. Ticks infesting the endangered Italian hare (*Lepus corsicanus*) and their habitat in an ecological park in southern Italy. Experimental and Applied Acarology 53:95–102.

Davis, W. B. 1944. Notes on Mexican mammals. Journal of Mammalogy 25:370–403.

Davis, W. B., and P. W. Lukens, Jr. 1958. Mammals of the Mexican state of Guerrero, exclusive of Chiroptera and Rodentia. Journal of Mammalogy 39:347–367.

Davis, W. B., and R. J. Russell. 1954. Mammals of the Mexican state of Morelos. Journal of Mammalogy 35:63–80.

Dawson, M. R. 1981. Evolution of the modern lagomorphs. Pp. 1–8. In: K. Myers and C. D. MacInnes (editors). Proceedings of the World Lagomorph Conference. University of Guelph: Guelph, Ontario, Canada.

Dawson, T. J., and K. Schmidt-Nielsen. 1966. Effect of thermal conductance on water economy in the antelope jackrabbit, *Lepus alleni*. Journal of Cell Physiology 67:463–472.

Daxner, G., and O. Fejfar. 1967. Uber die Gattungen *Alilepus* Dice, 1931 und *Pliopentalagus* Gureev, 1964 (Lagomorpha, Mammalia). Annalen des Naturhistorischen Museums in Wien 71:37–55.

Dayrat, B. 2005. Towards integrative taxonomy. Biological Journal of the Linnean Society 85:407–415.

Dearing, M. D. 1997a. The manipulations of plant toxins by a food-hoarding herbivore, *Ochotona princeps*. Ecology 78:774–781.

———. 1997b. The function of haypiles of pikas (*Ochotona princeps*). Journal of Mammalogy 78:1156–1163.

De Battisti, R., S. Mangiafico, S. Migliore, L. Masutti, V. Trocchi, and A. M. De Marinis. 2004. The diet of the Italian hare *Lepus corsicanus* on Etna Mountain, Sicily. Poster, 2nd World Lagomorph Conference. Research Center in Biodiversity and Genetic Resources: Vairao, Portugal.

Debelica, A., and M. L. Thies. 2009. Atlas and key to the hair of terrestrial Texas mammals. Special Publications, The Museum, Texas Tech University 55:1–102.

Delgadillo-Quezada, G. 2011. Distribución, Selección de Hábitat y Densidad de la Liebre Torda (*Lepus callotis*, Wagler, 1830) en el Valle de Perote. Unpublished Ph.D. dissertation, Instituto de Ecología, A. C. Xalapa, México.

Delibes, R., and M. Delibes-Mateos. 2015. Linking historical ecology and invasion biology: some lessons from European rabbit introductions into the new world before the 19th century. Biological Invasions 17:2505–2515.

Delibes-Mateos, M., M. Delibes, P. Ferreras, and R. Villafuerte. 2008. Key role of European rabbits in the conservation of the Western Mediterranean Basin Hotspot. Conservation Biology 17:559–574.

Delibes-Mateos, M., C. Ferreira, C. Rouco, R. Villafuerte, and I. C. Barrio. 2014. Conservationists, hunters and farmers: the European rabbit *Oryctolagus cuniculus* management conflict in the Iberian Peninsula. Mammal Review 44:190–203.

Delibes-Mateos, M., P. Ferreras, and R. Villafuerte. 2009. European rabbit population trends and associated factors: a review of the situation in the Iberian Peninsula. Mammal Review 39:124–140.

Delibes-Mateos, M., S. E. Redpath, E. Angulo, P. Ferreras, and R. Villafuerte. 2007. Rabbits as a keystone species in southern Europe. Biological Conservation 137:149–156.

Delibes-Mateos, M., A. T. Smith, C. N. Slobodchikoff, and J. E. Swenson. 2011. The paradox of keystone species persecuted as pests: a call for the conservation of abundant small mammals in their native range. Biological Conservation 144:1335–1346.

De Marinis, A., V. Trocchi, and S. Mangiafico. 2007a. First data on reproductive biology of Italian hare *Lepus corsicanus*. Hystrix, the Italian Journal of Mammalogy Supp.:78.

De Marinis, A., V. Trocchi, S. Mangiafico, and C. E. Mallia Fassò. 2007b. Reproductive strategies of three species of hare in Italy. Pp. 75–82. In: G. de Philip, D. L. Rice, F. Row, V. Trocchi, and S. R. Troisi (editors). 2007. Conservation of Corsican Hare de Winton, 1898: The State of Knowledge. IGF Publishing: Naples, Italy. [In Italian].

DeMay, S. M., P. A. Becker, L. P. Waits, T. R. Johnson, and J. L. Rachlow. 2016. Consequences for conservation: population density and genetic effects on reproduction of an endangered lagomorph. Ecological Applications 26:784–795.

Denisman, L. G., N. K. Kiseleva, and A. V. Kniazev. 1989. The History of Ecosystems of the Mongolian People's Republic. Biological Resources and Natural Conditions of the Mongolian People's Republic. Volume 32. Nauka: Moscow-Leningrad. [In Russian].

De Queiroz, K. 2007. Species concepts and species delimitation. Systematic Biology 56:879–886.

Desmond, M. J. 2004. Habitat associations and co-occurrence of Chihuahuan Desert hares (*Lepus californicus* and *L. callotis*). The American Midland Naturalist 151:414–419.

DeVos, A. 1964. Range changes of mammals in the Great Lakes Region. The American Midland Naturalist 71:210–231.

Dice, L. R. 1929. The phylogeny of the Leporidae, with description of a new genus. Journal of Mammalogy 10:340–344.

Dickenson, V. 2014. Rabbit. Reaktion Books: London.

Diersing, V. E. 1981. Systematic status of *Sylvilagus brasiliensis* and *S. insonus* from North America. Journal of Mammalogy 62:539–556.

Diersing, V. E., and D. E. Wilson. 1980. Distribution and systematics of the rabbits (*Sylvilagus*) of west-central Mexico. Smithsonian Contributions in Zoology 297:1–34.

Dinets, V. 2010. Observation of Sumatran striped rabbit (*Nesolagus netscheri*) in the wild. Mammalia 74:1.

Dingerkus, S. K., and W. I. Montgomery. 2002. A review of the status and decline in abundance of the Irish Hare (*Lepus timidus hibernicus*) in Northern Ireland. Mammal Review 32:1–11.

Dixon, J. E. W. 1975. A note on the burrows used by the hare *Lepus capensis*. Madoqua 9:45–46.

Dixon, K. R., J. A. Chapman, G. R. Willner, D. E. Wilson, and W. López-Forment. 1983. The New World jackrabbits and hares (genus *Lepus*)—2: numerical taxonomic analysis. Acta Zoológica Fennica 174:53–56.

Dobson, F. S., A. T. Smith, and X. G. Wang. 1998. Social and ecological influences on dispersal and philopatry in the plateau pika (*Ochotona curzoniae*). Behavioral Ecology 9:622–635.

———. 2000. The mating system and gene dynamics of plateau pikas. Behavioural Processes 51:101–110.

Donoho, H. S. 1971. Dispersion and dispersal of white-tailed and black-tailed jackrabbits, Pawnee National Grasslands. US/IBP Grassland Biome Technical Report No. 96.

Dooley, M. 1988. Abstract of the results for the Oxford University expedition to the Islas Marias, Mexico, 1987. Lagomorph Newsletter 7: 4–5.

Drew, C., D. O'Donovan, G. Simkins, M. Al Dosary, A. M. Al Khaldi,

O. B. Mohammed, et al. 2008. *Lepus capensis*. The IUCN Red List of Threatened Species 2008: e.T41277A10429185. http://dx.doi.org/10.2305/IUCN.UK.2008.RLTS.T41277A10429185.en.

Duarte, J. 2000. Liebre Iberica, *Lepus granatensis* Rosenhauer, 1856. Galemys 12:3–14.

Duckworth, J. W. 1996. Bird and mammal records from the Sangthong district, Vientiane municipality, Laos in 1996. Natural History Bulletin of the Siam Society 44:217–242.

Duckworth, J. W., R. Steinmetz, and A. Pattanavibool. 2008. *Lepus peguensis*. The IUCN Red List of Threatened Species 2008: e.T41284A10433206. http://dx.doi.org/10.2305/IUCN.UK.2008.RLTS.T41284A10433206.en.

Duckworth, J. W., R. J. Timmins, R. C. M. Thewlis, T. D. Evans, and G. Q. A. Anderson. 1994. Field observations of mammals in Laos, 1992–1993. Natural History Bulletin of the Siam Society 42:177–205.

Durant, P. 1980. Estudio ecológico del conejo Silvestre, *Sylvilagus brasiliensis meridensis* (Lagomorpha: Leporidae) en los páramos de los Andes Venezolanos. Caribbean Journal of Science 19:21–29.

Durant, P., and M. A. Guevara. 2000a. Habitat of a Venezuelan lowland rabbit, *Sylvilagus varynaensis* (Lagomorpha: Leporidae). Revista de Ecología Latino-Americana 7:1–10.

———. 2000b. Reproduction and productivity in *Sylvilagus varynaensis*, a lowland Venezuelan rabbit. Zoocriaderos 3:1–10.

———. 2001. A new rabbit species (*Sylvilagus*, Mammalia: Leporidae) from the lowlands of Venezuela. Revista de Biologia Tropical 49:369–381.

Duranthon, V., N. Beaujean, M. Brunner, K. E. Odening, A. N. Santos, I. Kacskovics, et al. 2012. On the emerging role of rabbit as human disease model and the instrumental role of novel transgenic tools. Transgenic Research 21:699–713.

Durette-Desset, M. C., K. E. Galbreath, and E. P. Hoberg. 2010. Discovery of new *Ohbayashinema* spp. (Nematoda: Heligmosomoidea) in *Ochotona princeps* and *Ochotona cansus* (Lagomorpha: Ochotonidae) from western North America and Central Asia, with considerations of historical biogeography. Journal of Parasitology 96:569–579.

Durrant, S. D. 1952. Mammals of Utah: Taxonomy and Distribution. University of Kansas Press: Lawrence, Kansas.

Duthie, A. G. 1989. The Ecology of the Riverine Rabbit, *Bunolagus monticularis*. Unpublished M.S. thesis, University of Pretoria, Pretoria, South Africa.

Duthie, A. G. 1997. Order Lagomorpha. Pp. 114–119. In: G. Mills and L. Hes (editors). The Complete Book of Southern African Mammals. Struik: Cape Town, South Africa.

Duthie, A. G., and T. J. Robinson. 1990. The African rabbits. Pp. 121–127. In: J. A. Chapman and J. E. C. Flux (editors). Rabbits, Hares and Pikas: Status Survey and Conservation Action Plan. International Union for Conservation of Nature: Gland, Switzerland.

Duthie, A. G., J. D. Skinner, and T. J. Robinson. 1989. The distribution and status of the riverine rabbit, *Bunolagus monticularis*, in South Africa. Biological Conservation 47:195–202.

Ecke, H. 1955. The reproductive cycle of the Mearns cottontail in Illinois. The American Midland Naturalist 53:294–311.

Eisenberg, J. F. 1989. Mammals of the Neotropics; the Northern Neotropics, Volume 1: Panama, Colombia, Venezuela, Guyana, Suriname, French Guyana. University of Chicago Press: Chicago, Illinois.

Eisenberg, J. F., and M. Lockhart. 1972. An ecological reconnaissance of Wilpattu National Park, Ceylon. Smithsonian Contributions to Zoology 101:1–128.

Eisenberg, J. F., P. O. Nameer, and A. J. T. Johnsingh. 2015. Little known mammals. Pp. 653–693. In: A. J. T. Johnsingh and N. Manjrekar (editors). Mammals of South Asia. Volume 2. Universities Press: Hyderabad, India.

Elbroch, M. 2006. Animal Skulls: A Guide to North American Species. Stackpole Books: Mechanicsburg, Pennsylvania.

Eldrige, D. J., and R. Simpson. 2002. Rabbit (*Oryctolagus cuniculus* L.) impacts on vegetation and soils, and implications for management of wooded rangelands. Basic and Applied Ecology 3:19–29.

Elias, B. A., L. A. Shipley, S. McCusker, R. D. Sayler, and T. R. Johnson. 2013. Effects of genetic management on reproduction, growth, and survival in captive endangered pygmy rabbits (*Brachylagus idahoensis*). Journal of Mammalogy 94:1282–1292.

Elias, B. A., L. A. Shipley, R. D. Sayler, and R. S. Lamson. 2006. Mating and parental care in captive pygmy rabbits. Journal of Mammalogy 87:921–928.

Ellerman, J. R., and T. C. S. Morrison-Scott. 1951. Checklist of Palaearctic and Indian Mammals 1758 to 1946. British Museum (Natural History): London.

Ellerman, J. R., T. C. S. Morrison-Scott, and R. W. Hayman. 1953. Southern African Mammals 1758 to 1951: A Reclassification. British Museum (Natural History): London.

Erb, L. P., C. Ray, and R. Guralnick. 2014. Determinants of pika population density vs. occupancy in the southern Rocky Mountains. Ecological Applications 24:429–435.

Erbajeva, M., L. J. Flynn, and N. Alexeeva. 2015. Late Cenozoic Asian Ochotonidae: taxonomic diversity, chronological distribution and biostratigraphy. Quaternary International 355:18–23.

Erbajeva, M. A. 1986. Lagomorpha, Mammalia: Cenozoic Lagomorpha: a systematic review. Rodents and lagomorphs of the Pozdov Cenozoic. Proceedings of the Zoological Institute, Academy of Sciences of the USSR 156:157–165. [In Russian].

Erbajeva, M. A., and Y. Ma. 2006. A new look at the taxonomic status of *Ochotona argentata* Howell, 1928. Acta Zoologica Cracoviensia 49A:135–149.

Erbajeva, M. A., and S. Zheng. 2005. New data on Late Miocene–Pleistocene ochotonids (Ochotonidae, Lagomorpha) from North China. Acta Zoologica Cracoviensia 48:93–117.

Erbajeva, M. A., J. I. Mead, N. V. Alexeeva, C. Angelone, and S. L. Swift. 2011. Taxonomic diversity of Late Cenozoic Asian and North American ochotonids. Palaeontologica Electronica 14:1–9.

Estes-Zumpf, W. A., and J. L. Rachlow. 2009. Natal dispersal by the pygmy rabbit (*Brachylagus idahoensis*). Journal of Mammalogy 90:363–372.

Estes-Zumpf, W. A., J. L. Rachlow, L. P. Waits, and K. I. Warheit. 2010. Dispersal, gene flow, and population genetic structure in the pygmy rabbit (*Brachylagus idahoensis*). Journal of Mammalogy 91:208–219.

Esteves, P. J., J. Abrantes, S. Bertagnoli, P. Cavadini, D. Gavier-Widen, J. S. Guitton, et al. 2015. Emergence of pathogenicity in lagoviruses: evolution from pre-existing nonpathogenic strains or through a species jump? PLoS Pathogens 11:e1005087.

Esteves, P. J., A. M. Lopes, M. J. Magalhães, A. Pinheiro, D. Gonçalves, and J. Abrantes. 2014. Rabbit hemorrhagic disease virus detected in Pico, Azores, Portugal, revealed a unique endemic strain with more than 17 years of independent evolution. Viruses 6:2698–2707.

Estonba, A., A. Solís, M. Iriondo, M. J. Sanz-Martín, G. Perez-Suarez,

G. Harkov, et al. 2006. The genetic distinctiveness of the three Iberian hare species: *Lepus europaeus, L. granatensis*, and *L. castroviejoi*. Mammalian Biology-Zeitschrift für Säugetierkunde 71:52–59.

Evans, G. E., and D. Thomson. 1974. The Leaping Hare. Country Book Club: Newton Abbot, United Kingdom.

Evans, T. D., J. W. Duckworth, and R. J. Timmins. 2000. Field observations of larger mammals in Laos, 1994–1995. Mammalia 64:55–100.

EWT (Endangered Wildlife Trust). 2014. Drylands conservation programme [Riverine Rabbit project]. Retrieved July 2015 from http://www.ewt.org.za/DCP/dcp.html.

Fa, J. E., and D. J. Bell. 1990. The volcano rabbit *Romerolagus diazi*. Pp. 43–146. In: J. A. Chapman and J. E. C. Flux (editors). Rabbits, Hares and Pikas: Status Survey and Conservation Action Plan. International Union for Conservation of Nature: Gland, Switzerland.

Fa, J. E., J. R. Stewart, L. Lloveras, and J. Mario Vargas. 2013. Rabbits and hominin survival in Iberia. Journal of Human Evolution 64:233–241.

Fagerstone, K. A., G. K. Lavoie, and R. E. Griffith, Jr. 1980. Black-tailed jackrabbit diet and density on rangeland and near agricultural crops. Journal of Range Management 33:229–233.

Fan, N. C., W. Y. Zhou, W. H. Wei, Q. Y. Wang, and Y. J. Jiang. 1999. Rodent pest management in the Qinghai-Tibet alpine meadow ecosystem. Pp. 285–304. In: G. Singleton, L. Hinds, H. Liers, and Z. B. Zhang (editors). Ecologically-Based Rodent Management. Australian Centre for International Agricultural Research: Canberra, Australia.

Farfán, M. A., J. Duarte, J. M. Vargas, and J. E. Fa. 2012. Effects of human induced land use changes on the distribution of the Iberian hare. Journal of Zoology 286:258–265.

Farfán, M. A., J. M. Vargas, R. Real, J. Palomo, and J. Duarte. 2004. Population parameters and reproductive biology of the Iberian hare *Lepus granatensis* in southern Iberia. Acta Theriologica 49:319–335.

Farías, V. 2004. Spatio-Temporal Ecology and Habitat Selection of the Critically Endangered Tropical Hare (*Lepus flavigularis*) in Oaxaca, Mexico. Unpublished Ph.D. dissertation, University of Massachusetts, Amherst, Massachusetts.

Farías, V., T. K. Fuller, F. A. Cervantes, and C. Lorenzo. 2006. Home range and social behavior of the endangered Tehuantepec jackrabbit (*Lepus flavigularis*) in Oaxaca, Mexico. Journal of Mammalogy 87:748–756.

Faulhaber, C. A., N. D. Perry, N. J. Silvy, R. R. Lopez, P. A. Frank, P. T. Hughes, et al. 2007. Updated distribution of the Lower Keys marsh rabbit. Journal of Wildlife Management 71:208–212.

Faulhaber, C. A., N. J. Silvy, R. R. Lopez, D. H. Lafever, P. A. Frank, and M. J. Peterson. 2008. Diurnal habitat use by Lower Keys marsh rabbits. Journal of Wildlife Management 72:1161–1167.

Feilden, H. W. 1877. On the mammalia of North Greenland and Grinnell land. The Zoologist 1:353–361.

Fenderson, L. E., A. I. Kovach, J. A. Litvaitis, and M. K. Litvaitis. 2011. Population genetic structure and history of fragmented remnant populations of the New England cottontail (*Sylvilagus transitionalis*). Conservation Genetics 12:943–958.

Fenderson, L. E., A. I. Kovach, J. A. Litvaitis, K. O'Brien, K. Boland, and W. Jakubas. 2014. Genetic structure and connectivity of the New England cottontail (*Sylvilagus transitionalis*) in a highly-fragmented landscape. Ecology and Evolution 4:1853–1875.

Feng, Z. J. 1973. A new species of *Ochotona* (Ochotonidae, Mammalia) from Mount Jolmo-Iungma area. Acta Zoologica Sinica 19:69–75.

Feng, Z. J., and Y. T. Kao. 1974. Taxonomic notes on the Tibetan pika and allied species—including a new subspecies. Acta Zoologica Sinica 20:76–88. [In Chinese].

Feng, Z. J., and C. L. Zheng. 1985. Studies on the pikas (genus *Ochotona*) of China taxonomic notes and distribution. Acta Theriologica Sinica 5:269–289.

Feng, Z. J., G. Q. Cai, and C. L. Zheng. 1986. The Mammals of Xizang. Science Press, Academia Sinica: Beijing, PRC. [In Chinese].

Fenner, F., and B. Fantini. 1999. Biological Control of Vertebrate Pests. The History of Myxomatosis—an Experiment in Evolution. CAB International: New York.

Fenner, F., and F. N. Ratcliffe. 1965. Myxomatosis. Cambridge University Press: London.

Ferrand, N. 2008. Inferring the evolutionary history of the European rabbit (*Oryctolagus cuniculus*) from molecular markers. Pp. 47–63. In: P. C. Alves, N. Ferrand, and K. Hackländer (editors). Lagomorph Biology: Evolution, Ecology, and Conservation. Springer: Berlin.

Ferreira, C., and P. C. Alves. 2009. Influence of habitat management on the abundance and diet of wild rabbit (*Oryctolagus cuniculus algirus*) populations in Mediterranean ecosystems. Wildlife Research 55:487–496.

Ferreira, C., F. Castro, V. Piorno, I. Barrio, M. Delibes-Mateos, C. Rouco, et al. 2015. Biometrics reveal major differences between the two European rabbit subspecies. Biological Journal of the Linnean Society 116:106–116.

Ferreira, C., J. Touza, C. Rouco, F. Díaz-Ruiz, J. Fernández de Simón, C. A. Ríos-Saldaña, et al. 2014. Habitat management as a generalized tool to boost European rabbit *Oryctolagus cuniculus* populations in the Iberian Peninsula: a cost-effectiveness analysis. Mammal Review 44:30–43.

Ferrusquía-Villafranca, I., J. Arroyo-Cabrales, E. Martínez-Hernández, J. Gama-Castro, J. Ruiz-González, O. J. Polaco, et al. 2010. Pleistocene mammals of Mexico: a critical review of regional chronofaunas, climate change response and biogeographic provinciality. Quaternary International 217:53–104.

Fischer, G. 1817. Adversaria zoologica. Mémoires de la Société Impériale des Naturalistes de Moscou 5:357–428.

Fisher, C. T., and D. W. Yalden. 2004. The steppe pika *Ochotona pusilla* in Britain, and a new northerly record. Mammal Review 34:320–324.

Fisher, J. L. 2012. Shifting prehistoric abundances of leporids at Five Finger Ridge, a central Utah archaeological site. Western North American Naturalist 72:60–68.

Fitzgerald, S. M., and L. B. Keith. 1990. Intra- and inter-specific dominance relationships among arctic and snowshoe hares. Canadian Journal of Zoology 68:457–464.

Flinders, J. T., and J. A. Chapman. 2003. Black-tailed jackrabbit (*Lepus californicus* and allies). Pp. 126–146. In: G. A. Feldhamer, B. C. Thompson, and J. A. Chapman (editors). Wild Mammals of North America. Johns Hopkins University Press: Baltimore, Maryland.

Flux, J. E. C. 1967. Reproduction and body weights of the hare *Lepus europaeus* Pallas, in New Zealand. New Zealand Journal of Science 10:357–401.

———. 1969. Current work on the reproduction of the African hare, *Lepus capensis* L., in Kenya. Journal of Reproduction and Fertility Supplement 6:225–227.

Flux, J. E. C. 1970. Life history of the mountain hare (*Lepus timidus scoticus*) in north-east Scotland. Journal of Zoology (London) 161:75–123.

———. 1981a. Field observations of behaviour in the genus *Lepus*. Pp. 377–394. In: K. Myers and C. D. MacInnes (editors). Proceedings of the World Lagomorph Conference. University of Guelph: Guelph, Ontario, Canada.

———. 1981b. Reproductive strategies in the genus *Lepus*. Pp. 155–174. In: K. Myers and C. D. MacInnes (editors). Proceedings of the World Lagomorph Conference. University of Guelph: Guelph, Ontario, Canada.

———. 1983. Introduction to the taxonomic problems in hares. Acta Zoologica Fennica 174:7–10.

———. 1987. Moult, condition and body weight in mountain hares (*Lepus timidus*). Journal of Zoology (London) 212:365–367.

———. 1990a. Brown hare. Pp. 161–172. In: C. M. King (editor). The Handbook of New Zealand Mammals. Oxford University Press: Oxford.

———. 1990b. The Sumatran rabbit *Nesolagus netscheri*. Pp. 137–139. In: J. A. Chapman and J. E. C. Flux (editors). Rabbits, Hares and Pikas: Status Survey and Conservation Action Plan. International Union for Conservation of Nature: Gland, Switzerland.

———. 1994. World distribution. Pp. 8–17. In: H. V. Thompson and C. M. King (editors). The European Rabbit: The History and Biology of a Successful Colonizer. Oxford University Press: Oxford.

Flux, J. E. C., and R. Angermann. 1990. The hares and jackrabbits. Pp. 61–94. In: J. A. Chapman and J. E. C. Flux (editors). Rabbits, Hares and Pikas: Status Survey and Conservation Action Plan. International Union for Conservation of Nature: Gland, Switzerland.

Flux, J. E. C., and M. M. Flux. 1983. Taxonomy and distribution of East African hares. Acta Zoologica Fennica 174:41–43.

Flux, J. E. C., and P. Fullagar. 1992. World distribution of the rabbit *Oryctolagus cuniculus* on islands. Mammal Review 22:151–205.

Flux, J. E. C., and J. U. M. Jarvis. 1970. Growth rates of two African hares, *Lepus capensis*. Journal of Mammalogy 51:798–799.

Fontanesi, L., F. Di Palma, P. Flicek, A. T. Smith, C.-G. Thulin, P. C. Alves, and Lagomorph Genomics Consortium. 2016. LaGomiCs-Lagomorph Genomics Consortium: an international collaborative effort for sequencing the genomes of an entire mammalian order. Journal of Heredity 107:295–308.

Formozov, N. A. 1997. Pikas (*Ochotona*) of the world: systematics and conservation. Giber Faune Sauvage (Game and Wildlife) 14: 506–507.

Formozov, N. A., and I. Yu. Baklushinskaya. 1999. The species status of a new pika (*Ochotona hoffmanni* Formozov et al., 1996) and its inclusion in the list of Russian fauna. Byulleten' Moskovskovo Obshchestva Ispytalelei Prirodym, Otdel Biologicheskii 104:68–72. [In Russian].

———. 2011. Manchurian pika (*Ochotona mantchurica scorodumovi*) from the interfluve of the Shilka and Argun rivers: karyotype and problem of pika taxonomy in Amurland and adjacent territories. Zoologicheskii Zhurnal 90:490–497. [In Russian].

Formozov, N. A., and N. S. Proskurina. 1980. Spatial structure of the settlements, and elements of the behavior of closely related forms of pikas. Pp. 293–294. In: Rodents. Materials of the Fifth All-Union Meeting. Nauka: Moscow-Leningrad. [In Russian].

Formozov, N. A., and E. L. Yakhontov. 2003. Sympatry zone of alpine (*Ochotona alpina*) and northern (*O. hyperborea*) pikas on the Putorana plateau with description of new subspecies (*Ochotona hyperborea naumovi* spp. n.). Zoologicheskii Zhurnal 82:485–496.

Formozov, N. A., I. Yu. Baklushinskaya, and Y. Ma. 2004. Taxonomic status of the Helan-shan pika, *Ochotona argentata*, from the Helan-shan ridge (Ningxia, China). Zoologicheskii Zhurnal 83:995–1007. [In Russian].

Formozov, N. A., T. V. Grigoreva, and V. L. Surin. 2006. Molecular systematics of pikas of the subgenus *Pika* (*Ochotona*, Lagomorpha). Zoologicheskii Zhurnal 85:1465–1473. [In Russian].

Formozov, N. A., A. A. Lissovsky, and I. Yu. Baklushinskaya. 1999. Karyological diagnostics of pikas (*Ochotona*, Lagomorpha) from the Putorana Plateau (Northern Central Siberia). Zoologicheskii Zhurnal 78: 606–612. [In Russian].

Formozov, N. A., E. L. Yakhontov, and P. P. Dmitriev. 1996. New form of alpine pika (*Ochotona alpina hoffmanni* ssp. n.) from Hentiyn Nuruu ridge (Mongolia) and probable natural history of this species. Byulleten' Moskovskovo Obshchestva Ispytatelei Prirodym, Otdel Biologicheskii 101:28–36. [In Russian].

Foronda, P. R., E. O. Figueruelo, A. R. Ortega, N. A. Abreu, and J. C. Casanova. 2005. Parasites (viruses, coccidia and helminths) of the wild rabbit (*Oryctolagus cuniculus*) introduced to Canary Islands from Iberian Peninsula. Acta Parasitologica 50:80–84.

Forys, E. A. 1995. Metapopulations of Marsh Rabbits: A Population Viability Analysis for the Lower Keys Marsh Rabbit (*Sylvilagus palustris hefneri*). Unpublished Ph.D. dissertation, University of Florida, Gainesville, Florida.

———. 1999. Food habits of the Lower Florida Keys marsh rabbit (*Sylvilagus palustris hefneri*). Florida Scientist 62:106–110.

Forys, E. A., and S. R. Humphrey. 1996. Home range and movement of the Lower Keys marsh rabbit in a highly fragmented environment. Journal of Mammalogy 77:1042–1048.

———. 1999. Use of population viability analysis to evaluate management options for the endangered Lower Keys marsh rabbit. Journal of Wildlife Management 63:251–260.

Fostowicz-Frelik, L., and J. Meng. 2013. Comparative morphology of premolar foramen in lagomorphs (Mammalia: Glires) and its functional and phylogenetic implications. PLoS ONE 8:e79794.

Fostowicz-Frelik, L., G. J. Frelik, and M. Gasparik. 2010. Morphological phylogeny of pika (Lagomorpha: *Ochotona*), with a description of a new species from the Pliocene/Pleistocene transition of Hungary. Proceedings of the Academy of Natural Sciences of Philadelphia 159:97–118.

Fowler, A., and R. E. Kissell, Jr. 2007. Winter relative abundance and habitat associations of swamp rabbits in eastern Arkansas. Southeastern Naturalist 6:247–258.

Fragulione, D. 1962–1964. Introduction artificielle et tentatives d'acclimatation du lievre commun (*Lepus europaeus* Pallas 1778) (de par le monde). Revues du Royal Saint Hubert Club de Belgique, juin 1962 a fevrier 1964. 16 pp.

Franken, R. J., and D. S. Hik. 2004. Interannual variation in timing of parturition and growth of collared pikas (*Ochotona collaris*) in the southwest Yukon. Integrative and Comparative Biology 44:186–193.

Fredsted, T., T. Wincentz, and O. Villesen. 2006. Introgression of mountain hare (*Lepus timidus*) mitochondrial DNA into wild brown hares (*Lepus europaeus*) in Denmark. BMC Ecology 6:17.

French, N. R., R. McBride, and J. Detmer. 1965. Fertility and population density of the black-tailed jackrabbit. Journal of Wildlife Management 29:14–26.

Freschi, P., S. Fascetti, M. Musto, C. Cosentino, R. Paolino, and V. Valentini. 2016. Seasonal variation in food habits of the Italian hare in a south Apennine semi-natural landscape. Ethology, Ecology and Evolution 28:148–162.

Freschi, P., S. Fascetti, M. Musto, E. Mallia, A. C. Blasi, C. Cosentino, et al. 2014. Diet of the Apennine hare in a southern Italy Regional Park. European Journal of Wildlife Research 60:423–430.

Freschi, P., S. Fascetti, M. Musto, E. Mallia, C. Cosentino, and R. Paolino. 2015. Diet of the Italian hare (*Lepus corsicanus*) in a semi-natural landscape of southern Italy. Mammalia 1:51–59.

Frey, J. K., R. D. Fisher, and L. A. Ruedas. 1997. Identification and restriction of the type locality of the Manzano Mountains cottontail, *Sylvilagus cognatus* Nelson, 1907 (Mammalia: Lagomorpha: Leporidae). Proceedings of the Biological Society of Washington 110:329–331.

Fujimoto, K., N. Yamaguti, and M. Takahashi. 1986. Ecological studies on ixodid ticks: ixodid ticks on vegetations and wild animals at the low mountain zone lying south-western part of Saitama Prefecture. Japanese Journal of Sanitary Zoology 37:325–331. [In Japanese].

Fujita, H. 2004. Tularemia. Modern Media 50:99–103. [In Japanese].

Fukasawa, K., T. Hashimoto, M. Tatara, and S. Abe. 2013. Reconstruction and prediction of invasive mongoose population dynamics from history of introduction and management: a Bayesian state-space modeling approach. Journal of Applied Ecology 50:469–478.

Fulk, G. W., and A. R. Khokhar. 1980. Observations on the natural history of a pika (*Ochotona rufescens*) from Pakistan. Mammalia 44: 51–58.

Fuller, S., and A. Tur. 2012. Conservation strategy for the New England cottontail (*Sylvilagus transitionalis*). U.S. Fish and Wildlife Publications. Paper 320.

Galbreath, K. E., and E. P. Hoberg. 2012. Return to Beringia: parasites reveal cryptic biogeographical history of North American pikas. Proceedings of the Royal Society B 279:371–378.

Galbreath, K. E., D. J. Hafner, and K. R. Zamudio. 2009. When cold is better: climate-driven elevation shifts yield complex patterns of diversification and demography in an alpine specialist (American pika, *Ochotona princeps*). Evolution 63:2848–2863.

Galbreath, K. E., D. J. Hafner, K. R. Zamudio, and K. Agnew. 2010. Isolation and introgression in the Intermountain West: contrasting gene genealogies reveal the complex biogeographic history of the American pika (*Ochotona princeps*). Journal of Biogeography 37: 344–362.

Gao, Y. T., and Z. J. Feng. 1964. On the subspecies of the Chinese gray-tailed hare, *Lepus oiostolus* Hodgson. Acta Zootaxonomica Sinica 1: 19–30. [In Chinese].

Gastil, G., J. Minch, and R. P. Phillips. 1983. The Geology and Ages of Islands. Pp. 13–25. In: T. J. Case and M. L. Cody (editors). Island Biogeography in the Sea of Cortez. University of California Press: Berkeley, California.

Gavier-Widen, D., and T. Mörner. 1993. Descriptive epizootiological study of European brown hare syndrome in Sweden. Journal of Wildlife Diseases 29:15–20.

Ge, D., A. A. Lissovsky, L. Xia, C. Cheng, A. T. Smith, and Q. Yang. 2012. Reevaluation of several taxa of Chinese lagomorphs (Mammalia: Lagomorpha) described on the basis of pelage phenotype variation. Mammalian Biology 77:113–123.

Ge, D., Z. Wen, L. Xia, Z. Zhang, M. Erbajeva, C. M. Huang, et al. 2013.

Evolutionary history of lagomorphs in response to global environmental change. PLoS ONE 8:e59668.

Ge, D., L. Yao, L. Xia, Z. Zhang, and Q. Yang. 2015. Geometric morphometric analysis of skull morphology reveals loss of phylogenetic signal at the generic level in extant lagomorphs (Mammalia: Lagomorpha). Contributions to Zoology 84:267–284.

Geraldes, A., N. Ferrand, and M. W. Nachman. 2006. Contrasting patterns of introgression at X-linked loci across the hybrid zone between subspecies of the European rabbit (*Oryctolagus cuniculus*). Genetics 173:919–933.

Ghose, R. K. 1971. Field observations on the habits of the Pachmarhi hare, *Lepus nigricollis mahadeva* Wroughton & Ryley, in Madhya Pradesh, India. Journal of the Zoological Society of India 23: 167–169.

Gibb, J. A. 1990. The European rabbit. Pp. 116–120. In: J. A. Chapman and J. E. C. Flux (editors). Rabbits, Hares and Pikas: Status Survey and Conservation Action Plan. International Union for Conservation of Nature: Gland, Switzerland.

Gidley, J. W. 1912. The lagomorphs an independent order. Science, New Series 36:285–286.

Gilcrease, K. 2014. The Mexican cottontail (*Sylvilagus cunicularius*): a historical perspective of hunting and grazing and implications for conservation planning. Acta Zoológica Mexicana (nueva serie) 30: 32–40.

Girma, Z., A. Bekele, and H. Graham. 2012. Large mammals and mountain encroachments on Mount Kaka and Hunkolo fragments, southeast Ethiopia. Asian Journal of Applied Sciences 5:279–289.

Gliwicz, J., S. Pagacz, and J. Witczuk. 2006. Strategy of food plant selection in the Siberian northern pika, *Ochotona hyperborea*. Arctic, Antarctic, and Alpine Research 38:54–59.

Gliwicz, J., J. Witczuk, and S. Pagacz. 2005. Spatial behavior of the rock-dwelling pika (*Ochotona hyperborea*). Journal of Zoology, London 267:113–120.

Gómez-Nísino, A. 2006. Ficha técnica de *Sylvilagus mansuetus*. In: R. Medellín (editor). Los Mamíferos Mexicanos en Riesgo de Extinción Según el PROY-NOM-059-ECOL-2000. Instituto de Ecología, Universidad Nacional Autónoma de México. Bases de datos SNIB-CONABIO. Proyecto No. W005: Ciudad de México, México.

Gonçalves, H., P. C. Alves, and A. Rocha. 2002. Seasonal variation in the reproductive activity of the wild rabbit (*Oryctolagus cuniculus algirus*) in a Mediterranean ecosystem. Wildlife Research 29:165–173.

González, J., C. Lara, J. Vázquez, and M. Martínez-Gómez. 2007. Demography, density, and survival of an endemic and near threatened cottontail *Sylvilagus cunicularius* in central Mexico. Acta Theriologica 52:299–305.

González-Cózatl, F. X., R. M. Vallejo, and F. A. Cervantes. 2007. Avances en el estudio de la sistemática de lagomorfos utilizando marcadores moleculares: filogenia del género *Sylvilagus* basada en secuencias del gen 16S. Pp. 31–46. In: G. Sánchez-Rojas and A. Rojas-Martínez (editors). Tópicos en Sistemática, Biogeografía, Ecología y Conservación de Mamíferos. Universidad Autónoma del Estado de Hidalgo: Hidalgo, México.

Grajales-Tam, K. M., and A. González-Romero. 2014. Determinación de la dieta estacional del coyote (*Canis latrans*) en la región norte de la Reserva de la Biosfera Mapimí, México. Revista Mexicana de Biodiversidad 85:553–564.

Gray, D. R. 1993. Behavioural adaptations to arctic winter: shelter seeking by arctic hare (*Lepus arcticus*). Arctic 46:340–353.

Gray, J. E. 1821. On the natural arrangement of the vertebrose animals. London Medical Repository 15(part 1):296–310.

Gray, J. P. 1977. Population Studies of Sonoran Desert Lagomorphs. Unpublished M.S. thesis, University of Arizona, Tucson, Arizona.

Grayson, D. K. 1977. On the Holocene history of some Great Basin lagomorphs. Journal of Mammalogy 58:507–513.

Green, J. S., and J. T. Flinders. 1980a. *Brachylagus idahoensis*. Mammalian Species 125:1–4.

———. 1980b. Habitat and dietary relationships of the pygmy rabbit. Journal of Range Management 22:136–142.

Guberti, V., M. A. De Marco, F. Riga, A. Lavazza, V. Trocchi, and L. Capucci. 2000. Virology and species conservation: the case of EBHSV and the Italian hare (*Lepus corsicanus*). Proceedings of the 5th International Congress of the European Society for Veterinary Virology (ESVV) 2000:198–199.

Guilday, J. E., and M. S. Bender. 1958. A recent fissure deposit in Bedford County, Pennsylvania. Annals of Carnegie Museum 35:127–138.

Gunther, K. A., R. A. Renkin, J. C. Halfpenny, S. M. Gunther, T. Davis, P. Schullelry, et al. 2009. Presence and distribution of white-tailed jackrabbits in Yellowstone National Park. Yellowstone Science 17:24–32.

Gurung, K. K., and R. Singh. 1996. Field Guide to the Mammals of the Indian Subcontinent. Academic Press: San Diego, California.

Gustavsson, I. 1971. Mitotic and meiotic chromosomes of the variable hare (*Lepus timidus* L.), the common hare (*Lepus europaeus* Pall.) and their hybrids. Hereditas 67:27–34.

Guthrie, R. D. 1973. Mummified pika (*Ochotona*) carcass and dung pellets from Pleistocene deposits in interior Alaska. Journal of Mammalogy 54:970–971.

Gyldenstolpe, A. 1917. Zoological results of the Swedish zoological expeditions to Siam 1911–1912 and 1914–1915: V. Mammals II. Kungliga Svenska vetenskapsakademiens handlingar 57:1–9.

Hackländer, K., W. Arnold, and T. Ruf. 2002. Postnatal development and thermoregulation in the precocial European hare (*Lepus europaeus*). Journal of Comparative Physiology B 172:183–190.

Hackländer, K., N. Ferrand, and P. C. Alves. 2008. Overview of Lagomorph research: what we have learned and what we still need to do. Pp. 381–394. In: P. C. Alves, N. Ferrand, and K. Hackländer (editors). Lagomorph Biology: Evolution, Ecology, and Conservation. Springer: Berlin.

Hackländer, K., C. Frisch, E. Klansek, T. Steineck, and T. Ruf. 2001. Die Fruchtbarkeit weiblicher Feldhasen (*Lepus europaeus*) aus Revieren mit unterschiedlicher Populationsdichte. Zeitschrift für Jagdwissenschaft 47:100–110.

Hackländer, K., C. Zeitlhofer, T. Ceulemans, and F. Suchentrunk. 2011. Continentality affects body condition and size but not yearly reproductive output in female European hares (*Lepus europaeus*). Mammalian Biology 76:662–664.

Hafner, D. J. 1994. Pikas and permafrost: post-Wisconsin historical zoogeography of *Ochotona* in the southern Rocky Mountains, U.S.A. Arctic and Alpine Research 26:375–382.

Hafner, D. J., and D. S. Sullivan. 1995. Historical and ecological biogeography of Nearctic pikas (Lagomorpha: Ochotonidae). Journal of Mammalogy 76:302–321.

Haga, R. 1960. Observation on the ecology of the Japanese pika. Journal of Mammalogy 41:200–212.

Hal, S. 2012. The Geographical Distribution of Animal Viral Diseases. Academic Press: New York.

Halanych, K. M., and T. J. Robinson. 1997. Phylogenetic relationships of cottontails (*Sylvilagus*, Lagomorpha): congruence of 12S rDNA and cytogenetic data. Molecular Phylogenetics and Evolution 7:294–302.

Halanych, K. M., J. R. Demboski, B. J. van Vuuren, D. R. Klein, and J. A. Cook. 1999. Cytochrome *b* phylogeny of North American hares and jackrabbits (*Lepus*, Lagomorpha) and the effects of saturation in outgroup taxa. Molecular Phylogenetics and Evolution 11:213–221.

Hall, E. R. 1951. A synopsis of the North American Lagomorpha. University of Kansas Press: Lawrence, Kansas.

———. 1981. The Mammals of North America. 2nd edition. John Wiley and Sons: New York.

Hall, E. R., and B. Villa. 1949. An annotated check list of the mammals of Michoacán, Mexico. University of Kansas Museum, Natural History Publication 1:431–472.

Hall, M. C. 1916. Nematode parasites of mammals of the orders Rodentia, Lagomorpha, and Hyracoidea. Proceedings of the United States National Museum 50:1–258.

Hamill, R. M., D. Doyle, and E. J. Duke. 2007. Microsatellite analysis of mountain hares (*Lepus timidus hibernicus*): low genetic differentiation and possible sex-bias in dispersal. Journal of Mammalogy 88:784–792.

Hamilton, L. P., P. A. Kelly, D. F. Williams, D. Kelt, and H. U. Wittmer. 2010. Factors associated with survival of reintroduced riparian brush rabbits in California. Biological Conservation 143:999–1007.

Happold, D. C. D. (editor). 2013. Mammals of Africa. Volume III: Rodents, Hares and Rabbits. Bloomsbury Publishing: New York.

Happold, D. C. D., and W. Wendelen. 2006. The distribution of *Poelagus marjorita* (Lagomorpha: Leporidae) in central Africa. Mammalian Biology 71:377–383.

Harada, M., T. H. Yoshida, S. Hattori, and S. Takada. 1985. Karyotypes and chromosome banding patterns of a rare Leporids species, *Pentalagus furnessi* (Lagomorpha, Leporidae). Proceedings of the Japan Academy Series B 61:319–321.

Harington, C. R. 2011. Pleistocene vertebrates of the Yukon Territory. Quaternary Science Reviews 30:2341–2354.

Harris, A. H., and J. Hearst. 2012. Late Wisconsin mammalian fauna from Dust Cave, Guadalupe Mountains National Park, Culberson County, Texas. The Southwestern Naturalist 57:202–206.

Harris, W. P., Jr. 1932. Four new mammals from Costa Rica. Occasional Papers of the Museum of Zoology, University of Michigan 248:1–6.

Hartman, A. C., and R. E. Barry. 2010. Survival and winter diet of *Sylvilagus obscurus* (Appalachian cottontail) at Dolly Sods, West Virginia. Northeastern Naturalist 17:505–516.

Hassinger, J. D. 1973. A survey of the mammals of Afghanistan: resulting from the 1965 Street Expedition (excluding bats). Fieldiana: Zoology 60:1–195.

Hatt, R. T. 1940. Lagomorpha and Rodentia other than Sciuridae, Anomaluridae and Idiuridae collected by the American Museum Congo Expedition. Bulletin of the American Museum of Natural History 76:457–604.

He, H. 1958. Report on Mammalian Survey in Northeast China. Science Press: Beijing, PRC. [In Chinese].

Hearn, B. J., L. B. Keith, and O. J. Rongstad. 1987. Demography and ecology of the arctic hare (*Lepus arcticus*) in southwestern Newfoundland. Canadian Journal of Zoology 65:852–861.

Henke, S. E., and F. C. Bryant. 1999. Effects of coyote removal on the faunal community in western Texas. Journal of Wildlife Management 63:1066–1081.

Henry, P., and M. A. Russello. 2013. Adaptive divergence along environmental gradients in a climate-change-sensitive mammal. Ecology and Evolution 3:3906–3917.

Hershkovitz, P. 1938. A review of the rabbits of the *andinus* group and their distribution in Ecuador. Occasional Papers of the Museum of Zoology, University of Michigan 393:1–15.

———. 1950. Mammals of northern Colombia preliminary report no. 6: rabbits (Leporidae), with notes on the classification and distribution of the South American forms. Proceedings of the United States National Museum 100:327–375.

Hewson, R. 1962. Food and feeding habits of the mountain hare *Lepus timidus scoticus* Hilzheimer. Proceedings of the Zoological Society of London 139:415–426.

———. 1985. Long-term fluctuations in populations of mountain hares (*Lepus timidus*). Journal of Zoology (London) 206:269–273.

Hewson, R., and D. C. Hinge. 1990. Characteristics of the home range of mountain hares *Lepus timidus*. Journal of Applied Ecology 27: 651–666.

Hibbard, C. M. 1963. The origin of the P3 pattern of *Sylvilagus*, *Caprolagus*, *Oryctolagus*, and *Lepus*. Journal of Mammalogy 44:1–15.

Hinds, D. S. 1973. Acclimatization of thermoregulation in the desert cottontail, *Sylvilagus audubonii*. Journal of Mammalogy 54:708–728.

———. 1977. Acclimatization of thermoregulation in desert-inhabiting jackrabbits (*Lepus alleni* and *Lepus californicus*). Ecology 58: 246–264.

Hirakawa, H. 2001. Coprophagy in leporids and other mammalian herbivores. Mammal Review 31:61–80.

Hirakawa, H., T. Kuwahata, Y. Shibata, and E. Yamada. 1992. Insular variation of the Japanese hare (*Lepus brachyurus*) on the Oki Island, Japan. Journal of Mammalogy 73:672–679.

Hobbs, R. P., and W. M. Samuel. 1974. Coccidia (Protozoa, Eimeriidae) of the pikas *Ochotona collaris*, *O. princeps*, and *O. hyperborea yesoensis*. Canadian Journal of Zoology 52:1079–1085.

Hodges, K. E. 2000. The ecology of snowshoe hares in northern boreal forests. Pp. 117–161. In: L. F. Ruggiero, K. B. Aubry, S. W. Buskirk, G. M. Koehler, C. J. Krebs, K. S. McKelvey, et al. (editors). Ecology and Conservation of Lynx in the United States. University Press of Colorado: Denver, Colorado.

Hodges, K. E., C. J. Krebs, D. S. Hik, C. I. Stefan, E. A. Gillis, and C. E. Doyle. 2001. Snowshoe hare demography. Pp. 141–178. In: C. J. Krebs, S. Boutin, and R. Boonstra (editors). Ecosystem Dynamics of the Boreal Forest: The Kluane Project. Oxford University Press: New York.

Hoffmann, R. S., and A. T. Smith. 2005. Order Lagomorpha. Pp. 185–211. In: D. E. Wilson and D. M. Reeder (editors). Mammal Species of the World: A Taxonomic and Geographic Reference. 3rd edition. Johns Hopkins University Press: Baltimore, Maryland.

Hoffmeister, D. F. 1984. Mammals of Arizona. University of Arizona Press: Tucson, Arizona.

Hoffmeister, D. F., and M. R. Lee. 1963. Taxonomic review of cottontails, *Sylvilagus floridanus* and *Sylvilagus nuttallii*, in Arizona. The American Midland Naturalist 70:138–148.

Höglund, N. H. 1957. Reproduction in the mountain hare (*Lepus t. timidus* Lin.). Viltrevy 1:267–282.

Holler, N. R., and C. H. Conaway. 1979. Reproduction of the marsh rabbit (*Sylvilagus palustris*) in South Florida. Journal of Mammalogy 60:769–777.

Holley, A. J. F. 2001. The daily activity period of the brown hare (*Lepus europaeus*). Mammalian Biology 66:357–364.

Horak, I. G., and L. Fourie. 1991. Parasites of domestic and wild animals in South Africa. XXIX. Ixodid ticks on hares in the Cape Province and on hares and red rock rabbits in the Orange Free State. Onderstepoort Journal of Veterinary Research 58:261–270.

Hoth, J., A. Velázquez, F. J. Romero, L. León, M. Aranda, and D. J. Bell. 1987. The volcano rabbit—a shrinking distribution and a threatened habitat. Oryx 21:85–91.

Howell, A. B. 1928. New Asiatic mammals collected by F. R. Wulsin. Proceedings of the Biological Society of Washington 32:105–110.

Huey, L. M. 1942. A vertebrate faunal survey of the Organ-pipe Cactus National Monument. Transactions of the San Diego Society of Natural History 9:353–375.

Hughes, G. O., W. Thuiller, G. F. Midgley, and K. Collins. 2008. Environmental change hastens the demise of the critically endangered riverine rabbit (*Bunolagus monticularis*). Biological Conservation 141:23–34.

Humphreys, A. M., and T. G. Barraclough. 2014. The evolutionary reality of higher taxa in mammals. Proceedings of the Royal Society B 281:20132750.

Hunt, T. P. 1959. Breeding habits of the swamp rabbit with notes on its life history. Journal of Mammalogy 40:82–91.

Huntly, N. J., A. T. Smith, and B. L. Ivins. 1986. Foraging behavior of the pika (*Ochotona princeps*) with comparisons of grazing versus haying. Journal of Mammalogy 67:139–148.

Hwang, H. S., H. S. Son, H. Kang, and S. J. Rhim. 2014. Ecological factors influencing the winter abundance of mammals in temperate forest. Folia Zoologica 63:296–300.

Iason, G. R. 1989. Growth and mortality in mountain hares: the effect of sex and date of birth. Oecologia 81:540–546.

———. 1990. The effect of size, age and a cost of early breeding on reproduction in female mountain hares. Holarctic Ecology 13:81–89.

Iason, G. R., and B. Boag. 1988. Do intestinal helminths affect condition and fecundity of adult mountain hares? Journal of Wildlife Diseases 24:599–605.

Ichikawa, T. (editor). 1999. We want to hear pikas call: the story of the struggle of a small NGO. Nakiusagi Fan Club: Sapporo, Japan.

Illiger, J. K. W. 1811. Prodromus systematis mammalium et avium; additis terminis zoographicis utriusque classis, eorumque versione germanica. Sumptibus C. Salfeld: Berlin.

Imaizumi, Y. 1960. Coloured Illustrations of the Mammals of Japan. Hoikusha Publishing Co., Ltd.: Osaka, Japan. [In Japanese].

Ingles, L. G. 1941. Natural history observations on the Audubon cottontail. Journal of Mammalogy 22:227–250.

Ivins, B. L., and A. T. Smith. 1983. Responses of pikas (*Ochotona princeps*: Lagomorpha) to naturally occurring terrestrial predators. Behavioral Ecology and Sociobiology 13:277–285.

Jacobson, E. 1921. Notes on some mammals from Sumatra. Journal of the Federated Malay States Museum 10:235–241.

Jacobson, E., and C. B. Kloss. 1919. Notes on the Sumatran hare. Journal of the Federated Malay States Museum 7:293–297.

Janzen, D. H. 1967. Why mountain passes are higher in the tropics. The American Naturalist 101:233–249.

Jeffress, M. R., T. J. Rodhouse, C. Ray, S. Wolff, and C. W. Epps. 2013. The idiosyncrasies of place: geographic variation in the climate-distribution relationships of the American pika. Ecological Applications 23:864–878.

Jennings, N. V., R. K. Smith, K. Hackländer, S. Harris, and P. C. L. White. 2006. Variation in demography, condition and dietary quality of hares *Lepus europaeus* from high-density and low-density populations. Wildlife Biology 12:179–189.

Jensen, A. 2003. Movements and Natal Dispersal of the Appalachian Cottontail, *Sylvilagus obscurus*. Unpublished M.S. thesis, Frostburg State University, Frostburg, Maryland.

Jiang, Y. J., and Z. W. Wang. 1981. Social behavior of *Ochotona cansus*: adaptation to the alpine environment. Acta Theriologica Sinica 11: 23–40. [In Chinese].

Jiang, Z. G., and W. P. Xia. 1987. The niches of yak, Tibetan sheep, and plateau pikas in the alpine meadow ecosystem. Acta Biologica Plateau Sinica 8:115–146. [In Chinese].

Jiang, Z. G., J. Jiang, Y. Wang, E. Zhang, Y. Zhang, L. Li, et al. 2016. Red List of China's vertebrates. Biodiversity Science 24:500–551. [In Chinese].

Jiménez-Almaraz, T., J. Juárez Gómez, and L. León Paniagua. 1993. Mamíferos. Pp. 503–549. In: I. Luna Vega and J. Llorente Bousquets (editors). Historia Natural del Parque Ecológico Estatal Omiltemi, Chilpancingo, Guerrero, México. Comisión para el Conocimiento y Uso de la Biodiversidad y Universidad Nacional Autónoma de México: Ciudad de México, México.

Jiménez-Ruiz, F. A., S. L. Gardner, F. A. Cervantes, and C. Lorenzo. 2004. A new species of *Pelecitus* (Filarioidea: Onchocercidae) from the endangered Tehuantepec jackrabbit *Lepus flavigularis*. Journal of Parasitology 90:803–807.

Jin, C. Z., Y. Tomida, Y. Wang, and Y. Q. Zhang. 2010. First discovery of fossil *Nesolagus* (Leporidae, Lagomorpha) from Southeast Asia. Science China: Earth Science 53:1134–1140.

Johnson, R. D., and J. E. Anderson. 1984. Diets of black-tailed jack rabbits in relation to population density and vegetation. Journal of Range Management 37:79–83.

Jones, J. K., and D. H. Johnson. 1965. Synopsis of the lagomorphs and rodents of Korea. University of Kansas Publications, Museum of Natural History 16:357–407.

Joseph, J., M. Collins, J. Holechek, R. Valdez, and R. Steiner. 2003. Conservative and moderate grazing effects on Chihuahuan Desert wildlife sightings. Western North American Naturalist 63:43–49.

Joyce, T. L. 2002. Impact of Hunting on Snowshoe Hares in Newfoundland. Unpublished M.S. thesis, Department of Zoology, University of British Columbia, Vancouver, British Columbia.

Katzner, T. E., and K. L. Parker. 1997. Vegetative characteristics and size of home ranges used by pygmy rabbits (*Brachylagus idahoensis*) during winter. Journal of Mammalogy 78:1063–1072.

Katzner, T. E., K. L. Parker, and H. H. Harlow. 1997. Metabolism and thermal response in winter-acclimatized pygmy rabbits (*Brachylagus idahoensis*). Journal of Mammalogy 78:1053–1062.

Kawamichi, T. 1968. Winter behaviour of the Himalayan pika, *Ochotona roylei*. Journal of the Faculty of Science, Hokkaido University, Series VI Zoology 16:582–594.

———. 1969. Behaviour and daily activities of the Japanese pika, *Ocho-tona hyperborea yesoensis*. Journal of the Faculty of Science, Hokkaido University, Series VI Zoology 17:127–151.

———. 1970. Social pattern of the Japanese pika, *Ochotona hyperborea yesoensis*, preliminary report. Journal of the Faculty of Science, Hokkaido University, Series VI Zoology 17:462–473.

———. 1971a. Annual cycle of behavior and social pattern of the Japanese pika, *Ochotona hyperborea yesoensis*. Journal of the Faculty of Science, Hokkaido University, Series VI Zoology 18:173–185.

———. 1971b. Daily activities and social pattern of two Himalayan pikas, *Ochotona macrotis* and *O. roylei*, Observed at Mt. Everest. Journal of the Faculty of Science, Hokkaido University, Series VI Zoology 17:587–609.

Kawamichi, T., and S. Dawanyam. 1997. Structure of a breeding nest of the Daurian pika, *Ochotona daurica*, in Mongolia. Mammal Study 22:89–93.

Kawamura, Y., T. Kamei, and H. Taruno. 1989. Middle and Late Pleistocene mammalian faunas in Japan. The Quaternary Research 28: 317–326. [In Japanese].

Keith, L. B. 1963. Wildlife's Ten-year Cycle. University of Wisconsin Press: Madison, Wisconsin.

———. 1990. Dynamics of snowshoe hare populations. Current Mammalogy 4:119–195.

Keith, L. B., J. R. Cary, T. M. Yuill, and I. M. Keith. 1985. Prevalence of helminths in a cyclic snowshoe hare population. Journal of Wildlife Diseases 21:233–253.

Kelt, D. A., P. A. Kelly, S. E. Phillips, and D. F. Williams. 2014. Home range size and habitat selection of reintroduced *Sylvilagus bachmani riparius*. Journal of Mammalogy 95:516–524.

Khalilipour, O., H. R. Rezaei, A. A. Shabani, M. Kaboli, and S. Ashrafi. 2014. Genetic structure and differentiation of four populations of Afghan pika (*Ochotona rufescens*) in Iran based on mitochondrial cytochrome *b* gene. Zoology in the Middle East 60:288–298.

Khanal, B. 2007. New report on the symbiotic relation of *Ochotona roylei* (Lagomorpha: Ochotonidae) and scaly breasted wren babbler (*Pnoepyge albiventer*) at Ganesh Himalaya area of central Nepal. Our Nature 5:37–40.

Kim, S. W., and W. K. Kim. 1974. Avi-mammalian fauna of Korea. Wildlife population census in Korea #5. Office of Forestry, Forest Research Institute: Seoul, South Korea. [In Korean].

Kingdon, J. 1974. East African Mammals. Volume II, Part B: Hares and Rodents. Academic Press: London.

Kirk, D. A., and G. M. Bathe. 1994. Population size and home range of black-napped hares *Lepus nigricollis* on Cousin Island (Seychelles Indian Ocean). Mammalia 58:557–562.

Kishida, K. 1930. Diagnosis of a new piping hare from Yeso. Lansania 2: 45–47. [In Japanese].

———. 1932. A new species of piping hares from Saghalien. Lansania 4: 149–152.

Kishida, K., and T. Mori. 1930. Summer pelage of the Corean piping hare, *Ochotona coreana*. Lansania 2:49–52. [In Japanese].

Kjolhaug, M. S., and A. Woolf. 1988. Home range of the swamp rabbit in southern Illinois. Journal of Mammalogy 69:194–197.

Klein, D. R. 1995. Tundra or arctic hares. P. 359. In: E. T. LaRoe (editor). Our Living Resources: A Report to the Nation on the Distribution, Abundance, and Health of U.S. Plants, Animals, and Ecosystems. U.S. Department of the Interior, National Biological Service: Washington, D.C.

Kline, P. D. 1963. Notes on the biology of jackrabbits in Iowa. Proceedings of the Iowa Academy of Science 70:196–204.

Kloss, C. B. 1919. On mammals collected in Siam. Journal of the Natural History Society of Siam 4:333–407.

Kniazev, A. V., and A. B. Savinetski. 1988. Changes of populations of small mammals of the Tsaagan-Bogdo ridge (Transaltai Gobi) in the Late Holocene. Zoologichesky Zhurnal 67:297–300.

Knick, S. K., and D. L. Dyer. 1997. Distribution of black-tailed jackrabbit habitat determined by GIS in southwestern Idaho. Journal of Wildlife Management 61:75–85.

Koh, H. S., and K. H. Jang. 2010. Genetic distinctness of the Korean hare, Lepus coreanus (Mammalia, Lagomorpha), revealed by nuclear thyroglobulin gene and mtDNA control region sequences. Biochemical Genetics 48:706–710.

Koh, H. S., T. Y. Chun, H. S. Yoo, Y. Zhang, J. Wang, M. Zhang, et al. 2001. Mitochondrial cytochrome b gene sequence diversity in the Korean hare, Lepus coreanus Thomas (Mammalia, Lagomorpha). Biochemical Genetics 39:417–429.

Koh, S., and D. S. Hik. 2007. Herbivory mediates grass-endophyte relationships. Ecology 88:2752–2757.

Komonen, M., A. Komonen, and A. Otgonsuren. 2003. Daurian pikas (Ochotona daurica) and grassland condition in eastern Mongolia. Journal of Zoology (London) 259:281–288.

Kong, L., W. Wang, H. Cong, T. Son Nguyen, Q. Yang, Y. Wu, et al. 2016. Molecular evidence revealed Lepus hainanus and L. peguensis have a conspecific relationship. Mitochondrial DNA 27:265–269.

Koutsogiannouli, E. A., K. A. Moutou, C. Stamatis, and Z. Mamuris. 2012. Analysis of MC1R genetic variation in Lepus species in Mediterranean refugia. Mammalian Biology 77:428–433.

Kraatz, B. P., E. Sherratt, N. Bumacod, and M. J. Wedel. 2015. Ecological correlates to cranial morphology in Leporids (Mammalia, Lagomorpha). PeerJ 3:e844.

Krebs, C. J., R. Boonstra, S. Boutin, and A. R. E. Sinclair. 2001. What drives the 10-year cycle of snowshoe hares? BioScience 51:25–35.

Krebs, C. J., S. Boutin, R. Boonstra, A. R. E. Sinclair, J. N. M. Smith, M. R. T. Dale, et al. 1995. Impact of food and predation on the snowshoe hare cycle. Science 269:1112–1115.

Krebs, C. J., J. Bryant, M. O'Donoghue, K. Kielland, F. Doyle, C. McIntyre, et al. 2014. What factors determine cyclic amplitude in the snowshoe hare cycle? Canadian Journal of Zoology 92:1039–1048.

Krebs, C. J., B. S. Gilbert, S. Boutin, A. R. E. Sinclair, and J. N. M. Smith. 1986. Population biology of snowshoe hares. I. Demography of food-supplemented populations in the southern Yukon, 1976–84. Journal of Animal Ecology 55:963–982.

Krebs, C. J., K. Kielland, J. Bryant, M. O'Donoghue, F. Doyle, C. McIntyre, et al. 2013. Synchrony in the snowshoe hare cycle in northwestern North America, 1970–2012. Canadian Journal of Zoology 91:562–572.

Kryger, U. 2002. Genetic Variation Among Southern African Hares (Lepus spec.). Unpublished Ph.D. dissertation, University of Pretoria, Pretoria, South Africa.

Kryger, U., T. J. Robinson, and P. Bloomer. 2004. Population structure and history of southern African scrub hares, Lepus saxatilis. Journal of Zoology (London) 263:121–133.

Kubo, M., T. Nakashima, T. Honda, Y. Kochi, Y. Ito, S. Hattori, et al. 2013. Histopathological examination of spontaneous lesions in Amami rabbits (Pentalagus furnessi): a preliminary study using formalin-fixed archival specimens. Japanese Journal of Zoo and Wildlife Medicine 18:65–70.

Kürten, B., and E. Anderson. 1980. Pleistocene Mammals of North America. Columbia University Press: New York.

Lai, C. H., and A. T. Smith. 2003. Keystone status of plateau pikas (Ochotona curzoniae): effect of control on biodiversity of native birds. Biodiversity and Conservation 12:1901–1912.

Lanier, H. C., and L. E. Olson. 2009. Inferring divergence times within pikas (Ochotona spp.) using mtDNA and relaxed molecular dating techniques. Molecular Phylogenetics and Evolution 53:1–2.

———. 2013. Deep barriers, shallow divergences: reduced phylogeographic structure in the collared pika (Ochotona collaris). Journal of Biogeography 40:466–478.

Lanier, H. C., A. M. Gunderson, M. Weksler, V. B. Fedorov, and L. E. Olson. 2015a. Comparative phylogeography of eastern Beringian mammals highlights the double-edged sword of climate change faced by arctic- and alpine-adapted species. PLoS ONE 10: e0118396.

Lanier, H. C., R. Massatti, Q. He, L. E. Olson, and L. L. Knowles. 2015b. Colonization from divergent ancestors: glaciation signatures on contemporary patterns of genomics variation in collared pikas (Ochotona collaris). Molecular Ecology 24:3688–705.

Larrucea, E. S., and P. F. Brussard. 2008. Habitat selection and current distribution of the pygmy rabbit in Nevada and California, USA. Journal of Mammalogy 89:691–699.

———. 2009. Diel and seasonal activity patterns of pygmy rabbits (Brachylagus idahoensis). Journal of Mammalogy 90:1176–1183.

Larter, N. C. 1999. Seasonal changes in arctic hare, Lepus arcticus, diet composition and differential digestibility. The Canadian Field-Naturalist 113:481–486.

Laseter, B. R. 1999. Estimates of Population Density for the Appalachian Cottontail (Sylvilagus obscurus) in Eastern Tennessee. Unpublished M.S. thesis, University of Memphis, Memphis, Tennessee.

Lavazza, A., P. Cavadini, I. Barbieri, P. Tizzani, A. Pinheiro, J. Abrantes, et al. 2015. Field and experimental data indicate that the eastern cottontail (Sylvilagus floridanus) is susceptible to infection with European brown hare syndrome (EBHS) virus and not with rabbit haemorrhagic disease (RHD) virus. Veterinary Research 46:13.

Lavazza, A., M. T. Scicluna, and L. Capucci. 1996. Susceptibility of hares and rabbits to the European brown hare syndrome virus (EBHSV) and rabbit haemorrhagic disease virus (RHDV) under experimental conditions. Journal of Veterinary Medicine 43: 401–410.

Lavocat, R. 1978. Rodentia and Lagomorpha. Pp. 84–89. In: V. J. Maglio and H. B. S. Coole (editors). Evolution of African Mammals. Harvard University Press: Cambridge, Massachusetts.

Lawes, T. J., R. G. Anthony, W. D. Robinson, J. T. Forbes, and G. A. Lorton. 2012. Homing behavior and survival of pygmy rabbits after experimental translocation. Western North American Naturalist 72: 569–581.

———. 2013. Movements and settlement site selection of pygmy rabbits after experimental translocation. Journal of Wildlife Management 77:1170–1181.

Lay, D. M. 1967. A study of the mammals of Iran resulting from the Street Expedition of 1962–63. Fieldiana: Zoology 54:1–282.

Lazell, J., W. H. Lu, W. Xia, S. Y. Li, and A. T. Smith. 1995. Status of the Hainan hare (Lepus hainanus). Species 25:61–62.

Lazell, J. D., Jr. 1984. A new marsh rabbit (*Sylvilagus palustris*) from Florida's Lower Keys. Journal of Mammalogy 65:26–33.

Lechleitner, R. R. 1958a. Movements, density, and mortality in a black-tailed jackrabbit population. Journal of Wildlife Management 22: 371–384.

———. 1958b. Certain aspects of behavior of the black-tailed jackrabbit. The American Midland Naturalist 60:145–155.

———. 1959. Sex ratio, age classes, and reproduction of the black-tailed jackrabbit. Journal of Mammalogy 40:63–81.

Lee, D. N., R. S. Pfau, and L. K. Ammerman. 2010. Taxonomic status of the Davis Mountains cottontail, *Sylvilagus robustus*, revealed by amplified fragment length polymorphism. Journal of Mammalogy 91:1473–1483.

Lee, J. E., R. T. Larsen, J. T. Flinders, and D. L. Eggett. 2010. Daily and seasonal patterns of activity at pygmy rabbit burrows in Utah. Western North American Naturalist 70:189–197.

Lees, A. C., and D. J. Bell. 2008. A conservation paradox for the 21st century: the European wild rabbit *Oryctolagus cuniculus*, an invasive alien and an endangered native species. Mammal Review 30: 304–320.

Le Gall-Reculé, G., A. Lavazza, S. Bertagnoli, F. Zwingelstein, P. Cavadini, N. Martinelli, et al. 2013. Emergence of a new lagovirus related to rabbit haemorrhagic disease virus. Veterinary Research 44:81.

Le Gall-Reculé, G., F. Zwingelstein, S. Boucher, B. Le Normand, G. Plassiart, Y. Portejoie, et al. 2011. Detection of a new variant of rabbit haemorrhagic disease virus in France. Veterinary Record 168: 137–138.

Le Gall-Reculé, G., F. Zwingelstein, S. Laurent, C. De Boisseson, Y. Portejoie, and D. Rasschaert. 2003. Phylogenetic analysis of rabbit haemorrhagic disease virus in France between 1993 and 2000, and the characterisation of RHDV antigenic variants. Archives of Virology 148:65–81.

Lekagul, B., and J. A. McNeely. 1977. Mammals of Thailand. Association for the Conservation of Wildlife: Kurusapha, Bangkok, Thailand.

Lenghaus, C., H. Westbury, B. Collins, N. Ratnamoban, and C. Morrissy. 1994. Overview of the RHD Project in Australia. Pp. 104–129. In: R. Williams and R. Munro (editors). Rabbit Haemorrhagic Disease: Issues in Assessment for Biological Control. Australian Government Printing Service: Canberra, Australia.

Leopold, A. 1966. A Sand County Almanac with other Essays on Conservation from Round River. Oxford University Press: New York.

Leopold, A. S. 1959. Wildlife of Mexico: The Game Birds and Mammals. University of California Press: Berkeley, California.

———. 1972. Wildlife of Mexico: The Game Birds and Mammals. 2nd edition. University of California Press: Berkeley, California.

Lewis, R. E. 1971. A new species of *Chaetopsylla kohaut*, 1903, infesting pikas in Nepal (Siphonaptera: Vermipsyllidae). Journal of Parasitology 57:1344–1348.

Li, C.-K., and S.-Y. Ting. 1985. Possible phylogenetic relationship of Asiatic eurymylids and rodents, with comments on mimotonids. Pp. 35–58. In: W. P. Luckett and J.-L. Hartenberger (editors). Evolutionary Relationships among Rodents: A Multidisciplinary Analysis. Plenum Press: New York.

Li, C.-K., J. Meng, and Y.-Q. Wang. 2007. *Dawsonolagus antiquus*, a primitive lagomorph from the Eocene Arshanto formation, Nei Mongol, China. Pp. 97–110. In: K. C. Beard and Z.-X. Luo (editors). Mammalian Paleontology on a Global Stage: Papers in Honor of Mary R. Dawson. Bulletin of Carnegie Museum of Natural History 39.

Li, D. H. (editor). 1989. Qinghai Fauna Economica. Qinghai People's Press: Xining, PRC. [In Chinese].

Li, W. (editor). 2013. Comprehensive Scientific Investigation of Arjinshan National Nature Reserve in Xinjiang. Xinjiang Science and Technology Press: Ürümqi, PRC. [In Chinese].

Li, W., H. Zhang, and Z. Liu. 2006. Brief report on the status of Koslov's pika, *Ochotona koslowi* (Büchner), in the east Kunlun mountains of China. Integrative Zoology 1:22–24.

Li, W. D. 1997. An endangered species of Lagomorpha—Ili pika (*Ochotona iliensis*). Chinese Biodiversity 5 (supplement):23–28. [In Chinese].

Li, W. D. 2003. The comparative research on status of Ili Pika in the past ten years. Chinese Journal of Zoology 38:64–68. [In Chinese].

Li, W. D. 2004. The discovery and research on Ili Pika (*Ochotona iliensis*). Chinese Journal of Zoology 39:106–111. [In Chinese].

Li, W. D., and Y. Ma. 1986. A new species of Ochotonidae, Lagomorpha. Acta Zoologica Sinica 32:375–379. [In Chinese].

Li, W. D., and A. T. Smith. 2005. Dramatic decline of the threatened Ili pika *Ochotona iliensis* (Lagomorpha: Ochotonidae) in Xinjiang, China. Oryx 39:30–34.

———. 2015. In search of the illusive and iconic Ili pika (*Ochotona iliensis*). Mountain Views (CIRMOUNT) 9:21–27.

Li, W. D., and W. Zhao. 1991. The component species of Genus *Ochotona* and its distribution areas in Xinjiang. Chinese Journal of Vector Biology and Control 2:305–308. [In Chinese].

Li, W. D., H. Li, X. Hamit, and J. Ma. 1991. A preliminary study on the distribution and habitat of *Ochotona iliensis*. Chinese Journal of Zoology 26:28–30. [In Chinese].

Li, W. D., H. Li, X. Hamit, J. Ma, and W. Zhao. 1993a. Preliminary research on the daily activity rhythm of the Ili pika. Arid Zone Research 1:54–57. [In Chinese].

Li, W. D., H. Li, and J. Ma. 1993b. A preliminary exploration of the reproductive biology of the Ili pika. Chinese Journal of Vector Biology and Control 4:120–122. [In Chinese].

Li, Z. C., L. Xia, Y. M. Li, Q. S. Yang, and M. Y. Liang. 2006. Mitochondrial DNA variation and population structure of the Yarkand hare *Lepus yarkandensis*. Acta Theriologica 51:243–253.

Li, Z. C., L. Xia, Q. S. Yang, and M. Y. Liang. 2005. Population genetic structure of the Yarkand hare (*Lepus yarkandensis*). Acta Theriologica Sinica 25:224–228.

Liao, J., Z. Zhang, and N. Liu. 2007. Effects of altitudinal change on the auditory bulla in *Ochotona daurica* (Mammalia, Lagomorpha). Journal of Zoological Systematics and Evolutionary Research 45: 151–154.

Lightfoot, D. C., A. D. Davidson, C. M. McGlone, and D. G. Parker. 2010. Rabbit abundance relative to rainfall and plant production in northern Chihuahuan Desert grassland and shrubland habitats. Western North American Naturalist 70:490–499.

Lim, B. K. 1987. *Lepus townsendii*. Mammalian Species 288:1–6.

Lin, G. H., H. X. Xi, S. J. Thirgood, T. Z. Zhang, and J. P. Su. 2010. Genetic variation and molecular evolution of endangered Kozlov's pika (*Ochotona koslowi* Büchner) based on mitochondrial cytochrome *b* gene. Polish Journal of Ecology 58:563–568.

Lincoln, G. A. 1974. Reproduction and "March madness" in the brown hare, *Lepus europaeus*. Journal of Zoology, London 174:1–14.

Lind, E. A. 1963. Observations on the mutual relationship between the snow hare (*Lepus timidus*) and the field hare (*L. europaeus*). Suomen Riista 16:128–135.

Lindström, E., H. Andren, P. Angelstam, and P. Widen. 1986. Influence of predators on hare populations in Sweden: a critical review. Mammal Review 16:151–156.

Linnaeus, C. 1758. Systema naturæ per regna tria naturæ, secundum classes, ordines, genera, species, cum characteribus, differentiis, synonymis, locis; Editio decima, reformata; Tomus I. Holmiæ, Laurentii Salvii.

Linzey, D. W. 1998. The Mammals of Virginia. The McDonald & Woodward Publishing Company: Blacksburg, Virginia.

Lissovsky, A. A. 2003. Geographical variation of skull characters in pikas (Ochotona, Lagomorpha) of the alpina-hyperborea group. Acta Theriologica 48:11–24.

———. 2005. Comparative analyses of the vocalizations of pikas (Ochotona, Mammalia) from the alpina-hyperborea group. Byulleten Moskovskogo Obshchestva Ispytatelei Prirody. Otdel Biologicheskii 110:12–26. [In Russian].

———. 2014. Taxonomic revision of pikas Ochotona (Lagomorpha, Mammalia) at the species level. Mammalia 78:199–216.

———. 2015. A new subspecies of Manchurian pika Ochotona mantchurica (Lagomorpha, Ochotonidae) from the Lesser Khinggan Range, China. Russian Journal of Theriology 14:145–152.

Lissovsky, A. A., and E. V. Lissovskaya. 2002. Diagnostics of pikas (Lagomorpha, Ochotonidae, Ochotona) from the Putorana Plateau, eastern Siberia. Russian Journal of Theriology 1:37–42.

Lissovsky, A. A., N. V. Ivanova, and A. V. Borisenko. 2007. Molecular phylogenetics and taxonomy of the subgenus Pika (Ochotona, Lagomorpha). Journal of Mammalogy 88:1195–1204.

Lissovsky, A. A., Q. S. Yang, and A. E. Pilnikov. 2008. Taxonomy and distribution of the pikas (Ochotona, Lagomorpha) of alpina-hyperborea group in north-east China and adjacent territories. Russian Journal of Theriology 7:5–16.

Lissovsky, A. A., S. P. Yatsentyuk, and D. Ge. 2016. Phylogeny and taxonomic reassessment of pikas Ochotona pallasii and O. argentata (Mammalia, Lagomorpha). Zoologica Scripta 45:583–594.

Litvaitis, J. A. 1993. Response of early successional vertebrates to historic changes in land use. Conservation Biology 7:866–873.

———. 2001. Importance of early successional habitats to mammals in eastern forests. Wildlife Society Bulletin 29:466–473.

Litvaitis, J. A., and R. Villafuerte. 1996. Factors affecting the persistence of New England cottontail metapopulations: the role of habitat management. Wildlife Society Bulletin 24:686–693.

Litvaitis, J. A., M. S. Barbour, A. L. Brown, A. I. Kovach, J. D. Oehler, B. L. Probert, et al. 2007. Testing multiple hypotheses to identify the causes of the decline of a lagomorph species: the New England cottontail as a case study. Pp. 167–185. In: P. C. Alves, N. Ferrand, and K. Hackländer (editors). Biology of Lagomorphs: Evolution, Ecology and Conservation. Springer: Berlin.

Litvaitis, M. K., J. A. Litvaitis, W-J. Lee, and T. Kocher. 1997. Variation in the mitochondrial DNA of the Sylvilagus complex occupying the northeastern United States. Canadian Journal of Zoology 75:595–605.

Litvaitis, J. A., J. P. Tash, M. K. Litvaitis, M. N. Marchand, A. I. Kovach, and R. Jenkins. 2006. A range-wide survey to determine the current distribution of New England cottontails. Wildlife Society Bulletin 34:1190–1197.

Liu, J., P. Chen, L. Yu, S. F. Wu, Y. P. Zhang, and X. L. Jiang. 2011a. The taxonomic status of Lepus melainus (Lagomorpha: Leporidae) based on nuclear DNA and morphological analyses. Zootaxa 3010:47–57.

Liu, J., L. Yu, M. L. Arnold, C. H. Wu, S. F. Wu, X. Lu, et al. 2011b. Reticulate evolution: frequent introgressive hybridization among Chinese hares (genus Lepus) revealed by analyses of multiple mitochondrial and nuclear DNA loci. BMC Evolutionary Biology 11:223.

Liu, W., Y. Zhang, X. Wang, J. H. Zhao, Q. M. Xu, and L. Zhou. 2009. Caching selection by plateau pika and its biological significance. Acta Theriologica Sinica 29:152–159. [In Chinese].

Lombardi, L., N. Fernandez, and S. Moreno. 2007. Habitat use and spatial behaviour in the European rabbit in three Mediterranean environments. Basic and Applied Ecology 8:453–463.

Long, J. L. 2003. Introduced Mammals of the World: Their History, Distribution and Influence. CSIRO Publishing: Collingwood, Victoria, Australia.

Lopes, A. M., S. Marques, E. Silva, M. Magalhães, A. Pinheiro, P. C. Alves, et al. 2014. Detection of RHDV strains in the Iberian hare (Lepus granatensis): earliest evidence of rabbit lagovirus cross-species infection. Veterinary Research 45:94.

López, J. A., C. Lorenzo, F. Barragán, and J. Bolaños. 2009. Mamíferos terrestres de la zona lagunar del Istmo de Tehuantepec, Oaxaca, México. Revista Mexicana de Biodiversidad 80:491–505.

López-Forment, W., I. E. Lira, and C. Müdespacher. 1996. Mamíferos: Su Biodiversidad en Islas Mexicanas. AGT Editor, S. A.: México.

López-Martínez, N., 1989. Revisión sistemática y bioestratigráfica de los lagomorfos (Mammalia) del Terciario y Cuaternario de España. Memorias del Museo Paleontológico, Universidad of Zaragoza 3:1–343.

———. 2008. The lagomorph fossil record and the origin of the European rabbit. Pp. 27–46. In: P. C. Alves, N. Ferrand, and K. Hackländer (editors). Lagomorph Biology: Evolution, Ecology, and Conservation. Springer: Berlin.

López-Vidal, J. C., C. Elizalde-Arellano, L. Hernández, J. W. Laundré, A. González-Romero, and F. A. Cervantes. 2014. Foraging of the bobcat (Lynx rufus) in the Chihuahuan Desert: generalist or specialist? The Southwestern Naturalist 59:157–166.

Lorenzo, C., S. T. Álvarez-Castañeda, and J. Vázquez. 2011. Conservation status of the threatened, insular San Jose brush rabbit (Sylvilagus mansuetus). Western North American Naturalist 71:10–16.

Lorenzo, C., D. E. Brown, S. Amirsultan, and M. García. 2014a. Evolutionary history of the antelope jackrabbit, Lepus alleni. Journal of the Arizona-Nevada Academy of Science 45:70–75.

Lorenzo, C., A. Carrillo-Reyes, M. Gómez-Sánchez, A. Velázquez, and E. Espinoza. 2011. Diet of the endangered Tehuantepec jackrabbit, Lepus flavigularis. Therya 2:67–76.

Lorenzo, C., A. Carrillo-Reyes, T. Rioja-Paradela, and M. de La Paz-Cuevas. 2012. Estado actual de conservación de liebres insulares en Baja California Sur, México. Therya 3:185–206.

Lorenzo, C., A. Carrillo-Reyes, T. M. Rioja-Paradela, M. de la Paz-Cuevas, J. Bolaños-Citalán, and S. T. Álvarez-Castañeda. 2014b. Estado actual de conservación de liebres y conejos en categoría de riesgo en México. Technical Report. Project HK052. Comisión Nacional para el Conocimiento y Uso de la Biodiversidad: Ciudad de México, México.

Lorenzo, C., F. A. Cervantes, and M. A. Aguilar. 1993. The karyotypes of some Mexican cottontail rabbits of the genus Sylvilagus. Pp. 129–136. In: R. A. Medellín and G. Ceballos (editors). Avances en el Estudio de los Mamíferos de México. Publicaciones Especiales, Asociación Mexicana de Mastozoología: México City, México.

Lorenzo, C., F. A. Cervantes, F. Barragán, and J. Vargas. 2006. New records of the endangered Tehuantepec jackrabbit (*Lepus flavigularis*) from Oaxaca, Mexico. The Southwestern Naturalist 51:116–119.

Lorenzo, C., P. Cortes-Calva, G. Ruiz-Campos, and S. T. Álvarez-Castañeda. 2013. Current distributional status of two subspecies of *Sylvilagus bachmani* on the Baja California Peninsula, Mexico. Western North American Naturalist 73:219–223.

Lorenzo, C., T. M. Rioja, A. Carrillo, and F. A. Cervantes. 2008. Population fluctuations of *Lepus flavigularis* (Lagomorpha: Leporidae) at Tehuantepec Isthmus, Oaxaca, Mexico. Acta Zoológica Mexicana (nueva serie) 24:207–220.

Lorenzo, C., T. M. Rioja-Paradela, and A. Carrillo-Reyes. 2015. State of knowledge and conservation of endangered and critically endangered lagomorphs worldwide. Therya 6:11–30.

Lorusso, V., R. P. Lia, F. Dantas-Torres, E. Mallia, S. Ravagnan, G. Capelli, et al. 2011. Ixodid ticks of road-killed wildlife species in southern Italy: new tick-host associations and locality records. Experimental and Applied Acarology 55:293–300.

Loukashkin, A. S. 1940. On the pikas of north Manchuria. Journal of Mammalogy 21:402–405.

———. 1943. On the hares of northern Manchuria. Journal of Mammalogy 24:73–81.

Lowe, C. E. 1958. Ecology of the swamp rabbit in Georgia. Journal of Mammalogy 39:116–127.

Lozano-Garcial, S., and L. Vazquez-Selem. 2005. A high-elevation Holocene pollen record from Iztaccíhuatl volcano, central Mexico. The Holocene 15:329–338.

Lu, J., G. Lu, and X. Li. 1997. Rodents of Henan. Henan Science and Technology Press: Zhengzhou, PRC. [In Chinese].

Lu, X. 2010. Demographic data on the woolly hare *Lepus oiostolus* near Lhasa, Tibet. Mammalian Biology 75:572–576.

———. 2011. Habitat use and abundance of the woolly hare *Lepus oiostolus* in the Lhasa mountains. Mammalia 75:35–40.

Lumpkin, S., and J. Seidensticker. 2011. Rabbits: The Animal Answer Guide. Johns Hopkins University Press: Baltimore, Maryland.

Lyman, R. L. 1991. Late Quaternary biogeography of the pygmy rabbit (*Brachylagus idahoensis*) in eastern Washington. Journal of Mammalogy 72:110–117.

———. 2004. Biogeographic and conservation implications of Late Quaternary pygmy rabbits. Western North American Naturalist 64:1–6.

Lynch, A. J., D. W. Duszynski, and J. A. Cook. 2007. Species of coccidia (Apicomplexa: Eimeriidae) infecting pikas from Alaska, U.S.A. and Northeastern Siberia, Russia. Journal of Parasitology 93:1230–1234.

MacArthur, R. A., and L. C. H. Wang. 1974. Behavioral thermoregulation in the pika, *Ochotona princeps*: a field study using radiotelemetry. Canadian Journal of Zoology 52:353–358.

MacDonald, S. O., and C. Jones. 1987. *Ochotona collaris*. Mammalian Species 281:1–4.

MacKenzie, D. I., and W. L. Kendall. 2002. How should detection probability be incorporated into estimates of relative abundance? Ecology 83:2387–2393.

Maddison, W. P. 1997. Gene trees in species trees. Systematic Biology 46:523–536.

Madriñan, S., A. J. Cortés, and J. E. Richardson. 2013. Páramo is the world's fastest evolving and coolest biodiversity hotspot. Frontiers in Genetics 4:192.

Madsen, R. L. 1974. The Influence of Rainfall on the Reproduction of Sonoran Desert Lagomorphs. Unpublished M.S. thesis, University of Arizona, Tucson, Arizona.

Maheswaran, G. 2002. Status and Ecology of Endangered Hispid Hare *Caprolagus hispidus* in Jaldapara Wildlife Sanctuary, West Bengal, India. Bombay Natural History Society and Wildlife Conservation Society: New York.

Mallon, D. 1991. Lagomorphs in Ladakh. Manchester, United Kingdom.

Mamuris, Z., A. I. Sfougaris, and C. Stamatis. 2001. Genetic structure of Greek brown hare (*Lepus europaeus*) populations as revealed by mtDNA RFLP-PCR analysis: implications for conserving genetic diversity. Biological Conservation 101:187–196.

Manakadan, R., and A. R. Rahmani. 1999. Population densities of the blacknaped hare *Lepus nigricollis* at Rollapadu Wildlife Sanctuary, Kurnool District, Andhra Pradesh. Journal of the Bombay Natural History Society 96:221–224.

Mankin, P. C., and R. E. Warner. 1999a. A regional model of the eastern cottontail and land use changes in Illinois. Journal of Wildlife Management 63:956–963.

Mankin, P. C., and R. E. Warner. 1999b. Responses of eastern cottontails to intensive row-crop farming. Journal of Mammalogy 80:940–949.

Marboutin, E., Y. Bray, R. Péroux, B. Mauvy, and A. Lartiges. 2003. Population dynamics in European hare: breeding parameters and sustainable harvest rates. Journal of Applied Ecology 40:580–591.

Marchandeau, S., G. Le Gall-Recule, S. Bertagnoli, J. Aubineau, G. Botti, and A. Lavazza. 2005. Serological evidence for a non-protective RHDV-like virus. Veterinary Research 36:53–62.

Marcström, V., L. B. Keith, A. Engren, and J. R. Cary. 1989. Demographic responses of arctic hares (*Lepus timidus*) to experimental reductions of red foxes (*Vulpes vulpes*) and martens (*Martes martes*). Canadian Journal of Zoology 67:658–668.

Margono, B. A., S. Turubanova, I. Zhuravleva, P. Potapov, A. Tyukavina, A. Baccini, et al. 2012. Mapping and monitoring deforestation and forest degradation in Sumatra (Indonesia) using Landsat time series data sets from 1990 to 2010. Environmental Research Letters 7:1–16.

Marín, A. I., L. Hernández, and J. W. Laundré. 2003. Predation risk and food quantity in the selection of habitat by black-tailed jackrabbit (*Lepus californicus*): an optimal foraging approach. Journal of Arid Environments 55:101–110.

Markham, K. W., and W. D. Webster. 1993. Diets of marsh rabbits, *Sylvilagus palustris* (Lagomorpha, Leporidae), from coastal islands in southeastern North Carolina. Brimleyana 19:147–154.

Marsden, H. M., and N. R. Holler. 1964. Social behavior in confined populations of the cottontail and the swamp rabbit. Wildlife Monographs 13:3–39.

Martins, H., J. A. Milne, and F. Rego. 2002. Seasonal and spatial variation in the diet of the wild rabbit (*Oryctolagus cuniculus* L.) in Portugal. Journal of Zoology 258:395–404.

Matson, J. O., and R. H. Baker. 1986. Mammals of Zacatecas. Special Publication, Museum Texas Tech University 24:1–88.

Matsuzaki, T., H. Suzuki, and M. Kamiya. 1989. Laboratory rearing of the Amami rabbits (*Pentalagus furnessi* Stone, 1900) in captivity. Experimental Animals 38:65–69.

Matthee, C. A. 1993. Mitochondrial DNA Variability and Geographic Population Structure in *Pronolagus rupestris* and *P. randensis* (Mammalia: Lagomorpha). Unpublished M.Sc. thesis, University of Pretoria, Pretoria, South Africa.

Matthee, C. A., and T. J. Robinson. 1996. Mitochondrial DNA differentiation among geographical populations of *Pronolagus rupestris*, Smith's red rock rabbit (Mammalia: Lagomorpha). Heredity 76: 514–523.

Matthee, C. A., K. Collins, and M. Keith. 2004a. *Pronolagus crassicaudatus*. Pp. 418–419. In: Y. Friedman and B. Daly (editors). Red Data Book of the Mammals of South Africa: A Conservation Assessment. CBSG Southern Africa, Conservation Breeding Specialist Group (SSC/IUCN), Endangered Wildlife Trust.

———. 2004b. *Pronolagus randensis*. Pp. 420–421. In: Y. Friedman and B. Daly (editors). Red Data Book of the Mammals of South Africa: A Conservation Assessment. CBSG Southern Africa, Conservation Breeding Specialist Group (SSC/IUCN), Endangered Wildlife Trust.

———. 2004c. *Pronolagus rupestris*. Pp. 422–423. In: Y. Friedman and B. Daly (editors). Red Data Book of the Mammals of South Africa: A Conservation Assessment. CBSG Southern Africa, Conservation Breeding Specialist Group (SSC/IUCN), Endangered Wildlife Trust.

———. 2004d. *Pronolagus saundersiae*. Pp. 424–425. In: Y. Friedman and B. Daly (editors). Red Data Book of the Mammals of South Africa: A Conservation Assessment. CBSG Southern Africa, Conservation Breeding Specialist Group (SSC/IUCN), Endangered Wildlife Trust.

Matthee, C. A., B. J. Van Vuuren, D. Bell, and T. J. Robinson. 2004e. A molecular supermatrix of the rabbits and hares (Leporidae) allows for the identification of five intercontinental exchanges during the Miocene. Systematic Biology 53:433–447.

Mayr, E. 1942. Systematics and the Origin of Species. Columbia University Press: New York.

McCarthy, J. L., T. K. Fuller, K. P. McCarthy, H. T. Wibisono, and M. C. Livolsi. 2012. Using camera trap photos and direct sightings to identify possible refugia for the Vulnerable Sumatran striped rabbit *Nesolagus netscheri*. Oryx 46:438–441.

McCleery, R. A., A. Sovie, R. N. Reed, M. W. Cunningham, M. E. Hunter, and K. M. Hart. 2015. Marsh rabbit mortalities tie pythons to the precipitous decline of mammals in the Everglades. Proceedings of the Royal Society B 282:20150120.

McKenna, M. C., and S. K. Bell. 1997. Classification of Mammals Above the Species Level. Columbia University Press: New York.

Meaney, C. 1987. Cheek-gland odors in pikas (*Ochotona princeps*): discrimination of individual and sex differences. Journal of Mammalogy 68:391–395.

Mearns, E. A. 1890. Descriptions of supposed new species and subspecies of mammals from Arizona. Bulletin of the American Museum of Natural History 2:277–307.

———. 1895. Preliminary description of a new subgenus and six species and subspecies of hares, from the Mexican border of the United States. Proceedings of the United States National Museum 18: 551–565.

Mekonnen, T., M. Yaba, A. Bekele, and J. Malcolm. 2011. Food selection and habitat association of Starck's hare (*Lepus starcki* Petter, 1963) in Bale Mountains National Park, Ethiopia. Asian Journal of Applied Sciences 4:728–734.

Melo-Ferreira J., P. C. Alves, H. Freitas, N. Ferrand, and P. Boursot. 2009. The genomic legacy from the extinct *Lepus timidus* to the three hare species of Iberia: contrast between mtDNA, sex chromosomes and autosomes. Molecular Ecology 18:2643–2658.

Melo-Ferreira J., P. C. Alves, J. Rocha, N. Ferrand, and P. Boursot. 2011.

Interspecific X-chromosome and mitochondrial DNA introgression in the Iberian hare: selection or allele surfing? Evolution 65: 1956–1968.

Melo-Ferreira J., P. Boursot, M. Carneiro, P. J. Esteves, L. Farelo, and P. C. Alves. 2012. Recurrent introgression of mitochondrial DNA among hares (*Lepus* spp.) revealed by species-tree inference and coalescent simulation. Systematic Biology 61:367–381.

Melo-Ferreira, J., P. Boursot, E. Randi, A. Kryukov, F. Suchentrunk, N. Ferrand, et al. 2007. The rise and fall of the mountain hare (*Lepus timidus*) during Pleistocene glaciations: expansion and retreat with hybridization in the Iberian Peninsula. Molecular Ecology 16: 605–618.

Melo-Ferreira, J., P. Boursot, F. Suchentrunk, N. Ferrand, and P. C. Alves. 2005. Invasion from the cold past: extensive introgression of mountain hare (*Lepus timidus*) mitochondrial DNA into three other hare species in northern Iberia. Molecular Ecology 14:2459–2464.

Melo-Ferreira, J., L. Farelo, H. Freitas, F. Suchentrunk, P. Boursot, and P. C. Alves. 2014c. Home-loving boreal hare mitochondria survived several invasions in Iberia: the relative roles of recurrent hybridisation and allele surfing. Heredity 112:265–273.

Melo-Ferreira, J., A. Lemos de Matos, H. Areal, A. A. Lissovsky, M. Carneiro, and P. J. Esteves. 2015. The phylogeny of pikas (*Ochotona*) inferred from a multilocus coalescent approach. Molecular Phylogenetics and Evolution 84:240–244.

Melo-Ferreira J., F. A. Seixas, E. Cheng, L. S. Mills, and P. C. Alves. 2014a. The hidden history of the snowshoe hare, *Lepus americanus*: extensive mitochondrial DNA introgression inferred from multilocus genetic variation. Molecular Ecology 23:4617–4630.

Melo-Ferreira, J., J. Vilela, M. M. Fonseca, R. R. da Fonseca, P. Boursot, and P. C. Alves. 2014b. The elusive nature of adaptive mitochondrial DNA evolution of an arctic lineage prone to frequent introgression. Genome Biology and Evolution 6:886–896.

Meng, J., and Y.-M. Hu. 2004. Lagomorphs from the Yihesubu Late Eocene of Nei Mongol (Inner Mongolia). Vertebrata PalAsiatica 42: 261–275.

Meng, J., and A. R. Wyss. 2005. Glires (Lagomorpha, Rodentia). Pp. 145–158. In: K. D. Rose and J. D. Archibald (editors). The Rise of Placental Mammals: Origins and Relationships of Major Extant Clades. Johns Hopkins University Press: Baltimore, Maryland.

Meng, J., G. J. Bowen, J. Ye, P. L. Koch, S. Ting, Q. Li, et al. 2004. *Gomphos elkema* (Glires, Mammalia) from the Erlian Basin: evidence for the Early Tertiary Bumbanian Land Mammal Age in Nei-Mongol, China. American Museum Novitates 3425:1–24.

Meng, J., Y.-M. Hu, and C.-K. Li. 2003. The osteology of *Rhombomylus* (Mammalia, Glires): implications for phylogeny and evolution of Glires. Bulletin of the American Museum of Natural History 275: 1–247.

———. 2005. *Gobiolagus* (Lagomorpha, Mammalia) from Eocene Ula Usu, Inner Mongolia, and comments on Eocene lagomorphs of Asia. Palaeontologia Electronica 8:7A.

Mengoni, C., N. Mucci, and E. Randi. 2015. Genetic diversity and no evidences of recent hybridization in the endemic Italian hare (*Lepus corsicanus*). Conservation Genetics 16:477–489.

Menon, V. 2009. Mammals of India. Princeton University Press: Princeton, New Jersey.

———. 2014. Indian Mammals—A Field Guide. Hachette India: Guragon, India.

Mercer, W. E., B. J. Hearn, and C. Finlay. 1981. Arctic hare populations in insular Newfoundland. Pp. 450–468. In: K. Myers and C. D. Mac-Innes (editors). Proceedings of the World Lagomorph Conference. University of Guelph: Guelph, Ontario, Canada.

Meredith, R. W., J. E. Janecka, J. Gatesy, O. A. Ryder, C. A. Fisher, E. C. Teeling, et al. 2011. Impacts of the Cretaceous Terrestrial Revolution and KPg extinction on mammal diversification. Science 334: 521–524.

Merritt, J. F. 1987. Guide to the Mammals of Pennsylvania. University of Pittsburgh Press: Pittsburgh, Pennsylvania.

Millar, C. I., R. D. Westfall, and D. L. Delany. 2013. New records of marginal locations for American pika (*Ochotona princeps*) in the western Great Basin. Western North American Naturalist 73:457–476.

Millar, J. S. 1974. Success of reproduction in pikas, *Ochotona princeps* (Richardson). Journal of Mammalogy 55:527–542.

Millar, J. S., and F. C. Zwickel. 1972. Characteristics and ecological significance of hay piles of pikas. Mammalia 36:657–667.

Mitchell, R. M. 1978. The *Ochotona* (Lagomorpha: Ochotonidae) of Nepal. Säugetierkundliche Mitteilungen 3:208–214.

———. 1981. The *Ochotona* (Lagomorpha: Ochotonidae) of Asia. Pp. 1031–1038. In: D. Liu (editor). Geological and Ecological Studies of Qinghai-Xizang Plateau. Science Press: Beijing, PRC.

Mitchell-Jones, A. J., G. Amori, W. Bogdanowicz, B. Kryštufek, P. J. H. Reijnders, F. Spitzenberger, et al. 1999. The Atlas of European Mammals. T. & A. D. Poyser: London.

Mohr, W. P., and C. O. Mohr. 1936. Recent jackrabbit populations at Rapidan, Minnesota. Journal of Mammalogy 17:112–114.

Molur, S., C. Srinivasulu, B. Srinivasulu, S. Walker, P. O. Nameer, and L. Ravikumar (editors). 2005. Status of South Asian Non-volant Small Mammals: Conservation Assessment and Management Plan (C.A.M.P.), workshop report. Zoo Outreach Organisation / CBSG-South Asia: Coimbatore, India.

Montes, C., A. Cardona, C. Jaramillo, A. Pardo, J. C. Silva, V. Valencia, et al. 2015. Middle Miocene closure of the Central American seaway. Science 348:226–229.

Mora, J. M. 2000. Mamíferos Silvestres de Costa Rica. Editorial UNED: San José, Costa Rica.

Morgan, G. S., and R. S. White, Jr. 2005. Miocene and Pliocene vertebrates from Arizona. Pp. 115–136. In: A. B. Heckert and S. G. Lucas (editors). Vertebrate Paleontology in Arizona. New Mexico Museum of Natural History and Science Bulletin 29:115–136.

Morrison, S. F., and D. S. Hik. 2007. Demographic analysis of a declining pika *Ochotona collaris* population: linking survival to broad-scale climate patterns via spring snowmelt patterns. Journal of Animal Ecology 76:899–907.

Morrison, S. F., L. Barton, P. Caputa, and D. S. Hik. 2004. Forage selection by collared pikas, *Ochotona collaris*, under varying degrees of predation risk. Canadian Journal of Zoology 82:533–540.

Morrison, S. F., G. Pelchat, A. Donahue, and D. S. Hik. 2009. Influence of food hoarding behavior on the over-winter survival of pikas in strongly seasonal environments. Oecologia 159:107–116.

Mossman, A. S. 1955. Reproduction of the brush rabbit in California. Journal of Wildlife Management 19:177–184.

Mucina, L., and M. C. Rutherford (editors). 2006. The Vegetation of South Africa, Lesotho and Swaziland. Strelitzia 19, South African National Biodiversity Institute: Pretoria, South Africa.

Murphy, W. J., T. H. Pringle, T. A. Crider, M. S. Springer, and W. Miller.

2007. Using genomic data to unravel the root of the placental mammal phylogeny. Genome Research 17:413–421.

Murray, D., and A. T. Smith. 2008. *Lepus othus*. The IUCN Red List of Threatened Species 2008: e.T11795A3308465. http://dx.doi.org/10.2305/IUCN.UK.2008.RLTS.T11795A3308465.en.

Murray, D. L. 1999. An assessment of overwinter food limitation in a snowshoe hare population at a cyclic low. Oecologia 120:50–58.

———. 2000. A geographic analysis of snowshoe hare population demography. Canadian Journal of Zoology 78:1207–1217.

———. 2003. Snowshoe hare and other hares. Pp. 147–175. In: G. A. Feldhamer and B. Thompson (editors). Wild Mammals of North America. Volume II. Johns Hopkins University Press: Baltimore, Maryland.

Murray, D. L., L. B. Keith, and J. R. Cary. 1998. Do parasitism and nutritional status interact to affect production in snowshoe hares? Ecology 79:1209–1222.

Nagata, J., Y. Sonoda, K. Hamaguchi, N. Ohnishi, S. Kobayashi, K. Sugimura, et al. 2009. Isolation and characterization of microsatellite loci in the Amami rabbit (*Pentalagus furnessi*). Conservation Genetics 10:1121–1123.

Nalls, A. V., L. K. Ammerman, and R. C. Dowler. 2012. Genetic and morphologic variation in the Davis mountains cottontail (*Sylvilagus robustus*). The Southwestern Naturalist 57:1–7.

Nath, N. K., and K. Machary. 2015. An ecological assessment of hispid hare *Caprolagus hispidus* (Mammalia: Lagomorpha) in Manas National Park. Journal of Threatened Taxa 7:8195–8204.

Naumov, R. L. 1974. Ecology of *Ochotona alpina* in the west Sayan. Zoologicheskii Zhurnal 53:1524–1529. [In Russian].

Nelson, E. W. 1899. Mammals of the Tres Marías Islands. North American Fauna 14:14–19.

———. 1904. Description of seven new rabbits from Mexico. Proceedings of the Biological Society of Washington 17:103–110.

———. 1907. Descriptions of new North American rabbits. Proceedings of the Biological Society of Washington 20:81–84.

———. 1909. The rabbits of North America. North American Fauna 29: 9–287.

Nelson, T., J. L. Holechek, R. Valdez, and M. Cardenas. 1997. Wildlife numbers on late and mid seral Chihuahuan Desert rangelands. Journal of Range Management 50:593–599.

Newey, S., and S. Thirgood. 2009. Parasite-mediated reduction in fecundity of mountain hares. Proceeding of the Royal Society B-Biological Sciences 271:S413–S415.

Newmark, W. D. 1995. Extinction of mammal populations in western North American national parks. Conservation Biology 9:512–526.

NIER (National Institute of Environmental Research). 2006. Study on the Outbreak and Prevention of Wildlife Diseases. National Institute of Environmental Research: Incheon, South Korea. [In Korean].

Nikolskii, A. A., and T. D. Mukhamediev. 1997. Territoriality in the Altai pika (*Ochotona alpina*). Giber Faune Sauvage 143:359–383.

Nowak, R. M. 1991. Walker's Mammals of the World. Johns Hopkins University Press: Baltimore, Maryland.

Nunome, M., G. Kinoshita, M. Tomozawa, H. Torii, R. Matsuki, F. Yamada, et al. 2014. Lack of association between winter coat colour and genetic population structure in the Japanese hare, *Lepus brachyurus* (Lagomorpha: Leporidae). Biological Journal of the Linnean Society 111:761–776.

Nunome, M., H. Torii, R. Matsuki, G. Kinoshita, and H. Suzuki. 2010.

The influence of Pleistocene refugia on the evolutionary history of the Japanese hare, *Lepus brachyurus*. Zoological Science 27:746–754.

O'Donoghue, M. 1994. Early survival of juvenile snowshoe hares. Ecology 75:1582–1592.

O'Donoghue, M., and C. M. Bergman. 1992. Early movements and dispersal of juvenile snowshoe hares. Canadian Journal of Zoology 70:1787–1791.

O'Donoghue, M., and S. Boutin. 1995. Does reproductive synchrony affect juvenile survival rates of northern mammals? Oikos 74: 115–120.

O'Donoghue, M., and C. J. Krebs. 1992. Effects of supplemental food on snowshoe hare reproduction and juvenile growth at a cyclic population peak. Journal of Animal Ecology 61:631–641.

Ogen-Odoi, A., and T. G. Dilworth. 1984. Effects of grassland burning on the savanna hare-predator relationships in Uganda. African Journal of Ecology 22:101–106.

———. 1985. Ectozoochory by hares (*Lepus crawshayi*) in Queen Elizabeth National Park, Uganda. The East Africa Natural History Society 75:1–6.

Ognev, S. I. 1966. Mammals of the U.S.S.R. and Adjacent Countries. Mammals of Eastern Europe and Northern Asia. Volume IV: Rodents. Israel Program for Scientific Translations: Jerusalem, Israel.

Olcott, S. P., and R. E. Barry. 2000. Environmental correlates of geographic variation in body size of the eastern cottontail (*Sylvilagus floridanus*). Journal of Mammalogy 81:986–998.

O'Leary, M. A., J. I. Bloch, J. J. Flynn, T. J. Gaudin, A. Giallombardo, N. P. Giannini, et al. 2013. The placental mammal ancestor and the post-K-Pg radiation of placentals. Science 339:662–667.

Oliveira e Silva, J. R., and P. M. Dellias. 1973. Biologia do *Sylvilagus brasiliensis tapetillus* (Lagomorpha) tapeti em cativeiro. Contribuição para o estudo. Revista da Faculdade de Odontologia de São José dos Campos 2:27–31.

Oliver, W. L. R. 1979. The doubtful future of the pygmy hog and the hispid hare. Pygmy hog survey report—Part 1. Journal of Bombay Natural History Society 75:337–341.

———. 1980. The Pygmy Hog: The Biology and Conservation of the Pygmy Hog, *Sus salvanius* and the Hispid Hare, *Caprolagus hispidus*. Special Scientific Report #1. Jersey Wildlife Preservation Trust: Jersey, United Kingdom.

Olsen, J., B. Cooke, S. Trost, and D. Judge. 2014. Is wedge-tailed eagle, *Aquila audax*, survival and breeding closely linked to the abundance of introduced wild rabbits *Oryctolagus cuniculus*? Wildlife Research 41:95–105.

Ong, D. M. 1998. The Convention on International Trade in Endangered Species (CITES, 1973): implications of recent developments in international and EC environmental law. Journal of Environmental Law 10:291–314.

Orlov, G. I. 1983. The neck gland of pikas (Lagomorpha, Ochotonidae) and the scent marking of *Ochotona alpina* related to its functioning. Zoologicheskii Zhurnal 62:1709–1717. [In Russian].

Orr, R. T. 1940. The rabbits of California. Occasional Papers of the California Academy of Science 19:1–227.

———. 1960. An analysis of the recent land mammals. Systematic Zoology 9:47–90.

Otsu, S. 1974. The ecology and control of the Tohoku hare *Lepus brachyurus angustidens* Hollister. Bulletin of the Yamagata Prefectural Forest Experiment Station 5:1–94. [In Japanese].

Padgett, T. M. 1989. A range extension of the marsh rabbit, *Sylvilagus palustris*, from southeastern Virginia. Virginia Journal of Science 40:177.

Pak, U. I., M. S. Kim, et al. 2002. Red data book of DPR Korea (Animal). MAB National Committee: Pyongyang, DPR Korea.

Palacios, F. 1976. Descripción de una nueva especie de liebre (*Lepus castroviejoi*) endémica de la cordillera Cantábrica. Doñana Acta Vertebrata 3:205–223.

———. 1983. On the taxonomic status of the genus *Lepus* in Spain. Acta Zoologica Fennica 174:27–30.

———. 1989. Biometric and morphologic features of the species of the genus *Lepus* in Spain. Mammalia 53:227–263.

———. 1995. Systematics of the indigenous hares of Italy traditionally identified as *Lepus europaeus* Pallas, 1778 (Mammalia: Leporidae). Bonner Zoologische Beiträge 46:59–91.

Palacios, F., and J. Fernandez. 1992. A new subspecies of hare from Majorca (Baleartic Islands). Mammalia 56:71–85.

Palacios, F., and N. Lopez. 1980. Morfologia dentaria de las liebres europeas (Lagomorpha, Leporidae). Doñana Acta Vertebrata 7:61–91.

Palacios, F., and M. Meijide. 1979. Distribución Geográfica y Hábitat de las Liebres en la Península Ibérica. Instituto Nacional para la Conservación de la Naturaleza: Madrid, Spain.

Palmer, T. S. 1897. The jack rabbits of the United States. USDA Biological Survey Bulletin 8:1–88.

Pan, Q. H., Y. X. Wang, and K. Yan (editors). 2007. A Field Guide to the Mammals of China. China Forestry Publishing House: Beijing, PRC. [In Chinese].

Panek, M., and R. Kamieniarz. 1999. Relationships between density of brown hare *Lepus europaeus* and landscape structure in Poland in the years 1981–1995. Acta Theriologica 44:67–75.

Pante, E., C. Schoelinck, and N. Puillandre. 2015. From integrative taxonomy to species description: one step beyond. Systematic Biology 64:152–160.

Parker, G. R. 1977. Morphology, reproduction, diet, and behavior of the arctic hare (*Lepus arcticus monstrabilis*) on Axel Heiberg Island, Northwest Territories. The Canadian Field-Naturalist 91:8–18.

Paupério, J., and P. C. Alves. 2008. Diet of the Iberian hare (*Lepus granatensis*) in a mountain ecosystem. European Journal of Wildlife Research 54:571–579.

Peacock, M. M., and A. T. Smith. 1997. The effect of habitat fragmentation on dispersal patterns, mating behavior, and genetic variation in a pika (*Ochotona princeps*) metapopulation. Oecologia 112:524–533.

Pech, R. P., A. D. Arthur, Y. M. Zhang, and L. Hui. 2007. Population dynamics and responses to management of plateau pikas *Ochotona curzoniae*. Journal of Applied Ecology 44:615–624.

Pedlar, R. D., R. Brandle, J. Read, R. Southgate, P. Bird, and K. Moseby. 2016. Rabbit biocontrol and landscape-scale recovery of threatened desert mammals. Conservation Biology 30:774–782.

Pehrson, Å. 1983. Caecotrophy in caged mountain hares (*Lepus timidus*). Journal of Zoology (London) 199:563–574.

Pehrson, Å., and B. Lindlöf. 1984. Impact of winter nutrition on reproduction in captive mountain hares (*Lepus timidus*) (Mammalia: Lagomorpha). Journal of Zoology (London) 204:201–209.

Pen, H. S., Y. T. Kao, C. K. Lu, Z. C. Feng, and C. X. Chen. 1962. Report on mammals from southwestern Szechwan and northwestern Yunnan. Acta Zoologica Sinica 14 (supplement):105–133. [In Chinese].

Peng, J., and X. Zhong. 2005. A Guide to Identification and Conservation of Wild Mammals in Ganzi Tibetan Autonomous Prefecture of Sichuan Province. Sichuan Science and Technology Press: Chengdu, PRC.

Pérez, V. A., J. Arrollo-Cabrales, and A. Santos. 2008. Generalidades de los mamíferos del Pleistoceno tardío de Oaxaca. Naturaleza y Desarrollo 6:5–11.

Pérez-Amador, M. C., A. García, A. Velázquez, and H. Granados. 1985. Variaciones cuantitativas de la melanina durante la muda del pelaje del conejo de los volcanes (*Romerolagus diazi*). Archivos de Investigación Médica 16:41–46.

Petter, F. 1959. Eléments d'une révision des lièvres africains du sous-genre *Lepus*. Mammalia 23:41–67.

———. 1961. Eléments d'une révision des lievres européens et asiatiques du sous-genre *Lepus*. Zeitschrift für Säugetierkunde 26:30–40.

———. 1963. Nouveaux éléments d'une révision des lièvres africains. Mammalia 27:238–255.

———. 1972. Order Lagomorpha. Pp. 1–7. In: J. Meester and H. W. Setzer (editors). The Mammals of Africa: An Identification Manual. Part 5. Smithsonian Institution Press: Washington, D.C.

Pfaffenberger, G. S., and V. B. Valencia. 1988. Helminths of sympatric black-tailed jack rabbits (*Lepus californicus*) and desert cottontails (*Sylvilagus audubonii*) from the high plains of eastern New Mexico. Journal of Wildlife Diseases 24:375–377.

Pfeffer, P. 1969. Considerations sur l'écologie des forêts claires du Cambodge oriental. Terre et la Vie 23:3–24.

Pielowski, Z. 1972. Home range and degree of residence of the European hare. Acta Theriologica 17:93–103.

Pierce, J. E., R. T. Larsen, J. T. Flinders, and J. C. Whiting. 2011. Fragmentation of sagebrush communities: does an increase in habitat edge impact pygmy rabbits? Animal Conservation 14:314–321.

Pierpaoli, M., F. Riga, V. Trocchi, and E. Randi. 1999. Species distinction and evolutionary relationships of the Italian hare (*Lepus corsicanus*) as described by mitochondrial DNA sequencing. Molecular Ecology 8:1805–1817.

———. 2003. Hare populations in Europe: intra- and interspecific analysis of mtDNA variation. Comptes Rendus Biologies 326:80–84.

Pietri, C. 2015. Range and status of the Italian hare *Lepus corsicanus* in Corsica. Hystrix, the Italian Journal of Mammalogy 26:166–168.

Pietri, C., P. C. Alves, and J. Melo-Ferreira. 2011. Hares in Corsica: high prevalence of *Lepus corsicanus* and hybridization with introduced *L. europaeus* and *L. granatensis*. European Journal of Wildlife Research 57:313–321.

Pinheiro, A., F. Neves, A. L. Matos, J. Abrantes, W. van der Loo, R. Mage, and P. J. Esteves. 2015. An overview of the lagomorph immune system and its genetic diversity. Immunogenetics 68:83–107.

Platt, S. G., and M. Bunch. 2000. Distribution and status of the swamp rabbit in South Carolina. Proceedings of the Annual Conference for the Association of Southeastern Fish and Wildlife Agencies 54:408–414.

Posada, D. 2016. Phylogenomics for systematic biology. Systematic Biology 65:353–356.

Prakash, I., and G. C. Taneja. 1969. Reproduction biology of the Indian desert hare, *Lepus nigricollis dayanus* Blanford. Mammalia 33:102–117.

Price, A. J., and J. L. Rachlow. 2011. Development of an index of abundance for pygmy rabbits. Journal of Wildlife Management 75:929–937.

Price, A. J., W. A. Estes-Zumpf, and J. L. Rachlow. 2010. Survival of juvenile pygmy rabbits. Journal of Wildlife Management 74:43–47.

Pringle, J. A. 1974. The distribution of mammals in Natal. Part 1. Primates, Hyracoidea, Lagomorpha (except *Lepus*), Pholidota and Tubulidentata. Annals of the Natal Museum 22:173–186.

Probert, B. L., and J. A. Litvaitis 1996. Behavioral interactions between invading and endemic lagomorphs: implications for conserving a declining species. Biological Conservation 76:289–295.

Proskurina, N. S. 1991. Interrelationship of locomotion and acoustic activity in *Ochotona daurica* (Mammalia, Lagomorpha). Zoologichesky Zhurnal 70:91–96.

Puggioni, G., P. Cavadini, C. Maestrale, R. Scivoli, G. Botti, C. Ligios, et al. 2013. The new French 2010 variant of the rabbit hemorrhagic disease virus causes an RHD-like disease in the Sardinian Cape hare (*Lepus capensis mediterraneus*). Veterinary Research 44:96.

Qu, J. P., M. Liu, M. Yang, and Y. M. Zhang. 2012. Reproduction of plateau pika (*Ochotona curzoniae*) on the Qinghai-Tibetan plateau. European Journal of Wildlife Research 58:269–277.

Rachlow, J. L., D. M. Sanchez, and W. A. Estes-Zumpf. 2005. Natal burrows and nests of free-ranging pygmy rabbits (*Brachylagus idahoensis*). Western North American Naturalist 65:136–139.

Ramírez-Albores, J. E., L. León-Paniagua, and A. G. Navarro-Sigüenza. 2014. Mamíferos silvestres del parque ecoturístico Piedra Canteada y alrededores, Tlaxcala, México; con notas sobre algunos registros notables para el área. Revista Mexicana de Biodiversidad 85:48–61.

Ramírez-Pulido, J., A. Martínez, and G. Urbano. 1977. Mamíferos de la costa grande de Guerrero. Anales del Instituto de Biología, Universidad Nacional Autónoma de México, Serie Zoología 48:243–292.

Ramírez-Silva, J. P., F. X. González-Cózatl, E. Vázquez-Domínguez, and F. A. Cervantes. 2010. Phylogenetic position of Mexican jackrabbits within the genus *Lepus* (Mammalia: Lagomorpha): a molecular perspective. Revista Mexicana de Biodiversidad 81:721–731.

Rausch, R. L. 1961. Notes on the collared pika, *Ochotona collaris* (Nelson), in Alaska. The Murrlet 42:22–24.

———. 1963. *Schizorchis yamashitai* sp. n. (Cestoda: Anoplocephalidae) from the northern pika *Ochotona hyperborea* Pallas in Hokkaido. Journal of Parasitology 49:479–482.

Rausch, R. L., and M. Ohbayashi. 1974. On some anoplocephaline cestodes from pikas, *Ochotona* spp. (Lagomorpha), in Nepal, with the description of *Ectopocephalium abei* gen. et sp. n. Journal of Parasitology 60:596–604.

Ray, C., E. Beever, and S. Loarie. 2012. Retreat of the American pika: up the mountain or into the void? Pp. 245–270. In: J. F. Brodie, E. Post, and D. F. Doak (editors). Wildlife Conservation in a Changing Climate. University of Chicago Press: Chicago, Illinois.

Reading, R. R., H. Mix, B. Lhagvasuren, and N. Tseveenmyadag. 1998. The commercial harvest of wildlife in Dorod Aimag, Mongolia. Journal of Wildlife Management 62:59–71.

Reid, F. A. 1997. A Field Guide to the Mammals of Central America and Southeast Mexico. Oxford University Press: New York.

———. 2006. A Field Guide to Mammals of North America North of Mexico. 4th edition. Houghton Mifflin Company: Boston, Massachusetts.

Reid, N. 2011. European hare (*Lepus europaeus*) invasion ecology: implication for the conservation of the endemic Irish hare (*Lepus timidus hibernicus*). Biological Invasions 13:559–569.

Reid, N., R. A. McDonald, and W. I. Montgomery. 2007. Mammals and agri-environment schemes: hare haven or pest paradise? Journal of Applied Ecology 44:1200–1208.

Rice, B. L., and M. Westoby. 1978. Vegetation responses of some Great Basin shrub communities protected against jackrabbits and domestic stock. Journal of Range Management 31:28–34.

Richard, M., and B. Daniel. 2013. African savannah hare *Lepus microtis* near Tantan—a new mammal species for North Moroccan Atlantic Sahara. Go-South Bulletin 10:263–265.

Richerson, J. V., J. F. Scudday, and S. P. Tabor. 1992. An ectoparasite survey of mammals in Brewster County, Texas, 1982–1985. Southwestern Entomologist 17:7–15.

Rico, Y., C. Lorenzo, F. X. González-Cózatl, and E. Espinoza. 2008. Phylogeography and population structure of the endangered Tehuantepec jackrabbit *Lepus flavigularis*: implications for conservation. Conservation Genetics 9:1467–1477.

Riga, F., V. Trocchi, E. Randi, and S. Toso. 2001. Morphometric differentiation between the Italian hare (*Lepus corsicanus* de Winton, 1898) and the European brown hare (*Lepus europaeus* Pallas, 1778). Journal of Zoology, London 253:241–252.

Rioja, T., A. Carrillo-Reyes, and C. Lorenzo. 2012. Análisis de población viable para determinar el riesgo de extinción de la liebre de Tehuantepec (*Lepus flavigularis*) en Santa María del Mar, Oaxaca. Therya 3: 137–150.

Rioja, T., C. Lorenzo, E. Naranjo, L. Scott, and A. Carrillo. 2008. Polygynous mating behavior in the endangered Tehuantepec jackrabbit (*Lepus flavigularis*). Western North American Naturalist 68: 343–349.

———. 2011. Breeding and parental care in the endangered Tehuantepec jackrabbit (*Lepus flavigularis*). Western North American Naturalist 71:56–66.

Roberts, A. 1951. The Mammals of South Africa. Central News Agency: Cape Town, South Africa.

Roberts, T. J. 1977. The Mammals of Pakistan. Ernest Benn Limited: London.

Robinson, T. J. 1981a. The bushman hare. Pp. 934–938. In: K. Myers and C. D. MacInnes (editors). Proceedings of the World Lagomorph Conference. University of Guelph: Guelph, Ontario, Canada.

———. 1981b. Systematics of the South African Leporidae. D.Sc. dissertation, University of Pretoria, Pretoria, South Africa.

———. 1981c. Chromosome homologies in South African Lagomorphs. Pp. 56–63. In: K. Myers and C. D. MacInnes (editors). Proceedings of the World Lagomorph Conference. University of Guelph: Guelph, Ontario, Canada.

———. 1986. Incisor morphology as an aid in the systematics of the South African Leporidae. South African Journal of Zoology 21: 297–302.

Robinson, T. J., and N. J. Dippenaar. 1983. Observations on the status of *Lepus saxatilis*, *L. whytei* and *L. crawshayi* in southern Africa. Acta Zoologica Fennica 174:35–39.

———. 1987. Morphometrics of the Southern African Leporidae II: *Lepus* Linnaeus 1758, and *Bunolagus* Thomas, 1929. Annals of the Transvaal Museum 34:379–404.

Robinson T. J., and C. A. Matthee. 2005. Phylogeny and evolutionary origins of the Leporidae: a review of cytogenetics, molecular analyses and a supermatrix analysis. Mammal Review 35:231–247.

Robinson, T. J., and J. D. Skinner. 1983. Karyology of the riverine rabbit, *Bunolagus monticularis*, and its taxonomic implications. Journal of Mammalogy 64:678–681.

Robinson, T. J., F. F. B. Elder, and J. A. Chapman. 1983a. Karyotypic conservatism in the genus *Lepus* (Order Lagomorpha). Canadian Journal of Genetics and Cytology 25:540–544.

———.1983b. Evolution of chromosomal variation in cottontails, genus *Sylvilagus* (Mammalia, Lagomorpha): *S. aquaticus*, *S. floridanus*, and *S. transitionalis*. Cytogenetics and Cell Genetics 35:216–222.

———. 1984. Evolution of chromosomal variation in cottontails, genus *Sylvilagus* (Mammalia: Lagomorpha). Cytogenetic and Genome Research 38:282–289.

Robinson, T. J., F. F. B. Elder, and W. López-Forment. 1981. Banding studies in the volcano rabbit *Romerolagus diazi* and Crawshay's hare *Lepus crawshayi*: evidence of the leporid ancestral karyotype. Canadian Journal of Genetics and Cytology 23:469–474.

Robinson, T. J., F. Yang, and W. R. Harrison. 2002. Chromosome painting refines the history of genome evolution in hares and rabbits (Order Lagomorpha). Cytogenetics and Genome Research 96:223–227.

Rodriguez, M., J. Palacios, J. Martin, T. Yanes, C. Sánchez, M. Navesco, et al. 1997. La liebre. Ediciones Mundi-Prensa: Madrid, Spain.

Rodríguez-Martínez, L. 2015. Descripción de la Conducta Materna y Desarrollo de las Crías del Conejo Montés *Sylvilagus cunicularius* en Condiciones de Semicautiverio. Unpublished Ph.D. dissertation, Universidad Veracruzana, México.

Rodríguez-Martínez, L., R. Hudson, M. Martínez-Gómez, and A. Bautista. 2014. Description of the nursery burrow of the Mexican cottontail rabbit *Sylvilagus cunicularius* under seminatural conditions. Acta Theriologica 59:193–201.

Rogowitz, G. L. 1990. Seasonal energetics of the white-tailed jackrabbit (*Lepus townsendii*). Journal of Mammalogy 71:277–285.

———. 1992. Reproduction of white-tailed jackrabbits on semi-arid range. Journal of Wildlife Management 56:676–684.

———. 1997. Locomotor and foraging activity of the white-tailed jackrabbit (*Lepus townsendii*). Journal of Mammalogy 78:1172–1181.

Rogowitz, G. L., and M. L. Wolfe. 1991. Intraspecific variation on life-history traits of the white-tailed jackrabbit (*Lepus townsendii*). Journal of Mammalogy 72:796–806.

Romero-Rodriguez, J. 1976. Contribución al conocimiento de las coccidiopatias de los Lagomorpha: Estudio de los Protozoa-Eimeridae, *E. leporis* y *E. europaea*, parasitas de *Lepus granatensis*. Revista Iberica Parasitologia 36:131–143.

Rondinini, C., F. Chiozza, and L. Boitani. 2006. High human density in the irreplaceable sites for African vertebrate conservation. Biological Conservation 133:358–363.

Rose, K. D., V. Burke DeLeon, P. Missiaen, R. S. Rana, A. Sahni, L. Singh, et al. 2008. Early Eocene lagomorph (Mammalia) from Western India and the early diversification of Lagomorpha. Proceedings of the Royal Society B 275:1203–1208.

Rosell, J. M. (editor). 2000. Enfermedades del Conejo (Diseases of the Rabbit). Mundi-Prensa Libros SA: Madrid, Spain.

Ross, J., and A. M. Tittensor. 1986. Influence of myxomatosis in regulating rabbit numbers. Mammal Review 16:163–168.

Ross, S., B. Munkhtsog, and S. Harris. 2010. Dietary composition, plasticity, and prey selection of Pallas's cats. Journal of Mammalogy 91: 811–817.

Roth, E. L., and E. L. Cockrum. 1976. A survey of the mammals of the Fort Bowie National Historic Site. In: Survey of Vertebrate Fauna of

Fort Bowie Historic Site, Arizona. Technical Report No. 2. Cooperative National Park Resources Studies Unit, University of Arizona: Tucson, Arizona.

Roth, H. H., and G. Merz. 1997. Wildlife Resources: A Global Account of Economic Use. Springer: Berlin.

Rouco, C., R. Villafuerte, F. Castro, and P. Ferreras. 2011. Effect of artificial warren size on a restocked European wild rabbit population. Animal Conservation 14:117–123.

Roy Nielsen, C. L., S. M. Wakamiya, and C. K. Nielsen. 2008. Viability and patch occupancy of the state-endangered swamp rabbit metapopulation in southwestern Indiana. Biological Conservation 141: 1043–1054.

Ruedas, L. A. 1998. Systematics of Sylvilagus Gray, 1867 (Lagomorpha: Leporidae) from southwestern North America. Journal of Mammalogy 79:1355–1378.

Ruedas, L. A., and J. Salazar-Bravo. 2007. Morphological and chromosomal taxonomic assessment of Sylvilagus brasiliensis gabbi (Leporidae). Mammalia 71:63–69.

Ruedas, L. A., R. C. Dowler, and E. Aita. 1989. Chromosomal variation in the New England cottontail, Sylvilagus transitionalis. Journal of Mammalogy 70:860–864.

Ruedas, L. A., S. M. Silva, J. H. French, R. N. Platt, II, J. Salazar-Bravo, J. M. Mora, and C. W. Thompson. 2017. A prolegomenon to the systematics of South American cottontail rabbits (Mammalia, Lagomorpha, Leporidae: Sylvilagus): designation of a neotype for S. brasiliensis (Linnaeus, 1758), and restoration of S. andinus (Thomas, 1897) and S. tapetillus Thomas, 1913. Miscellaneous Publications, Museum of Zoology, University of Michigan 205:1–67.

Ruiz-Fons, F., E. Ferroglio, and C. Gortázar. 2013. Leishmania infantum in free-ranging hares, Spain, 2004–2010. Euro Surveill 18:30.

Sako, T., M. Uchimura, and Y. Koreeda. 1991. Keeping and breeding of the Amami rabbit in the Kagoshima Hirakawa Zoo. Animals and Zoo 43:272–274. [In Japanese].

Sanchez, D. M., and J. L. Rachlow. 2008. Spatio-temporal factors affecting space use by pygmy rabbits. Journal of Wildlife Management 72: 1304–1310.

Sántiz, E., A. González-Romero, C. Lorenzo, S. Gallina-Tessaro, and F. A. Cervantes. 2012. Uso y selección de asociaciones vegetales por la liebre de Tehuantepec (Lepus flavigularis) en Oaxaca, México. Therya 3:127–136.

Saprageldyev, M. 1987. Ecology of the Afghan pika in Turkmenistan. Ylym: Ashkhabad, Turkmenistan. [In Russian].

Scalera, R., and F. M. Angelici. 2003. Rediscovery of the Apennine hare Lepus corsicanus in Corsica. Bollettino del Museo Regionale di Scienze Naturali, Torino 20:161–166.

Scandura, M., L. Iacolina, H. Ben Slimen, F. Suchentrunk, and M. Apollonio. 2007. Mitochondrial CR-1 variation in Sardinian hares and its relationship with other Old World hares (genus Lepus). Biochemical Genetics 45:305–323.

Schai-Braun, S. C., and K. Hackländer. 2014. Home range use by the European hare (Lepus europaeus) in a structurally diverse agricultural landscape analysed at a fine temporal scale. Acta Theriologica 59:277–287.

Schai-Braun, S. C., T. S. Reichlin, T. Ruf, E. Klansek, F. Tataruch, W. Arnold, et al. 2015. The European hare (Lepus europaeus): a picky herbivore searching for plant parts rich in fat. PLoS 10:e0134278.

Schai-Braun, S. C., H. G. Rödel, and K. Hackländer. 2012. The influence of daylight regime on diurnal locomotor activity patterns of the European hare (Lepus europaeus) during summer. Mammalian Biology 77:434–440.

Schai-Braun, S. C., D. Weber, and K. Hackländer. 2013. Spring and autumn habitat preferences of active European hares (Lepus europaeus) in an agricultural area with low hare density. European Journal of Wildlife Research 59:387–397.

Scharine, P. D., C. K. Nielsen, E. M. Schauber, and L. Rubert. 2009. Swamp rabbits in floodplain ecosystems: influence of landscape- and stand-level habitat on relative abundance. Wetlands 29: 615–623.

Scharine, P. D., C. K. Nielsen, E. M. Schauber, L. Rubert, and J. C. Crawford. 2011. Occupancy, detection, and habitat associations of sympatric lagomorphs in early-successional bottomland hardwood forests. Journal of Mammalogy 92:880–890.

Schauber, E. M., P. D. Scharine, C. K. Nielsen, and L. Rubert. 2008. An artificial latrine log for swamp rabbit studies. Journal of Wildlife Management 72:561–563.

Scheibe, J. S., and R. Henson. 2003. The distribution of swamp rabbits in southeast Missouri. Southeastern Naturalist 2:327–334.

Schipper, J., J. S. Chanson, F. Chiozza, N. A. Cox, M. Hoffmann, V. Katariya, et al. 2008. The status of the world's land and marine mammals: diversity, threat, and knowledge. Science 322:225–230.

Schmalz, J. M., B. Wachocki, M. Wright, S. I. Zeveloff, and M. M. Skopec. 2014. Habitat selection by the pygmy rabbit (Brachylagus idahoensis) in northeastern Utah. Western North American Naturalist 74: 456–466.

Schmidt, J. A., R. A. McCleery, J. R. Seavey, S. E. Cameron Devitt, and P. M. Schmidt. 2012. Impacts of a half century of sea-level rise and development on an endangered mammal. Global Change Biology 18:3536–3542.

Schmidt, P. M., R. A. McCleery, R. R. Lopez, N. J. Silvy, and J. A. Schmidt. 2010. Habitat succession, hardwood encroachment and raccoons as limiting factors for Lower Keys marsh rabbits. Biological Conservation 143:2703–2710.

Schmidt-Nielsen, K., T. J. Dawson, H. T. Hammel, D. Hinds, and D. C. Jackson. 1965. The jack rabbit—a study in its desert survival. Hvalradets Skrifter 48:125–142.

Schmitz, M. M., J. E. Foley, R. W. Kasten, B. B. Chomel, and R. S. Larsen. 2014. Prevalence of vector-borne bacterial pathogens in riparian brush rabbits (Sylvilagus bachmani riparius) and their ticks. Journal of Wildlife Diseases 50:369–373.

Schneider, E. 1990. Lagomorphs. Pp. 244–326. In: S. P. Parker (editor). Grzimek's Encyclopedia of Mammals. Volume 4. McGraw-Hill: New York.

Schnell, I. B., P. F. Thomsen, N. Wilkinson, M. Rasmussen, L. R. D. Jensen, E. Willerslev, et al. 2012. Screening mammal biodiversity using DNA from leeches. Current Biology 22:R262–R263.

Schwartz, C. W., and E. R. Schwartz. 2001. The Wild Mammals of Missouri. University of Missouri Press: Columbia, Missouri.

Schwensow, N., B. Cooke, J. Fickel, W. Lutz, and S. Sommer. 2012. Changes in liver gene expression indicate pathways associated with rabbit haemorrhagic disease expression in wild rabbits. The Open Immunology Journal 5:20–26.

Seixas, F. A., J. Juste, P. F. Campos, M. Carneiro, N. Ferrand, P. C. Alves, et al. 2014. Colonization history of Mallorca Island by the European rabbit, Oryctolagus cuniculus, and the Iberian hare, Lepus granatensis (Lagomorpha: Leporidae). Biological Journal of the Linnean Society 111:748–760.

SEMARNAT (Secretaría de Medio Ambiente y Recursos Naturales). 2010. Norma Oficial Mexicana NOM-059-SEMARNAT-2010. Protección ambiental, especies nativas de flora y fauna silvestres de México, categorías de riesgo y especificaciones para su inclusión, exclusión o cambio, y lista de especies en riesgo. Diario Oficial de la Federación, 30 de diciembre de 2010:1–78.

Shafi, M. M., S. W. A. Rizvi, A. Pervez, R. Ali, and S. Z. Shah. 1992. Some observations on the reproductive biology of collared pika *Ochotona rufescens* Gray, 1942 from Ziarat Valley, Baluchistan, Pakistan. Acta Theriologica 37:423–427.

Shan, W. J., and Y. G. Liu. 2015. The complete mitochondrial DNA sequence of the cape hare *Lepus capensis pamirensis*. Mitochondrial DNA 7:1–2.

Shan, W. J., J. Liu, L. Yu, R. W. Murphy, H. Mahmut, and Y. P. Zhang. 2011. Genetic consequences of postglacial colonization by the endemic Yarkand hare (*Lepus yarkandensis*) of the arid Tarim Basin. Chinese Science Bulletin 56:1370–1382.

Shayilawu, C. M., Y. F. Ni, Muhetaer, and Bolati. 2009. New record of *Ochotona pusilla* in Jayer Mountain, Xinjiang. Chinese Journal of Zoology 44:152–154.

Sheriff, M. J., C. J. Krebs, and R. Boonstra. 2009. The sensitive hare: sublethal effects of predator stress on reproduction in snowshoe hares. Journal of Animal Ecology 78:1249–1258.

Sheriff, M. J., E. McMahon, C. J. Krebs, and R. Boonstra. 2015. Predator-induced maternal stress and population demography in snowshoe hares: the more severe the risk, the longer the generational effect. Journal of Zoology 296:305–310.

Shimizu, R., and K. Shimano. 2010. Food and habitat selection of *Lepus brachyurus lyoni* Kishida, a near-threatened species on Sado Island, Japan. Mammal Study 35:169–177.

Shipley, L. A., T. B. Davila, N. J. Thines, and B. A. Elias. 2006. Nutritional requirements and diet choices of the pygmy rabbit (*Brachylagus idahoensis*): a sagebrush specialist. Journal of Chemical Ecology 32:2455–2474.

Shipley, L. A., E. M. Davis, L. A. Felicetti, S. McLean, and J. S. Forbey. 2012. Mechanisms for eliminating monoterpenes of sagebrush by specialist and generalist rabbits. Journal of Chemical Ecology 38:1178–1189.

Shou, Z., and Z. J. Feng. 1984. A new subspecies of the Tibetan pika from China. Acta Theriologica Sinica 2:151–154.

Sikes, R. K., and R. W. Chamberlain. 1954. Laboratory observations on three species of bird mites. Journal of Parasitology 40:691–697.

Silva, S. M., C. Ferreira, J. Paupério, R. M. Silva, P. C. Alves, and A. Lemos. 2015. Coccidiosis in European rabbit (*Oryctolagus cuniculus algirus*) populations in the Iberian Peninsula. Acta Parasitologica 60:350–355.

Simes, M. T., K. M. Longshore, K. E. Nussear, G. L. Beatty, D. E. Brown, and T. C. Esque. 2015. Black-tailed and white-tailed jackrabbits in the American west: history, ecology, ecological significance, and survey methods. Western North American Naturalist 75:491–519.

Simpson, G. G. 1945. The principles of classification and a classification of mammals. Bulletin of the American Museum of Natural History 85:1–350.

Sinclair, A. R. E., D. Chitty, C. I. Stefan, and C. J. Krebs. 2003. Mammal population cycles: evidence for intrinsic differences during snowshoe hare cycles. Canadian Journal of Zoology 81:216–220.

Sinclair, A. R. E., C. J. Krebs, J. N. M. Smith, and S. Boutin. 1988. Population biology of snowshoe hares III. Nutrition, plant secondary compounds and food limitation. Journal of Animal Ecology 57:787–806.

Skalon, V. N. 1934. Some zoological discoveries in South-Eastern Transbaikalia. Bulletin of Antiplague Organisation of Eastern Siberian Territory 1:83–100. [In Russian].

Skinner, J. D., and C. T. Chimimba. 2005. The Mammals of the South African Subregion. 3rd edition. Cambridge University Press: Cape Town, South Africa.

Slamečka, J., P. Hell, and R. Jurčík. 1997. Brown hare in the westslovak lowland. Acta Scientarum Naturalium Academiae Scientarum Bohemicae Brno 31:1–115.

Slough, B. G., and T. S. Jung. 2007. Diversity and distribution of the terrestrial mammals of the Yukon Territory: a review. The Canadian Field-Naturalist 121:119–127.

Sludskii, A. A., and E. I. Strautman (editors). 1980. Mammals of Kazakhstan. Volume 2: Lagomorpha. Nauka of Kazakh SSR: Alma-Ata, Kazakhstan. [In Russian].

Smirnov, P. K. 1974. Biotopic distribution and territorial relationship of the steppe and Pallas' pikas in the sympatric zone of their ranges. Byulleten' Moskovskovo Obshchestva Ispytatelei Prirodym, Otdel Biologicheski 75:72–80. [In Russian].

———. 1986. Habitat preference in sympatric species of *Ochotona*. Vestn. Leningrad University, Biology 3:10–15. [In Russian].

Smith, A. T. 1974. The distribution and dispersal of pikas: influences of behavior and climate. Ecology 55:1368–1376.

———. 1978. Comparative demography of pikas: effect of spatial and temporal age-specific mortality. Ecology 59:133–139.

———. 1987. Population structure of pikas: dispersal versus philopatry. Pp. 128–142. In: B. D. Chepko-Sade and Z. T. Halpin (editors). Mammalian Dispersal Patterns: The Effects of Social Structure on Population Genetics. University of Chicago Press: Chicago, Illinois.

———. 1994. Lagomorph specialist group. Species 23:68–69.

———. 2008. Conservation of endangered lagomorphs. Pp. 297–315. In: P. C. Alves, N. Ferrand, and K. Hackländer (editors). Lagomorph Biology: Evolution, Ecology, and Conservation. Springer: Berlin.

Smith, A. T., and A. F. Boyer. 2008a. *Oryctolagus cuniculus*. The IUCN Red List of Threatened Species 2008: e.T41291A10415170. http://dx.doi.org/10.2305/IUCN.UK.2008.RLTS.T41291A10415170.en.

———. 2008b. *Sylvilagus dicei*. The IUCN Red List of Threatened Species 2008: e.T21209A9256840. http://dx.doi.org/10.2305/IUCN.UK.2008.RLTS.T21209A9256840.en.

Smith, A. T., and J. M. Foggin. 1999. The plateau pika (*Ochotona curzoniae*) is a keystone species for biodiversity on the Tibetan plateau. Animal Conservation 2:235–240.

Smith, A. T., and B. L. Ivins. 1983a. Colonization in a pika population: dispersal versus philopatry. Behavioral Ecology and Sociobiology 13:37–47.

———. 1983b. Reproductive tactics of pikas: why have two litters? Canadian Journal of Zoology 61:1551–1559.

———. 1984. Spatial relationships and social organization in adult pikas: a facultatively monogamous mammal. Zeitschrift für Tierpsychologie 66:289–308.

Smith, A. T., and J. D. Nagy. 2015. Population resilience in an American pika (*Ochotona princeps*) metapopulation. Journal of Mammalogy 96:394–404.

Smith, A. T., and X. G. Wang. 1991. Social relationships of adult black-lipped pikas (*Ochotona curzoniae*). Journal of Mammalogy 72:231–247.

Smith, A. T., and M. L. Weston. 1990. *Ochotona princeps*. Mammalian Species 352:1–8.

Smith, A. T., and Y. Xie (editors). 2008. A Guide to the Mammals of China. Princeton University Press: Princeton, New Jersey.

Smith, A. T., N. A. Formozov, R. S. Hoffmann, C. L. Zheng, and M. A. Erbajeva. 1990. The pikas. Pp. 14–60. In: J. A. Chapman and J. E. C. Flux (editors). Rabbits, Hares and Pikas: Status Survey and Conservation Action Plan. International Union for Conservation of Nature: Gland, Switzerland.

Smith, A. T., H. J. Smith, X. G. Wang, X. C. Yin, and J. Liang. 1986. Social behavior of the steppe-dwelling black-lipped pika. National Geographic Research 2:57–74.

Smith, D. F., and J. A. Litvaitis. 2000. Foraging strategies of sympatric lagomorphs: implications for differential success in fragmented landscapes. Canadian Journal of Zoology 78:2134–2141.

Smith, G., J. L. Holechek, and M. Cardenas. 1996. Wildlife numbers on excellent and good condition Chihuahuan Desert rangelands: an observation. Journal of Range Management 49:489–493.

Smith, G. W. 1990. Home range and activity patterns of black-tailed jackrabbits. Great Basin Naturalist 50:249–256.

Smith, G. W., and N. C. Nydegger. 1985. A spotlight, line-transect method for surveying jack rabbits. Journal of Wildlife Management 49:699–702.

Smith, G. W., L. C. Stoddart, and F. F. Knowlton. 2002. Long-distance movements of black-tailed jackrabbits. Journal of Wildlife Management 66:463–469.

Smith, J. A., and L. P. Erb. 2013. Patterns of selective caching behavior of a generalist herbivore, the American pika (*Ochotona princeps*). Arctic, Antarctic and Alpine Research 45:396–403.

Smith, J. N. M., C. J. Krebs, A. R. E. Sinclair, and R. Boonstra. 1988. Population biology of snowshoe hares II. Interactions with winter food plants. Journal of Animal Ecology 57:269–286.

Smith, R. K., N. V. Jennings, and S. Harris. 2005a. A quantitative analysis of the abundance and demography of European hares *Lepus europaeus* in relation to habitat type, intensity of agriculture and climate. Mammal Review 35:1–24.

Smith, R. K., N. V. Jennings, F. Tataruch, K. Hackländer, and S. Harris. 2005b. Vegetation quality and habitat selection by European hares *Lepus europaeus* in a pastural landscape. Acta Theriologica 50: 391–404.

Smithers, R. H. N. 1971. The Mammals of Botswana. Museum Memoir No. 4. National Museums of Rhodesia: Salisbury, Rhodesia.

Sokolov, V. E., E. Yu. Ivanitskaya, V. V. Gruzdev, and V. G. Heptner. 2009. Lagomorphs: Mammals of Russia and Adjacent Regions. English translation (A. T. Smith Scientific Editor). Smithsonian Institution Libraries: Washington, D.C.

Soltis, D. E., A. B. Morris, J. S. McLachlan, P. S. Manos, and P. S. Soltis. 2006. Comparative phylogeography of unglaciated eastern North America. Molecular Ecology 15:4261–4293.

Sommer, M. A. 1997. Distribution, Habitat, and Home Range of the New England Cottontail (*Sylvilagus transitionalis*) in Western Maryland. Unpublished M.S. thesis, Frostburg State University, Frostburg, Maryland.

South African Mammal CAMP Workshop. 2013. *Bunolagus monticularis*. The IUCN Red List of Threatened Species: e.T3326A43710964. http://dx.doi.org/10.2305/IUCN.UK.2013–1.RLTS.T3326A437109 64.en.

Sovie, A. 2015. Evaluating the Impact of Burmese Pythons on Marsh Rabbits in the Greater Everglades Ecosystem. Unpublished M.S. thesis, University of Florida, Gainesville, Florida.

Springer, M. S., W. J. Murphy, E. Eizirik, and S. J. O'Brien. 2003. Placental mammal diversification and the Cretaceous-Tertiary boundary. Proceedings of the National Academy of Sciences of the United States of America 100:1056–1061.

St. Leger, J. 1929. An interesting collection of mammals, with a remarkable new species of hare from Uganda. Annals and Magazine of Natural History Series 10, 4:290–294.

———. 1932. A new genus for the Uganda hare (*Lepus marjorita*). Proceedings of the Zoological Society of London 1932:119–123.

Stefan, C. I., and C. J. Krebs. 2001. Reproductive changes in a cyclic population of snowshoe hares. Canadian Journal of Zoology 79: 2101–2108.

Stevens, M. A., and R. E. Barry. 2002. Selection, size, and use of home range of the Appalachian cottontail, *Sylvilagus obscurus*. The Canadian Field-Naturalist 116:529–535.

Stewart, D. R. M. 1971. Seasonal food preferences of *Lepus capensis* in Kenya. East African Wildlife Journal 9:163–166.

Stodart, E., and I. Parer. 1988. Colonisation of Australia by the Rabbit. CSIRO Division of Wildlife and Ecology Project Report No. 6.

Stoddart, L. C. 1985. Severe weather related mortality of black-tailed jackrabbits. Journal of Wildlife Management 49:696–698.

Stone, W. 1900. Descriptions of a new rabbit from the Liu Kiu Islands and a new flying squirrel from Borneo. Proceedings of the Academy of Natural Sciences of Philadelphia 52:460–463.

———. 1914. On a collection of mammals from Ecuador. Proceedings of the Academy of Natural Sciences of Philadelphia 66:9–19.

Storer, J. E. 1984. Mammals of the Swift Current Creek Local Fauna (Eocene: Uintan), Saskatchewan. Natural History Contributions No. 7. Museum of Natural History: Regina, Saskatchewan, Canada.

Stout, G. G. 1970. The breeding biology of the desert cottontail in the Phoenix region, Arizona. Journal of Wildlife Management 34: 47–51.

Strive, T., J. D. Wright, and A. J. Robinson. 2009. Identification and partial characterisation of a new lagovirus in Australian wild rabbits. Virology 38:97–105.

Stucky, R. K., and M. C. McKenna. 1993. Mammalia. Pp. 739–771. In: M. J. Benton (editor). The Fossil Record 2. Chapman and Hall: London.

Su, J. P. 2001. A Comparative Study on the Habitat Selection of Plateau Pika (*Ochotona curzoniae*) and Gansu Pika (*Ochotona cansus*). Unpublished Ph.D. dissertation, Northwest Plateau Institute of Biology: Xining, Qinghai. [In Chinese].

Suchentrunk, F. 2004. Phylogenetic relationships between Indian and Burmese hares (*Lepus nigricollis* and *L. peguensis*) inferred from epigenetic dental characters. Mammalian Biology 69:28–45.

Suchentrunk, F., and M. Davidovic. 2004. Evaluation of the classification of Indian hares (*Lepus nigricollis*) into the genus *Indolagus* Gureev, 1953 (Leporidae, Lagomorpha). Mammalian Biology 69:46–57.

Suchentrunk, F., and J. E. C. Flux. 1996. Minor dental traits in East African cape hares and savanna hares (*Lepus capensis* and *Lepus victoriae*): a study of intra- and interspecific variability. Journal of Zoology 238:495–511.

Suchentrunk, F., J. E. C. Flux, M. M. Flux, and H. Ben Slimen. 2007. Multivariate discrimination between East African cape hares (*Lepus capensis*) and savanna hares (*L. victoriae*) based on occipital bone shape. Mammalian Biology 72:372–383.

Sucke, A. C. 2002. Survival, Winter Diet, Density and Macrohabitat of the Appalachian Cottontail, *Sylvilagus obscurus*, in West Virginia. Unpublished M.S. thesis, Frostburg State University, Frostburg, Maryland.

Sugimura, K., and F. Yamada. 2004. Estimating population size of the Amami rabbit *Pentalagus furnessi* based on fecal pellet counts on Amami Island, Japan. Acta Zoologica Sinica 50:519–526.

Sugimura, K., K. Ishida, S. Abe, N. Nagai, Y. Watari, M. Tatara, et al. 2014. Monitoring the effects of forest clear-cutting and mongoose *Herpestes auropunctatus* invasion on wildlife diversity on Amami Island, Japan. Oryx 48:241–249.

Sugimura, K., S. Sato, F. Yamada, S. Abe, H. Hirakawa, and Y. Handa. 2000. Distribution and abundance of the Amami rabbit *Pentalagus furnessi* in the Amami and Tokuno Islands, Japan. Oryx 34:198–206.

Surridge, A. K., R. J. Timmins, G. M. Hewett, and D. J. Bell. 1999. Striped rabbits in Southeast Asia. Nature 400:726.

Syrjälä, P., M. Nylund, and S. Heinikainen. 2005. European brown hare syndrome in free-living mountain hares (*Lepus timidus*) and European brown hares (*Lepus europaeus*) in Finland 1990–2002. Journal of Wildlife Diseases 41:42–47.

Tablado, Z., E. Revilla, and F. Palomares. 2009. Breeding like rabbits: global patterns of variability and determinants of European wild rabbit reproduction. Ecography 32:310–320.

Takada, N., and T. Yamaguchi. 1974. Studies on ixodid fauna in the northern part of Honshu, Japan: ixodid ticks (Ixodidae) parasitic on wild mammals and some cases of human infestation. Japanese Journal of Sanitary Zoology 25:35–40. [In Japanese].

Tandan, P., B. Dhakal, K. Karki, and A. Aryal. 2013. Tropical grasslands supporting the endangered hispid hare (*Caprolagus hispidus*) population in Bardia National Park, Nepal. Current Science 105:691–694.

Taniguchi, A. 1986. Studies on the forest damage caused by the Japanese hare *Lepus brachyurus* in Kagoshima Prefecture. Bulletin of Kagoshima Prefectural Forest Experiment Station 2:1–38. [In Japanese].

Tapia, L., J. Dominguez, and L. Rodríguez. 2010. Modelling habitat use by Iberian hare *Lepus granatensis* and European wild rabbit *Oryctolagus cuniculus* in a mountainous area in northwestern Spain. Acta Theriologica 55:73–79.

Tapper, S. C., and R. F. W. Barnes. 1986. Influence of farming practice on the ecology of the brown hare (*Lepus europaeus*). Journal of Applied Ecology 23:39–52.

Tash, J. P., and J. A. Litvaitis. 2007. Characteristics of occupied habitats and identification of sites for restoration and translocation of New England cottontail populations. Biological Conservation 137:584–598.

Tate, G. H. H. 1933. Taxonomic history of the Neotropical hares of the genus *Sylvilagus*, subgenus *Tapeti*. American Museum Novitates 661:1–10.

Taylor, P. J. 1998. The smaller mammals of KwaZulu-Natal. University of Natal Press: Pietermaritzburg, South Africa.

Tefft, B. C., and J. A. Chapman. 1983. Growth and development of nestling New England cottontails, *Sylvilagus transitionalis*. Acta Theriologica 28:317–337.

Temminck, E. S. 1844. *Lepus brachyurus*. In: Philipp Franz Bathazar von Siebold's Fauna Japonica, Mammalia. Lugduni Batavorum: Leiden, The Netherlands.

Thapa, A., B. V. Dahal, N. P. Koju, and S. Thapa. 2011. A Review on Pikas of Nepal. Small Mammals Conservation and Research Foundation: New Baneshwor, Kathmandu, Nepal.

Thimmayya, A. C., and S. W. Buskirk. 2012. Genetic connectivity and diversity of pygmy rabbits (*Brachylagus idahoensis*) in southern Wyoming. Journal of Mammalogy 93:29–37.

Thines, N. J., L. A. Shipley, and R. D. Sayler. 2004. Effects of cattle grazing on ecology and habitat of Columbia Basin pygmy rabbits (*Brachylagus idahoensis*). Biological Conservation 119:525–534.

Thomas, H. H., and T. L. Best. 1994a. *Sylvilagus mansuetus*. Mammalian Species 464:1–2.

———. 1994b. *Lepus insularis*. Mammalian Species 465:1–3.

Thomas, O. 1890. On a collection of mammals from Central Vera Cruz, Mexico. Proceedings of the Zoological Society of London 11:71–76.

———. 1892. Diagnosis of a new subspecies of hare from the Corea. Annals and Magazine of Natural History Series 6, 9:146–147.

———. 1903. On a remarkable new hare from the Cape Colony. Annals and Magazine of Natural History Series 7, 11:78–79.

———. 1909. A collection of mammals from northern and central Mantchuria. Annals and Magazine of Natural History Series 8, 4:500–505.

———. 1911a. New rodents from Sze-chwan collected by Capt. F. M. Bailey. Annals and Magazine of Natural History Series 8, 8:727–729.

———. 1911b. The Duke of Bedford's zoological exploration in eastern Asia—XIV. On mammals from southern Shen-si, central China. Proceedings of the Zoological Society of London 1911:687–695.

———. 1912. On mammals from Central Asia, collected by Mr. Douglas Carruthers. Annals and Magazine of Natural History Series 8, 9:391–408.

———. 1922. On some new forms of *Ochotona*. Annals and Magazine of Natural History Series 9, 9:187–193.

———. 1923. On mammals from the Li-kiang Range, Yunnan, being a further collection obtained by Mr. George Forrest. Annals and Magazine of Natural History Series 9, 11:655–663.

———. 1929. On mammals from the Kaokoveld, South-West Africa, obtained during Captain Shortridge's fifth Percy Sladen and Kaffrarian Museum Expedition. Proceedings of the Zoological Society of London 1:109–110.

Thompson, H. V., and C. M. King. 1994. The European Rabbit: The History of a Successful Colonizer. Oxford University Press: Oxford.

Thulin, C. G. 2003. The distribution of mountain hares *Lepus timidus* in Europe: a challenge from brown hares *L. europaeus*? Mammal Review 33:29–42.

Thulin, C. G., M. Jaarola, and H. Tegelstrom. 1997. The occurrence of mountain hare mitochondrial DNA in wild brown hares. Molecular Ecology 6:463–467.

Timm, R. M., D. E. Wilson, B. L. Clauson, R. K. LaVal, and C. S. Vaughan. 1989. Mammals of the La Selva-Braulio Carrillo Complex, Costa Rica. North American Fauna 75:1–162.

Tizzani, P., S. Catalano, L. Rossi, P. J. Duignan, and P. G. Meneguz. 2014. Invasive species and their parasites: eastern cottontail rabbit *Sylvilagus floridanus* and *Trichostrongylus affinis* (Graybill, 1924) from Northwestern Italy. Parasitology Research 113:1301–1303.

Tizzani, P., A. Menzano, S. Catalano, L. Rossi, and P. G. Meneguz. 2011.

First report of *Obeliscoides cuniculi* in European brown hare (*Lepus europaeus*). Parasitology Research 109:963–966.

Tolesa, Z. 2014. Evolutionary Relationships among Hares (*Lepus* spp.) from Ethiopia: Multivariate Morphometry, Molecular Phylogenetics and Population Genetics. Unpublished Ph.D. dissertation, Addis Ababa University (Ethiopia) and Christian Albrecht's University of Kiel (Germany).

Tomida, Y., and C. Jin. 2002. Morphological evolution of the genus *Pliopentalagus* based on the fossil material from Anhui Province, China: a preliminary study. In: S. Kubodera, M. Higuchi, and R. Miyawaki (editors). Proceedings of the Third and Fourth Symposia on Collection Building and Natural History Studies in Asia and the Pacific Rim. National Science Museum Monographs 22:97–107.

Tomida, Y., and H. Otsuka. 1993. First discovery of fossil Amami rabbit (*Pentalagus furnessi*) from Tokunoshima, Southern Japan. Bulletin of the National Museum of Nature and Science Series C 19:73–79.

Tomiya, S., J. L. McGuire, R. W. Dedon, S. D. Lerner, R. Setsuda, A. N. Lipps, et al. 2011. A report on late Quaternary vertebrate fossil assemblages from the eastern San Francisco Bay region, California. PaleoBios 30:50–71.

Tomkins, I. R. 1935. The marsh rabbit: an incomplete life history. Journal of Mammalogy 16:201–205.

Torii, H. 1989. Mammals in Shizuoka Prefecture. Daiichihouki: Tokyo, Japan.

———. 1990. Survey of hare tracks on the snow. Journal of the Japanese Society for Hares 17:21–28. [In Japanese].

Townsend, C. H. 1912. Mammals collected by the "Albatross" expedition in Lower California in 1911, with descriptions of new species. Bulletin of the American Museum of Natural History 30:117–130.

Traphagen, M. B. 2011. Final Report on the Status of the White-sided Jackrabbit (*Lepus callotis gaillardi*) in New Mexico. New Mexico Department of Game and Fish: Santa Fe, New Mexico.

Trefry, S. A., and D. S. Hik. 2009. Eavesdropping on the neighborhood: collared pika (*Ochotona collaris*) responses to playback calls of conspecifics and heterospecifics. Ethology 115:928–938.

Tsuchiya, K. 1979. A contribution to the chromosome study in Japanese mammals. Proceedings of the Japan Academy, Series B 55:191–195.

Tufts, D. M., C. Natarajan, I. G. Revsbech, J. Projecto-Garcia, F. G. Hoffmann, R. E. Weber, et al. 2015. Epistasis constrains mutational pathways of hemoglobin adaptations in high-altitude pikas. Molecular Biology and Evolution 32:287–298.

Turkowski, F. J. 1975. Dietary adaptability of the desert cottontail. Journal of Wildlife Management 39:748–756.

Tursi, R. M., P. T. Hughes, and E. A. Hoffman. 2012. Taxonomy versus phylogeny: evolutionary history of marsh rabbits without hopping to conclusions. Diversity and Distributions 19:120–133.

Twain, M. 1962. Roughing It. Signet Classics: New York.

Uribe-Alcocer, M., F. A. Cervantes, C. Lorenzo, and L. Gúereña-Gándara. 1989. Karyotype of the tropical hare (*Lepus flavigularis*, Leporidae). The Southwestern Naturalist 34:304–306.

Usai, F., R. Rinnovati, V. Trocchi, and L. Stancampiano. 2012. *Lepus corsicanus* gastro-intestinal helminths: first report. Helminthologia 49:71–77.

USFWS (U.S. Fish and Wildlife Service). 2016. Great Thicket National Wildlife Refuge (Proposed): Draft Land Protection Plan / Environmental Assessment. U.S. Fish and Wildlife Service: Hadley, Massachusetts.

Vakurin, A. A., V. P. Korablev, X. L. Jiang, and T. V. Grigor'eva. 2012. The chromosomes of Tsing-Ling pika, *Ochotona huangensis* Matschie, 1908 (Lagomorpha, Ochotonidae). Comparative Cytogenetics 6:347–358.

Vale, K. B., and R. E. Kissell, Jr. 2010. Male swamp rabbit (*Sylvilagus aquaticus*) habitat selection at multiple scales. Southeastern Naturalist 9:547–562.

Van Devender, T. R., and G. L. Bradley. 1990. Late Quaternary mammals from the Chihuahuan Desert: paleoecology and latitudinal gradients. Pp. 350–362. In: J. L. Betancourt, T. R. Van Devender, and P. S. Martin (editors). Packrat Middens, the Last 40,000 Years of Biotic Change. University of Arizona Press: Tucson, Arizona.

Van Peenen, P. F. D. 1969. Preliminary Identification Manual for Mammals of South Vietnam. United States National Museum, Smithsonian Institution: Washington, D.C.

Vargas, J. 2000. Distribución, Abundancia y Hábitat de la Liebre Endémica *Lepus flavigularis* (Mammalia: Lagomorpha). Unpublished M.S. thesis, Facultad de Ciencias, Universidad Nacional Autónoma de México, México City, México.

Vargas Cuenca, J., and F. A. Cervantes. 2005. *Sylvilagus audubonii* (Baird, 1858) conejo del desierto. Pp. 838–839. In: G. Ceballos and G. Oliva (editors). Los Mamíferos Silvestres de México. Comisión Nacional para el Conocimiento y Uso de la Biodiversidad. Fondo de Cultura Económica: México, D.F.

Vaughan, C., and M. A. Rodriguez. 1986. Comparación de los hábitos alimentarios del coyote (*Canis latrans*) en dos localidades en Costa Rica. Vida Silvestre Neotropical 1:6–11.

Vázquez, J., V. Farías, L. Rodríguez-Martínez, A. Bautista, G. Palacios-Roque, and M. Martínez-Gómez. 2013. Ámbito hogareño del conejo mexicano (*Sylvilagus cunicularius*) en un bosque templado del centro del México. Therya 4:581–595.

Vázquez J., A. J. Martínez, R. Hudson, L. Rodríguez-Martínez, and M. Martínez-Gómez. 2007. Seasonal reproduction in Mexican cottontail rabbits *Sylvilagus cunicularius* in La Malinche National Park, central Mexico. Acta Theriologica 52:361–369.

Velázquez, A. 1993. Man-made and ecological habitat fragmentation: study case of the volcano rabbit (*Romerolagus diazi*). Zeitschrift für Säugetierkunde 58:54–61.

———. 1994. Distribution and population size of *Romerolagus diazi* on El Pelado volcano, México. Journal of Mammalogy 75:743–749.

———. 2012. El contexto geográfico de los lagomorfos de México. Therya 3:223–238.

Velázquez, A., and F. J. Romero (editors). 1999. Biodiversidad de la Región de Montaña del Sur de la Cuenca de México. Universidad Autonoma Metropolitana: Ciudad de México, México.

Velázquez, A., G. Bocco, F. J. Romero, and A. P. Vega. 2003. A landscape perspective on biodiversity conservation: the case of central Mexico. Mountain Research and Development 23:240–246.

Velázquez, A., F. A. Cervantes, and C. Galindo-Leal. 1993. The volcano rabbit (*Romerolagus diazi*) a peculiar lagomorph. Lutra 36:62–70.

Velázquez, A., A. Larrazabal, and F. Romero. 2011. Del conocimiento específico a la conservación de todos los niveles de organización biológica: el caso del zacatuche y los paisajes que denotan su hábitat. Investigación Ambiental 3:59–62.

Velázquez, A., F. J. Romero, H. Cordero-Rangel, and G. Heil. 2001. Effects of landscape changes on mammalian assemblages at Izta-Popo volcanoes, México. Biodiversity and Conservation 10:1059–1075.

Velázquez, A., F. J. Romero, and J. López-Paniagua (editors). 1996. Ecología y Conservación del Conejo Zacatuche y su Hábitat. Universidad Nacional Autónoma de México: Ciudad de México, México.

Verheyen, W., and J. Verschuren. 1966. Rongeurs et Lagomorphs: Exploration du Parc National de la Garamba. Institut des Parcs Nationaux du Congo, Bruxelles 50:1–71.

Verts, B. J., and L. N. Carraway. 1998. Land Mammals of Oregon. University of California Press: Berkeley, California.

Verts, B. J., S. D. Gehman, and K. J. Hundertmark. 1984. *Sylvilagus nuttallii*: a semiarboreal lagomorph. Journal of Mammalogy 65: 131–135.

Vestal, A. L. 2005. Genetic Variation in the Davis Mountains Cottontail (*Sylvilagus robustus*). Unpublished M.S. thesis, Angelo State University, San Angelo, Texas.

Vidus-Rosin, A., N. Gilio, and A. Meriggi. 2008. Introduced lagomorphs as a threat to "native" lagomorphs: the case of the eastern cottontail (*Sylvilagus floridanus*) in northern Italy. Pp. 153–165. In: P. C. Alves, N. Ferrand, and K. Hackländer (editors). Lagomorph Biology: Evolution, Ecology, and Conservation. Springer: Berlin.

Vidus-Rosin, A., A. Meriggi, E. Cardarelli, S. Serrano-Perez, M.-C Mariani, C. Coraddelli, and A. Barba. 2011. Habitat overlap between sympatric European hares (*Lepus europaeus*) and Eastern cottontails (*Sylvilagus floridanus*) in northern Italy. Acta Theriologica 56:53–61.

Vigne, J. D. 1988. Les Mammifères Post-glaciaires de Corse: Etude Archéozoologique. Éditions du Centre National de la Recherche Scientifique: Paris.

Vila, T., J. Casanova, J. Miquel, and C. Feliu. 1999. Sobre las Helmintofaunas de *Lepus castroviejoi* y *L. capensis* (Lagomorpha; Leporidae) Ibérica. VI. Congreso Ibérico de Parasitología. Córdoba: 27.

Villafuerte, R. and S. Moreno. 1997. Predation risk, cover type, and group size in European rabbits in Doñana (SW Spain). Acta Theriologica 42:225–230.

Villafuerte R., A. Lazo, and S. Moreno. 1997. Influence of food abundance and quality on rabbit fluctuations: conservation and management implications in Doñana National Park (SW Spain). Rev d'écologie 52:345–356.

Villafuerte, R., J. A. Litvaitis, and D. F. Smith. 1997. Physiological responses by lagomorphs to resource limitations imposed by habitat fragmentation: implications to condition-sensitive predation. Canadian Journal of Zoology 75:148–151.

Vorhies, C. H., and W. P. Taylor. 1933. The life histories and ecology of jack rabbits, *Lepus alleni* and *Lepus californicus* ssp. in relation to grazing in Arizona. University of Arizona College of Agriculture Technical Bulletin 49:467–510.

Wainwright, M. 2002. The Natural History of Costa Rican Mammals. Zona Tropical, S.A: Miami, Florida.

Walhovd, H. 1965. Age criteria of the mountain hare (*Lepus timidus* L.) with analyses of age and sex ratios, body weights and growth in some Norwegian populations. Papers of Norwegian State Game Research Institute 22:1–57.

Waltari, E., and J. A. Cook. 2005. Hares on ice: phylogeography and historical demographics of *Lepus arcticus*, *L. othus*, and *L. timidus* (Mammalia: Lagomorpha). Molecular Ecology 14:3005–3016.

Waltari, E., J. R. Demboski, D. R. Klein, and J. A. Cook. 2004. A molecular perspective on the historical biogeography of the northern high latitudes. Journal of Mammalogy 85:591–600.

Wang, C. M., H. X. He, M. Li, F. M. Lei, J. J. Root, Y. Y. Wu, et al. 2009. Parasite species associated with wild plateau pika (*Ochotona curzoniae*) in southeastern Qinghai province, China. Journal of Wildlife Diseases 45:288–294.

Wang, L. C. H., D. L. Jones, R. A. MacArthur, and W. A. Fuller. 1973. Adaptation to cold: energy metabolism in an atypical lagomorph, the arctic hare (*Lepus arcticus*). Canadian Journal of Zoology 51: 841–846.

Wang, X. G., and A. T. Smith. 1988. On the natural winter mortality of the plateau pika (*Ochotona curzoniae*). Acta Theriologica Sinica 8: 152–156. [In Chinese].

———. 1989. Studies on the mating system in plateau pikas (*Ochotona curzoniae*). Acta Theriologica Sinica 9:210–215. [In Chinese].

Wangdwei, M., B. Steele, and R. B. Harris. 2013. Demographic responses of plateau pikas to vegetation cover and land use in the Tibet Autonomous Region, China. Journal of Mammalogy 94: 1077–1086.

Warren, A., J. A. Litvaitis, and D. Keirstead. 2016. Developing a habitat suitability index to guide management of New England cottontail habitats. Wildlife Society Bulletin 40:69–77.

Watari, Y., S. Nishijima, M. Fukasawa, F. Yamada, S. Abe, and T. Miyashita. 2013. Evaluating the "recovery-level" of endangered species without prior information before alien invasion. Ecology and Evolution 3:4711–4721.

Waterhouse, G. R. 1839. Observations on the Rodentia, with a view to point out the groups, as indicated by the structure of the Crania, in this order of mammals. Magazine of Natural History, new series 3: 90–96, 184–188, 274–279, 593–600.

Watkins, L. C., and R. M. Nowak. 1973. The white-tailed jackrabbit in Missouri. The Southwestern Naturalist 18:352–354.

Watland, A. M., E. M. Schauber, and A. Woolf. 2007. Translocation of swamp rabbits in southern Illinois. Southeastern Naturalist 6: 259–270.

Watson, A., and R. Hewson. 1973. Population densities of mountain hares (*Lepus timidus*) on western Scottish and Irish moors and on Scottish hills. Journal of Zoology (London) 170:151–159.

Weidman, T., and J. A. Litvaitis. 2011. Can supplemental food increase winter survival of a threatened cottontail rabbit? Biological Conservation 144:2054–2058.

Werdelin, L., and W. J. Sanders (editors). 2010. Cenozoic Mammals of Africa. University of California Press: Berkeley, California.

Wesche, K., K. Nadrowski, and V. Retzer. 2007. Habitat engineering under dry conditions: the impact of pikas (*Ochotona pallasi*) on vegetation and site conditions in southern Mongolian steppes. Journal of Vegetation Science 18:665–674.

Weston, M. L. 1981. The *Ochotona alpina* complex: a statistical re-evaluation. Pp. 73–89. In: K. Myers and C. D. MacInnes (editors). Proceedings of the World Lagomorph Conference. University of Guelph: Guelph, Ontario, Canada.

Whitaker, J. O., Jr., and W. J. Hamilton, Jr. 1998. Mammals of the Eastern United States. 3rd edition. Cornell University Press: Ithaca, New York.

Whitaker, J. O., Jr., and J. B. Morales-Malacara. 2005. Ectoparasites and other associates (Ectodytes) of mammals of México. Pp. 535–66. In: V. Sánchez-Cordero and R. Medellín (editors). Contribuciones Mastozoológicas en Homenaje a Bernardo Villa. Instituto de Biología, Universidad Nacional Autónoma de México, CONABIO: México City, México.

White, J. A., 1991. North American Leporinae (Mammalia, Lagomorpha) from late Miocene (Clarendonian) to latest Pliocene (Blancan). Journal of Vertebrate Paleontology 11:67–89.

Whiteford, C. E. M. 1995. Molecular phylogeny of the genus *Pronolagus* (Mammalia: Lagomorpha) and the use of morphological and molecular characters in the delineation of *P. rupestris*. M.Sc. thesis, University of Pretoria: Pretoria, South Africa.

Whitworth, M. R. 1984. Maternal care and behavioural development in pikas *Ochotona princeps*. Animal Behaviour 32:743–752.

Wible, J. R. 2007. On the cranial osteology of the Lagomorpha. Bulletin of the Carnegie Museum of Natural History 39:213–234.

Williams, D. F., P. A. Kelly, L. P. Hamilton, M. R. Lloyd, E. A. Williams, and J. J. Youngblom. 2008. Recovering the endangered riparian brush rabbit (*Sylvilagus bachmani riparius*): reproduction and growth in confinement and survival after translocation. Pp. 349–361. In P. C. Alves, N. Ferrand, and K. Hackländer (editors). Lagomorph Biology: Evolution, Ecology, and Conservation. Springer: Berlin.

Wilson, D. E. 1991. Mammals of the Tres Marías Islands. Pp. 214–250. In: T. A. Griffiths and D. Klingener (editors). Contributions to Mammalogy in Honor of Karl F. Koopman. Bulletin of the American Museum of Natural History 206:1–432.

Wilson, D. E., and D. M. Reeder. 2005. Mammal Species of the World: A Taxonomic and Geographic Reference. 3rd edition. Volume 1. Johns Hopkins University Press: Baltimore, Maryland.

Wilson, D. E., and S. Ruff (editors). 1999. The Smithsonian Book of North American Mammals. UBC Press: Vancouver, Canada.

Wilson, M. C., and A. T. Smith. 2015. The pika and the watershed: the impact of small mammal poisoning on the ecohydrology of the Qinghai-Tibetan Plateau. Ambio 44:16–22.

Wilson, T. L., F. P. Howe, and T. C. Edwards. 2011. Effects of sagebrush treatments on multi-scale resource selection by pygmy rabbits. Journal of Wildlife Management 75:393–398.

Wing, E. S. 1980. Faunal remains. Pp. 149–172. In: T. F. Lynch (editor). Guitarrero Cave: Early Man in the Andes. Academic Press: New York.

Winkler, A. J., L. F. Flynn, and Y. Tomida. 2011. Fossil lagomorphs from the Potwar Plateau, northern Pakistan. Palaeontologia Electronica 14:38A.

Won, C. M., and K. G. Smith. 1999. History and current status of mammals of the Korean Peninsula. Mammal Review 29:3–33.

Won, H. K. 1968. The Mammals of Korea. Institute of Science Press: Pyeongyang, North Korea.

Won, P. H. 1967. Illustrated Encyclopedia of Fauna and Flora of Korea. Volume 7: Mammals. Ministry of Education: Seoul, South Korea.

Wood, A. E. 1940. The mammalian fauna of the White River Oligocene. Part III: Lagomorpha. Transactions of the American Philosophical Society, New Series 28:271–362.

———. 1957. What, if anything, is a rabbit? Evolution 11:417–425.

Woods, B. A., J. L. Rachlow, C. Bunting, T. R. Johnson, and K. D. Bockting. 2013. Managing high-elevation sagebrush steppe: do conifer encroachment and prescribed fire affect habitat for pygmy rabbits? Journal of Rangeland Ecology and Management 66:462–471.

Woolsey, N. G. 1956. The vanishing antelope jackrabbit. Arizona Wildlife Sportsman November:24–27.

Worthington, D. H. 1970. The karyotype of the brush rabbit, *Sylvilagus bachmani*. Mammalian Chromosome Newsletter 11:21–22.

Wroughton, R. C. 1915. Bombay Natural History Society's mammal survey of India, Burma and Ceylon: Report no. 16. Journal of the Bombay Natural History Society 23:460–480.

Wu, C. H., H. P. Li, Y. X. Wang, and Y. P. Zhang. 2000. Low genetic variation of the Yunnan hare (*Lepus comus* G. Allen 1927) as revealed by mitochondrial cytochrome *b* gene sequences. Biochemical Genetics 38:147–153.

Wu, C. H, J. P. Wu, T. D. Bunch, Q. W. Li, Y. X. Wang, and Y. P. Zhang. 2005. Molecular phylogenetics and biogeography of *Lepus* in eastern Asia based on mitochondrial DNA sequences. Molecular Phylogenetics and Evolution 37:45–61.

Wu, Y. H., L. Xia, Q. Zhang, and Q. S. Yang. 2010. Habitat fragmentation affects genetic diversity and differentiation of the Yarkand hare. Conservation Genetics 11:183–194.

Wu, Y. H., L. Xia, Q. Zhang, Q. S. Yang, and X. X. Meng. 2011. Bidirectional introgressive hybridization between *Lepus capensis* and *Lepus yarkandensis*. Molecular Phylogenetics and Evolution 59:545–555.

Wywialowski, A. P., and L. C. Stoddart. 1988. Estimation of jack rabbit density: methodology makes a difference. Journal of Wildlife Management 52:57–59.

Xiao, N., T. Y. Li, J. M. Qiu, M. Nakao, X. W. Chen, K. Nakaya, et al. 2004. The Tibetan hare *Lepus oiostolus*: a novel intermediate host for *Echinococcus multilocularis*. Parasitology Research 92:352–353.

Xie, L., X. Z. Zhang, D. L. Qi, X. Y. Guo, B. Pang, Y. R. Du, et al. 2014. Inhibition of inducible nitric oxide synthase and nitric oxide production in plateau pika (*Ochotona curzoniae*) at high altitude on Qinghai-Tibet plateau. Nitric Oxide-Biology and Chemistry 38: 38–44.

Xu, K. F. 1986. Analysis on the karyotypes of *Lepus yarkandensis*. Acta Theriologica Sinica 6:249–253.

Yadhav, B. P., S. Sathyakumar, R. K. Koirala, and C. Pokharel. 2008. Status, distribution and habitat use of hispid hare (*Caprolagus hispidus*) in Royal Suklaphanta Wildlife Reserve, Nepal. Tiger Paper 35:8–14.

Yalden, D. W., and M. J. Largen. 1992. The endemic mammals of Ethiopia. Mammal Review 22:115–150.

Yalden, D. W., M. J. Largen, and D. Kock. 1986. Catalogue of the Mammals of Ethiopia. 6. Perissodactyla, Proboscidea, Hyracoidea, Lagomorpha, Tubulidentata, Sirenia and Cetacea. Monitore Zoologico Italiano, N.S. Suppl. XXI 4:31–103.

Yalden, D. W., M. J. Largen, D. Kock, and J. C. Hillman. 1996. Catalogue of the mammals of Ethiopia and Eritrea. 7. Revised checklist, zoogeography and conservation. Tropical Zoology 9:73–164.

Yamada, F. 1987. Reproductive behavior in the *Lepus* and its properties. Journal of the Japanese Society for Hares 14:17–21. [In Japanese].

———. 1990. Habitat selection and feeding habits of the Japanese hare and its damage to seedlings. Pp. 111–113. In: N. Maruyama (editor). Wildlife Conservation, Present Trends and Perspectives for the 21st Century. Japan Wildlife Research Center: Tokyo, Japan.

———. 2002. Impacts and control of introduced small Indian mongoose on Amami Island, Japan. Pp. 389–392. In: C. R. Veitch and M. N. Clout (editors). Turning the Tide: The Eradication of Invasive Species. International Union for Conservation of Nature: Gland, Switzerland.

———. 2008. A review of the biology and conservation of the Amami rabbit (*Pentalagus furnessi*). Pp. 369–377. In: P. C. Alves, N. Ferrand, and P. C. Hackländer (editors). Lagomorph Biology: Evolution, Ecology, and Conservation. Springer: Berlin.

———. 2014. *Lepus brachyurus lyoni* Kishida, 1937. P. 82. In: Ministry of the Environment (editor). Red Data Book 2014—Threatened Wildlife of Japan. Volume 1: Mammalia. Gyosei Corporation: Tokyo, Japan.

———. 2015a. *Lepus brachyurus* Temminck, 1844. Pp. 216–217. In: S. D. Ohdachi, Y. Ishibashi, M. A. Iwasa, D. Fukui, and T. Saitoh (editors). The Wild Mammals of Japan. 2nd edition. Shoukadoh Book Sellers: Kyoto, Japan.

———. 2015b. *Pentalagus furnessi* (Stone, 1900). Pp. 212–213. In: S. D. Ohdachi, Y. Ishibashi, M. A. Iwasa, D. Fukui, and T. Saitoh (editors). The Wild Mammals of Japan. 2nd edition. Shoukadoh Book Sellers: Kyoto, Japan.

Yamada, F., and F. A. Cervantes. 2005. *Pentalagus furnessi*. Mammalian Species 782:1–5.

Yamada, F., S. Shiraishi, A. Taniguchi, T. Mori, and T. A. Uchida. 1989. Follicular growth and timing of ovulation after coitus in the Japanese hare, *Lepus brachyurus*. Journal of the Mammalogical Society of Japan 14:1–9.

Yamada, F., S. Shiraishi, A. Taniguchi, and T. A. Uchida. 1990. Growth, development and age determination of the Japanese hare, *Lepus brachyurus*. Journal of the Mammalogical Society of Japan 14:65–77.

Yamada, F., S. Shiraishi, and T. A. Uchida. 1988. Parturition and nursing behaviours of the Japanese hare, *Lepus brachyurus*. Journal of the Mammalogical Society of Japan 13:59–68.

Yamada, F., K. Sugimura, S. Abe, and Y. Handa. 2000. Present status and conservation of the endangered Amami rabbit *Pentalagus furnessi*. Tropics 10:87–92.

Yamada, F., M. Takaki, and H. Suzuki. 2002. Molecular phylogeny of Japanese Leporidae, the Amami rabbit *Pentalagus furnessi*, the Japanese hare *Lepus brachyurus*, and the mountain hare *Lepus timidus*, inferred from mitochondrial DNA sequences. Genes & Genetic Systems 77:107–116.

Yamaguchi, M., H. Torii, and S. Higuchi. 2008. Habitat utilization of the Japanese hare (*Lepus brachyurus*) in hilly areas in the southern part of Kyoto Prefecture during the winter season. Journal of the Japanese Wildlife Research Society 33:12–19. [In Japanese].

Yamaguti, S. 1935. Studies on the helminth fauna of Japan. Part 7. Cestodes of mammals and snakes. Japanese Journal of Zoology 6: 233–246.

Yang, D. Y., J. R. Woiderski, and J. C. Driver. 2005. DNA analysis of archaeological rabbit remains from the American Southwest. Journal of Archaeological Science 32:567–578.

Yang, J., T. G. Bromage, Q. Zhao, B. H. Xu, W. L. Gao, H. F. Tian, et al. 2011. Functional evolution of leptin of *Ochotona curzoniae* in adaptive thermogenesis driven by cold environmental stress. PLoS ONE 6:e19833.

Yatake, H., M. Nashimoto, R. Matsuki, T. Takeuchi, S. Abe, and T. Ishii. 2003. Density estimation of Japanese hare *Lepus brachyurus* by fecal pellet count and INTGEP in Akita-komagatake mountains area. Mammalian Science 43:99–111. [In Japanese].

Yates, T. L., H. H. Genoways, and J. K. Jones, Jr. 1979. Rabbits (genus *Sylvilagus*) of Nicaragua. Mammalia 43:113–124.

Ye, R. Y., W. Li, B. Shi, et al. 2000. Six new records of Gamasid mite in Xinjiang and the deutonymph morphology of *Hirstionyssus meridianus* Zemskaja, 1955 (Acari). Endemic Diseases Bulletin 15:48–50. [In Chinese].

Yong, M., F. G. Wang, S. K. Jin, and S. H. Li. 1987. Glires (Rodents and Lagomorphs) of Northern Xinjiang and their Geographical Distribution. Science Press, Academia Sinica: Beijing, PRC.

Yu, F. H., S. P. Li, W. C. Kilpatrick, P. M. McGuire, K. He, and W. H. Wei. 2012. Biogeographical study of plateau pikas *Ochotona curzoniae* (Lagomorpha, Ochotonidae). Zoological Science 29:518–526.

Yu, H. Q., S. Y. Liu, W. Zhao, and J. H. Ran. 1996. Glover's pika control in plantation of arid valley in Maoxian County. Forest Pest and Disease Control 20:67–69. [In Chinese].

Yu, N., and C. L. Zheng. 1992. A taxonomic revision of Nubra pika (*Ochotona nubrica* Thomas, 1922). Acta Theriologica Sinica 12: 132–138. [In Chinese].

Yu, N., C. L. Zheng, Y. P. Zhang, and W. H. Li. 2000. Molecular systematics of pikas (genus *Ochotona*) inferred from mitochondrial DNA sequences. Molecular Phylogenetics and Evolution 16:85–95.

Yu, X. 2004. Molecular Systematics of the Genus *Lepus* in China. Unpublished M.S. thesis, Chinese Academy of Sciences, Beijing. [In Chinese].

Zeilinga de Boer, J., M. S. Drummond, M. J. Bordelon, M. J. Defant, H. Bellon, and R. C. Maury. 1995. Cenozoic magmatic phases of the Costa Rican island arc (Cordillera de Talamanca). Pp. 35–56. In: P. Mann (editor). Geologic and Tectonic Development of the Caribbean Plate Boundary in Southern Central America. Geological Society of America (Special Paper 295): Boulder, Colorado.

Zgurski, J. M., and D. S. Hik. 2012. Polygynandry and even-sexed dispersal in a population of collared pikas, *Ochotona collaris*. Animal Behavior 83:1075–1082.

Zhang, D. C., N. C. Fan, and H. Yin. 2001. Comparative analysis of behavior between *Ochotona daurica* and *Ochotona curzoniae* in the sympatric coexistence. Journal of Hebei University, Natural Science Edition 21:71–77.

Zheng, C. L. 1986. Recovery of Koslow's pika (*Ochotona koslowi* Büchner) in Kunlun Mountains of Xinjiang Uygur Autonomous Region, China. Acta Theriologica Sinica 5:285.

Zhong, W., G. Wang, Q. Zhou, X. Wan, and G. Wang. 2008. Effects of winter food availability on the abundance of Daurian pikas (*Ochotona dauurica*) in Inner Mongolian grasslands. Journal of Arid Environments 72:1383–1387.

Zollner, P. A., W. P. Smith, and L. A. Brennan. 1996. Characteristics and adaptive significance of latrines of swamp rabbits (*Sylvilagus aquaticus*). Journal of Mammalogy 77:1049–1058.

———. 2000a. Microhabitat characteristics of sites used by swamp rabbits. Wildlife Society Bulletin 28:1003–1011.

———. 2000b. Home range use by swamp rabbits (*Sylvilagus aquaticus*) in a frequently inundated bottomland forest. The American Midland Naturalist 143:64–69.

Zörner, H. 1996. Der Feldhase. Spektrum: Heidelberg, Germany.

INDEX